Spoil to Soil
Mine Site Rehabilitation and Revegetation

Spoil to Soil
Mine Site Rehabilitation and Revegetation

Edited by
N.S. Bolan, M.B. Kirkham, and Y.S. Ok

CRC Press
Taylor & Francis Group
Boca Raton London New York

CRC Press is an imprint of the
Taylor & Francis Group, an **informa** business

CRC Press
Taylor & Francis Group
6000 Broken Sound Parkway NW, Suite 300
Boca Raton, FL 33487-2742

First issued in paperback 2021

ISBN 13: 978-1-03-209641-4 (pbk)
ISBN 13: 978-1-4987-6761-3 (hbk)

Contents

Foreword ix
Preface... xi
Editors xiii
Contributors xv

SECTION I   Mine Site Characterization

Chapter 1   Characterization and Improvement in Physical, Chemical, and Biological
Properties of Mine Wastes 3

*Binoy Sarkar, Hasintha Wijesekara, Sanchita Mandal, Mandeep Singh,
and N.S. Bolan*

Chapter 2   Chemical Characterization of Mine Sites 17

Xinni Xiong, Daniel C.W. Tsang, and Y.S. Ok

Chapter 3   Sources and Management of Acid Mine Drainage 33

*S.R. Gurung, Hasintha Wijesekara, Balaji Seshadri, R.B. Stewart,
P.E.H. Gregg, and N.S. Bolan*

SECTION II   Mine Site Rehabilitation Practices

Chapter 4   Use of Biowaste for Mine Site Rehabilitation: A Meta-Analysis
on Soil Carbon Dynamics 59

*Hasintha Wijesekara, N.S. Bolan, Kim Colyvas, Balaji Seshadri,
Y.S. Ok, Yasser M. Awad, Yilu Xu, Ramesh Thangavel, Aravind Surapaneni,
Christopher Saint, and Meththika Vithanage*

Chapter 5   Rehabilitation of Biological Characteristics in Mine Tailings 75

Longbin Huang and Fang You

Chapter 6   Nanoscale Materials for Mine Site Remediation 95

Tapan Adhikari and Rajarathnam Dharmarajan

SECTION III   Post Mine Site Land-Use Practices

Chapter 7   Profitable Beef Cattle Production on Rehabilitated Mine Lands 111

Dee Murdoch and Rajasekar Karunanithi

Chapter 8   Restoring Forests on Surface Coal Mines in Appalachia: A Regional
Reforestation Approach with Global Application 123

*Christopher D. Barton, Kenton Sena, Teagan Dolan,
Patrick Angel, and Carl Zipper*

Chapter 9   Recreating a Headwater Stream System on a Valley Fill in Appalachia, USA 147

Carmen T. Agouridis, Christopher D. Barton, and Richard C. Warner

Chapter 10   Key Issues in Mine Closure Planning for Pit Lakes.. 175

Jerry A. Vandenberg and Cherie D. McCullough

Chapter 11   Carbon Sequestration Potential on Mined Lands... 189

Sally Brown, Andrew Trlica, John Lavery, and Mark Teshima

SECTION IV   Mine Site Revegetation Potential

Chapter 12   Phytotechnologies for Mine Site Rehabilitation 203

*Ramesh Thangavel, Rajasekar Karunanithi, Hasintha Wijesekara, Yubo Yan,
Balaji Seshadri, and N.S. Bolan*

Chapter 13   Phytocapping of Mine Waste at Derelict Mine Sites in New South Wales 215

*Dane Lamb, Peter Sanderson, Liang Wang,
Mohammed Kader, and Ravi Naidu*

Chapter 14   Rehabilitation of an Abandoned Mine Site with Biosolids 241

*Abdulaziz Alghamdi, M.B. Kirkham, Deann R. Presley, Ganga Hettiarachchi,
and Leigh Murray*

Chapter 15   Dynamics of Heavy Metal(loid)s in Mine Soils.. 259

*Anitha Kunhikrishnan, N.S. Bolan, Saikat Chowdhury, Jin Hee Park,
Hyuck Soo Kim, Girish Choppala, Bhupinder Pal Singh, and Won Il Kim*

SECTION V   Case Studies of Successful Mine Site Rehabilitation

Chapter 16 Mine Site Reclamation in Canada: Overview and Case Studies..............................291

Jin-Hyeob Kwak, Abimbola Ojekanmi, Min Duan, Scott X. Chang, and M. Anne Naeth

Chapter 17 Case Studies of Successful Mine Site Rehabilitation: Malaysia.............................309

Soon Kong Yong and Suhaimi Abdul-Talib

Chapter 18 Mine Rehabilitation in New Zealand: Overview and Case Studies 335

Robyn C. Simcock and Craig W. Ross

Index...363

Foreword

When we plant trees, we plant the seeds of peace and seeds of hope.

(Wangari Maathai, 2004 Nobel Laureate)

Restoring soils and vegetation on mine-impacted lands is no easy task. Soils possessing site-specific biological, physical, and chemical attributes can be extensively altered during the mining process. Soil microfauna and microflora, plant seeds and rhizomes, earthworms and insects, and even the mineral, clay, and organic carbon fractions of soil can be buried or severely compromised by mining disturbances. Spoil is the material that remains after mining, which provides the foundation for new life post rehabilitation. Unfortunately, spoil is not a universal media that exhibits similar characteristics across the globe. Spoil can be stockpiled topsoil and subsoil from the pre-mining landscape that has been replaced. Spoil can be a topsoil substitute consisting of weathered or unweathered overburden or tailings resulting from mineral extraction. Spoil can be a mixture of all of these and can oftentimes contain amendments that were imported from outside the mining landscape. Regardless, from a pedogenic perspective, spoil represents time zero in the soil formation process in disturbed sites including mine sites.

Rehabilitation, reclamation, restoration, revegetation, and remediation are all terms that have been used to describe the process of transforming spoil to soil and the subsequent recovery of ecosystem processes. In many parts of the world, this process is intended to create a post-mining land use that is equal or better than the pre-mining condition. Unfortunately, this is not always achieved. Even with decades of reclamation research, more information is needed on how to successfully rehabilitate mine-impacted lands. Spoil stabilization and vegetation establishment techniques have been well documented on mined lands, but the return of a landscape for long-term uses and benefits has not always been the goal of many reclamation endeavors.

Mining provided the raw materials needed for global industrialization and development. The legacy of these activities is at the forefront of conversations on climate change, food and water security, and economic sustainability. Mine-impacted lands should have great value even after the resources have been extracted if proper reclamation techniques are employed. Society at large can benefit by restoring ecosystem services and productivity to these lands, as they constitute an "environmental infrastructure" of tangible value to local and global communities. For example, rehabilitated forested landscapes help to maintain clean water and reduce flooding in rivers and streams. Forests are responsible for cleaning air and sequestering carbon, which will help offset emissions from mining activities and burning of fossil fuels. Targeted reforestation can reduce forest fragmentation from mining, as needed to restore habitat for wildlife species that depend on large expanses of unbroken forest. Reforestation with native species can also improve landscape aesthetics, thus enhancing the capacity of communities in mining areas to serve as tourist destinations and to support tourism-related businesses and jobs.

Beyond the forest, former surface mines may also provide opportunities for renewable energy applications (wind, solar, bioenergy including growing energy crops). These lands often contain infrastructure (transmission lines, roads, railroads) that can be utilized for a variety of other industrial and agricultural uses. Opportunities abound to stimulate job creation and economic growth from these types of post-mining land uses. However, we must ensure that this growth is sustainably developed and that the foundation for all of these endeavors to occur, the soil, is healthy and resilient to the many threats it faces. These threats are environmental and biological, such as climate change and exotic invasive pests and diseases, as well as threats associated with benign neglect and exploitive overuse of the land and its related resources.

Successful rehabilitation and revegetation of mine-impacted land are vital for the current and future prosperity of mining regions across the globe. Information in the following chapters outlines the scientific knowledge and best practices for protecting the environment and reestablishing productivity in post-mining landscapes. A commitment to improving conditions on reclaimed mined land for the future seems like a worthwhile investment. By improving our ability to rehabilitate mined land, we create new opportunities for lands that are often considered marginal and we stand a chance of contributing significantly to the development of a sustainable and economically viable future.

Christopher D. Barton
Professor of Forest Hydrology and Watershed Management
University of Kentucky
Lexington, Kentucky

Preface

The need to ensure environmental sustainability for current and future generations is a critical challenge for humanity. The forces driving increased environmental pressure include population growth, increased urbanization, and demand for human consumption. Thus, demands for energy resources, such as coal and base metals (iron and copper) used for industrial products and infrastructure, have been rising.

Mining can cause significant environmental impacts that include soil erosion, formation of mine voids and sinkholes, loss of biodiversity, and contamination of soil, groundwater, and surface water by chemicals from mining processes. Besides creating environmental damage, the contamination resulting from leakage of chemicals also affects the health of the local population. For example, acid mine drainage resulting from mine wastes can cause toxic metal mobilization. Furthermore, conflict can arise at the local level where mining is perceived as competing with agriculture and livestock grazing or other traditional land uses. Displacement and resettlement of land owners and farmers from mining areas also can cause conflict where comparable land cannot be obtained. Mining companies in many countries are required to follow environmental and rehabilitation codes, ensuring the area mined is returned close to its original state through rehabilitation and revegetation.

This book covers both the fundamental and practical aspects of remediation and revegetation of mine sites. It includes three major themes: (1) characterization of mine site spoils; (2) remediation of chemical, physical, and biological constraints of mine site spoils, including post mine site land-use practices; and (3) revegetation of remediated mine site spoils. Each theme includes chapters covering case studies involving mine sites around the world. The last section focuses specifically on case studies with successful mine site rehabilitation. This book provides a narrative of *how inert spoil can be converted to live soil*. The purpose of this book is to give students, scientists, and professional personnel in the mining industry sensible, science-based information needed to rehabilitate sustainably areas disturbed by mining activities. The key features of this book include the following:

- It provides a fundamental understanding of mine site spoil properties.
- It provides various approaches for remediation of mine site contaminants.
- It outlines the potential value of risk-based approaches for remediation.
- It includes case studies of revegetation of mine sites for various land-use practices.
- It is suitable for undergraduate and graduate students majoring in environmental, earth, and soil sciences; environmental and soil scientists; and mine site environmental engineers and regulators.

N.S. Bolan
M.B. Kirkham
Y.S. Ok

Preface

Editors

Professor N.S. Bolan completed his PhD in soil science and plant nutrition at the University of Western Australia and is currently working as a professor of environmental science at the University of Newcastle. His teaching and research interests include agronomic value of manures, fertilizers and soil amendments, soil acidification, nutrient and carbon cycling, pesticide and metal pollutant interactions in soils, greenhouse gas emission, soil remediation, and waste and wastewater management. Professor Bolan is a fellow of the Soil Science Society of America, the American Society of Agronomy, and the New Zealand Society of Soil Science and was awarded the Communicator of the Year Award by the New Zealand Institute of Agricultural Sciences. He has supervised more than 50 postgraduate students and was awarded the Massey University Research Medal for excellence in postgraduate students' supervision. He has published more than 300 book chapters and journal papers and was awarded the M.L. Leamy Award by the New Zealand Society of Soil Science in recognition of the most meritorious contribution to soil science.

Professor M.B. Kirkham is a professor in the Department of Agronomy at Kansas State University and a graduate of Wellesley College (BA) and the University of Wisconsin, Madison (MS and PhD). Professor Kirkham's research deals with water movement in the soil–plant–atmosphere continuum and uptake of heavy metals by plants grown on contaminated soil. Professor Kirkham has written three textbooks dealing with soil–plant–water relations and is the author or coauthor of more than 250 contributions to scientific publications and is on the editorial boards of 14 journals. Professor Kirkham is on the executive committee of the International Union of Soil Sciences and has received recognitions, including being named fellow of five societies and elected honorary member of the International Union of Soil Sciences.

Dr. Y.S. Ok is a full professor in the Division of Environmental Science and Ecological Engineering, Korea University, Seoul, Korea, where he also serves as director of the Korea Biochar Research Center. His positions include adjunct professor at the University of Wuppertal, Germany; adjunct professor at the University of Southern Queensland, Australia; guest professor at China Jiliang University, China; and guest professor at Ghent University Global Campus, Ghent University, Belgium. Professor Ok holds a B.S. (1998), an M.Sc. (2000), and a Ph.D. (2003) from the Division of Environmental Science and Ecological Engineering, Korea University. He was a postdoctoral fellow in the Department of Renewable Resources, University of Alberta, Canada, and held visiting professorships in the Department of Renewable Resources (University of Alberta), in the Faculty of Bioscience Engineering (Ghent University), in the Department of Civil and Environmental Engineering (Hong Kong Polytechnic University), and in the Department of Chemical and Biomolecular Engineering (National University of Singapore). Professor Ok's research interests include waste management, bioavailability of emerging contaminants, bioenergy, and value-added products such as biochar. Professor Ok also has experience in fundamental soil science and remediation of various contaminants in soils and sediments. Together with graduate students and colleagues, Professor Ok has published more than 400 research papers, 21 of which were ranked as ESI top papers (19 nominated as "highly cited papers" and 2 nominated as "hot papers"). Professor Ok maintains a worldwide professional network through his service as associate editor for *Environmental Pollution* and *Critical Reviews in Environmental Science and Technology* and as member of the editorial boards of *Chemosphere*, *Journal of Analytical and Applied Pyrolysis*, and several other international scientific journals. In addition, Professor Ok has served as guest editor for many leading journals, such as *Journal of Hazardous Materials*, *Bioresource Technology*, *Science of the Total Environment*,

Chemosphere, *Plant and Soil*, *Journal of Environmental Management*, *Applied Geochemistry*, *Environmental Geochemistry and Health*, *Environmental Science and Pollution Research*, and *Geoderma*. In professional service, Professor Ok has organized many international conferences to date, such as 2nd International Conference on Contaminated Land, Ecological Assessment, and Remediation (Korea, 2014); 3rd Asia Pacific Biochar Conference (Korea, 2016), 2nd International Conference on Biological Waste as Resource (Hong Kong, 2017); and others. Professor Ok has been involved in numerous national and international commissions and committees as an active member and regularly serves as an external reviewer for national and international funding agencies.

Contributors

Suhaimi Abdul-Talib
Faculty of Civil Engineering
Universiti Teknologi MARA
Selangor, Malaysia

Tapan Adhikari
Environmental Soil Science
Indian Institute of Soil Science
Bhopal, Madhya Pradesh, India

Carmen T. Agouridis
Department of Biosystems and Agricultural
    Engineering
University of Kentucky
Lexington, Kentucky

Abdulaziz Alghamdi
Department of Agronomy
Throckmorton Plant Sciences Center
Kansas State University
Manhattan, Kansas

and

Department of Soil Science
King Saud University
Riyadh, Saudi Arabia

Patrick Angel
Office of Surface Mining Reclamation and
    Enforcement
U.S. Department of Interior
London, Kentucky

Yasser M. Awad
Korea Biochar Research Centre
Kangwon National University
Chuncheon, Gangwon, Republic of Korea

Christopher D. Barton
Department of Forestry
University of Kentucky
Lexington, Kentucky

N.S. Bolan
Global Centre for Environmental Remediation
and
Faculty of Science
Cooperative Research Centre for
    Contamination Assessment and Remediation
    of the Environment
Advanced Technology Centre
The University of Newcastle
Callaghan, New South Wales, Australia

Sally Brown
School of Environmental and Forest Sciences
University of Washington
Seattle, Washington

Scott X. Chang
Department of Renewable Resources
University of Alberta
Edmonton, Alberta, Canada

Girish Choppala
Southern Cross GeoScience
Southern Cross University
Lismore, New South Wales, Australia

Saikat Chowdhury
Department of Soil Science
Sher-e-Bangla Agricultural University
Dhaka, Bangladesh

Kim Colyvas
Faculty of Science
School of Mathematical and Physical Sciences
The University of Newcastle
Callaghan, New South Wales, Australia

Rajarathnam Dharmarajan
Faculty of Science
Global Centre for Environmental Remediation
University of Newcastle
Callaghan, New South Wales, Australia

Teagan Dolan
Department of Forestry
University of Kentucky
Lexington, Kentucky

Min Duan
College of Chemical and Environmental
   Science
Shaanxi University of Technology
Hanzhong, Shaanxi, People's Republic of
   China

P.E.H. Gregg
Institute of Agriculture and Environment
Massey University
Palmerston North, New Zealand

S.R. Gurung
Department of Primary Industries, Parks,
   Water and Environment
Tasmanian Government
Hobart, Tasmania, Australia

Ganga Hettiarachchi
Department of Agronomy
Throckmorton Plant Sciences Center
Kansas State University
Manhattan, Kansas

Longbin Huang
Sustainable Minerals Institute
The University of Queensland
Brisbane, Queensland, Australia

Mohammed Kader
Faculty of Science
Global Centre for Environmental Research
The University of Newcastle
Callaghan, New South Wales, Australia

and

Cooperative Research Centre for
   Contamination Assessment and Remediation
   of the Environment
University of South Australia
Mawson Lakes, South Australia, Australia

Rajasekar Karunanithi
Faculty of Science
Global Centre for Environmental Remediation
Advanced Technology Centre
The University of Newcastle
Callaghan, New South Wales, Australia

Hyuck Soo Kim
Department of Biological Environment
Kangwon National University
Chuncheon, Gangwon, Republic of Korea

Won Il Kim
Department of Agro-Food Safety
National Institute of Agricultural Science
Wanju-gun, Jeollabuk, Republic of Korea

M.B. Kirkham
Department of Agronomy
Throckmorton Plant Sciences Center
Kansas State University
Manhattan, Kansas

Anitha Kunhikrishnan
NSW Department of Primary Industries
Elizabeth Macarthur Agricultural Institute
Menangle, New South Wales, Australia

Jin-Hyeob Kwak
Department of Renewable Resources
University of Alberta
Edmonton, Alberta, Canada

Dane Lamb
Faculty of Science
Global Centre for Environmental Research
The University of Newcastle
Callaghan, New South Wales, Australia

and

Cooperative Research Centre for
   Contamination Assessment and Remediation
   of the Environment
University of South Australia
Mawson Lakes, South Australia, Australia

John Lavery
Sylvis Environmental
Vancouver, British Columbia, Canada

Sanchita Mandal
Future Industries Institute
University of South Australia
Mawson Lakes, South Australia, Australia

Cherie D. McCullough
Golder Associates Pty Ltd
and
University of Western Australia
Perth, Western Australia, Australia

Dee Murdoch
AECOM Australia Pty. Ltd
Singleton, New South Wales, Australia

Leigh Murray
Department of Statistics
Kansas State University
Manhattan, Kansas

M. Anne Naeth
Department of Renewable Resources
University of Alberta
Edmonton, Alberta, Canada

Ravi Naidu
Faculty of Science
Global Centre for Environmental Research
The University of Newcastle
Callaghan, New South Wales, Australia

and

Cooperative Research Centre for
    Contamination Assessment and Remediation
    of the Environment
University of South Australia
Mawson Lakes, South Australia, Australia

Abimbola Ojekanmi
Department of Renewable Resources
University of Alberta
Edmonton, Alberta, Canada

Y.S. Ok
O-Jeong Eco-Resilience Institute
and
Division of Environmental Science and
    Ecological Engineering
Korea University
Seoul, Republic of Korea

Jin Hee Park
School of Crop Science and Agricultural
    Chemistry
Chungbuk National University
Cheongju, Chungbuk, Republic of Korea

Deann R. Presley
Department of Agronomy
Throckmorton Plant Sciences Center
Kansas State University
Manhattan, Kansas

Craig W. Ross
Landcare Research (Manaaki Whenua)
Manawatu Mail Centre
Palmerston North, New Zealand

Christopher Saint
Natural and Built Environments Research
    Centre
University of South Australia
Mawson Lakes, South Australia, Australia

Peter Sanderson
Faculty of Science
Global Centre for Environmental Research
The University of Newcastle
Callaghan, New South Wales, Australia

and

Cooperative Research Centre for
    Contamination Assessment and Remediation
    of the Environment
University of South Australia
Mawson Lakes, South Australia, Australia

Binoy Sarkar
Future Industries Institute
University of South Australia
Mawson Lakes, South Australia, Australia

and

Department of Animal and Plant Sciences
The University of Sheffield
Sheffield, United Kingdom

Kenton Sena
Department of Forestry
University of Kentucky
Lexington, Kentucky

Balaji Seshadri
Faculty of Science
Global Centre for Environmental Remediation
Advanced Technology Centre
The University of Newcastle
Callaghan, New South Wales, Australia

Robyn C. Simcock
Landcare Research (Manaaki Whenua)
Auckland Mail Centre
Auckland, New Zealand

Bhupinder Pal Singh
NSW Department of Primary Industries
Elizabeth Macarthur Agricultural Institute
Menangle, New South Wales, Australia

Mandeep Singh
Future Industries Institute
University of South Australia
Mawson Lakes, South Australia, Australia

R.B. Stewart
Institute of Agriculture and Environment
Massey University
Palmerston North, New Zealand

Aravind Surapaneni
South East Water
Frankston, Victoria, Australia

Mark Teshima
Sylvia Environmental
Vancouver, British Columbia, Canada

Ramesh Thangavel
Faculty of Science
Global Centre for Environmental Remediation
Advanced Technology Centre
The University of Newcastle
Callaghan, New South Wales, Australia

and

Division of Natural Resource Management
Indian Council of Agricultural Research
Umiam, Meghalaya, India

Andrew Trlica
Department of Geography
Boston University
Boston, Massachusetts

Daniel C.W. Tsang
Department of Civil and Environmental
    Engineering
The Hong Kong Polytechnic University
Hung Hom, Hong Kong, People's Republic of
    China

Jerry A. Vandenberg
Golder Associates Ltd
Kelowna, British Columbia, Canada

Meththika Vithanage
National Institute of Fundamental Studies
Kandy, Sri Lanka

and

Faculty of Applied Sciences
University of Sri Jayewardenepura
Nugegoda, Sri Lanka

Liang Wang
Faculty of Science
Global Centre for Environmental Research
The University of Newcastle
Callaghan, New South Wales, Australia

and

Cooperative Research Centre for
    Contamination Assessment and Remediation
    of the Environment
University of South Australia
Mawson Lakes, South Australia, Australia

Richard C. Warner
Department of Biosystems and Agricultural
    Engineering
University of Kentucky
Lexington, Kentucky

Hasintha Wijesekara
Faculty of Science
Global Centre for Environmental Remediation
Advanced Technology Centre
The University of Newcastle
Callaghan, New South Wales, Australia

Xinni Xiong
Department of Civil and Environmental
    Engineering
The Hong Kong Polytechnic University
Hung Hom, Hong Kong, People's Republic of
    China

Yilu Xu
Faculty of Science
Global Centre for Environmental Remediation
Advanced Technology Centre
The University of Newcastle
Callaghan, New South Wales, Australia

Yubo Yan
Faculty of Science
Global Centre for Environmental Remediation
Advanced Technology Centre
The University of Newcastle
Callaghan, New South Wales, Australia

Soon Kong Yong
Faculty of Applied Sciences
Universiti Teknologi MARA
Selangor, Malaysia

Fang You
Sustainable Minerals Institute
The University of Queensland
Brisbane, Queensland, Australia

Carl Zipper
Department of Crop and Soil Environmental
   Sciences
Virginia Polytechnic Institute and State University
Blacksburg, Virginia

Section I

Mine Site Characterization

1 Characterization and Improvement in Physical, Chemical, and Biological Properties of Mine Wastes

Binoy Sarkar, Hasintha Wijesekara, Sanchita Mandal, Mandeep Singh, and Nanthi S. Bolan

CONTENTS

1.1 Introduction 3
1.2 Nature of Mine Wastes 4
1.3 Mine Waste Characteristics 4
    1.3.1 Physical Properties 4
    1.3.2 Chemical Properties 5
    1.3.3 Biological Properties 6
1.4 Improving Mine Waste Characteristics 7
    1.4.1 Physical Characteristics 8
    1.4.2 Chemical Characteristics 9
    1.4.3 Biological Characteristics 10
1.5 Summary and Conclusions 11
References 11

1.1 INTRODUCTION

Degradation of land resources as a result of mining activities poses serious threat to the environment. It has been estimated that around 0.4×10^6 km^2 area of land is impacted by mining activities around the world (Hooke and Martín-Duque 2012). Unfortunately, a significant percentage of this area has never been reclaimed, which poses health risks to ecosystems and humans. Often, these wastes contain hazardous substances such as heavy metals, organic contaminants, radionuclides, and crushed limestone, where the latter could become a potential source of atmospheric CO_2 emission. Thus, they not only pose serious risk to the groundwater and surface water, but also to the atmosphere (Wijesekara et al. 2016). In order to tackle the issues related to mine wastes and manage the affected sites sustainably, an appropriate physical, chemical, and biological characterization of waste materials becomes very prudent. Due to the lack of both above- and below-ground biodiversity, mine waste sites are very poor in organic matter content. This in return leads to poor seed germination, plant growth, and vegetation establishment. In many cases, the associated toxic contaminants also seriously compromise the soil health, microbial life, and plant growth (Castillejo and Castelló 2010, Larney and Angers 2012). This chapter describes the physicochemical characteristics of mine wastes, including spoil, tailings, and overburden, by underpinning their source–property relationships. The value of readily available biowaste resources, including biosolids, composts, and manures, in improving such physicochemical properties of mining-impacted soils/sites is also discussed.

1.2   NATURE OF MINE WASTES

In general, two major types of wastes are produced during mining activities: mine spoil and mine tailings. The waste materials which are produced as a result of underground mining, quarries, or open-cast excavations are generally known as mine spoil (Wijesekara et al. 2016). Similarly, mine tailings refer to the concentrated admixture of crushed rocks and ore processing ingredients/fluids following the extraction of the valuable resource from ores (Kossoff et al. 2014). Due to the drastic disturbances during their processing and production, these wastes become extremely poor in available nutrients, including carbon, and hence unsuitable for any microbial habitat development and plant growth (Singh et al. 2004).

In addition to the two types of wastes mentioned, the removed material from the upper segments of a mountain, ridge, or valley is known as overburden. The overburden soil is not only unproductive, but also very much prone to erosion and landslides (Johnson 2003, Sopper 1992).

1.3   MINE WASTE CHARACTERISTICS

Mine wastes can impede seed germination, seedling establishment, and plant growth through a range of physical, chemical, and biological constraints. These negative attributes could ruin the revegetation attempt of a mine waste site either partially or completely. Depending on the sources of wastes and prevailing agroclimatic and environmental conditions, one type of constraint may become predominant over the other. Therefore, understanding these characteristics is of utmost importance for making effective recommendations to revegetate/rejuvenate an impacted site.

1.3.1   PHYSICAL PROPERTIES

The physical properties of mine wastes can largely be different depending on their forms and ore processing procedure (Table 1.1). For example, the overburden waste, which is produced during the excavation in order to uncover the main ore body, consists of comparatively less-weathered rocks. These rocks are very coarse in texture, having particle diameters ranging from 2 to 20 cm (Tordoff et al. 2000). Their degree of weathering can be predicted approximately from the color; usually bright red or brown color indicates a higher degree of oxidation or weathering (Sheoran et al. 2010). Depending on the variation in the rock hardness, blasting techniques, and handling processes, the content of coarse fragments in a mine spoil can vary from <30% to >70% (Sheoran et al. 2010). Such a large particle size is naturally an obstacle to particle aggregation. The larger particle size, together with a severe shortage of organic matter, gives rise to a poor soil texture and structure (Castillejo and Castelló 2010). The overburden or spoil composed of large rock particles prevents seed germination and seedling development mainly due to the extremely poor water-holding capacity. These wastes are also very prone to erosion by earthquakes, landslides, wind, and water (Dawson et al. 1998, Ulusay et al. 1995). The fine-grain fraction of the waste (e.g., tailings; <2 mm particle diameter) can form soil crusts and cracks, which also adversely influence the soil structure (Hossner and Hons 1992, Sheoran et al. 2010).

The mine tailings when pumped into the slime dams can cause surface soil compaction and even cementation, which can potentially create an impenetrable barrier to plant roots (Bradshaw et al. 1978). At the very beginning of the mining activity itself, the soil aggregate structure is destroyed due to the removal of successive soil layers, which are stockpiled at another location. This could also promote soil compaction and reduce the water-holding capacity and aeration. These parameters are prerequisites for a good plant establishment and growth. The bulk density of a soil optimum for plant growth ranges between 1.1 and 1.5 g cm^{-3}. The bulk density of a mine overburden can reach as high as 1.9 g cm^{-3} (Maiti and Ghose 2005), whereas a soil having value above 1.7 g cm^{-3} is considered as severely compacted. Understandably, this kind of soil becomes unfit for plant growth.

TABLE 1.1
Major Properties of Mine Wastes in Comparison with Agricultural Soils

Characteristics	Agricultural Soil	Mine Overburden	Mine Spoil	Mine Tailings
Texture[a]	Clay loam/sandy loam	Coarse (particles > 4 mm)	Medium (particles about 2–4 mm)	Fine (particles < 2 mm)
Structure	Strong	Structureless	Weak to moderate	Moderate to strong
Water-holding capacity	High	Very poor	Moderate	Moderate to high
Permeability	Moderate	Rapid to very rapid	Moderately rapid to moderate	Moderately slow to very slow
Crusting	Special situation	Rare	Often	Very often
Soil penetration	Easy	Hard	Moderate	Moderate to hard
Bulk density[b]	1.1–1.5 g cm^{-3}	Up to 1.9 g cm^{-3}	Up to 1.7 g cm^{-3}	1.4–1.8 g cm^{-3}
pH[c]	5–8	NA	As low as 1	As low as –3
Electrical conductivity[d]	0.2–2 dS m^{-1}	NA	2–12 dS m^{-1}	2–12 dS m^{-1}
Organic matter	Moderate	Extremely low	Very low	Very low
Nutrient contents	Moderate	Extremely low	Extremely low	Extremely low
Heavy metals	No concern	Extreme concern	Extreme concern	Extreme concern
Organic contaminants	No concern	Medium concern	Extreme concern	Extreme concern
Radionuclides	No concern	Extreme concern	Extreme concern	Extreme concern
Microbial community	Ambient proliferation	Establishment severely impacted	Establishment severely impacted	Establishment severely impacted
Enzymatic activity	Ambient activity	Activity severely impacted	Activity severely impacted	Activity severely impacted

[a] Sheoran et al. (2010) and Tordoff et al. (2000).
[b] Maiti and Ghose (2005).
[c] Nordstrom et al. (2000) and Gerke et al. (1998).
[d] Arocena et al. (2012) and Gardner et al. (2010).

1.3.2 CHEMICAL PROPERTIES

The chemical properties of mine wastes can be evaluated mainly by the pH of the waste matrix (Table 1.1). The pH value is largely controlled by the chemical nature of the waste itself. For example, weathering of pyrite-based rocks can lower the pH to a level which is unsuitable for plant growth, microbial proliferation, and water quality (Blechschmidt et al. 1999, Lindsay et al. 2015, Moncur et al. 2015). In mine waters in California, a highly negative pH value (pH = –3.6) was reported (Nordstrom et al. 2000). Similarly, the groundwater which was contaminated by acid mine drainage from overburden spoil piles at an open-pit lignite mine in Germany recorded a pH value as low as 1 (Gerke et al. 1998). The alteration in pH directly affects the mobility and availability of many major and micro-nutrient elements (Sheoran et al. 2010). Due to the predominance of intrinsically low pH in mine wastes, elevated heavy metal concentration is a common feature of most of the mine spoils. For example, Zn and Pb concentrations in mining- and smelting-contaminated soils could be as high as 14,700 and 27,000 mg kg^{-1}, respectively (Brown et al. 2003). Studies on some tailings of polymetallic mines in Morocco reported extremely high concentrations of heavy metals such as Zn (38,000–108,000 mg kg^{-1}), Pb (20,412–30,100 mg kg^{-1}), Cu (2,019–8,635 mg kg^{-1}), and Cd (148–228 mg kg^{-1}) (Boularbah et al. 2006). Similarly, metal concentrations (7384 mg kg^{-1} As, 7200 mg kg^{-1} Cu, 3450 mg kg^{-1} Pb, and 6270 mg kg^{-1} Zn) in soils around a semi-arid mining area in Mexico were many times greater than the permissible limits (Razo et al. 2004). Semi-arid mine tailings from Spain contained Pb, Zn, Cd, Cu, and As as high as 8.3 g kg^{-1}, 12.5 g kg^{-1}, 40.9 mg kg^{-1},

332.1 mg kg^{-1}, and 314.7 mg kg^{-1}, respectively (Navarro et al. 2008). The elevated concentration of metals in mine wastes imposes multiple adverse impacts, such as groundwater and surface water contamination, plant toxicity, and destruction of biodiversity and microbial activity. As a result of poor microbial proliferation, the decomposition of organic matter and subsequent humus formation could be deeply impacted (O'Reilly 1997). The conversion of soil organic matter and humus formation are of utmost importance for structural development (aggregate formation) in mine wastes. The elevated heavy metal levels in soils and waters around mining-impacted areas can potentially cause toxic effects in organisms, including humans, and thus exacerbate serious human health issues (Bolan et al. 2003, Lindsay et al. 2015). The altered pH can also adversely affect the availability of essential nutrients in mine wastes, which can prevent plant establishment and growth.

Wastes generated from uranium mining can often contain radioactive material and substances, which is also a serious issue to the environmental quality (Lottermoser 2010, Lottermoser and Ashley 2008). Uranium mine wastes can potentially impose risks to the environment and human health through the associated ambient radiation doses and contamination with other radionuclides such as ^{238}U (^{234}Th), ^{226}Ra, and ^{210}Pb (Carvalho et al. 2007). Plants grown on mine spoils from uranium mining were reported to take up a greater level of lithophile or rare earth elements such as Ce, Cr, La, Lu, Rb, Sc, Th, U, V, Y, and Yb (Li et al. 2013b, Lottermoser 2015, Lottermoser and Ashley 2006). Moreover, groundwater and surface waters can be polluted by enhanced transportation of radionuclides through sediments and acid drainage (Neiva et al. 2014, 2016).

In mine sites, organic contaminants such as petroleum and polycyclic aromatic hydrocarbons (PAHs) can also be a menace to the quality of the surrounding soil and surface water and groundwater. These organic contaminants are mainly derived from the mining activities, including digging, extraction, processing, and transport. For example, PAHs, mostly dominated by phenanthrene, fluoranthene, and pyrene, both of petrogenic and of pyrolytic origin (due to burning of coal waste piles), were reported in the Douro Coalfield, which is the largest outcrop of terrestrial Carboniferous coal-bearing strata in Portugal (Ribeiro et al. 2012, 2013). Similar results were also reported from the Tiefa coal mine district in China, where a study on the vertical distribution of the compounds showed their penetration into the deeper layers of the soil (Liu et al. 2012). Another study at the coal gangue dump of Jiulong Coal Mine in China showed that the wastewater stream could carry organic substances at least 1800 m far from the original dump (Sun et al. 2009). In addition, spontaneous combustion of coal and coal waste gobs is an important source of atmospheric pollutants, including toxic volatile organic compounds (Fabiańska et al. 2013, Querol et al. 2011). The occurrence of a mixture of inorganic and organic contaminants in abandoned mine sites can make the reclamation attempt extremely challenging. Such kinds of a complex legacy of contamination with benzo(a) pyrene, other PAHs, and unexpected mercury (Hg) organocompounds (phenylmercury propionate) were also reported from abandoned Hg–As mining and metallurgy sites in Spain (Gallego et al. 2015).

1.3.3 BIOLOGICAL PROPERTIES

Mine wastes are usually characterized as a poor habitat for microorganisms (Table 1.1). Many adverse physical and chemical conditions prevent the development of microbial communities in mine wastes. These include extreme pH, lack of organic matter, lack of available nutrients, heavy metal toxicity, and poor particle aggregation. The organic matter content in soils is directly correlated to the soil microbial life and health, which is prerequisite for plant growth (Castillejo and Castelló 2010, Larney and Angers 2012). Since plants cannot grow in most of the mine wastes, the chances of organic matter addition through plant litter and root exudates are very little (Asensio et al. 2013a). Enzymatic activity, which is considered as an early and effective indicator of soil quality change, in mine waste–impacted soils is very poor (Asensio et al. 2013b, Humphrey et al. 1995). It is also a vital index of soil fertility because various enzymes take part in the biogeochemical cycling of nutrients (Sarkar et al. 2016). Soil enzymes such as arylsulfatase and phosphatase are known to be highly sensitive to metal toxicity (Kandeler et al. 1996), which is a common occurrence in mine waste–impacted soils. Moreover, mine waste, especially from pyrite mines, which contain a very acidic reaction, can significantly influence the prevalence of a particular microbial community. It is because soil microbial populations (e.g., bacteria, fungi, archaea) are highly sensitive to the soil pH (Aciego Pietri and Brookes 2009). Another possible way of altering the diversity of micro- and macroorganisms in mine wastes could be through stockpiling near mining sites, which could create anaerobic conditions in subsurface soils. It was reported that such stockpiling could prohibit the growth and proliferation of earthworms in mine soils (Boyer et al. 2011).

1.4 IMPROVING MINE WASTE CHARACTERISTICS

Having understood the physical, chemical, and biological characteristics of mine wastes, various rehabilitation and revegetation strategies can be undertaken. The choice of a particular mine site rehabilitation strategy would largely depend on the degree of disturbance, level of contamination, prevailing agroclimatic conditions, and the cost of the concerned technology. The level of remediation could also be decided based upon the future use of the impacted land. For example, if the impacted land would be used for recreational purposes (e.g., to develop a parkland), the degree of reclamation required would be less intensive than in the soils that would be used for agricultural crop production. A risk-based rehabilitation strategy (e.g., in order to minimize the risk just up to the level it is needed) in this case could be implemented to confine the cost to a profitable limit (Naidu 2013, Naidu et al. 2008, Sanderson et al. 2015).

Several approaches have been tried to revegetate mine waste–contaminated soils that contain heavy metals. The key strategies include immobilization of heavy metals through addition of organic or inorganic amendments (i.e., sewage sludge, compost, fly ash, or biochar), application of a cap or liner over the toxic materials and initiation of growth of vegetation by using uncontaminated soils and heavy metal–tolerant genotypes (Lamb et al. 2014, O'Reilly 1997, Wijesekara et al. 2016). Peat materials and forest floor, which are rich sources of natural organic matter, can also be used for mine site rehabilitation (Beasse 2012). In addition, mine wastes can be stored in isolated impoundments under water or behind dams (Kossoff et al. 2014). However, this is risky because the impoundments may frequently fail, which may cause unwanted release of the contaminants in the surrounding environment. Recent scientific advancements in mine site reclamation include amending with nanoenhanced materials such as iron oxides, phosphate-based nanoparticles, iron sulfide nanoparticles, and carbon nanotubes (Liu and Lal 2012, Wijesekara et al. 2016). Such materials can improve the mine spoil quality by enhancing the usage efficiency of other coamendments, immobilizing heavy metals, and preventing soil erosion (Liu and Lal 2012).

Among various types of amendments, copious and readily available biowaste resources have been found to be very popular in successful rehabilitation of mining-impacted lands (Gardner et al. 2010, Sopper 1992, Stolt et al. 2001). The key features of biowastes that have made this technology popular are their easy local availability, cheaper costs, and ease in practical applications. Different types of biowastes include, but are not limited to, biosolids, sawdust, sewage sludge, municipal solid waste, yard waste, composts, manures, and biowaste-derived biochar. Biowastes are able to bring about beneficial changes in the physical, chemical, and biological properties of mine wastes and mine waste–impacted soils (Figure 1.1). Some of these effects are evident directly (e.g., lowering of the bulk density), while several indirect chemical and biological properties may take some time to become apparent. Numerous laboratory and field-scale experiments and case studies have successfully demonstrated the efficiency of biowastes in mine site rehabilitation (Larney and Angers 2012, Wijesekara et al. 2016). In the long run, many additional behavioral, nutritional, or growth responses of flora and fauna and groundwater quality improvements could be achieved through biowaste application to mine waste sites and surrounding areas (Sopper 1992). Figure 1.1 depicts the inherent mechanisms of how biowastes could rehabilitate mine wastes and/or mine waste–impacted soils.

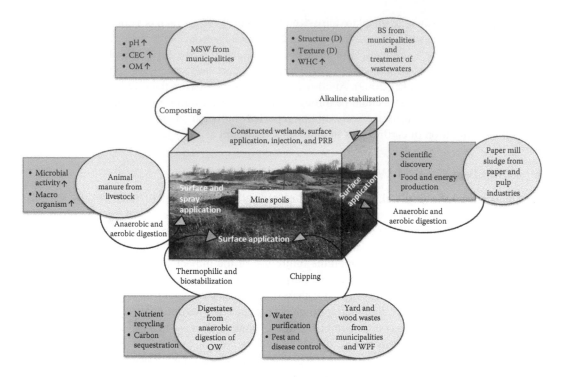

FIGURE 1.1   Various approaches in utilization of different biowastes for mine waste rehabilitation and potential benefits. MWS, municipal solid waste; BS, Biosolids; D, Develop; CEC, Cation exchange capacity; WHC, Water-holding capacity; OM, Organic matter; PRB, Permeable reactive barriers; OW, Organic waste; WPF, Wood processing facilities.

1.4.1 PHYSICAL CHARACTERISTICS

The key component in biowastes which helps in improving the physical characteristics of mine wastes is organic matter. Upon application, biowaste can decrease the bulk density, increase the water-holding capacity and hydraulic conductivity, improve porosity and particle aggregation, reduce soil erosion, and improve soil temperature and aeration regimes (Larney and Angers 2012, Sopper 1992). For example, application of pig manure and sewage sludge to acidic mine tailings increased the aggregate stability from 19% to 30% and 19% to 28%, respectively (Zanuzzi et al. 2009). Similarly, application of urban municipal solid waste to a gypsum spoil quarry improved the available water capacity from 13.7% to 17.1%, which consequently decreased the permanent wilting point from 14.1% to 11.8% (Castillejo and Castelló 2010). Improvements in the physical characteristics bring about an improvement in the clay content, overall structure, and texture of impacted soils. At mine waste dumps where soil incrustation is a problem, application of biowastes improves water infiltration and aeration. Since most of the physical properties of soils are interrelated, application of biowastes can simultaneously enhance multiple properties at a time such as bulk density, porosity, hydraulic conductivity, and infiltration. For example, the comparative positive effect of saw dust and municipal sewage sludge amendment to a mine site in Virginia, the United States, on aggregate stability, bulk density, and porosity was apparent following 5 years of application (Bendfeldt et al. 2001). However, the effect saw no lasting soil quality improvement after 16 years of amendment application (Bendfeldt et al. 2001). Similarly, the water-holding capacity increased and bulk density decreased as a result of wood compost application to an open-cast mine contaminated with tertiary and quaternary substrates of coal spoils (Nada et al. 2012).

1.4.2 CHEMICAL CHARACTERISTICS

Like the physical characteristics, biowastes can improve various chemical characteristics of mining-impacted soils, and many of these characteristics are interrelated in their functions. Chemical properties that can be directly improved through biowaste application include pH, electrical conductivity (EC), cation exchange capacity (CEC), and nutrient and organic matter contents (Wijesekara et al. 2016). The availability and toxicity of heavy metals in mine spoils can be ameliorated directly through adsorption/immobilization on the applied biowastes or indirectly through alternation in the pH value. Heavy metal mobility and availability in mine waste–impacted soils having very low pH values are a critical issue. Biowaste amendments have a direct role in increasing the pH values, thereby reducing metal mobility (Nada et al. 2012, Sopper 1992). For example, application of municipal solid waste increased the pH of a mining soil from 6.8 to 7.4 (Bendfeldt et al. 2001). While this pH change was small, a more prominent pH change in an acidic tailing (from 2.9 to 7.3) was achieved through the application of pig manure (Zanuzzi et al. 2009). In another example, pig manure application to mine tailings raised the pH from 2.8 to 5.9 (Arocena et al. 2012). Sewage sludge also brought about a similar remarkable rise in the pH value (2.8–5.8) of mine tailings (Arocena et al. 2012).

Biowaste-derived biochar has received tremendous popularity in controlling heavy metal pollution in mining-impacted soils (Abdelhafez et al. 2014, Fellet et al. 2011, Kelly et al. 2014, Puga et al. 2016). Biochar products are also well known to improve the nutrient retention capacity of amended soils (Mandal et al. 2016a,b), and thus can facilitate plant establishment and growth in mining-impacted soils by supplying nutrients. Biosolids, sewage sludge, and municipal solid waste are similar amendments that can also provide the dual function of nutrient supply and heavy metal immobilization (Bolan et al. 2014, Park et al. 2011). These biowastes are rich in macronutrients such as N and P and micronutrients such as B, Cu, Fe, Mo, Mn, and Zn. Application of sewage sludge to a calcareous soil from bauxite mining increased the total N and Olsen P contents from 100 to 5000 mg kg^{-1} and from 24.7 to 255.3 mg kg^{-1}, respectively (Brofas et al. 2000). Similarly, application of municipal solid waste to mine soils in Spain increased water-extractable P and Fe contents from 6.6 to 29.6 mg kg^{-1} and from 2.8 to 5.4 mg kg^{-1}, respectively (de Varennes et al. 2010). Owing to these nutritional values, biowaste application improves the fertility of mine spoil soils (Jones et al. 2010). The fertility effect can be further enhanced by combining biowastes with other amendments and fertilizer materials prior to application to mine sites (Li et al. 2013a).

Often, lime is added to stabilize biowastes, which increases the pH value of the final product and enhances heavy metal immobilization (Amuda et al. 2008, Basta et al. 2001, Feagley et al. 1994, Lim et al. 2016). In highly acidic sulfur-bearing mine tailings, however, the pH may decrease gradually over the course of time and heavy metals may potentially remobilize (Sopper 1992). Both biotic and abiotic phenomena may be responsible for this slow pH reversal. Biowastes when applied to mine waste undergo slow decomposition, which generates organic acid. Also, during the nitrification of ammonium-bearing compounds in biowaste, numerous protons are generated. These effects, together with acid production as a result of pyrite mineral oxidation, may lower the pH of the matrix. Therefore, for an effective management of mine wastes, care should be taken to decide the application rate, incorporation depth, and addition of coamendments (Castillejo and Castelló 2010). For example, co-composting of biowastes with alkaline materials such as lime, fluidized bed boiler ash, flue gas desulfurization gypsum, and red mud improved carbon sequestration and the revegetation potential of disturbed soils (Chowdhury et al. 2016). Care should also be taken that application of biowastes does not increase the EC (i.e., soluble salt contents) of the plant growth matrix to an alarming level. Excess soluble salts may promote the formation of metal–inorganic complexes, which are far more readily available to plants than other chemical species. As a result of such excess metal uptake, phytotoxicity may appear, which can potentially impact the mine site rehabilitation strategy through phytoremediation, along with the application of biowastes (Feagley et al. 1994, Smolders et al. 1998, Wilden et al. 1999). Excess salts can also directly prevent plant establishment and growth by impacting the osmotic regulation of water and nutrients uptake (Rodgers and Anderson 1995). However, some instances are also available where the EC dropped as a result of biowaste application to mine wastes. For example, application of pig manure and sewage sludge to acid tailings decreased the EC from 2.8 to 2.2 and 2.3, respectively (Arocena et al. 2012). When heavy metal cations are immobilized by the organic matter of biowastes, other soluble salts are leached down, which may consequently lower the EC value (Clapp et al. 1986).

Another key chemical parameter that is shown to change as a result of biowaste application to mine wastes is the CEC (Wijesekara et al. 2016). CEC is highly dependent on the content of clay minerals and sesquioxides in mine wastes (Gardner et al. 2010). The contribution from biowastes to raise the CEC comes through the addition of organic colloids in the form of humic substances. The degree of humification of organic matter in biowastes largely depends on time, climatic conditions, and the availability of biological activities (Fierro et al. 1999, Larney and Angers 2012).

1.4.3 Biological Characteristics

In order to thrive in the soil environment, microorganisms require energy, which they gain from soil organic matter. Through the application of biowaste, a huge amount of organic matter is added to mine wastes, which facilitates the development and growth of microbial communities. Biological soil health parameters such as microbial biomass carbon (MBC), enzymatic activities, and microbial populations can be monitored following the application of biowastes to mine wastes. In general, biowaste application enhances these quality parameters in both short- and long-run experiments. For example, application of compost from mixed municipal solid waste significantly enhanced activities of six different soil enzymes, namely acid phosphatase, β-glucosidase, cellulose, urease, protease, and dehydrogenase, in a soil developed on pyrite mine wastes (de Varennes et al. 2010). Similarly, dehydrogenase, β-glucosidase, and phosphatase activities were enhanced by the application of horticultural waste compost to abandoned metalliferous mine wastes (Melgar-Ramírez et al. 2012). Both of these examples reported the biological effects in a less-than-a-year study. A-long term and sustainable improvement in the microbial population and activities was also reported from a copper mine tailing site in the United States which received class A biosolids treatment 10 years ago (Pepper et al. 2012). It was found that microbial functions such as nitrification and sulfur oxidation and dehydrogenase activity were sustained after 10 years of biosolids application (Pepper et al. 2012). In order to observe a significant increase in the population of soil microorganisms (e.g., bacteria and fungi), a large application rate of biowaste might be needed, whereas changes in enzymatic activities might be evident even at a smaller application rate (Baker et al. 2011). The application of composted beef manure to a heavy metal–contaminated mine waste increased arylsulfatase, phosphatise, and β-glucosidase enzyme activities when coapplied with bentonite (Baker et al. 2011). Other coamendments may include liming material and chemical fertilizers. For example, when sewage sludge was applied in combination with chemical fertilizers, it increased the populations of fungi, bacteria, and actinomycetes and the activity of soil urease in abandoned open-cast mining areas (Li et al. 2005, 2013a). It was reported that co-composting of biowaste may provide an almost similar benefit for rehabilitating a mine waste compared with the incorporation of matured compost (Tandy et al. 2009).

In addition to microorganisms, application of biowastes to mine wastes can improve the population and diversity of many beneficial macroorganisms. For example, a higher population density of earthworms was observed in open-cast coal mining spoil soils by the application of biowastes (i.e., biosolids, compost, and coal ash) in comparison with mineral fertilizers (Emmerling and Paulsch 2001). On the contrary, biosolids application may sometime cause earthworm mortality. Seasonal influences, along with ecological factors such as the diet of earthworms, are important factors that may control the survival and proliferation of the organisms in a given habitat (Ireland 1983, Ireland and Wooton 1976).

1.5 SUMMARY AND CONCLUSIONS

Mining activities produce a big amount of wastes in the form of mine spoils, tailings, and overburden. Often, these wastes are highly contaminated with heavy metals, organic contaminants, and even radionuclides. In addition, these wastes have very poor physical, chemical, and biological characteristics, which are unsuitable for plant establishment and growth. While the contaminants in these wastes may create a menace to the surrounding environment and impact surface water and groundwater quality, other poor characteristics provide obstruction to revegetation and rehabilitation attempts.

Lack of organic matter and microbial life is one of the key reasons that may spoil the revegetation attempt of a mining-impacted site. Therefore, application of biowastes, which are a rich source of organic matter, has been found to be very effective. Unlike many other inorganic amendments, biowastes are readily available and cheap in cost. Along with immobilizing toxic heavy metals and organic contaminants, biowastes are able to improve numerous physical, chemical, and biological properties of impacted soils. Thus, they improve the overall soil quality and fertility. However, care should be taken in deciding the appropriate application rate and application method of these biowastes in a rehabilitation program. The longevity and effectiveness of biowaste amendments should be predicted prior to their real application in a contaminated site. Some biowastes might carry secondary pollutants and harmful microorganisms, which also need special regulatory attention. Application of biowastes for mine site rehabilitation may also provide additional advantages such as recycling of energy for plant production, carbon sequestration, and greenhouse gas mitigation. More scientific research is required under different agroclimatic conditions to understand metal immobilization and microbial community development mechanisms (atomic and molecular levels) in mine spoils through biowaste application.

REFERENCES

Abdelhafez, A.A., J. Li, and M.H.H. Abbas. 2014. Feasibility of biochar manufactured from organic wastes on the stabilization of heavy metals in a metal smelter contaminated soil. *Chemosphere* 117:66–71.

Aciego Pietri, J.C., and P.C. Brookes. 2009. Substrate inputs and pH as factors controlling microbial biomass, activity and community structure in an arable soil. *Soil Biol. Biochem.* 41(7):1396–1405.

Amuda, O., A. Deng, A. Alade, and Y.-T. Hung. 2008. Conversion of sewage sludge to biosolids. In *Biosolids Engineering and Management*, L. Wang, N. Shammas, and Y.-T. Hung (Eds.), pp. 65–119. New York: Humana Press.

Arocena, J.M., J.M. van Mourik, and A. Faz Cano. 2012. Granular soil structure indicates reclamation of degraded to productive soils: A case study in southeast Spain. *Can. J. Soil Sci.* 92(1):243–251.

Asensio, V., E.F. Covelo, and E. Kandeler. 2013a. Soil management of copper mine tailing soils—Sludge amendment and tree vegetation could improve biological soil quality. *Sci. Total Environ.* 456–457:82–90.

Asensio, V., F.A. Vega, B.R. Singh, and E.F. Covelo. 2013b. Effects of tree vegetation and waste amendments on the fractionation of Cr, Cu, Ni, Pb and Zn in polluted mine soils. *Sci. Total Environ.* 443:446–453.

Baker, L.R., P.M. White, and G.M. Pierzynski. 2011. Changes in microbial properties after manure, lime, and bentonite application to a heavy metal-contaminated mine waste. *Appl. Soil Ecol.* 48(1):1–10.

Basta, N.T., R. Gradwohl, K.L. Snethen, and J.L. Schroder. 2001. Chemical immobilization of lead, zinc, and cadmium in smelter-contaminated soils using biosolids and rock phosphate. *J. Environ. Qual.* 30(4):1222–1230.

Beasse, M.L. 2012. Microbial communities in organic substrates used for oil sands reclamation and their link to boreal seedling growth. MSc thesis, Department of Renewable Resources, University of Alberta, Edmonton, Alberta, Canada.

Bendfeldt, E.S., J.A. Burger, and W.L. Daniels. 2001. Quality of amended mine soils after sixteen years. *Soil Sci. Soc. Am. J.* 65(6):1736–1744.

Blechschmidt, R., W. Schaaf, and R.F. Hüttl. 1999. Soil microcosm experiments to study the effects of waste material application on nitrogen and carbon turnover of lignite mine spoils in Lusatia (Germany). *Plant Soil* 213(1):23–30.

Bolan, N.S., D.C. Adriano, and D. Curtin. 2003. Soil acidification and liming interactions with nutrient and heavy metal transformation and bioavailability. *Adv. Agron.* 78:215–272.

Bolan, N.S., A. Kunhikrishnan, R. Thangarajan, J. Kumpiene, J.H. Park, T. Makino, M.B. Kirkham, and K.G. Scheckel. 2014. Remediation of heavy metal(loid)s contaminated soils—To mobilize or to immobilize? *J. Hazard. Mater.* 266:141–166.

Boularbah, A., C. Schwartz, G. Bitton, and J.L. Morel. 2006. Heavy metal contamination from mining sites in South Morocco: 1. Use of a biotest to assess metal toxicity of tailings and soils. *Chemosphere* 63(5):802–810.

Boyer, S., S. Wratten, M. Pizey, and P. Weber. 2011. Impact of soil stockpiling and mining rehabilitation on earthworm communities. *Pedobiologia* 54(Suppl):S99–S102.

Bradshaw, A.D., M.O. Humphreys, and M.S. Johnson. 1978. The value of heavy metal tolerance in the reveg- etation of metalliferous mine wastes. In *Environmental Management of Mineral Wastes*, G.T. Goodman and M.J. Chadwick (Eds.), pp. 311–334. Alphen aan den Rijn, the Netherlands: Sijthoff & Noordhoff.

Brofas, G., P. Michopoulos, and D. Alifragis. 2000. Sewage sludge as an amendment for calcareous bauxite mine spoils reclamation. *J. Environ. Qual.* 29(3):811–816.

Brown, S.L., C.L. Henry, R. Chaney, H. Compton, and P.S. DeVolder. 2003. Using municipal biosolids in com- bination with other residuals to restore metal-contaminated mining areas. *Plant Soil* 249(1):203–215.

Carvalho, F.P., M.J. Madruga, M.C. Reis, J.G. Alves, J.M. Oliveira, J. Gouveia, and L. Silva. 2007. Radioactivity in the environment around past radium and uranium mining sites of Portugal. *J. Environ. Radioact.* 96(1–3):39–46.

Castillejo, J.M., and R. Castelló. 2010. Influence of the application rate of an organic amendment (Municipal Solid Waste [MSW] compost) on gypsum quarry rehabilitation in semiarid environments. *Arid Land Res. Manage.* 24(4):344–364.

Chowdhury, S., N.S. Bolan, B. Seshadri, A. Kunhikrishnan, H. Wijesekara, Y. Xu, J.Y. Yang, G.H. Kim, D.L. Sparks, and C. Rumpel. 2016. Co-composting solid biowastes with alkaline materials to enhance carbon stabilization and revegetation potential. *Environ. Sci. Pollut. Res.* 23(8):7099–7110.

Clapp, C.E., S.A. Stark, D.E. Clay, and W.E. Larson. 1986. Sewage sludge organic matter and soil properties. In *The Role of Organic Matter in Modern Agriculture*, Y. Chen and Y. Avnimelech (Eds.), pp. 209–253. Dordrecht, the Netherlands: Springer.

Dawson, R.F., N.R. Morgenstern, and A.W. Stokes. 1998. Liquefaction flowslides in Rocky Mountain coal mine waste dumps. *Can. Geotech. J.* 35(2):328–343.

de Varennes, A., M.M. Abreu, G. Qu, and C. Cunha-Queda. 2010. Enzymatic activity of a mine soil varies according to vegetation cover and level of compost applied. *Int. J. Phytorem.* 12(4):371–383.

Emmerling, C., and D. Paulsch. 2001. Improvement of earthworm (Lumbricidae) community and activ- ity in mine soils from open-cast coal mining by the application of different organic waste materials. *Pedobiologia* 45(5):396–407.

Fabiańska, M.J., J. Ciesielczuk, Ł. Kruszewski, M. Misz-Kennan, D.R. Blake, G. Stracher, and I. Moszumańska. 2013. Gaseous compounds and efflorescences generated in self-heating coal-waste dumps—A case study from the Upper and Lower Silesian Coal Basins (Poland). *Int. J. Coal Geol.* 116–117:247–261.

Feagley, S.E., M.S. Valdez, and W.H. Hudnall. 1994. Bleached, primary papermill sludge effect on bermudag- rass grown on a mine soil. *Soil Sci.* 157(6):389–397.

Fellet, G., L. Marchiol, G. Delle Vedove, and A. Peressotti. 2011. Application of biochar on mine tailings: Effects and perspectives for land reclamation. *Chemosphere* 83(9):1262–1267.

Fierro, A., D.A. Angers, and C.J. Beauchamp. 1999. Restoration of ecosystem function in an abandoned sand- pit: Plant and soil responses to paper de-inking sludge. *J. Appl. Ecol.* 36(2):244–253.

Gallego, J.R., N. Esquinas, E. Rodríguez-Valdés, J.M. Menéndez-Aguado, and C. Sierra. 2015. Comprehensive waste characterization and organic pollution co-occurrence in a Hg and As mining and metallurgy brown- field. *J. Hazard. Mater.* 300:561–571.

Gardner, W.C., K. Broersma, A. Naeth, D. Chanasyk, and A. Jobson. 2010. Influence of biosolids and fertilizer amendments on physical, chemical and microbiological properties of copper mine tailings. *Can. J. Soil Sci.* 90(4):571–583.

Gerke, H.H., J.W. Molson, and E.O. Frind. 1998. Modelling the effect of chemical heterogeneity on acidifica- tion and solute leaching in overburden mine spoils. *J. Hydrol.* 209(1–4):166–185.

Hooke, R.L., and J.F. Martín-Duque. 2012. Land transformation by humans: A review. *GSA Today* 22:4–10.

Hossner, L.R., and F.M. Hons. 1992. Reclamation of mine tailings. In *Soil Restoration*, R. Lal and B.A. Stewart (Eds.), pp. 311–350. New York: Springer.

Humphrey, C.L., D.P. Faith, and P.L. Dostine. 1995. Baseline requirements for assessment of mining impact using biological monitoring. *Aust. J. Ecol.* 20(1):150–166.

Ireland, M.P. 1983. Heavy metal uptake and tissue distribution in earthworms. In *Earthworm Ecology*, J.E. Satchell (Ed.), pp. 247–265. Dordrecht, the Netherlands: Springer.

Ireland, M.P., and R.J. Wooton. 1976. Variations in the lead, zinc and calcium content of *Dendrobaena rubida* (oligochaeta) in a base metal mining area. *Environ. Pollut.* 10(3):201–208.

Johnson, D.B. 2003. Chemical and microbiological characteristics of mineral spoils and drainage waters at abandoned coal and metal mines. *Water Air Soil Pollut. Focus* 3(1):47–66.

Jones, B.E.H., R.J. Haynes, and I.R. Phillips. 2010. Effect of amendment of bauxite processing sand with organic materials on its chemical, physical and microbial properties. *J. Environ. Manage.* 91(11):2281–2288.

Kandeler, F., C. Kampichler, and O. Horak. 1996. Influence of heavy metals on the functional diversity of soil microbial communities. *Biol. Fertil. Soils* 23(3):299–306.

Kelly, C.N., C.D. Peltz, M. Stanton, D.W. Rutherford, and C.E. Rostad. 2014. Biochar application to hardrock mine tailings: Soil quality, microbial activity, and toxic element sorption. *Appl. Geochem.* 43:35–48.

Kossoff, D., W.E. Dubbin, M. Alfredsson, S.J. Edwards, M.G. Macklin, and K.A. Hudson-Edwards. 2014. Mine tailings dams: Characteristics, failure, environmental impacts, and remediation. *Appl. Geochem.* 51:229–245.

Lamb, D.T., K. Venkatraman, N.S. Bolan, N. Ashwath, G. Choppala, and R. Naidu. 2014. Phytocapping: An alternative technology for the sustainable management of landfill sites. *Crit. Rev. Environ. Sci. Technol.* 44(6):561–637.

Larney, F.J., and D.A. Angers. 2012. The role of organic amendments in soil reclamation: A review. *Can. J. Soil Sci.* 92(1):19–38.

Li, S., X. Di, D. Wu, and J. Zhang. 2013a. Effects of sewage sludge and nitrogen fertilizer on herbage growth and soil fertility improvement in restoration of the abandoned opencast mining areas in Shanxi, China. *Environ. Earth Sci.* 70(7):3323–3333.

Li, S., D. Wu, and J. Zhang. 2005. Effects of vegetation and fertilization on weathered particles of coal gob in Shanxi mining areas, China. *J. Hazard. Mater.* 124(1–3):209–216.

Li, X., Z. Chen, Z. Chen, and Y. Zhang. 2013b. A human health risk assessment of rare earth elements in soil and vegetables from a mining area in Fujian Province, Southeast China. *Chemosphere* 93(6):1240–1246.

Lim, J.E., J.K. Sung, B. Sarkar, H. Wang, Y. Hashimoto, D.C.W. Tsang, and Y.S. Ok. 2017. Impact of natural and calcined starfish (*Asterina pectinifera*) on the stabilization of Pb, Zn and As in contaminated agricultural soil. *Environ. Geochem. Health.* 39:431–441.

Lindsay, M.B.J., M.C. Moncur, J.G. Bain, J.L. Jambor, C.J. Ptacek, and D.W. Blowes. 2015. Geochemical and mineralogical aspects of sulfide mine tailings. *Appl. Geochem.* 57:157–177.

Liu, J., G. Liu, J. Zhang, H. Yin, and R. Wang. 2012. Occurrence and risk assessment of polycyclic aromatic hydrocarbons in soil from the Tiefa coal mine district, Liaoning, China. *J. Environ. Monit.* 14(10):2634–2642.

Liu, R., and R. Lal. 2012. Nanoenhanced materials for reclamation of mine lands and other degraded soils: A review. *J. Nanotechnol.* 2012:18.

Lottermoser, B., and P. Ashley. 2008. Assessment of rehabilitated uranium mine sites, Australia. In *Uranium, Mining and Hydrogeology*, B.J. Merkel and A. Hasche-Berger (Eds.), pp. 335–340. Berlin, Germany: Springer.

Lottermoser, B.G. 2010. Radioactive wastes of uranium ores. In *Mine Wastes: Characterization, Treatment and Environmental Impacts*, B.G. Lottermoser (Ed.), pp. 263–312. Berlin, Germany: Springer.

Lottermoser, B.G. 2015. Rare earth elements in Australian uranium deposits. *Uranium—Past and Future Challenges: Proceedings of the Seventh International Conference on Uranium Mining and Hydrogeology*, pp. 25–30. Cham (ZG), Switzerland.

Lottermoser, B.G., and P.M. Ashley. 2006. Physical dispersion of radioactive mine waste at the rehabilitated Radium Hill uranium mine site, South Australia. *Aust. J. Earth Sci.* 53:485–499.

Maiti, S.K., and M.K. Ghose. 2005. Ecological restoration of acidic coalmine overburden dumps—An Indian case study. *Land Contam. Reclam.* 13:361–369.

Mandal, S., B. Sarkar, N. Bolan, J. Novak, Y.S. Ok, L. Van Zwieten, B.P. Singh et al. 2016a. Designing advanced biochar products for maximizing greenhouse gas mitigation potential. *Crit. Rev. Environ. Sci. Technol.* 46:1367–1401.

Mandal, S., R. Thangarajan, N.S. Bolan, B. Sarkar, N. Khan, Y.S. Ok, and R. Naidu. 2016b. Biochar-induced concomitant decrease in ammonia volatilization and increase in nitrogen use efficiency by wheat. *Chemosphere* 142:120–127.

Melgar-Ramírez, R., V. González, J.A. Sánchez, and I. García. 2012. Effects of application of organic and inorganic wastes for restoration of sulphur-mine soil. *Water Air Soil Pollut.* 223(9):6123–6131.

Moncur, M.C., C.J. Ptacek, M.B.J. Lindsay, D.W. Blowes, and J.L. Jambor. 2015. Long-term mineralogical and geochemical evolution of sulfide mine tailings under a shallow water cover. *Appl. Geochem.*
57:178–193.

Nada, W., O. Blumenstein, S. Claassens, and L. van Rensburg. 2012. Effect of wood compost on extreme soil
characteristics in the lusatian lignite region. *Open J. Soil Sci.* 2:347–352.

Naidu, R. 2013. Recent advances in contaminated site remediation. *Water Air Soil Pollut.* 224(12):1705.

Naidu, R., S.J.T. Pollard, N.S. Bolan, G. Owens, and A.W. Pruszinski. 2008. Chapter 4—Bioavailability: The
underlying basis for risk-based land management. In *Developments in Soil Science*, A.B. McBratney,
A.E. Hartemink, and R. Naidu (Eds.), pp. 53–72. Amsterdam, the Netherlands: Elsevier.

Navarro, M.C., C. Pérez-Sirvent, M.J. Martínez-Sánchez, J. Vidal, P.J. Tovar, and J. Bech. 2008. Abandoned
mine sites as a source of contamination by heavy metals: A case study in a semi-arid zone. *J. Geochem.
Explor.* 96(2–3):183–193.

Neiva, A.M.R., P.C.S. Carvalho, I.M.H.R. Antunes, M.M.V.G. Silva, A.C.T. Santos, M.M.S. Cabral Pinto, and
P.P. Cunha. 2014. Contaminated water, stream sediments and soils close to the abandoned Pinhal do
Souto uranium mine, central Portugal. *J. Geochem. Explor.* 136:102–117.

Neiva, A.M.R., P.C.S. de Carvalho, I.M.H.R. Antunes, M.M. da Silva Cabral Pinto, A.C.T. dos Santos, P.P.
Cunha, and M.M. Costa. 2016. Spatial variability of soils and stream sediments and the remediation
effects in a Portuguese uranium mine area. *Chem. Erde.* 76:501–518.

Nordstrom, D.K., C.N. Alpers, C.J. Ptacek, and D.W. Blowes. 2000. Negative pH and extremely acidic mine
waters from Iron Mountain, California. *Environ. Sci. Technol.* 34(2):254–258.

O'Reilly, J.L. 1997. An incubation study to assess the effect of waste sludge additions on some chemical characteristics of mine spoils. MSc thesis, Department of Soil Science, Massey University, Palmerston North,
New Zealand.

Park, J.H., D. Lamb, P. Paneerselvam, G. Choppala, N. Bolan, and J.W. Chung. 2011. Role of organic
amendments on enhanced bioremediation of heavy metal(loid) contaminated soils. *J. Hazard. Mater.*
185(2–3):549–574.

Pepper, I.L., H.G. Zerzghi, S.A. Bengson, B.C. Iker, M.J. Banerjee, and J.P. Brooks. 2012. Bacterial populations within copper mine tailings: Long-term effects of amendment with Class A biosolids. *J. Appl.
Microbiol.* 113(3):569–577.

Puga, A.P., L.C.A. Melo, C.A. de Abreu, A.R. Coscione, and J. Paz-Ferreiro. 2016. Leaching and fractionation
of heavy metals in mining soils amended with biochar. *Soil Tillage Res.* 164:25–33.

Querol, X., X. Zhuang, O. Font, M. Izquierdo, A. Alastuey, I. Castro, B. L. van Drooge et al. 2011. Influence
of soil cover on reducing the environmental impact of spontaneous coal combustion in coal waste gobs:
A review and new experimental data. *Int. J. Coal Geol.* 85(1):2–22.

Razo, I., L. Carrizales, J. Castro, F. Díaz-Barriga, and M. Monroy. 2004. Arsenic and heavy metal pollution of soil, water and sediments in a semi-arid climate mining area in Mexico. *Water Air Soil Pollut.*
152(1):129–152.

Ribeiro, J., T. Silva, J.G.M. Filho, and D. Flores. 2012. Polycyclic aromatic hydrocarbons (PAHs) in burning
and non-burning coal waste piles. *J. Hazard. Mater.* 199–200:105–110.

Ribeiro, J., S.R. Taffarel, C.H. Sampaio, D. Flores, and L.F.O. Silva. 2013. Mineral speciation and fate of
some hazardous contaminants in coal waste pile from anthracite mining in Portugal. *Int. J. Coal Geol.*
109–110:15–23.

Rodgers, C.S., and R.C. Anderson. 1995. Plant growth inhibition by soluble salts in sewage sludge-amended
mine spoils. *J. Environ. Qual.* 24(4):627–630.

Sanderson, P., R. Naidu, and N. Bolan. 2015. Effectiveness of chemical amendments for stabilisation of lead
and antimony in risk-based land management of soils of shooting ranges. *Environ. Sci. Pollut. Res.*
22(12):8942–8956.

Sarkar, B., H.L. Choi, K. Zhu, A. Mandal, B. Biswas, and A. Suresh. 2016. Monitoring of soil biochemical
quality parameters under greenhouse spinach cultivation through animal waste recycling. *Chem. Ecol.*
32(5):407–418.

Sheoran, V., A.S. Sheoran, and P. Poonia. 2010. Soil reclamation of abandoned mine land by revegetation: A
review. *Int. J. Soil Sediment Water* 3(2):Article 13.

Singh, A.N., A.S. Raghubanshi, and J.S. Singh. 2004. Comparative performance and restoration potential of
two Albizia species planted on mine spoil in a dry tropical region, India. *Ecol. Eng.* 22(2):123–140.

Smolders, E., R.M. Lambregts, M.J. McLaughlin, and K.G. Tiller. 1998. Effect of soil solution chloride on
cadmium availability to Swiss Chard. *J. Environ. Qual.* 27(2):426–431.

Sopper, W.E. 1992. Reclamation of mine land using municipal sludge. In *Soil Restoration*, R. Lal and B.A.
Stewart (Eds.), pp. 351–431. New York: Springer.

Stolt, M.H., J.C. Baker, T.W. Simpson, D.C. Martens, J.R. McKenna, and J.R. Fulcher. 2001. Physical reconstruction of mine tailings after surface mining mineral sands from prime agricultural land. *Soil Sci.* 166:29–37.

Sun, Y.Z., J.S. Fan, P. Qin, and H.Y. Niu. 2009. Pollution extents of organic substances from a coal gangue dump of Jiulong Coal Mine, China. *Environ. Geochem. Health* 31(1):81–89.

Tandy, S., J.R. Healey, M.A. Nason, J.C. Williamson, and D.L. Jones. 2009. Remediation of metal polluted mine soil with compost: Co-composting versus incorporation. *Environ. Pollut.* 157(2):690–697.

Tordoff, G.M., A.J.M. Baker, and A.J. Willis. 2000. Current approaches to the revegetation and reclamation of metalliferous mine wastes. *Chemosphere* 41(1–2):219–228.

Ulusay, R., F. Arikan, M.F. Yoleri, and D. Çağlan. 1995. Engineering geological characterization of coal mine waste material and an evaluation in the context of back-analysis of spoil pile instabilities in a strip mine, SW Turkey. *Eng. Geol.* 40(1):77–101.

Wijesekara, H., N.S. Bolan, M. Vithanage, Y. Xu, S. Mandal, S.L. Brown, G.M. Hettiarachchi et al. 2016. Utilization of biowaste for mine spoil rehabilitation. *Adv. Agron.* 138:97–173.

Wilden, R., W. Schaaf, and R.F. Hüttl. 1999. Soil solution chemistry of two reclamation sites in the Lusatian lignite mining district as influenced by organic matter application. *Plant Soil* 213(1):231–240.

Zanuzzi, A., J.M. Arocena, J.M. van Mourik, and A. Faz Cano. 2009. Amendments with organic and industrial wastes stimulate soil formation in mine tailings as revealed by micromorphology. *Geoderma* 154(1–2):69–75.

2 Chemical Characterization of Mine Sites

Xinni Xiong, Daniel C.W. Tsang, and Y.S. Ok

CONTENTS

2.1 Introduction...17
2.2 Environmental Impact...17
    2.2.1 Mine Tailings 17
    2.2.2 Acid Mine Drainage 19
2.3 Remediation Techniques...20
    2.3.1 Mine Tailings 20
    2.3.2 Acid Mine Drainage 21
2.4 Biochar Applications for Mine Site Remediation.......................................23
    2.4.1 Characteristics of Biochar..23
    2.4.2 Biochar Use in Mine Site Remediation 24
    2.4.3 Engineered Biochar for Enhancing Remediation 27
References...28

2.1 INTRODUCTION

Along with industrial and economic development, intensive mining activities have been causing detrimental environmental degradation, including elevated concentrations of toxic elements, a decrease in soil fertility and microbial activity, and generation of acid mine drainage (AMD) (Rodríguez-Vila et al. 2015, Yang et al. 2013). The adverse impact could remain several decades after the mining activity ceases.

Data are affected by technological, economic, and social issues, but, in general, the statistics illustrate that the extent of waste rock is increasing, while ore grades are declining. Nevertheless, there are economic resources for the development of various minerals, with considerable resources-to-production ratios (Mudd 2009). Therefore, chemical characterization of mine sites is important.

Hydrogeochemistry of abandoned mine sites is of great significance because a wide range of reactions, from oxidation to acidification, may occur, which may present an unstable status for remediation. Hydrogeochemistry and geochemical dynamics of mining wastes can be characterized by column leachate studies, tailing mineralogy observation, as well as soil respiration analysis (Li et al. 2013).

2.2 ENVIRONMENTAL IMPACT

2.2.1 MINE TAILINGS

Mining wastes include tailings, smelting ashes, and leaching tanks' residues. Sulfide minerals produce high concentrations of sulfate, protons, and toxic metals through a series of biogeochemical reactions, including hydration, oxidation, and hydrolysis. According to the mineral components and natural conditions, major geochemical reactions in mining wastes are summarized in Table 2.1,

TABLE 2.1

Geochemical Reactions in Mining Wastes

Metal	Mineral	Component	Reaction
Iron	Pyrite	FeS_2	$FeS_{2(s)} + H_2O + 7/2O_{2(aq)} \rightarrow Fe^{2+} + SO_4^{2-} + 2H^+$
			$FeS_{2(s)} + 8H_2O + 14Fe^{3+} \rightarrow 15Fe^{2+} + 2SO_4^{2-} + 16H^+$
			$2FeS_2 + 15H_2O_2 \rightarrow 2Fe^{2+} + 4SO_4^{2-} + 14H_2O + 2H^+$
	Arsenopyrite	FeAsS	$FeAsS + 7/2O_2 + H_2O \rightarrow Fe^{3+} + SO_4^{2-} + H_2AsO_4^-$
			$FeAsS + 14Fe^3 + 10H_2O \rightarrow 14Fe^{2+}$
			$\quad + FeAsSO_4 \cdot 2H_2O + SO_4^{2-} + 16H^+$
	Ferrous iron	Fe^{2+}	$2Fe^{2+} + 1/4\ O_{2(aq)} + H^+ \rightarrow 2Fe^{3+} + 1/2H_2O$
	Jarosite	$KFe_3(SO_4)_2(OH)_6$	$KFe_3(SO_4)_2(OH)_6 + 6H^+ \rightarrow K^+ + 2SO_4^{2-}$
			$\quad + 3Fe^{3+} \cdot 6H_2O$
	Schwertmannite	$Fe_8O_8(OH)_6SO_4$	$Fe_8O_8(OH)_6SO_4 + 22H^+ \rightarrow 8Fe^{3+} + SO_4^{2-} + 14H_2O$
	Scorodite	$FeAsO_4$	$FeAsO_4 \cdot 2H_2O \rightarrow Fe(OH)_2^+ + H_2AsO_4^-$
	Goethite	FeOOH	$FeOOH + 3H^+ \rightarrow Fe^{3+} + 2H_2O$
	Pyrrhotite	Fe_7S_8	$Fe_{(1-x)}S + (2-x/2)O_2 + xH_2O \rightarrow (1-x)Fe^{2+}$
			$\quad + SO_4^{2-} + 2xH^+$
			$Fe_{(1-x)}S + (8-2x)Fe^{3+} + 4H_2O \rightarrow (9-3x)Fe^{2+}$
			$\quad + SO_4^{2-} + 8H^+$
	Siderite	$FeCO_3$	$FeCO_3 + H^+ \rightarrow Fe^{2+}HCO_3^-$
Lead	Galena	PbS	$PbS + H^+ \rightarrow Pb^{2+} + HS^-$
	Beudantite	Pb^{2+}	$Pb^{2+} + 3Fe^{3+} + SO_4^{2-} + H_2AsO_4^- + 6H_2O \rightarrow$
			$\quad PbFe_3(AsO_4)(SO_4)(OH)_4 + 8H^+$
	Cerussite	$PbCO_3$	$PbCO_3 + H^+ \rightarrow Pb^{2+} + HCO_3^-$
Copper	Chalcopyrite	$CuFeS_2$	$CuFeS_{2(s)} + 4O_{2(aq)} \rightarrow Fe^{2+} + 2SO_4^{2-} + Cu^{2+}$
			$CuFeS_{2(s)} + 16Fe^{3+} + 8H_2O_{(aq)} \rightarrow 17Fe^{2+} + 2SO_4^{2-} +$
			$\quad Cu^{2+} + 16H^+$
	Chalcocite	Cu_2S	$Cu_2S + 2O_2 \rightarrow 2Cu^{2+} + SO_4^{2-}$
	Covellite	CuS	$CuS + 2O_2 \rightarrow Cu^{2+} + SO_4^{2-}$
Zinc	Sphalerite	ZnS	$ZnS_{(s)} + 2O_{2(aq)} \rightarrow Zn^{2+} + SO_4^{2-}$
			$ZnS_{(s)} + 8Fe^{3+} + 4H_2O \rightarrow Zn^{2+} + SO_4^{2-} + 8Fe^{2+} +$
			$\quad 8H^+$
	Smithsonite	$ZnCO_3$	$ZnCO_3 \rightarrow Zn^{2+} + CO_3^{2-}$
Nickel	Millerite	NiS	$NiS + 2O_2 \rightarrow Ni^{2+} + SO_4^{2-}$
Arsenic	Orpiment	As_2S_3	$As_2S_3 + 7O_2 + 6H_2O \rightarrow 2HAsO_4^{2-} + 3SO_4^{2-} + 10H^+$
	Realgar	AsS	$AsS + 2.75O_2 + 2.5H_2O \rightarrow HAsO_4^{2-} + SO_4^{2-} + 4H^+$
Antimony	Stibnite	Sb_2S_3	$Sb_2S_3 + 18H_2O \rightarrow 3SO_4^{2-} + 2SbO_3^- + 26e^- + 36H^+$
			$Sb_2S_3 + 3H_2O + 6O_2 \rightarrow 3SO_4^{2-} + 2Sb_2O_3 + 6H^+$
	Valentinite	Sb_2O_3	$Sb_2O_3 + 3H_2O \rightarrow 2SbO^{3-} + 4e^- + 6H^+$
Calcium, magnesium	Dolomite	$CaMg(CO_3)_2$	$2H^+ + CaMg(CO_3)_2 \rightarrow Ca^{2+} + Mg^{2+} + 2HCO_3^-$
Magnesium	Brucite	$Mg(OH)_2$	$Mg(OH)_2 + 2H^+ \rightarrow Mg^{2+} + 2H_2O, Fe^{3+} + 3H_2O \rightarrow$
			$\quad Fe(OH)_3 + 3H^+$
Manganese	Rhodochrosite	$MnCO_3$	$MnCO_3 \rightarrow Mn^{2+} + CO_3^{2-}$
	Pyrolusite	MnO_2	$MnO_2 + 2H^+ \rightarrow Mn^{2+} + 1/2O_2 + H_2O$
	Pyrochroite	MnOOH	$MnOOH + 2H^+ \rightarrow Mn^{2+} + 1/4O_2 + 3/2H_2O$

(Continued)

TABLE 2.1 (*Continued*)
Geochemical Reactions in Mining Wastes

Metal	Mineral	Component	Reaction
Aluminum	Anorthite	$CaAl_2Si_2O_8$	$CaAl_2Si_2O_8 + 2H^+ + H_2O \rightarrow Ca^{2+} + Al_2Si_2O_5(OH)_4$
	Albite	$NaAlSi_3O_8$	$2NaAlSi_3O_8 + 2H^+ + 9H_2O \rightarrow 2Na^+ + 4H_4SiO_4$ $+ Al_2Si_2O_5(OH)_4$
	K-Feldspar	$KAlSi_3O_8$	$KAlSi_3O_8 + H^+ + 9/2H_2O \rightarrow K^+ + 2H_4SiO_4$ $+ 1/2Al_2Si_2O_5(OH)_4$
	Muscovite	$KAl_3Si_3O_8$	$2KAl_3Si_3O_8 + 2H^+ + 3H_2O \rightarrow 3Al_2Si_2O_5(OH)_4$ $+ 2K^+$
	Chlorite	$AlSi_3O_{10}(OH)_8$	$AlSi_3O_{10}(OH)_8 + 16H^+ \rightarrow 4.5Mg^{2+} + 0.2Fe^{3+}$ $+ 0.2Fe^{2+} + 2Al^{3+} + 3SiO_2 + 12H_2O$
	Biotite	$KMg_{1.5}Fe_{1.5}AlSi_3O_{10}$ $(OH)_2$	$KMg_{1.5}Fe_{1.5}AlSi_3O_{10}(OH)_2 + 7H^+ + 1/2H_2O \rightarrow$ $K^+ + 1.5Mg^{2+} + 1.5Fe^{2+} + 2H_4SiO_4$ $+ 1/2Al_2Si_2O_5(OH)_4$
	Kaolinite	$Al_2Si_2O_5(OH)_4$	$Al_2Si_2O_5(OH)_4 + 6H^+ \rightarrow 2Al^{3+} + 2Si(OH)_4 + H_2O$
	Boehmite	$AlOOH$	$AlOOH + 3H^+ \rightarrow Al^{3+} + 2H_2O$
	Gibbsite	$Al(OH)_3$	$Al(OH)_3 + 3H^+ \rightarrow Al^{3+} + 3H_2O$
	Zeolite	$NaAlSi_2O_6$	$2NaAlSi_2O_6 \cdot H_2O + Ca^{2+} \rightarrow Ca(AlSi_2O_6)_2 \cdot H_2O$ $+ 2Na^+$
			$2NaAlSi_2O_6 \cdot H_2O + Mg^{2+} \rightarrow Mg(AlSi_2O_6)_2 \cdot H_2O$ $+ 2Na^+$
Calcium	Calcite	$CaCO_3$	$H^+ + CaCO_3 \rightarrow Ca^{2+} + HCO_3^-$
	Gypsum	$CaSO_4$	$CaSO_4 \cdot 2H_2O \rightarrow Ca^{2+} + SO_4^{2-} + 2H_2O$
		$CaCO_3$	$CaCO_{3(s)} + 2H^+ + SO_4^{2-} + 2H_2O \rightarrow CaSO_4 \cdot 2H_2O_{(s)}$ $+ H_2CO_3$
	Anhydrite	$CaSO_4$	$CaSO_4 \rightarrow Ca^{2+} + SO_4^{2-}$
Calcium, arsenic	Ca-arsenate	Ca^{2+}, AsO_4^{3-}	$4Ca^{2+} + 2AsO_4^{3-} + 2OH^- + 4H_2O \rightarrow Ca_3(AsO_4) \cdot$ $Ca(OH)_2 \cdot 4H_2O$

with information from recent literature (Anawar 2015, Bea et al. 2010, Kruse and Younger 2009, Modabberi et al. 2013, Salmon and Malmström 2004).

Environmental conditions, including weathering, geochemical reactions, and contamination levels, have an effect on the various mining processes, from ore extraction to the concentrating process, which, in turn, will affect the characteristics of mine tailings. Warm climates enhance oxygen transport and diffusion, which result in rapid weathering of minerals. Hydrological changes such as rainfall and drought conditions also affect sulfate oxidation and contaminant transport of mine tailings by controlling temperature, pH, availability of oxygen and water, and natural attenuation, as well as biogeochemical processes (Anawar 2013, 2015, Bhattacharya et al. 2006).

2.2.2 Acid Mine Drainage

Another outcome of mining activities is acid mine drainage (AMD) formation, which is caused by abiotic or microbial oxidation of iron pyrite (FeS_2), chalcopyrite ($CuFeS_2$), galena (ZnS), arsenopyrite (FeAsS), and other sulfide minerals when exposed to oxygen and water during mining activities (Johnson 2003). Such acidic wastewater of mine sites contains high amounts of heavy metals and may be discharged into surface water and groundwater, threatening the health of the surrounding ecosystems. For instance, the severe contamination event in January 1992 at the Wheal Jane, a tin mine in West Cornwall, England, was caused by the discharge of strongly acidic wastewater from abandoned mines containing high concentrations of heavy metals into the river (Neal et al. 2005, Younger et al. 2005).

Bacteria present in mining sites significantly contribute to the fast oxidation process of pyrite weathering, which can be accelerated by up to 100 times by them (Kelly et al. 2012). For example, the dissolution process of pyrite produces sulfuric acid (H_2SO_4), which creates an environment conducive to metal solubilization. The pyrite oxidation initiated by oxygen and microorganisms is relatively slow due to the natural alkalinity of water. With decreasing pH, the oxidizing ability of microorganisms increases and plays a more dominant role. Consequently, in the extremely acidic environment, microorganisms oxidize ferrous iron to ferric iron, which participates in the oxidation of pyrite, thus resulting in high acidity, sulfate, and total iron in the solution. Both inorganic and biological mechanisms simultaneously occur and depend on the changes in pH and Eh values (Favas et al. 2016).

The most significant characteristics of AMD are the low pH and high heavy metal concentrations. Total acidity, which is measured by the consumption of hydroxyl ions when neutralizing the alkalinity of a solution, distinguishes AMD from other low pH wastewater. The change in acidity also influences the release of dissolved metals and precipitation of metal hydroxides. The concentrations of toxic metals, including Zn, Cu, Cd, Co, Ni, and Pb, are influenced by their natural content, solubilities, and rates of dissolution, which are responsible for ecotoxicological effects on the surrounding environment. Besides, the two major elements in AMD of extremely high concentrations are iron and aluminum, which result in the formation of coatings of oxyhydroxides and induce adsorption and coprecipitation of other dissolved metals. In addition, a high content of sulfate originating from sulfide oxidation is another remarkable feature of AMD, which serves as a reference to evaluate the behavior and mechanisms of other mineral components. Physical parameters such as turbidity and suspended solids also affect the health of an ecosystem, and they may be correlated to the transport of metals and metalloids (Favas et al. 2016, Han et al. 2015). For example, a previous study revealed that the metal concentrations of AMD in summer are higher than those in spring, and in the summer, they contain up to 95.5 mg/L Ca, 19.9 mg/L Mg, 302.8 mg/L Fe, 12.1 mg/L Mn, 51.3 mg/L Al, 0.4 mg/L As, 30.7 mg/L Cu, 27.8 mg/L Zn, and 1464.8 mg/L SO_4 (Oh and Yoon 2013).

2.3 REMEDIATION TECHNIQUES

2.3.1 MINE TAILINGS

Rehabilitation of mine sites should consider many issues, such as legislation, long-term effectiveness, assessment criteria, and remediation goals (Mudd 2009). Remediation options for mine sites vary from passive/containment treatment such as store-and-release cover design (Gatzweiler et al. 2001) and vegetation cover (Valente et al. 2012) to active treatment such as electrodialytic remediation (Hansen et al. 2005), washing remediation (Moutsatsou et al. 2006), and phosphate remediation (Tang and Yang 2012). The conditions and concerns are listed in Table 2.2.

Capping or covering is an effective technique, particularly applicable to hazardous waste management in remote areas, including simple soil caps, drainage layers, geotextiles, evapotranspiration covers, impermeable caps, hardened covers, vegetative covers, and phytostabilization. Capping remediation methods can be easily implemented with a permanent and rapid effect, and they are associated with a lower cost because leaching contamination at the source can be prevented. However, proper long-term monitoring and maintenance are mandated to guarantee the performance.

Phytostabilization is also a sustainable and inexpensive method of in situ rehabilitation by planting highly tolerant species, especially hyperaccumulators, to reduce wind or water erosion, promote water percolation, and extract heavy metals from contaminated soil (Fellet et al. 2014, Reeves 2006, van der Ent et al. 2012). However, phytostabilization is often limited by the degraded physical structure, toxic elements, and extreme pH values, as well as by the low nutrient and organic content in the mining sites, thus making it difficult for revegetation remediation on the field scale (Mendez and Maier 2008,

TABLE 2.2

Remediation Techniques for Mine Tailings

Remediation Technique		Advantages	Disadvantages	References
Barrier	Zero-valent iron	Reduces acidic leachates and contaminants; low cost	Decreases iron reactivity by contamination	Bartzas et al. (2006)
	Store-release cover	Stable; in situ remediation; sustainable	Susceptible to climate hazard	Gatzweiler et al. (2001)
Phytostabilization	Amended with municipal and sewage wastes	Commercially available sorbents; realistic and cost-effective	Low microbial and enzymatic activity due to metal toxicity	Ciccu et al. (2003) and Alvarenga et al. (2009)
	Vegetation cover	Long-term stable: ecologically safe	Slow and takes a long time	Valente et al. (2012)
Biomass amendments	Biochar-aided rehabilitation	Sustainable; effective for revegetation and phytostabilization	No significant disadvantage	Fellet et al. (2011)
	Algal/microbial biomass	Long-term means to remove uranium/radionuclides	No significant disadvantage	Kalin et al. (2004)
Inorganic amendments	Carbonate, lime, marble, fly ash	Realistic; moderately cost effective	Overdose may mobilize contaminants; need care	Pérez-López et al. (2009) and Zornoza et al. (2013)
	Phosphorus amendment	Significantly lower bioavailability of heavy metals	No significant disadvantage	Hettiarachchi et al. (2001)
Electrodialytic remediation	Copper mine tailings	Fast removal; effective	Expensive; not sustainable; damage to soil	Hansen et al. (2005)
Washing remediation	Washed with deionized water, HCl, H_2SO_4, HNO_3, and Na_2EDTA	Permanent solution; ex situ process	Expensive	Moutsatsou et al. (2006)

Rodríguez-Vila et al. 2015). Hence, a number of soil amendments, such as compost, lime, organic waste, sludge, and biochar, are proposed to enhance the remediation by improving soil carbon capacity, microbial composition, and nutrient availability (Anawar et al. 2015, Green and Renault 2008, Mendez et al. 2007, Ye et al. 1999). Chemical stabilization using buffered phosphate and biosolids also has attracted attention for both in situ and ex situ rehabilitation of mining wastes (Hettiarachchi et al. 2001).

Various species of acidophilic microorganisms favor an acidic environment in mine tailings, which offer a potential approach to extract and recover toxic heavy metals, also known as the biomining technology. Microbial consortia including iron/sulfur-oxidizing microorganisms and thermophilic bacteria can be employed to accelerate mineral dissolution and precipitation reactions (Anawar 2015, Johnson 2008).

2.3.2 ACID MINE DRAINAGE

Although source control measures are employed to prevent the formation of AMD by means of sealing underground mines and blending mineral wastes, contamination often happens and requires further remediation measures to mitigate the environmental impact. The options of AMD remediation can be divided into active and passive treatment processes (as illustrated in Figure 2.1). The former refers to

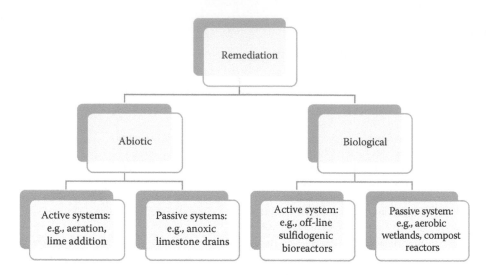

FIGURE 2.1   Technique classification on acid mine drainage treatment.

remediation with intensive operations, usually by applying alkaline agents to neutralize acidic drainage, while the latter is more based on the natural environment exploiting chemical and biological activities in the natural systems to mitigate the contamination (Johnson and Hallberg 2005).

The total acidity used for offsetting against alkalinity is regarded as net acidity, which consists of both proton acidity and mineral acidity. As for net alkaline mine waters, treatment strategies such as aerobic wetlands are often applied, where ferrous iron is oxidized to ferric iron, followed by hydrolysis and then the generation of acidity. Anoxic limestone drains produce alkalinity by maintaining the iron in its reduced form. Anaerobic wetlands/compost bioreactors mainly depend on microbially

TABLE 2.3
Remediation Techniques for Acid Mine Drainage

Classification	AMD Remediation Technique	Reference
Prevention	Sealing layer, which is made from clay, covers the spoil	Swanson et al. (1997)
	Underwater storage to prevent contact between the minerals and dissolved oxygen	Li et al. (1997)
	Blend acid-generating and acid-consuming materials to produce environmentally benign composites	Mehling et al. (1997)
	Surface protective coating using soluble phosphate with hydrogen peroxide	Evangelou (1998)
Abiotic remediation	Addition of chemical-neutralizing agents, including lime, slaked lime, calcium carbonate, sodium carbonate, sodium hydroxide, and magnesium oxide	Coulton et al. (2003)
	Anoxic limestone drains	Kleinmann et al. (1998)
Biological remediation	Aerobic wetlands	Coupland et al. (2004)
	Anaerobic wetlands/compost bioreactors	Vile and Wieder (1993)
	Composite aerobic and anaerobic wetlands	Kalin et al. (1991)
	Permeable reactive barriers	Benner et al. (1997)
	Iron-oxidizing bioreactors	Long et al. (2003)
	Sulfidogenic bioreactors	Johnson (2000)

catalyzed reactions to reduce sulfate to sulfide and generate alkalinity. Permeable reactive barriers incorporating reactive materials can also be employed to precipitate heavy metals with sulfides, hydroxides, and carbonates in a trench or pit. Passive biological systems such as iron-oxidizing bioreactors accelerate the oxidation of ferrous iron to ferric iron by certain bacteria and archaea (Johnson and Hallberg 2005).

Active systems are more frequently used on occasions where mine drainage pollution is severe. High-acidity and high-flow drainage is particularly suitable for the adoption of active systems due to its larger and faster consumption of neutralizing materials. Abandoned mines are often treated by passive systems, while operational mine sites usually require active systems, considering the space limitation and variations in chemical properties (Cavanagh et al. 2014). Despite the lower maintenance costs, passive systems are less economic to set up in the first place compared with active systems (Johnson and Hallberg 2005). Table 2.3 summarizes some examples of remediation techniques.

2.4 BIOCHAR APPLICATIONS FOR MINE SITE REMEDIATION

2.4.1 CHARACTERISTICS OF BIOCHAR

Biochar is a carbonaceous by-product obtained by pyrolysis of carbon-rich biomass under a limited oxygen condition. Biochar has attracted increasing attention worldwide in view of its multifunctional use in agricultural and environmental applications for mitigating greenhouse gas emissions, increasing crop growth and soil fertility, and immobilizing soil contaminants (Anawar et al. 2015, Beesley et al. 2011, Lehmann 2007). Biochar production from agricultural waste relieves the burden of landfill and incineration. Rich in carbon and nutrients, biochar effectively has enhanced soil properties and improved crop productivity by its high alkalinity, porosity, cation exchange capacity (CEC), surface area, and adsorption capacity of contaminants, as well as by stimulation to microbial activity (Lehmann et al. 2011, Mohan et al. 2014, Tang et al. 2013). Therefore, biochar serves as a good soil amendment, indicating its potential use in mine site rehabilitation and remediation.

The characteristics of biochar mainly depend on feedstock type and the pyrolysis conditions such as heating rate, residence time, and pyrolysis temperature (Keiluweit et al. 2010, Lehmann 2007, Mohan et al. 2014, Rajapaksha et al. 2014, Tang et al. 2013). The fraction of carbonized organic matter and noncarbonized organic matter is determined by the pyrolytic temperature, and it influences the structure and behavior of biochar (Chen et al. 2008). Biochar produced from high-temperature pyrolysis provides a larger surface area and a higher sorption capacity (Devi and Saroha 2015), while low-temperature pyrolysis produces a less condensed carbon structure and more available nutrients from the biochar, which are favorable for improving soil fertility (Anawar et al. 2015, Steinbeiss et al. 2009).

As for biochar feedstocks, the International Biochar Initiative has classified biomass material into two categories: (1) unprocessed feedstock types—rice hulls and straw, maize cobs and stover, nonmaize cereal straws, sugarcane bagasse and trash, switch grass, silver grass and bamboo, oil crop residues, leguminous crop residues, hemp residues, softwoods, and hardwoods; and (2) processed feedstock types—cattle manure, pig manure, poultry litter, sheep manure, horse manure, paper mill sludge, sewage sludge, distillers grain, anaerobic digester sludge, biomass fraction of municipal solid waste, and food industry waste (International Biochar Initiative 2012). The biomass material influences biochar nutrient composition and surface functional groups, which play an important role in biochar applications. For example, a study investigated three feedstocks, including pruning residues, fir tree pellets, and manure pellets, where biochar produced from manure performed best, with higher metal immobilization and higher plant biomass production (Fellet et al. 2014). Other factors, including heating rate, residence time, pyrolysis medium, and atmospheric conditions, have also been studied in the recent literature (Kwon et al. 2015, Luo et al. 2015, Tan et al. 2015b). A diverse set of key parameters determines the characteristics of biochar and thus influence its performance for contaminated site remediation.

2.4.2 Biochar Use in Mine Site Remediation

Various studies have proved the applicability of biochar for increasing the pH and decreasing the bioavailability of heavy metals in mine tailings. Algal biochar was found to affect strongly the establishment and growth of a native plant for mine soil rehabilitation (Roberts et al. 2015). A biochar amendment with acacia revegetation proved to affect soil properties and plant nutritional status as well as enhance photosynthetic efficiency in nitrogen utilization (Reverchon et al. 2015), making it favorable for revegetation in mine site rehabilitation. The addition of biochar in an acidic soil effectively ameliorated Al toxicity and P deficiency in maize plants (Zhu et al. 2014). The addition of biochar to an artificial soil cover was found to improve grass growth in contaminated colliery sites (Ryan et al. 2014). Softwood- and hardwood-derived biochars were found to adsorb Cu and Zn ions (Jiang et al. 2016), while sugarcane straw–derived biochar was found to reduce effectively metal mobility, as shown by leaching and fractionation tests (Puga et al. 2016).

Also, biochar is recommended to be applied in combination with compost to reduce soluble metal concentrations while enhancing nutrient availability for plant growth (Beesley et al. 2014). The mixture of waste and biochar as an amendment to mine tailings was found to have a high sorption capacity and to favor the retention of Cu, Pb, and Zn in contaminated soil (Forján et al. 2016). Biochar and compost application in a copper mine soil, along with phytostabilization, was also reported as a highly efficient method for reclamation of mining soils (Rodríguez-Vila et al. 2014). Biochar uses in remediation of mine sites as documented in various studies are summarized in Table 2.4.

Biochar's potential for treating AMD from abandoned mining sites has also been studied (Oh and Yoon 2013), and the results showed that biochar served as a neutralizer and sorbent and could remove toxic constituents from AMD. Research on sludge-derived biochar revealed that coordination with organic hydroxyl and carboxyl functional groups, as well as coprecipitation or complexation on mineral surfaces, accounts for its sorption of Pb in AMD (Lu et al. 2012). Surface complexation or coprecipitation involves electrostatic cation exchange or metal exchange reactions with Ca, Mg, K, and Na, which are released from the biochar during sorption of heavy metals such as Pb and Zn (Yang et al. 2016, Zhang et al. 2016) and oxyanions such as As and Cr (Fang et al. 2016, Zhang et al. 2015).

This adsorption capacity and pH-regulating ability are attributed to the physicochemical properties of biochar, including porosity, ash content, surface area, functional groups, buffering capacity, and CEC (Devi and Saroha 2015, Jiang et al. 2016). Metal immobilization mechanisms involve ion exchange, electrostatic attraction, physical adsorption, and carbonate or hydroxide precipitation (Tan et al. 2015a, Uchimiya et al. 2010). A column leaching experiment investigating the geochemical dynamics of tailing leachate found that biochar addition improved the seepage water quality and plant survival (Li et al. 2013). Another dynamic leaching study found that amending high-sulfur mine rejects with biochar inhibited the acid production rate, altered salt and carbon dissolution, and reduced the release of dissolved organic carbon (Jain et al. 2014), thus facilitating rehabilitation of mine sites.

Biological stability is the resistance and resilience of soil microbial communities responding after a disturbance or environmental change, which is crucial to measure bioremediation processes (Griffiths and Philippot 2013). As for soil biochemical and microbiological activities, soil enzymes (e.g., phosphatase, urease, dehydrogenase, cellulase, and phenol oxidases) play an important role in nutrient cycling and soil resilience during phytoremediation. Biochar amendments increase the availability of these indicators, resulting in high nutrient content and pH increase in biochar-amended mine soil (Ahmad et al. 2016, Jain et al. 2016).

Biochar not only enhances the activity of N-fixing bacteria (Quilliam et al. 2013) and arbuscular and ectomycorrhizal fungi (Warnock et al. 2007) but also protects soil-colonizing microbes from their predators and competitors because the porous structure of biochar serves as a refuge site or microhabitat (Thies and Rillig 2009). Biochar influences microbial community composition and enzyme activities by sorption and inactivation of growth-inhibiting substances in soil (Lehmann et al. 2011).

TABLE 2.4
Biochar Applications in Mine Site Remediation

Biochar Feedstock	Waste Type	Effects	Mechanism	Reference
Jarrah wood	Mine rejects: iron ore	Higher soil pH, C/N ratio, C and P content, and N use efficiency of the plants	Revegetation	Reverchon et al. (2015)
Algae	Stockpiled mine soils: a red basalt ferrosol and a saline–sodic sodosol	Enhances plant growth and contributes essential trace elements (K) to soil pore water	Revegetation	Roberts et al. (2015)
Orchard pruning residues	Mine tailings	Bioavailability decreases from 2.1 to 0.21 (Cd), from 39.2 to 24.2 (Pb), from 3.64 to 3.13 (Tl), from 898 to 683 (Zn) with 10% biochar (mg/kg)	Phytostabilization	Fellet et al. (2011)
Orchard pruning residues	Mine waste rock	Bioavailability decreases from 5.23 to 1.51 (Cd), from 0.032 to ND (Cr), from 79.5 to 50.9 (Pb), from 954 to 819 (Zn) (mg/kg)	Phytostabilization	Fellet et al. (2014)
Orchard pruning residues	Contaminated mine soil	Metal contents decrease from 7,490 to 13 (As), from 74 to ND (Cd), from 2,940 to 449 (Cu), from 4,170 to 17 (Pb), from 13,200 to 483 (Zn) (mg/kg)	Organic wastes remediation	Beesley et al. (2014)
Olive tree pruning	Contaminated mine soils	Bioavailability decreases from 2.4 to 1.0 (As), from 0.65 to ND (Pb), from 5 to 3.7 (Zn) (mg/kg)	Revegetation	Rosende et al. (2016)
Holm oak wood	Cu-contaminated mine soil	Mobility factors decrease from 19.4 to 2.4 (Cu), from 52.7 to 7.3 (Ni), from 29.9 to 4.6 (Pb)	Revegetation	Rodríguez-Vila et al. (2015)
Acacia dealbata biomass	Copper mine tailing	Metal concentrations decrease from 138 to ND (Cu), from 0.28 to ND (Pb), from 2.41 to 1.31 (Zn) (mg/kg)	Waste and biochar amendment	Forján et al. (2016)
Hardwood (jarrah)	Cu/Pb–Zn mine tailings	The max adsorption capacities of Cu and Zn are 4.39 and 2.31 mg/g, respectively	Adsorption	Jiang et al. (2016)
Fresh biogas slurry and residue	As-contaminated tailing mine sediment	Reduction of As(V) (10%–13%) and Fe(III) (12%–17%)	Biological remediation	Chen et al. (2016)

(Continued)

TABLE 2.4 (*Continued*)
Biochar Applications in Mine Site Remediation

Biochar Feedstock	Waste Type	Effects	Mechanism	Reference
Cymbopogon flexuous (lemongrass)	Overburden or mine spoils	Higher nutrient, pH, and availability of phosphatase, β-glucosidase, urease, and dehydrogenase	Phytoremediation	Jain et al. (2016)
Cymbopogon flexuosus (lemongrass)	High-sulfur mine rejects	Acid production rate decreases from 10.4 to 3.81 kg/Mt/h and alkali consumption rate increases from 9.7 to 13.9 kg/Mt/h	Neutralization	Jain et al. (2014)
Sugarcane straw	Zn-contaminated mining soil	Reduction of Cd (57%–73%), Pb (45%–55%), and Zn (46%) concentrations in leachate	Immobilization	Puga et al. (2016)
Holm oak wood	Cu mine soil	$CaCl_2$-extractable concentrations decrease from 7.15 to 0.1 (Co) from 152 to 0.86 (Cu), from 7.22 to 0.17 (Ni) (mg/kg)	Phytoremediation	Rodríguez-Vila et al. (2014)
Poultry litter	AMD from abandoned Cu mine	Complete removal of Fe, Al, Cu, and As, and 99%, 61%, 31% removal of Zn, Mn, SO_4^{2-}, respectively	Neutralization and sorption	Oh and Yoon (2013)

ND, Not detected.

Bioremediation analysis of iron and arsenic in abandoned tailings from mine sediment with realgar (which is an arsenic sulfide mineral) via DNA extraction and amplicon sequencing found that biochar stimulated the bioavailability of dissolved organic matter and shifted the microbial community composition by increasing the relative abundance of arsenic/iron-reducing bacteria and bridging a biochar–bacterial dissolved organic matter consortium for electron transfer (Chen et al. 2016). A study using oligonucleotide fingerprint grouping found that a biochar-enriched Terra Preta soil contained a higher number of bacterial communities than unmodified soil (Kim et al. 2007). Other widely used assessment methods of microbial abundance can be applied for measuring the success of remediation of mine sites, including phospholipid fatty acid extraction, microbial carbon, basal respiration, metabolic quotient of CO_2, and ratio of microbial carbon to organic carbon (Langer and Rinklebe 2009, Moche et al. 2015, Rinklebe and Langer 2013).

However, in some cases, biochar amendment has not been effective. Although applying biochar derived from beetle-killed pine to hard rock mine tailings increased soil pH and organic matter content as well as decreased bulk density and extractable salt content in two mining sites (Kelly et al. 2014), the microbial population and activity showed no difference upon biochar addition, and N availability increased only in one mining material. In another study, biochar was applied to mine tailings to improve soil nutrients and water retention, but the bioavailability of metals and their leachability showed variability and even some unfavorable results; that is, the bioavailability and leachability of Cu increased with biochar addition (Fellet et al. 2011). Similarly, biochar amendments in a multielement polluted soil revealed no effect on As and Cu, which increased in pore water (Beesley et al. 2010). In another study, heavy metals were divided into two categories: Cd, Ni, and Zn as high-mobility metals and Cu, Cr, and Pb as low-mobility metals (Kim et al.

2015). Mixed results on metal sorption of biochar were possibly due to the toxicity and mobility of the diverse metals as well as the experimental conditions (Beesley and Marmiroli 2011, Kelly et al. 2014).

Moreover, because biochar sorption of metals depends on the increase in soil pH (Melo et al. 2015, Rees et al. 2014), the long-term stability of biochar is likely to be affected by rainwater and introduction of N through fertilization or revegetation. For instance, addition of mineral N fertilizer in the ammonium form may result in short-term acidification, thereby increasing heavy metal mobilization and transfer to groundwater (Puga et al. 2016). Nutrient-enriched biochar, in combination with fertilizers, was found to reduce synthetic N use efficiency during mine site rehabilitation with plant revegetation (Reverchon et al. 2015). It was also reported that biochar amendments caused a reduction in crop yield, including yields of rice, wheat, maize, lettuce, and tomato, as well as enhanced gaseous emissions from biochar-amended soils (Mukherjee and Lal 2014).

Therefore, the performance of biochar amendments, especially in multimetal-contaminated wastewater and soils (e.g., in the case of remediation of mine sites), depends on many variables such as the feedstock type and the characteristics of contaminants of concern (Fellet et al. 2014). A statistical meta-analysis of biochar applications also found that the effects varied with the experimental setup, soil properties, and conditions (Jeffery et al. 2011). In order to achieve sustainable rehabilitation of mining waste and AMD, issues including geochemistry, mine type, mineralogy, texture, ore extraction, and climate knowledge should be assessed comprehensively upfront (Anawar 2015). Further investigations into tailoring the biochar functionalities for mine site remediation and monitoring, to understand the long-term fate and aging processes of biochar under field conditions, are recommended.

2.4.3 ENGINEERED BIOCHAR FOR ENHANCING REMEDIATION

Engineering biochar's surface chemistry and pore size distribution would distinctly improve the dynamics and functions in amended soil. Up-to-date available modification methods include chemical modifications, physical modifications, impregnation with mineral sorbents, and magnetic modifications (Rajapaksha et al. 2016a). Magnetic biochar (Chen et al. 2011), chitosan-modified biochar (Zhou et al. 2013), carbon nanotube–biochar nanocomposite (Inyang et al. 2014), MnO_x-loaded biochars (Song et al. 2014), clay–biochar composites (Yao et al. 2014), and biochar modified by steam activation (Rajapaksha et al. 2016b), oxidation (Huff and Lee 2016), acidification, and alkalization (Liu et al. 2012) show a better structure composition and functional characteristics for improving the performance of site remediation than unmodified biochar. Blending feedstocks with various minerals to produce a biochar–mineral complex (Joseph et al. 2015, Li et al. 2014, Nielsen et al. 2014) is another option to promote plant nutrient uptake for revegetation and long-term stability for biochar itself. Hence, engineered biochar is a potentially promising soil conditioner and amendment for enhancing the efficacy and efficiency of rehabilitation of mine sites.

Findings suggest that the performance of biochar application on remediation of mine sites may vary according to the properties of biochar and the characteristics of mine tailings. A systematic classification of biochar is needed to sort out application-oriented physicochemical properties and conditions. Presynthesis and postsynthesis modification methods of biochar potentially can tailor its physicochemical properties and enhance the performance for remediation of mine sites, which is of great significance for further research. As there is no one-size-fits-all solution to different mine sites that are multielement contaminated, it is necessary to better understand the amelioration mechanisms and tailor the biochar in view of the characteristics of each specific site, as well as conduct long-term investigations and evaluations on the effectiveness of methods used in remediation of mine sites.

REFERENCES

Ahmad, M., Lee, S. S., Lee, S. E., Al-Wabel, M. I., Tsang, D. C. W., and Ok, Y. S. 2016. Biochar-induced changes in soil properties affected immobilization/mobilization of metals/metalloids in contaminated soils. *Journal of Soils and Sediments* 17:717–730.

Alvarenga, P., Gonçalves, A. P., Fernandes, R. M. et al. 2009. Organic residues as immobilizing agents in aided phytostabilization: (I) Effects on soil chemical characteristics. *Chemosphere* 74:1292–1300.

Anawar, H. M. 2013. Impact of climate change on acid mine drainage generation and contaminant transport in water ecosystems of semi-arid and arid mining areas. *Physics and Chemistry of the Earth, Parts A/B/C* 58:13–21.

Anawar, H. M. 2015. Sustainable rehabilitation of mining waste and acid mine drainage using geochemistry, mine type, mineralogy, texture, ore extraction and climate knowledge. *Journal of Environmental Management* 158:111–121.

Anawar, H. M., Akter, F., Solaiman, Z. M., and Strezov, V. 2015. Biochar: An emerging panacea for remediation of soil contaminants from mining, industry and sewage wastes. *Pedosphere* 25:654–665.

Bartzas, G., Komnitsas, K., and Paspaliaris, I. 2006. Laboratory evaluation of Fe^0 barriers to treat acidic leachates. *Minerals Engineering* 19:505–514.

Bea, S. A., Ayora, C., Carrera, J., Saaltink, M. W., and Dold, B. 2010. Geochemical and environmental controls on the genesis of soluble efflorescent salts in coastal mine tailings deposits: A discussion based on reactive transport modeling. *Journal of Contaminant Hydrology* 111:65–82.

Beesley, L., Inneh, O. S., Norton, G. J., Moreno-Jimenez, E., Pardo, T., Clemente, R., and Dawson, J. J. 2014. Assessing the influence of compost and biochar amendments on the mobility and toxicity of metals and arsenic in a naturally contaminated mine soil. *Environmental Pollution* 186:195–202.

Beesley, L., and Marmiroli, M. 2011. The immobilisation and retention of soluble arsenic, cadmium and zinc by biochar. *Environmental Pollution* 159:474–480.

Beesley, L., Moreno-Jiménez, E., and Gomez-Eyles, J. L. 2010. Effects of biochar and greenwaste compost amendments on mobility, bioavailability and toxicity of inorganic and organic contaminants in a multi-element polluted soil. *Environmental Pollution* 158:2282–2287.

Beesley, L., Moreno-Jimenez, E., Gomez-Eyles, J. L., Harris, E., Robinson, B., and Sizmur, E. 2011. A review of biochars' potential role in the remediation, revegetation and restoration of contaminated soils. *Environmental Pollution* 159:3269–3282.

Benner, S. G., Blowes, D. W., and Ptacek, C. J. 1997. A full-scale porous reactive wall for prevention of acid mine drainage. *Groundwater Monitoring and Remediation* 17:99–107.

Bhattacharya, A., Routh, J., Jacks, G., Bhattacharya, P., and Mörth, M. 2006. Environmental assessment of abandoned mine tailings in Adak, Västerbotten district (northern Sweden). *Applied Geochemistry* 21:1760–1780.

Cavanagh, J. E., Pope, J., Harding, J. S. et al. 2014. A framework for predicting and managing water quality impacts of mining on streams: A user's guide, Landcare Research, New Zealand, 139–234.

Chen, B., Chen, Z., and Lv, S. 2011. A novel magnetic biochar efficiently sorbs organic pollutants and phosphate. *Bioresource Technology* 102:716–723.

Chen, B., Zhou, D., and Zhu, L. 2008. Transitional adsorption and partition of nonpolar and polar aromatic contaminants by biochars of pine needles with different pyrolytic temperatures. *Environmental Science and Technology* 42:5137–5143.

Chen, Z., Wang, Y., Xia, D. et al. 2016. Enhanced bioreduction of iron and arsenic in sediment by biochar amendment influencing microbial community composition and dissolved organic matter content and composition. *Journal of Hazardous Materials* 311:20–29.

Ciccu, R., Ghiani, M., Serci, A., Fadda, S., Peretti, R., and Zucca, A. 2003. Heavy metal immobilization in the mining-contaminated soils using various industrial wastes. *Minerals Engineering* 16:187–192.

Coulton, R., Bullen, C., and Hallett, C. 2003. The design and optimisation of active mine water treatment plants. *Land Contamination and Reclamation* 11:273–280.

Coupland, K., Battaglia-Brunet, F., Hallberg, K. B., Dictor, M. C., Garrido, F., and Johnson, D. B. 2004. Oxidation of iron, sulfur and arsenic in mine waters and mine wastes: An important role for novel *Thiomonas* spp., In: *Biohydrometallurgy: A Sustainable Technology in Evolution: Proceedings on the 15th International Biohydrometallurgy Symposium*, In press.

Devi, P., and Saroha, A. K. 2015. Effect of pyrolysis temperature on polycyclic aromatic hydrocarbons toxicity and sorption behaviour of biochars prepared by pyrolysis of paper mill effluent treatment plant sludge. *Bioresource Technology* 192:312–320.

Evangelou, V. P. 1998. Pyrite chemistry: The key for abatement of acid mine drainage. In *Acidic Mining Lakes*, Geller, A., Klapper, H., and Salomons, W. (Eds.), pp. 197–222. Springer, Berlin, Germany.

Fang, S., Tsang, D. C. W., Zhou, F., Zhang, W., and Qiu, R. 2016. Stabilization of cationic and anionic metal species in contaminated soils using sludge-derived biochar. *Chemosphere* 149: 263–271.

Favas, P. J. C., Sarkar, S. K., Rakshit, D., Venkatachalam, P., and Prasad, M. N. V. 2016. Acid mine drainages from abandoned mines: Hydrochemistry, environmental impact, resource recovery, and prevention of pollution. In *Environmental Materials and Waste: Resource Recovery and Pollution Prevention*, Prasad, M.N.V. and Shih, K. (Eds.), pp. 413–462, Elsevier-Academic Press.

Fellet, G., Marchiol, L., Delle Vedove, G., and Peressotti, A. 2011. Application of biochar on mine tailings: Effects and perspectives for land reclamation. *Chemosphere* 83:1262–1267.

Fellet, G., Marmiroli, M., and Marchiol, L. 2014. Elements uptake by metal accumulator species grown on mine tailings amended with three types of biochar. *Science of the Total Environment* 468–469:598–608.

Forján, R., Asensio, V., Rodríguez-Vila, A., and Covelo, E. F. 2016. Contribution of waste and biochar amendment to the sorption of metals in a copper mine tailing. *Catena* 137:120–125.

Gatzweiler, R., Jahn, S., Neubert, G., and Paul, M. 2001. Cover design for radioactive and AMD-producing mine waste in the Ronneburg area, Eastern Thuringia. *Waste Management* 21:175–184.

Green, S., and Renault, S. 2008. Influence of papermill sludge on growth of *Medicago sativa*, *Festuca rubra* and *Agropyron trachycaulum* in gold mine tailings: A greenhouse study. *Environmental Pollution* 151:524–531.

Griffiths, B. S., and Philippot, L. 2013. Insights into the resistance and resilience of the soil microbial community. *FEMS Microbiology Reviews* 37:112–129.

Han, Y. S., Youm, S. J., Oh, C., Cho, Y. C., and Ahn, J. S. 2015. Geochemical and eco-toxicological characteristics of stream water and its sediments affected by acid mine drainage. *Catena* 148:52–59.

Hansen, H. K., Rojo, A., and Ottosen, L. M. 2005. Electrodialytic remediation of copper mine tailings. *Journal of Hazardous Materials* 117:179–183.

Hettiarachchi, G. M., Pierzynski, G. M., and Ransom, M. D. 2001. In situ stabilization of soil lead using phosphorus. *Journal of Environmental Quality* 30:1214–1221.

Huff, M. D., and Lee, J. W. 2016. Biochar-surface oxygenation with hydrogen peroxide. *Journal of Environmental Management* 165:17–21.

International Biochar Initiative. 2012. Standardized product definition and product testing guidelines for biochar that is used in soil. IBI Biochar Standards.

Inyang, M., Gao, B., Zimmerman, A., Zhang, M., and Chen, H. 2014. Synthesis, characterization, and dye sorption ability of carbon nanotube–biochar nanocomposites. *Chemical Engineering Journal* 236:39–46.

Jain, S., Baruah, B. P., and Khare, P. 2014. Kinetic leaching of high sulphur mine rejects amended with biochar: Buffering implication. *Ecological Engineering* 71:703–709.

Jain, S., Mishra, D., Khare, P., Yadav, V., Deshmukh, Y., and Meena, A. 2016. Impact of biochar amendment on enzymatic resilience properties of mine spoils. *Science of the Total Environment* 544:410–421.

Jeffery, S., Verheijen, F. G., Van Der Velde, M., and Bastos, A. C. 2011. A quantitative review of the effects of biochar application to soils on crop productivity using meta-analysis. *Agriculture, Ecosystems and Environment* 144:175–187.

Jiang, S., Huang, L. Nguyen, T. A. et al. 2016. Copper and zinc adsorption by softwood and hardwood biochars under elevated sulphate-induced salinity and acidic pH conditions. *Chemosphere* 142:64–71.

Johnson, D. B. 2000. Biological removal of sulfurous compounds from inorganic wastewaters. In *Environmental Technologies to Treat Sulfur Pollution: Principles and Engineering*, Lens, P. and Hulshoff Pol, L. (Eds.), London, IWA Publishing, pp. 175–206.

Johnson, D. B. 2003. Chemical and microbiological characteristics of mineral spoils and drainage waters at abandoned coal and metal mines. *Water, Air and Soil Pollution: Focus* 3:47–66.

Johnson, D. B. 2008. Biodiversity and interactions of acidophiles: Key to understanding and optimizing microbial processing of ores and concentrates. *Transactions of Nonferrous Metals Society of China* 18:1367–1373.

Johnson, D. B., and Hallberg, K. B. 2005. Acid mine drainage remediation options: A review. *Science of the Total Environment* 338:3–14.

Joseph, S., Anawar, H. M., Storer, P. et al. 2015. Effects of enriched biochars containing magnetic iron nanoparticles on mycorrhizal colonisation, plant growth, nutrient uptake and soil quality improvement. *Pedosphere* 25:749–760.

Kalin, M., Cairns, J., and McCready, R. 1991. Ecological engineering methods for acid mine drainage treatment of coal wastes. *Resources, Conservation and Recycling* 5:265–275.

Kalin, M., Wheeler, W. N., and Meinrath, G. 2004. The removal of uranium from mining waste water using algal/microbial biomass. *Journal of Environmental Radioactivity* 78:151–177.

Keiluweit, M., Nico, P. S., Johnson, M. G., and Kleber, M. 2010. Dynamic molecular structure of plant biomass-derived black carbon (biochar). *Environmental Science and Technology* 44:1247–1253.

Kelly, C. N., Peltz, C. D., Stanton, M., Rutherford, D. W., and Rostad, C. E. 2014. Biochar application to hardrock mine tailings: Soil quality, microbial activity, and toxic element sorption. *Applied Geochemistry* 43:35–48.

Kelly, M., Allison, W. J., Garman, A. R., and Symon, C. J. 2012. *Mining and the Freshwater Environment.* Springer Science and Business Media, London.

Kim, J. S., Sparovek, G., Longo, R. M., De Melo, W. J., and Crowley, D. 2007. Bacterial diversity of terra preta and pristine forest soil from the Western Amazon. *Soil Biology and Biochemistry* 39:684–690.

Kim, R. Y., Yoon, J. K., Kim, T. S., Yang, J. E., Owens, G., and Kim, K. R. 2015. Bioavailability of heavy metals in soils: Definitions and practical implementation—A critical review. *Environmental Geochemistry and Health* 37:1041–1061.

Kleinmann, R. L. P., Hedin, R. S., and Nairn, R. W. 1998. Treatment of mine drainage by anoxic limestone drains and constructed wetlands. In *Acidic Mining Lakes*, Geller, A., Klapper, H., Salomons, W. (Eds.), pp. 303–319. Springer, Berlin, Germany.

Kruse, N. A., and Younger, P. L. 2009. Development of thermodynamically-based models for simulation of hydrogeochemical processes coupled to channel flow processes in abandoned underground mines. *Applied Geochemistry* 24:1301–1311.

Kwon, E. E., Oh, J. I., and Kim, K. H. 2015. Polycyclic aromatic hydrocarbons (PAHs) and volatile organic compounds (VOCs) mitigation in the pyrolysis process of waste tires using CO_2 as a reaction medium. *Journal of Environmental Management* 160:306–311.

Langer, U., and Rinklebe, J. 2009. Lipid biomarkers for assessment of microbial communities in floodplain soils of the Elbe River (Germany). *Wetlands* 29:353–362.

Lehmann, J. 2007. A handful of carbon. *Nature* 447:143–144.

Lehmann, J., Rillig, M. C., Thies, J., Masiello, C. A., Hockaday, W. C., and Crowley, D. 2011. Biochar effects on soil biota—A review. *Soil Biology and Biochemistry* 43:1812–1836.

Li, F., Cao, X., Zhao, L., Wang, J., and Ding, Z. 2014. Effects of mineral additives on biochar formation: Carbon retention, stability, and properties. *Environmental Science and Technology* 48:11211–11217.

Li, M. G., Aube, B., and St-Arnaud, L. 1997. Considerations in the use of shallow water covers for decommissioning reactive tailings. In *Proceedings of the Fourth International Conference on Acid Rock Drainage*, Vancouver, British Columbia, Canada.

Li, X., You, F., Huang, L., Strounina, E., and Edraki, M. 2013. Dynamics in leachate chemistry of Cu-Au tailings in response to biochar and woodchip amendments: A column leaching study. *Environmental Sciences Europe* 25:1.

Liu, P., Liu, W. J., Jiang, H., Chen, J. J., Li, W. W., and Yu, H. Q. 2012. Modification of biochar derived from fast pyrolysis of biomass and its application in removal of tetracycline from aqueous solution. *Bioresource Technology* 121:235–240.

Long, Z. E., Huang, Y., Cai, Z., Cong, W., and Ouyang, F. 2003. Biooxidation of ferrous iron by immobilized *Acidithiobacillus ferrooxidans* in poly (vinyl alcohol) cryogen carriers. *Biotechnology Letters* 25:245–249.

Lu, H., Zhang, W., Yang, Y., Huang, X., Wang, S., and Qiu, R. 2012. Relative distribution of Pb^{2+} sorption mechanisms by sludge-derived biochar. *Water Research* 46:854–862.

Luo, L., Xu, C., Chen, Z., and Zhang, S. 2015. Properties of biomass-derived biochars: Combined effects of operating conditions and biomass types. *Bioresource Technology* 192:83–89.

Mehling, P. E., Day, S. J., and Sexsmith, K. S. 1997. Blending and layering waste rock to delay, mitigate or prevent acid generation: A case review study. In *Proceedings of the Fourth International Conference on Acid Rock Drainage*, May 30 to June 6, 1997, Vancouver, British Columbia, Canada, vol. II, pp. 953–970.

Melo, L. C., Puga, A. P., Coscione, A. R., Beesley, L., Abreu, C. A., and Camargo, O. A. 2015. Sorption and desorption of cadmium and zinc in two tropical soils amended with sugarcane-straw-derived biochar. *Journal of Soils and Sediments* 16:226–234.

Mendez, M. O., Glenn, E. P., and Maier, R. M. 2007. Phytostabilization potential of quail bush for mine tailings. *Journal of Environmental Quality* 36:245–253.

Mendez, M. O., and Maier, R. M. 2008. Phytostabilization of mine tailings in arid and semiarid environments—An emerging remediation technology. *Environmental Health Perspectives* 116:278.

Moche, M., Gutknecht, J., Schulz, E., Langer, U., and Rinklebe, J. 2015. Monthly dynamics of microbial community structure and their controlling factors in three floodplain soils. *Soil Biology and Biochemistry* 90:169–178.

Modabberi, S., Alizadegan, A., Mirnejad, H., and Esmaeilzadeh, E. 2013. Prediction of AMD generation potential in mining waste piles, in the sarcheshmeh porphyry copper deposit, Iran. *Environmental Monitoring and Assessment* 185:9077–9087.

Mohan, D., Sarswat, A., Ok, Y. S., and Pittman, C. U. 2014. Organic and inorganic contaminants removal from water with biochar, a renewable, low cost and sustainable adsorbent—A critical review. *Bioresource Technology* 160:191–202.

Moutsatsou, A., Gregou, M., Matsas, D., and Protonotarios, V. 2006. Washing as a remediation technology applicable in soils heavily polluted by mining-metallurgical activities. *Chemosphere* 63:1632–1640.

Mudd, G. M. 2009. *The Sustainability of Mining in Australia: Key Production Trends and Environmental Implications*. Monash University and Mineral Policy Institute, Melbourne, Victoria, Australia.

Mukherjee, A., and Lal, R. 2014. The biochar dilemma. *Soil Research* 52:217–230.

Neal, C., Whitehead, P. G., Jeffery, H., and Neal, M. 2005. The water quality of the River Carron, west Cornwall, November 1992 to March 1994: The impacts of Wheal Jane discharges. *Science and the Total Environment* 338:23–39.

Nielsen, S., Minchin, T., Kimber, S. et al. 2014. Comparative analysis of the microbial communities in agricultural soil amended with enhanced biochars or traditional fertilisers. *Agriculture, Ecosystems and Environment* 191:73–82.

Oh, S. Y., and Yoon, M. K. 2013. Biochar for treating acid mine drainage. *Environmental Engineering Science* 30:589–593.

Pérez-López, R., Cama, J., Nieto, J. M., Ayora, C., and Saaltink, M. W. 2009. Attenuation of pyrite oxidation with a fly ash pre-barrier: Reactive transport modelling of column experiments. *Applied Geochemistry* 24:1712–1723.

Puga, A. P., Melo, L. C. A., de Abreu, C. A., Coscione, A. R., and Paz-Ferreiro, J. 2016. Leaching and fractionation of heavy metals in mining soils amended with biochar. *Soil and Tillage Research* 164:25–33.

Quilliam, R. S., DeLuca, T. H., and Jones, D. L. 2013. Biochar application reduces nodulation but increases nitrogenase activity in clover. *Plant and Soil* 366:83–92.

Rajapaksha, A. U., Chen, S. S., Tsang, D. C. W. et al. 2016a. Engineered/designer biochar for contaminant removal/immobilization from soil and water: Potential and implication of biochar modification. *Chemosphere* 148:276–291.

Rajapaksha, A. U., Vithanage, M., Lee, S. S., Seo, D. C., Tsang, D. C. W., and Ok, Y. S. 2016b. Steam activation of biochars facilitates kinetics and pH-resilience of sulfamethazine sorption. *Journal of Soils and Sediments* 16:889–895.

Rajapaksha, A. U., Vithanage, M., Zhang, M. et al. 2014. Pyrolysis condition affected sulfamethazine sorption by tea waste biochars. *Bioresource Technology* 166:303–308.

Rees, F., Simonnot, M. O., and Morel, J. L. 2014. Short-term effects of biochar on soil heavy metal mobility are controlled by intra-particle diffusion and soil pH increase. *European Journal of Soil Science* 65:149–161.

Reeves, R. D. 2006. Hyperaccumulation of trace elements by plants. *Phytoremediation of Metal-Contaminated Soils* 68:25–52.

Reverchon, F., Yang, H., Ho, T. Y. et al. 2015. A preliminary assessment of the potential of using an acacia–biochar system for spent mine site rehabilitation. *Environmental Science and Pollution Research International* 22:2138–2144.

Rinklebe, J., and Langer, U. 2013. Soil microbial biomass and phospholipid fatty acids. In *Methods in Biogeochemistry of Wetlands*, DeLaune, R. D., Reddy, K. R., Richardson, C. J., and Megonigal, J. P. (Eds.), pp. 331–348. American Society of Agronomy.

Roberts, D. A., Cole, A. J., Paul, N. A., and De Nys, R. 2015. Algal biochar enhances the re-vegetation of stockpiled mine soils with native grass. *Journal of Environmental Management* 161:173–180.

Rodríguez-Vila, A., Asensio, V., Forján, R., and Covelo, E. F. 2015. Chemical fractionation of Cu, Ni, Pb and Zn in a mine soil amended with compost and biochar and vegetated with *Brassica juncea* L. *Journal of Geochemical Exploration* 158:74–81.

Rodríguez-Vila, A., Covelo, E. F., Forján, R., and Asensio, V. 2014. Phytoremediating a copper mine soil with *Brassica juncea* L., compost and biochar. *Environmental Science and Pollution Research International* 21:11293–11304.

Rosende, M., Beesley, L., Moreno-Jimenez, E., and Miro, M. 2016. Automatic flow-through dynamic extraction: A fast tool to evaluate char-based remediation of multi-element contaminated mine soils. *Talanta* 148:686–693.

Ryan, A., Street-Perrott, A., Eastwood, D., and Brackenbury, S. 2014. The use of sustainable 'biochar compost' for remediation of contaminated land. In *EGU General Assembly Conference Abstracts*, Vienna, Austria.

Salmon, S. U. and Malmström, M. E. 2004. Geochemical processes in mill tailings deposits: Modelling of groundwater composition. *Applied Geochemistry* 19:1–17.

Song, Z., Lian, F., Yu, Z., Zhu, L., Xing, B., and Qiu, W. 2014. Synthesis and characterization of a novel MnOx-loaded biochar and its adsorption properties for Cu^{2+} in aqueous solution. *Chemical Engineering Journal* 242:36–42.

Steinbeiss, S., Gleixner, G., and Antonietti, M. 2009. Effect of biochar amendment on soil carbon balance and soil microbial activity. *Soil Biology and Biochemistry* 41:1301–1310.

Swanson, D. A., Barbour, S. L., and Wilson, G. W. 1997. Dry-site versus wet-site cover design. In *Proceedings of Fourth International Conference of Acid Rock Drainage*, Vancouver, British Columbia, Canada.

Tan, X., Liu, Y., Gu, Y. et al. 2015a. Immobilization of Cd (II) in acid soil amended with different biochars with a long term of incubation. *Environmental Science and Pollution Research* 22:12597–12604.

Tan, X., Liu, Y., Zeng, G. et al. 2015b. Application of biochar for the removal of pollutants from aqueous solutions. *Chemosphere* 125:70–85.

Tang, J., Zhu, W., Kookana, R., and Katayama, A. 2013. Characteristics of biochar and its application in remediation of contaminated soil. *Journal of Bioscience and Bioengineering* 116:653–659.

Tang, X., and Yang, J. 2012. Long-term stability and risk assessment of lead in mill waste treated by soluble phosphate. *Science of the Total Environment* 438:299–303.

Thies, J. E., and Rillig, M. C. 2009. Characteristics of biochar: Biological properties. In *Biochar for Environmental Management: Science and Technology*, Joseph, S. and Lehmann, J. (Eds.). Earthscan, London, U.K.

Uchimiya, M., Lima, I. M., Thomas Klasson, K., Chang, S., Wartelle, L. H., and Rodgers, J. E. 2010. Immobilization of heavy metal ions (CuII, CdII, NiII, and PbII) by broiler litter-derived biochars in water and soil. *Journal of Agricultural and Food Chemistry* 58:5538–5544.

Valente, T., Gomes, P., Pamplona, J., and de la Torre, M. L. 2012. Natural stabilization of mine waste-dumps— Evolution of the vegetation cover in distinctive geochemical and mineralogical environments. *Journal of Geochemical Exploration* 123:152–161.

van der Ent, A., Baker, A. J., Reeves, R. D., Pollard, A. J., and Schat, H. 2012. Hyperaccumulators of metal and metalloid trace elements: Facts and fiction. *Plant and Soil* 362:319–334.

Vile, M. A., and Wieder, R. K. 1993. Alkalinity generation by Fe (III) reduction versus sulfate reduction in wetlands constructed for acid mine drainage treatment. *Water, Air, and Soil Pollution* 69:425–441.

Warnock, D. D., Lehmann, J., Kuyper, T. W., and Rillig, M. C. 2007. Mycorrhizal responses to biochar in soil—Concepts and mechanisms. *Plant and Soil* 300:9–20.

Yang, S., Cao, J., Hu, W., Zhang, X., and Duan, C. 2013. An evaluation of the effectiveness of novel industrial by-products and organic wastes on heavy metal immobilization in Pb–Zn mine tailings. *Environmental Science: Processes and Impacts* 15:2059–2067.

Yang, X., Liu, J., McGrouther, K. et al. 2016. Effect of biochar on the extractability of heavy metals (Cd, Cu, Pb, and Zn) and enzyme activity in soil. *Environmental Science and Pollution Research* 23:974–984.

Yao, Y., Gao, B., Fang, J. et al. 2014. Characterization and environmental applications of clay–biochar composites. *Chemical Engineering Journal* 242:136–143.

Ye, Z. H., Wong, J. W. C., Wong, M. H., Lan, C. Y., and Baker, A. J. M. 1999. Lime and pig manure as ameliorants for revegetating lead/zinc mine tailings: A greenhouse study. *Bioresource Technology* 69:35–43.

Younger, P. L., Coulton, R. H., and Froggatt, E. C. 2005. The contribution of science to risk-based decision-making: Lessons from the development of full-scale treatment measures for acidic mine waters at Wheal Jane, UK. *Science of the Total Environment* 338:137–154.

Zhang, W., Huang, X., Rees, F., Tsang, D. C. W., Qiu, R., and Wang, H. 2016. Metal immobilization by sludge-derived biochar: Roles of mineral oxides and carbonized organic compartment. *Environmental Geochemistry and Health* 39:379–389.

Zhang, W., Zheng, J., Zheng, P., Tsang, D. C. W., and Qiu, R. 2015. Sludge-derived biochar for As(III) immobilization: Effects of solution chemistry on sorption behavior. *Journal of Environmental Quality* 44:1119–1126.

Zhou, Y., Gao, B., Zimmerman, A. R., Fang, J., Sun, Y., and Cao, X. 2013. Sorption of heavy metals on chitosan-modified biochars and its biological effects. *Chemical Engineering Journal* 231:512–518.

Zhu, Q., Peng, X., Huang, T., Xie, Z., and Holden, N. M. 2014. Effect of biochar addition on maize growth and nitrogen use efficiency in acidic red soils. *Pedosphere* 24:699–708.

Zornoza, R., Faz, Á., Carmona, D. M., Acosta, J. A., Martínez-Martínez, S., and de Vreng, A. 2013. Carbon mineralization, microbial activity and metal dynamics in tailing ponds amended with pig slurry and marble waste. *Chemosphere* 90:2606–2613.

3 Sources and Management of Acid Mine Drainage

S.R. Gurung, Hasintha Wijesekara, Balaji Seshadri, R.B. Stewart, P.E.H. Gregg, and N.S. Bolan

CONTENTS

3.1 Introduction..33
3.2 Sources of Acid Mine Drainage..34
3.3 Biochemical Aspects of Acid Mine Drainage Formation 35
    3.3.1 Generation of Acid from Pyrite Oxidation 35
    3.3.2 Accretion and Migration of Acid Mine Drainage..37
    3.3.3 By-Products of Acid Mine Drainage 38
        3.3.3.1 Metal Hydroxides..38
        3.3.3.2 Sulfate Salts 38
        3.3.3.3 Acidity..39
        3.3.3.4 Aluminum 40
3.4 Predictive Techniques for Acid Mine Drainage...40
    3.4.1 Static Tests 40
    3.4.2 Kinetic Tests 41
    3.4.3 Evaluation of the Predictive Techniques..42
3.5 Prevention and Control of Acid Mine Drainage 44
    3.5.1 Preventive Coatings 45
    3.5.2 Selective Handling..45
    3.5.3 Bactericides Treatment 45
    3.5.4 Oxidant Infiltration Barriers 46
3.6 Treatment of Acid Mine Drainage 47
    3.6.1 *In Situ* Neutralization...48
    3.6.2 Active Treatment...49
    3.6.3 Passive Treatment 50
3.7 Summary and Conclusions 50
Acknowledgment 51
References..51

3.1 INTRODUCTION

Acid mine drainage (AMD) from both active and abandoned mine sites is a major environmental issue for the mining industry in environmentally concerned regions of the world (Gray 1997, Lindsay et al. 2015). The term is used to describe any seepage, leachate, or drainage affected by the oxidation products of sulfide minerals in mine sites when exposed to air and water (Figure 3.1). Both chemical reactions and biological transformations are recognized as being responsible for generating AMD (Lindsay et al. 2015). AMD is typically characterized by low pH and high levels of dissolved metal salts, as well as high concentrations of acidity, sulfate, iron, and other metals (Gray 1997). Once the AMD process begins, it is difficult to control, often accelerates, and is likely to persist for decades or centuries. In the absence of natural or added neutralizing materials

FIGURE 3.1   Acid mine drainage in an abandoned open pit coal mine, Singleton, New South Wales, Australia.

(carbonate minerals such as calcite or dolomite), the AMD is likely to contain toxic levels of heavy metals such as Fe, Al, Mn, Cu, Pb, Zn, and Cd, which can cause serious environmental problems in soil and water systems (Sengupta 1994).

Regulations in many countries have developed ways to address the issue of AMD at the permitting stage rather than as an afterthought. There are signs that some operators, working in partnership with regulatory authorities and other stakeholders, are developing proactive methodologies based in part upon improvements in predictive techniques (Taylor et al. 1997). Many of the currently practiced preventive measures, however, still require extensive field validation before they can be prescribed as standard techniques in the environmental management of AMD. Since it is now a recognized fact that AMD generation is a site-specific phenomenon, effective control and treatment measures have been directed toward fulfilling the problem on site rather than to providing universal solutions.

This chapter describes the sources and chemical and biological reactions resulting in AMD. It also covers the measures to prevent AMD formation and the methods to manage their environmental impacts.

3.2   SOURCES OF ACID MINE DRAINAGE

There are five major sources of AMD, namely, drainage from underground workings, run-off and discharges from open pits, waste rock dumps, tailings, and ore stockpiles. The sources may be locally significant, for example, spent heap-leach piles, stockpiles of segregated sulfides, and natural seeps and springs in areas of sulfidic mineralization. In general terms, potential sources of AMD in mining activities are well understood and widely documented (Akcil and Koldas 2006, Klein et al. 2014, Robertson 1996).

Abandoned underground workings can result in the release of high concentrations of metal salts into the aqueous environment as the water table rebounds and the workings lead to flooding. These metal salts accumulate when the mine is pumped "dry" and in-place sulfides are exposed to oxygen and moisture. AMD generation may continue even after flooding if there is a persistent source of dissolved oxygen.

Open pit mining can expose very large areas of sulfide-bearing rock to air and water. Failure to control water flow into open pit slopes can result in large volumes of AMD. As oxidation of the sulfides proceeds, fresh sulfides may be exposed by spalling of the rock face, resulting in the constant renewal of the AMD source (Kuyucak et al. 1991, Morin and Hutt 1995).

Waste rock has become a more significant threat as open pit mining has replaced underground mining, particularly in developed countries, and the volumes produced have increased (Morin 1990). The highly permeable coarse nature of the waste rock facilitates rapid oxidation of sulfides after disposal. During extended dry periods, dumps may build up "reserved or stored" acid products and salts through evaporation and supersaturation processes, which are then released in the form of highly contaminated AMD during the subsequent significant precipitation event (USEPA 1994).

Tailings and/or mine spoils often have high sulfide content (mainly in the form of rejected pyrite, marcasite, and pyrrhotite) and are much finer than waste rock. Although tailings have a much higher specific surface area than waste rock, the uniform and fine particle size leads to a much lower permeability than that seen in waste rock piles. Therefore, the increased surface area available for oxidation and leaching reactions is balanced by reduced contact with oxygen due to saturation by a relatively static water body. Consequently, tailings often generate AMD more slowly than coarser, but more permeable, waste rock (Filipek et al. 1996).

3.3 BIOCHEMICAL ASPECTS OF ACID MINE DRAINAGE FORMATION

Numerous workers have studied the reaction kinetics of pyrite oxidation (Evangelou and Zhang 1995, Tao and Dongwei 2014). Briefly, acid generation involves a complex combination of organic and inorganic processes and reactions. This happens when reactive sulfide rock (e.g., pyrite) is initially exposed to air and water. The extent and duration of acid generation depend on the intrinsic geochemical characteristics of the sulfidic rocks.

The generation of acid in the oxidation of pyrite is considered to involve abiotic and biotic processes under aerobic environments. The former is catalyzed by ferric iron (Fe^{3+}) while the latter is catalyzed by acidophilic bacteria, chiefly *Thiobacillus ferrooxidans*. The role of bacteria in AMD generation from oxidation of pyrite has been recognized and widely studied (Haferburg and Kothe 2007, Tao and Dongwei 2014). Both abiotic and biotic processes are considered pH and temperature dependent. The net generation of acidity in any sulfidic substrate will ultimately depend on the host rock mineralogy, bio-geochemical, and environmental factors. The pyrite crystal morphology and degree of liberation of grains are important factors affecting oxidation rates (Mills 1996).

Generation and migration of AMD are complex subjects, and intimately linked with the nature of the ore body, host rock mineralogy, and local and regional hydrology and hydrogeology. The occurrence of AMD does not necessarily lead to its migration as there are a number of chemical processes that prevent the movement of AMD away from its source. The main AMD-forming processes are summarized briefly below.

3.3.1 GENERATION OF ACID FROM PYRITE OXIDATION

The dominant sulfide minerals involved in the development of AMD in sulfidic mine waste rocks are commonly pyrite (FeS_2), marcasite (FeS_2), and pyrrhotite (FeS). Other sulfide minerals, such as chalcopyrite ($CuFeS_2$), chalcocite (Cu_2S), covellite (CuS), pentlandite [$(Fe,Ni)_8S_9$], arsenopyrite ($FeAsS$), stibnite (Sb_2S_3), molybdenite (MoS_2), sphalerite (ZnS), and galena (PbS), are also likely to provide a secondary contribution to sulfide oxidation and acid generation (Gray 1997, Jambor and Blowes 1994, Mills 1997, MVTI 1994), although PbS, ZnS, and bornite (Cu_5FeS_4) are considered non-acid-generating (Kwong 1995). These non-pyritic sulfide minerals are, however, likely to be subjected to direct chemical reaction by Fe^{3+} generated from oxidation of associated pyrite minerals or indirect oxidation by *thiobacilli* resulting in the generation of significant amounts of acid (Kwong 1995, MVTI 1994).

The biochemistry of pyrite oxidation and its products has been studied in detail and described (Gray 1997, Tao and Dongwei 2014). The basic chemistry of AMD generation from oxidation of pyrite is summarized in the following equations:

$$2FeS_2 + 7O_2 + 2H_2O \rightarrow 2Fe^{2+} + 4SO_4^{2-} + 4H^+ \tag{3.1}$$

$$2Fe^{2+} + 1/2O_2 + 2H^+ \rightarrow 2Fe^{3+} + H_2O \tag{3.2}$$

$$FeS_2 + 14Fe^{3+} + 8H_2O \rightarrow 15Fe^{2+} + 2SO_4^{2-} + 16H^+ \tag{3.3}$$

$$Fe^{3+} + 3H_2O \rightarrow Fe(OH)_3 + 3H^+ \tag{3.4}$$

$$FeS_2 + 15/4O_2 + 7/2H_2O \rightarrow Fe(OH)_3 + 2H_2SO_4 \tag{3.5}$$

Ferrous iron (Fe^{2+}) is initially released by the oxidation of FeS_2 (Equation 3.1). After this reaction sequence has been initiated, an oxidation reaction cycle is established in which Fe^{2+} is oxidized by oxygen to Fe^{3+} (Equation 3.2). FeS_2 is subsequently oxidized by Fe^{3+}, generating additional Fe^{2+} and acid (Equation 3.3). Certain conditions can cause rapid oxidation and hydrolysis of Fe^{3+}, resulting in generation of further acidity (Equation 3.4). The forward reaction in Equation 3.4 depends on the rate of production of Fe^{3+} in Equation 3.2, and the reaction in Equation 3.2 is, therefore, the rate-limiting step in the generation of acid in sulfidic ores (Nordstrom 1982). The principal products of the overall pyrite oxidation are, therefore, ferric oxyhydroxide and sulfuric acid (Equation 3.5). The rate-controlling factor in pyrite oxidation is thus the availability of Fe^{3+} ions or the products of its hydrolysis and secondary reactions, $Fe(OH)_3$ and $Fe(SO_4)_3$. The hydrolysis of the soluble hydrous metal sulfates formed during the oxidation reactions and formation of ferric oxyhydroxides will result in the net production of acidity (Backes et al. 1986, Caruccio et al. 1988).

Pyrite oxidation in the absence of Fe^{3+} is relatively slow (Equation 3.1). The rate of abiotic oxidation of Fe^{2+} to Fe^{3+} (Equation 3.2) is also slow in acidic environments. However, the presence of certain catalyzing bacteria (e.g., *Thiobacillus ferrooxidans*) can increase the rate of Fe^{2+} oxidation by as much as a factor of 10^6 (Evangelou and Zhang 1995), accelerating Equation 3.2 to the right. The bacteria *Thiobacillus ferrooxidans* is considered to be the most important organism involved in the biochemical oxidation of sulfide minerals (Bruynesteyn and Hackel 1984). Biologically catalyzed oxidation of pyrite leads to the production of ferrous iron, sulfate, and hydrogen ions.

Most of the protons (H^+) generated from the oxidation of pyrite are spent on the subsequent oxidation of Fe^{2+} to Fe^{3+}. Catalytic oxidation of pyrite predominantly occurs at pH < 4 because Fe^{3+} is only soluble and *Thiobaccillus ferrooxidans* activity is optimum under these acidic conditions (Dent 1986). Both pH and temperature control the biological oxidation rate: the optimum pH range being 1.5–3.5 and temperature range of 30°C–35°C (Caruccio et al. 1988, Ritcey 1989). The pH also has a significant effect on the oxidation rate of ferrous ion. For example, Hutchison and Ellison (1989) indicate that at pH 7, the oxidation of Fe^{2+} to Fe^{3+} occurs within minutes, while it takes about 300 days at pH 4.5 and about 1000 days at pH 3.5 (Roman and Benner 1973).

Acid formation occurs due to hydrolysis of ferric sulfate (Nordstrom 1982). Ferric ions produced by the oxidation of pyrite usually either precipitate as ferric oxyhydroxides if the pH is sufficiently high or serve to oxidize other metal ions and become reduced again to supply additional substrate for microbial growth. The rate of pyrite oxidation by bacteria depends on the amount and form of pyrite, pyrite activity, O_2 and CO_2 availability, pH, temperature, and presence of inhibiting compounds (Nordstrom 1982, Ritcey 1989).

The Fe^{3+} generated during pyrite oxidation is a very strong oxidant, and will readily attack pyrite (generating acidity and Fe^{2+}) and oxidize other metal sulfides (releasing the metals into the acid aqueous phase) while itself being regenerated by biotic and abiotic reactions. The dissolution of non-pyritic minerals can be accelerated by galvanic interactions causing preferential dissolution in acid solutions (Doyle 1990, Kwong 1995), possibly enhanced by the presence of certain bacteria, which continuously oxidize the layer of elemental sulfur formed on the pyrite surface that would otherwise prevent galvanic action (Gray and Sullivan 1995). Sulfate concentrations can often also rise to high levels. Because of these reinforcing "feedback" loops, AMD generation is considered autocatalytic. Once the process has started, it can be very difficult to halt (Doyle 1990), indicating the importance of taking preventative measures to control initial oxidation of pyritic minerals.

AMD generation depends primarily on the ratio of acid-producing sulfide minerals to acid-neutralizing alkaline minerals. The presence or absence of alkaline material is generally regarded as the most critical element in determining AMD characteristics (Caruccio et al. 1988). The degree of acidity, and concentration and speciation of dissolved contaminants, varies according to a number of site-specific factors, but typical contaminants in AMD include Fe, Mn, Al, Cu, Pb, Zn, Cd, As, SO_4^{2-}, and Cl (Herr et al. 1996). Less commonly dissolved components may also be present, depending on localized and regional mineralogy. Dissolved concentrations can range from below the limits of detection up to thousands or tens of thousands of milligrams per liter, while pH can vary from near neutrality down to one and below.

3.3.2 ACCRETION AND MIGRATION OF ACID MINE DRAINAGE

The processes that occur in the accretion and migration of AMD are controlled by both physical and chemical factors. Local environmental conditions also play an important role in the accretion and migration of AMD. The degree of liberation of pyrite grains, the amount of rainfall infiltration, sulfide rock permeability, availability of pore water pressure, and migration mechanism of AMD (surface flow, capillary flow, discreet seepage) are important physical characteristics.

The migration and fate of many of the dissolved metals and metalloids present in AMD are significantly affected by adsorption by, or co-precipitation with, various iron compounds that may form under favorable conditions. In general, the iron species that precipitates is determined by the iron concentration, redox potential, pH, and concentration of complexing anions such as carbonate, sulfate, and sulfide, and partial pressure of CO_2 and O_2 (Schwertmann and Taylor 1989). For example, Fe^{3+} is rapidly hydrolyzed, even at relatively low pH, to form iron oxyhydroxides (ochre). These ochres can contain significant concentrations of metals through co-precipitation and adsorption (Bloomfield 1972). Similarly, aluminum hydroxide can play an important role in the adsorption and precipitation of other metals. In relatively dry environments, various iron sulfates may crystallize; if Fe^{2+} is oxidized to Fe^{3+}, basic sulfates and oxyhydroxides may form. In environments with low pH and high sulfate concentrations, jarosite ($KFe_3(SO_4)_2(OH)_6$) may form while at higher pH and lower sulfate concentrations iron oxide species such as goethite and hydroxides may precipitate (Herbert 1995). These minerals have a large capacity to adsorb non-ferrous metals and metalloids, and will also remove metals from solution by co-precipitation processes. Such reactions are considered capable of preventing the dispersion of dissolved contaminants (Sullivan and Yelton 1988).

The other main chemical control on AMD accretion is the neutralizing effect of minerals such as carbonates and silicates. However, even if the host rock contains significant quantities of acid-neutralizing minerals, the rate of acid generation may exceed that of neutralization, leading to AMD accretion. The rate of neutralization is influenced by pH, the partial pressure of CO_2, temperature, mineral composition and structure, redox conditions, and the presence of impurity ions in the neutralizing mineral's lattice (Sherlock et al. 1995). The most important factor in determining the extent of AMD is not the pH, but the total acidity (Erickson and Hedin 1988). Total acidity is a measure of the excess amount of H^+ over other ions present in the solution (Johnson and Hallberg 2005). This takes into account the neutralization of the acid by other materials present. Usually, however, a high acidity is accompanied by a low pH in the AMD. Consequently, a low pH has detrimental effects on the bicarbonate buffering system.

The concentrations and mobility of metal ions in AMD are important aspects in the assessment of the environmental impact of mining base metal sulfide deposits. Metal concentrations mobilized in soil and water exert a strong influence on vegetation since plant uptake of heavy metals is considered a function of element speciation (Bolan et al. 2014). The distinction between total metal and available metal (i.e., free metal ion concentration) is critical to the understanding of potential contamination/pollution problems. Although total metal concentrations in various forms can have little impact on the environment, bioavailability of dissolved metals in phytotoxic levels can be a limiting factor to plant growth (Bolan et al. 2008, Luo and Rimmer 1995).

A clear understanding of the accretion of AMD, its migration, and metal loading can be a powerful tool in planning AMD preventative or control measures. Natural processes such as attenuation and neutralization by alkaline minerals present *in situ* can go some way toward mitigating the impacts of AMD. Long-term preventative measures and effective treatment prescriptions adapted to site-specific criteria accordingly may be viable options for mitigating low pH environment created by AMD.

3.3.3 BY-PRODUCTS OF ACID MINE DRAINAGE

Pyrite oxidation and its oxidation by-products are as much environmental problems as AMD itself. Sulfide oxidation and migration of its by-products have been studied mainly on coal overburden using complex geochemical and numerical models (Vanberk and Wisokzky 1995, Wunderly et al. 1996). However, the overall products of pyrite oxidation would be similar for hard rock mine waste materials as well. The primary reaction products resulting from biochemical oxidation of pyrite are H^+, Fe^{2+}, Fe^{3+}, SO_4^{2-}, and $Fe(OH)_3$. The $Fe(OH)_3$ remains in solution as long as the pH of the AMD is <4. At this pH, most of the metals ions (Fe, Mn, Al, Cu, Zn, Pb, As, etc.) remain mobilized in AMD (Nordstrom 1982, Wunderly et al. 1996). If the pH of the AMD is increased by contact with carbonate minerals (if present) or entry into a water system of higher pH, a number of secondary products (metal hydroxides and sulfate salts) can form under surficial conditions resulting in considerable buildup of reserve or residual acidity (Wunderly et al. 1995).

3.3.3.1 Metal Hydroxides

When the pH AMD increases to > 4, metallic ions such as Fe^{3+}, Al^{3+}, Cu^{2+}, Zn^{2+}, Pb^{2+}, and As^{3+} will react with hydroxyl ions to eventually form hydroxides as precipitates (Equation 3.6):

$$M^{n+} + nOH^- \leftrightarrow M(OH)_n \tag{3.6}$$

where M = metal ion, OH^- = hydroxyl ion and $M(OH)_n$ = metal hydroxides, and n = oxidation state. Fe-oxyhydroxides [$(FeO \cdot OH)$ and $Fe(OH)_3$] precipitate directly by oxidation of dissolved Fe^{2+} (Dent 1986). These highly insoluble amorphous Fe-oxydroxide gels often precipitate as rusty-brown or yellowish surface expression in areas of intense sulfide mineral oxidation or as "ochre" in pore fillings. The formation of ironstone (ferricrete) from cementing of unconsolidated surficial material by oxyhydroxides in fracture zones and slope surfaces of weathering iron sulfides is generally indicative of natural AMD (Taylor and Thornber 1995).

3.3.3.2 Sulfate Salts

Associated with the acid environment are elevated levels of soluble salts, which solubilize during wet periods, and severely affect plant growth on the reclaimed mine waste. In the presence of carbonate minerals such as calcite ($CaCO_3$) and dolomite ($CaMgCO_3$), the sulfate salt most commonly formed is gypsum ($CaSO_4 \cdot 2H_2O$), which may precipitate copiously on surfaces during dry periods. Unlike iron, which remains immobilized in the waste rock substrate, most of the soluble sulfate salt is leached with the AMD and only a fraction of sulfate may be retained as gypsum, jarosite [$(KFe_3(SO_4)_2(OH)_6$], alunite [$(KAl_3(SO_4)_2(OH)_6$], and mallardite ($MnSO_4 \cdot 7H_2O$) (Herbert 1995).

Under very low pH (pH < 3.7) and strongly oxidizing environments (Eh > 400 mV), characteristic pale yellow deposits of minerals in the iron and aluminum jarosite–natrojarosite range [(K, Na) (Fe, Al)$_3$(SO$_4$)$_2$(OH)$_6$] commonly precipitate as pore fillings and coatings on exposed surfaces (Dent 1986). Jarosite is unstable at higher pH where it is hydrolyzed to iron oxide (goethite), releasing further acidity (Bloomfield 1972). Thus, neutralization with alkaline amendments is likely to affect dissolution of jarosite and formation of Fe-hydroxide. Since the precipitation of iron hydroxide leads to formation of free acidity, jarosite acts as a storage, or reserve sink, for the acidity formed by pyrite oxidation and weathering (Kerth and Wiggering 1990). The formation of efflorescence minerals such as jarosite (KFe(SO$_4$)$_2$·2Fe(OH)$_3$) and melanterite (FeSO$_4$·7H$_2$O), which usually occur as surface encrustation during dry periods, will, through hydrolysis, result in the release of acidity to the ambient AMD (Equation 3.7). This is considered to be the reason why, in the presence of jarosite, mine waste dumps remain acidic long after the cessation of pyrite oxidation (Dent 1986).

$$3Fe_2(SO_4)_3 + 2KOH + 10H_2O \rightarrow 2K[Fe(SO_4)_2 \cdot 2Fe(OH)_3] + 5H_2SO_4 \qquad (3.7)$$

$$2FeS_2 + 7O_2 + 16H_2O \rightarrow 2FeSO_4 \cdot 7H_2O + 2SO_4^{2-} + H_2SO_4 \qquad (3.8)$$

Accumulation of acid in pyritic waste materials also facilitates rapid chemical degradation, leading to progressive decomposition of clay minerals, and extreme acidification of the spoil can have a detrimental impact on the wider environment through salinization and contamination of ground waters (Equation 3.8).

3.3.3.3 Acidity

Acidity is the base-neutralizing capacity of the solution, which results from H$^+$, Al^{3+}, Fe^{3+}, Fe^{2+}, Mn^{2+}, and other hydrolyzable cations in a sample (Hedin and Erickson 1988). In polymetallic base metal deposits, Cu^{2+}, Zn^{2+}, and Pb^{2+} can also contribute significantly to total acidity of the material (Skousen 1995). Acidity data are therefore useful in defining AMD, which commonly contains substantial quantities of Fe^{2+}, Mn^{2+}, and Al^{3+} (and other hydrolyzable cations), the oxidation and hydrolysis of which can have significant effect on pH of the mine drainage. Oxidation and hydrolysis of Mn^{2+} is known to occur at circumneutral pH 6–7 and, therefore, can contribute significant acidity even in AMD with neutral pH (Hem 1963). In fact, Fe(OH)$_3$ can be directly involved in the heterogeneous adsorption and oxidation of Mn^{2+} to MnO$_2$. Hedin et al. (1994) have indicated that in the absence of acid-neutralizing materials, a theoretical total acidity of the mine drainage containing significant amounts of Fe^{2+}, Mn^{2+}, and Al^{3+} can be calculated from the Equation 3.9:

$$\text{Acidity (mg CaCO}_3\text{ L}^{-1}) = [50*(2*\text{Fe}/56) + (2*\text{Mn}/55) + (3*\text{Al}/27) + 10^3*(10\exp\text{-pH})] \qquad (3.9)$$

(where metal concentrations are in mg L^{-1}, converted to cmol$_c$ kg^{-1}).

Thus, the emerging AMD containing abundant Fe^{2+} and Mn^{2+} can have pH ~6, but it may also have significant acidity. Alternatively, if the AMD contains low concentrations of dissolved metals (hence low acidity), a very low pH does not necessarily mean high acidity. In other words, pH of the AMD alone does not provide much information on the inherent acidity of the solution. This has important implications for treatment and amendment of low pH conditions in oxidizing pyritic materials. For example, a material with a pH of 2, but very low metal concentrations, maybe neutralized more readily than a material with a pH of 5 and high metal concentrations. Acidity determination is, therefore, very important for assessing the buffering capacity of the material.

3.3.3.4 Aluminum

Dissolution of aluminum (Al) in acid soils is one of the most pronounced soil chemical effects of AMD. Because of its toxicity to plants, the chemical behavior of Al has been a major research topic for decades of agronomic and environmental studies (Bolan et al. 2003, Mulder et al. 1989). Low pH conditions promote solubility of Al and result in the concentration of phytotoxic levels of Al in the soil solution (Helyar et al. 1993). The distribution of Al in acid soil materials is dependent on the forms of inorganic (i.e., OH, SO_4, F, $Si(OH)_4$, and H_2PO_4) and organic anionic ligands present and the relative competition between Al and other cations for ligands (Ritchie 1989). Therefore, these ligands have been found to detoxify Al (i.e., reducing phytotoxicity) (Gurung et al. 1996). Soluble Al-hydroxy phosphate polymers and Al-hydroxy silicates are examples found in acidic soils (Blamey et al. 1983). The effect of Al adsorption on pH depends on the type of clay mineral present and the OH:Al ratio in the solution. Thus, removal of hydrolyzed Al species from solution would induce further hydrolysis of Al and, therefore, a lowering of pH is expected (Bache 1974).

Manganese (Mn) in soil is considered to behave similarly to Al in that it may be adsorbed onto the surface of hydrous oxides, clay particles, and organic matter or exist as discrete Mn compounds. However, it differs from Al because it exists in more than one oxidation state under conditions naturally found in soils. In very acid soils, however, Mn^{2+} may be the dominant species and competes with other cations for exchange sites (Salcedo et al. 1979).

3.4 PREDICTIVE TECHNIQUES FOR ACID MINE DRAINAGE

The ability to predict acid generation from mine waste materials is an important step in the process of preventing AMD. This allows the appropriate control measures to be taken to prevent/mitigate the formation of AMD. Predictive tests specifically designed for coal mine waste have been used for decades and significant advances in predictive techniques have been applied to hard rock metal mine waste. The prediction of acid-producing potential (APP) of materials begins with an understanding of the geology and geochemical properties of the rock type encountered during the mining process. Several laboratory and field test procedures to assess or predict the generations of AMD are in use or have been proposed. These include geochemical static tests, geochemical kinetic tests, mineralogical and petrologic studies, mathematical and geochemical modeling, and remote sensing.

Geochemical static and kinetic tests form the most commonly used techniques currently used for the prediction of acid generation from mine waste rock materials. To be useful, the techniques must firstly predict whether a particular mine waste will generate acid at some time and, if so, the rate at which it will occur; secondly, it should predict the characteristics of the drainage leaving the mine waste, from both controlled and uncontrolled sites. The current static and kinetic testing procedures used and their merits evaluated are described below.

3.4.1 STATIC TESTS

The static test is generally the first step in the analysis of acid generation potential. It is based on an acid–base accounting (ABA) procedure whereby the acid-neutralizing capacity (ANC) and APP of the samples are determined, and the difference, net acid producing—potential (NAPP) is calculated. It serves as a screening process to categorize materials into potentially acid-generating, potentially non-acid-generating, and uncertain groups.

ABA remains the most widely used screening test procedure for AMD prediction. It is based on the total sulfide S content and ANC of a sample. The APP of the material is estimated stoichiometrically by assuming ideal oxidation of pyrite (Equation 3.5) and the equivalent $CaCO_3$ required to neutralize the potential acidity produced from the oxidation of 1 mole of pyrite (Equation 3.10).

$$H_2SO_4 + CaCO_3 \leftrightarrow CaSO_4 + H_2O + CO_2 \qquad (3.10)$$

TABLE 3.1

Commonly Used ANC/APP Ratios to Screen AMD-Producing Samples

ANC/APP Ratio	Status of AMD Generation
<1	Likely to generate AMD unless sulfides are unreactive
1–2	Possible AMD generation if neutralizing minerals are preferentially depleted, coated, or otherwise unreactive
>2	AMD generation not expected

According to Equation 3.5, one mole of pyrite contains 64 g of pyritic S, which will theoretically produce 4 moles of acidity (H^+), equivalent to 2 moles of $CaCO_3$, or 200 g $CaCO_3$, equivalent acidity (i.e., 1 g S will potentially produce acidity equivalent to 3.125 g $CaCO_3$). A sample containing 1% pyritic S (sulfide-S), therefore, will have an APP of 31.25 kg $CaCO_3$ t^{-1} ("t^{-1}" stands for "per tonne of material"). The ABA of a sample is calculated according to the following relationship (Equation 3.11):

$$APP \ (kg \ CaCO_3 \ t^{-1}) = Total \ sulfide-S \ (\%) \times 31.25 \tag{3.11}$$

The ANC (kg $CaCO_3$ t^{-1}) of the pyritic material is determined by chemically digesting a pulverized sample with dilute acid (HCl) and the solution back-titrated to a predetermined endpoint (pH 7 or 8.3) with a standard base (NaOH) to determine the amount of acid consumed by the material. The ANC value is subtracted from the APP to derive NAPP where NAPP (kg $CaCO_3$ t^{-1}) = APP – ANC. Lime requirement for sulfidic mine waste is generally based on this NAPP value.

Alternatively, the net acid-generating (NAG) test procedure based on the accelerated oxidation of a sample by H_2O_2, and measuring the resultant pH, predicts the acid-generating nature of the material if the pH of the solution (NAGpH) is < 4 (Miller and Jeffrey 1995). However, the ratio ANC/APP is now more often used to determine the potential for AMD generation, with incorporation of a factor of safety into the ratio to reduce risks arising from unknown parameters. Table 3.1 shows the commonly used ANC/APP ratios to screen AMD-producing samples (Robertson and Ferguson 1995).

A still higher ANC/APP ratio may be warranted to cover the situation where the dissolution rate of acid-neutralizing minerals is generally low compared with the rate of pyrite oxidation. If an apparently "safe" ANC:APP ratio overestimates the neutralization capacity or rate, then there can be severe environmental and, ultimately, cost implications in terms of both site operation and closure (Sherlock et al. 1995).

In the Australasia and Pacific regions, both NAPP and NAG test procedures are commonly used as initial screening and monitoring tools for predicting acid generation from mine waste. Currently both the North American and the Australian terminology and units expressing the results of ABA testing are in use.

3.4.2 KINETIC TESTS

For materials where the potential for acid generation is uncertain, kinetic tests are performed to define acid generation characteristics. In kinetic test procedures, the acid generation (and metal mobilization and transport) characteristics of a sample are measured with respect to time. Kinetic test procedures include humidity cells, column tests, soxhlet reactors, Shake Flasks, field lysimeters, and test plots and barrel tests. Miller and Jeffrey (1995) suggested that the static NAG procedure could also be considered as a kinetic test as the data generated provide information on the kinetics of the reaction from the short-term simulation of the weathering process. The Minewall procedure is also being tested for determining the kinetic AMD characteristics of *in situ* rocks such as pitwalls and the rock surfaces of adits and stopes, and other underground workings (MEND 1995).

Most kinetics tests involve weathering of samples under laboratory-controlled conditions to simulate time-dependent chemical changes in the mine waste and determine the potential to generate net acidity, the rates of sulfide oxidation and neutralization, and the quality of the leachate/drainage. The most commonly used laboratory scale kinetics test procedures are humidity cells and columns.

Humidity cells are typically laboratory units in which samples are subjected to accelerated weathering by cyclic permeation of dry and humid air followed by flushing with water (ASTM 1996). The test usually determines if a given sample will generate acidity but not when the material will produce acidity since the cells undergo accelerated oxidation of sulfide minerals (Price 1997). The inducement of oxidation processes will thus result in an accelerated rate of generation of oxidation products as dissolved metals and/or precipitated metal compounds. The humidity cell tests are not designed to provide leachates that are similar to the actual leachate produced in field conditions and, therefore, are not intended to simulate site-specific leaching conditions (ASTM 1996). The humidity cell tests were originally developed for coal overburden in which the lag period is shorter and they run for only 10 weeks. For hard rock materials, the lag periods are longer and a 10-week period is not adequate to give accurate results (Miller and Jeffrey 1995).

Column tests of AMD generation are generally conducted to simulate the leaching effects of precipitation infiltration into, and drainage from, material exposed to the atmosphere. The aim is to monitor leachate quality with duration by cyclic flushing with water to simulate seasonal variations at a site. Unlike humidity cells, there is no standard testwork procedure, and the operation can be highly site or material specific with regard to material particle size range, sample mass, water infiltration rate, and degree of oxygenation (Mills 1995). The column is operated without active flushing so that oxidation products may accumulate at particle surfaces in addition to being removed in leachate. This behavior parallels field conditions and, as a result, leachate analyses from column test works are a better indicator of expected water quality than leachate analyses from humidity cells, particularly if column infiltration rate is varied to simulate site conditions (Mills 1996).

Column tests also allow treatments to be tested and compared, which is a major advantage over humidity cells (Miller and Jeffrey 1995). Column tests are intended to simulate natural conditions and are simple to construct, operate, and monitor. Various environmental factors can be assessed, as can the influence of various control measures such as cover systems. However, the kinetics of reaction may not be distinguishable from rate-limiting transport phenomena, and bacterial populations may differ from those found under field conditions. Column tests, on the other hand, can become water saturated and interpretations of the results become difficult (Miller and Jeffrey 1995).

Kinetic tests generally only attempt to predict what will happen in the early stages of acid generation processes (Sherlock et al. 1995) and the data interpretation and modeling are complex. Irrespective of the different types of kinetic tests, the overall objective of the tests is to provide data on the rate of acid generation and acid neutralization under laboratory-controlled conditions. In most kinetic tests, water is commonly added to a sample, the mixture is allowed to incubate for a certain period to allow acid–base reactions, and samples of leachate or extracts are collected and analyzed. The tests are required to run for a long time to generate overall acid generation information. The major parameters measured in kinetic tests are trends in pH, sulfate, acidity or alkalinity, and metals. The pH identifies the stage of the acid-generating process, sulfate production relates to rates of sulfide oxidation, acidity/alkalinity gives an indication of the rate of acid generation/neutralization, and metal levels evaluate metal solubility and leaching behavior.

3.4.3 EVALUATION OF THE PREDICTIVE TECHNIQUES

Although data generated from predictive techniques give a base for planning preventive, corrective, and remedial measures, so far they have been found to have restricted application as a universal predictive tool. Sherlock et al. (1995) considered that predictive techniques must be applied on a site-specific basis and took into account the mineralogy of the waste material. For example, minerals present, percent sulfides and their distribution within the rock mass and along fractures and other discontinuities, and the likely durability of the waste rock are likely to affect the predictive test results (Orava and Swider 1996). Moreover, predictive techniques alone do not account for the buildup of metal salts that may occur after disposal but prior to final reclamation. The tests also fail to incorporate an assessment of the coating of sulfide phases with unreactive phases that may occur naturally (Pratt et al. 1996).

Most of these tests also force oxidation or neutralization reactions that may never occur in the real situation. Acid generation (and neutralization) are time-dependent phenomenon, and until a test that takes into account the time dependency developed, there will never be an "exact" predictor. For example, a test that indicates the presence of neutralizers, inherent or added, does not mean that the net result will be no acid generation; that depends on the reactivity and kinetics. If the acid is generated faster than it can be neutralized, the net result will be an acid effluent regardless of how much neutralizer is available.

Short-term static tests, which are conducted to determine the acid-generating or acid-consuming potential, will usually provide only an indirect assessment of the net acid-generating potential. A longer-term kinetic test, which allows reactions to continue, will provide a more comprehensive assessment. However, even a kinetic test may not predict the net acid-generating potential accurately because the ongoing sulfide oxidation will continue to produce acid, and acid consumption by the alkalinity present in the system may be dominant only in the beginning. Conversely, the alkalinity in the system could overcome and exhaust the supply of sulfide-bearing rock present. It is therefore clear that both the static and kinetic tests must be designed to suit the mineralogy of the waste rock being tested. Both acid-generating and acid-consuming reaction rates in the oxidation of pyrite and reactions with carbonate and silicate minerals must be considered for predictive acid generation tests.

The issue of appropriate NP/AP (i.e., NP and AP refer to Neutralization Potential and Acid Generation Potential, respectively) ratios is a key area of debate among the regulatory agencies and the mining industry. To date there is no comprehensive compilation of case histories of mine sites with significant ABA data, NP/AP ratios, and AMD problems. The NP/AP ratio is the most significant of the variables that regulatory agencies are attempting to use as a prescriptive measure. Cravotta et al. (1990) suggested that NP/AP criterion separating potentially acid-generating and non-acid-generating samples could be about 2:1. However, in the data base presented by them, no sample with NP/AP > 1 produced acidic leachate in 166 laboratory leaching tests. There is also no field evidence of NP/AP > 1 producing AMD. The NP/AP ratio may be considered as a safety factor with a higher safety factor probably required for mines in wet climates where carbonate minerals may be preferentially leached from the mine waste relative to the oxidation of the contained sulfide minerals.

The first interpretative use published for ABA was an estimate of NNP > 5 kg $CaCO_3$ t^{-1}-producing alkaline conditions (Sobek et al. 1978). This screening criterion was selected based on soil quality and plant growth media considerations, not mine drainage prediction. ABA tests later began to be applied to coal mine drainage prediction, beginning in the late 1970s. At this time, it became apparent that ABA interpretation depended on whether the end use was mine drainage prediction or mine spoil/growth media suitability. ABA is still used for both purposes today and two sets of interpretative frameworks have been developed.

Several researchers have suggested that the standard ABA procedures may in fact substantially underestimate the neutralizing material requirement of the potentially acid-generating materials (Brady et al. 1994, Cravotta et al. 1990). It has been suggested that the currently used value of 3.125 g $CaCO_3$ equivalent to neutralize acidity from oxidation of 1 g S should in fact be 6.25 g $CaCO_3$ to assure a neutral AMD (Brady et al. 1994, Cravotta et al. 1990). Field studies conducted by Brady et al. (1994) found that the alkaline material requirement calculated from ABA analysis using sulfide S to $CaCO_3$ ratio of 3.125 was inadequate for controlling AMD. Only when the ratio was doubled to 6.25, was there an overall net alkalinity of water at 11 of 12 coal mine sites studied. Perry and Brady (1995) showed that material with an NP > 21 generally produced alkaline drainage, whereas material with NP < 10 produced acidic drainage. The APP, on the other hand, is considered adequate in predicting AMD only in materials that contained insignificant amounts of carbonates

(<1% $CaCO_3$), in which case a relationship between total sulfide S and acidity could be defined (Perry and Brady 1995). Other studies have shown that factors other than mine waste characteristics may be involved in the generation of AMD (diPretoro and Rauch 1988, Erickson and Hedin 1988). They found that there was a poor correlation among APP, NP, and NNP from ABA and net alkalinity from drainage water. O'Hagan and Caruccio (1986) found that addition of $CaCO_3$ at 5% by weight to a coal refuge containing 1% S produced alkaline drainage, whereas Lapakko (1988) indicated that $CaCO_3 > 3\%$ was needed to neutralize an overburden material with 1.17% S.

The discussions above indicate that there are still discrepancies in the use of ABA as screening tool in predicting AMD. It is also evident from several studies that lime requirement assessed from ABA analysis did not always produce neutral drainage from waste rock dumpsites. There was a need for further studies on the rate, application, and placement of alkaline materials in mine waste and mine sites with potential to generate AMD. There appears to be no universal set of threshold numbers for defining cutoffs on ABA. Instead, the data tend to group themselves in ranges (Cravotta et al. 1990, diPretoro and Rauch 1988). However, ABA testing procedures are firmly entrenched in the mining industry and it is likely to stay in use because it is familiar to industry, consultants, and regulators, and it is low cost and has rapid turnaround time.

As with the static tests, kinetic tests are also subject to queries about their accuracy in predicting real situation AMD conditions. Kinetic tests (humidity cells, columns, and soxhlets) not only produce a unique leachate but also modify the sample, and a significant variation in the accuracy of the results was observed. No data are yet available from weathering tests which were run long enough to see the sulfate generation rate begin to taper off. Most weathering test results are assessed as "acid or not acid" producing. In other words, there is not anything particularly kinetic about the data analysis. While static and kinetic tests and the associated models are far from perfect in their capacity to predict the generation or migration of AMD, they do allow for a more systematic approach to understanding the potential problems.

3.5 PREVENTION AND CONTROL OF ACID MINE DRAINAGE

Various physical, chemical, and biological control measures have been used to prevent, minimize, and treat AMD. Basically, there are two types of prevention and control, the first relates to the generation of AMD, while the second relates to its *in situ* mobilization and subsequent migration. Both are interrelated in as much as certain approaches to preventing and controlling generation can reduce migration and *vice versa* (Mills 1996).

Methods proposed for the prevention and control of AMD generation include treatment of sulfide surfaces via the formation of inert surface "coatings," the use of bactericides, the segregation of the principal AMD-generating waste fraction, and control of oxygen and/or water infiltration of the sulfide-bearing material. Currently, waste segregation and prevention or control of water/oxygen infiltration dominate, with the other methods having only limited application at full-scale. Certain approaches such as inert surface coatings are considered unproven at present but worthy of further research. In addition to the use of engineered covers, the most common approaches to preventing or controlling the migration of AMD are the re-routing of water away from the source or the use of subsurface seals and barriers to impede the movement of contaminated groundwater (Filipek et al. 1996). The greater the control achieved, the smaller the volume of AMD that is likely to require treatment. Just as waste minimization is typically more cost-effective than waste management, prevention or minimization of AMD is generally considered a cheaper option than long-term treatment (Filipek et al. 1996). The principle involved in the prevention and control of AMD from mining activities has been centered on removal and abatement of one or more of the essential components in the acid generation process. These components are the sulfide mineral, bacteria, air, and water. The prevention and control procedures currently practiced to abate acid generation are described in the following sections.

3.5.1 PREVENTIVE COATINGS

Several techniques have been experimented in the recent years, which inhibit pyrite and pyrrhotite oxidation using inert coatings. Most of the studies center on coating of isolated sulfide mineral grains with coating agents such as iron phosphate, acetyl acetone, humic acid, oxalic acid, sodium silicate, and lignin. Cathodic protection of the weathering sulfide ore body has also been attempted (Shelp et al. 1995). Preventive coating experiments have been done mainly on separated grains of pyrite crystals and their use in the field is impractical, because of different forms and types of pyrite present in the waste rock. Attempts were also made by Ahmed (1994) to form pyrrhotite hard pans on tailings surfaces to prevent atmospheric oxidation of pyrrhotite. The method involved electrochemical treatment of FeS-rich tailings with ferrous solution to form an oxyhydrate (goethite) matrix.

The use of fatty acid amines and silica derivatives, which suppress bacterial activity and chemical oxidation through a process of hydrophobic coating, has been researched (Diao et al. 2013, Nyavor et al. 1996). Treatment with the amine makes the pyrite highly hydrophobic and the pyrite surface consequently repels oxidizing ions (i.e., Fe^{3+}). It is, however, unclear whether the effect of the hydrophobicity prevents the bacteria from contacting the pyrite surface.

While some of these techniques have prevented further oxidation of pyrite minerals, their applicability to the field situation is not proven as yet. Further research is necessary to detail the economic costs and technical constraints of larger-scale applications, but this approach does look promising, possibly as a means of treating segregated high-sulfide wastes prior to disposal.

3.5.2 SELECTIVE HANDLING

Partitioning of wastes into sulfide-rich and sulfide-depleted fractions offers the chance to expand waste management options in the control and prevention of AMD. In theory, the low-volume, sulfide rich fraction can be disposed of at a highly engineered disposal site or at least isolated as buried "cells" within the bulk waste while the high-volume sulfide-depleted fraction can be disposed of as an inert waste. Blending of acid-generating and non acid-generating wastes can also be used to prevent or control AMD. However, as the latter waste is often exempt from permitting due to its inert nature, blending can sometimes result in the permitting of a much larger volume. This approach may, therefore, be environmentally attractive, but constrained by operator's reticence to extend or further complicate the permitting process.

Several mineral-processing techniques are available for separating acid-generating sulfides. Methods include gravity separation (e.g., centrifugal concentrator, shaking table, spiral concentrator), flotation, magnetic separation, and cyclone classification. The segregated pyrite can be substituted for elemental sulfur in the production of sulfuric acid. Research also shows that pyrite may have some use in the removal of dissolved arsenic species by adsorption (Humber 1995).

3.5.3 BACTERICIDES TREATMENT

Bactericides normally contain anionic surfactants that destroy the greasy outer cell membrane coating of the bacteria *Thiobacillus ferrooxidans*. The greasy film coating normally protects the bacteria from the surrounding acid environment and once it is destroyed, the bacteria cannot survive in the acid conditions (MVTI 1994). ProMac System, comprising commercial short-term and long-term slow release bactericides, has been widely used in the treatment of AMD (MVTI 1994, Sengupta 1994). Alternative bactericides, such as sodium dodecylbenzene sulfonate (SDS), sodium laurel sulfate (SLS), sodium triethylenetetramine bisdithiocarbamate (STB), and thiol-blocking agents have also been suggested for the inhibition of *Thiobacillus* bacteria (Sengupta 1994, Stichbury et al. 1995, Wei and Wolfe 2013).

3.5.4 OXIDANT INFILTRATION BARRIERS

Seals, grouting, cover layers, interception trenches, and subsurface barriers have been used to prevent infiltration of oxidants (air and water) to control AMD generation (Johnson and Hallberg 2005). Sealing mine openings, tunnels, and adits can prevent the infiltration of water into, and the migration of AMD out of, underground sites. Preventing the movement of water through such workings can minimize sulfide oxidation even if the sites are flooded, as static water will quickly become anoxic as oxygen is consumed by chemical and biological reactions. Cementitious grouts can also be applied to mine adit and pitwalls to prevent the infiltration of oxygenated water (Scheetz et al. 1995). Interception trenches for directing AMD to passive treatment systems (i.e., permeable reactive barriers) have also been successfully used to prevent the migration of AMD (Ayala-Parra et al. 2016, Mueller et al. 1996).

Although waste rock dump and pitwall geometry can be important in defining surface area exposure and air infiltration rates, engineered covers are effective at controlling oxygen and water infiltration and can be classified as oxygen barriers, oxygen consumers, or reaction inhibitors. Pitwalls generally have steep gradients and, therefore, placement of amendments becomes impractical. Often heavy engineering of the acid-generating pitwall area is required to implement stabilized placement of suitable cover system (Watson 1995). Engineered cover systems often include layers that promote lateral rather than vertical movement of water, as well as providing a substrate for vegetation and protective layers between the geofabric and the waste to reduce the risk of physical damage to plant root systems.

Synthetic geofabrics such as polyvinyl chloride (PVC) and high density polyethylene (HDPE) have been used to prevent and control water and oxygen infiltration into AMD-generating wastes (Ritcey 1989, Sengupta 1994). Geofabrics are expensive but if applied properly (i.e., to avoid punctures and rips) they are likely to have useful working lives in excess of 100 years (Filipek et al. 1996). *Clays* and clay mixtures such as kaolin amorphous derivatives (KAD) and sand–bentonite mixture have often been used as oxidant infiltration barriers because of their minimal permeability when compacted and relatively low cost (Mackinnon et al. 1997). However, there is a danger that if the clay cover dries, deep cracks can occur, allowing the rapid ingress of water and oxygen. Moreover, compacted clays provide poor growing media for plants.

Organic covers consisting of sewage sludge, paper mill sludge, topsoil, and wood bark have also been found to be effective in reducing oxygen infiltration in sulfidic mine wastes and tailings ponds (Pierce et al. 1994). Surface application of organic materials not only provides a physical barrier to oxygen and moisture but also provides leachate rich in soluble organic compounds that encourage the activity of sulfate-reducing bacteria (Pierce et al. 1994). The placement of a biologically active organic layer on top of tailings has been suggested as a means of reducing oxygen diffusion and to confine metal contaminants (Ricthie 1997). Oxygen infiltration into the tailings is controlled by its consumption in the organic layer (e.g., via conversion to carbon dioxide and water), while metal diffusion into the water can be further controlled by inoculation of the organic layer with sulfate-reducing bacteria capable of precipitating the metals as sulfides (Nicholson et al. 1989).

Organic ligands leached from decomposing organic matter are found to be effective in reducing phytotoxic levels of Al by readily complexing with the exchangeable Al (Gurung et al. 1996). One potential drawback to this approach has been highlighted by research that has shown that ferric oxyhydroxides present in weathered tailings dissolve when in contact with organic acids originating from carbon-rich oxygen-consuming covers (Ribet et al. 1995). The dissolution of the oxyhydroxide phase can result in the release of adsorbed or co-precipitated non-ferrous metals into the aqueous phase.

Although reclamation of mine waste materials by municipal sludge application has been shown to be beneficial to plant growth by providing a growth medium and nutrient reserves (Wijesekara et al. 2016), there are serious environmental problems associated with such organic-based amendments (Sopper 1992). Composted organic sludge materials commonly contain high concentrations of heavy metals such as Cd, Cu, Mo, Mn, Pb, and Zn, which can be hazardous contaminants in soils, water, and plants. Surface applications of sludge materials can mobilize heavy metals through reductive dissolution as well as mobilizing nitrate into the ground water system (Ribet et al. 1995, Sopper 1992). Field and laboratory studies have shown that surface-applied sewage sludge to promote revegetation at reclaimed mines resulted in increased down gradient concentrations of sulfate and acidity, indicating that surficially applied sludge was not an effective barrier to O_2 entering into underlying zones (Cravotta 1996). Laboratory leaching tests also indicated that sewage sludge additions produced significant increases in microbial pyrite oxidation and that only when $CaCO_3$ was added was there a reduction in bacterial oxidation of pyrite (Cravotta 1996).

A composite cover system consisting of layers of non-acid-generating geologic materials has been developed with an aim to establish vegetation as well as isolating sulfidic waste materials (Lamb et al. 2014). Composite soil covers are often up to 1 m thick and they can be applied only on low gradient rehabilitation sites. Such a cover system has been used in the rehabilitation of Rum Jungle mine site in Australia. Long-term monitoring of the effectiveness of the composite cover system at Rum Jungle has shown significant reduction in metal loading in the AMD from the rehabilitated dump sites (Sengupta 1994).

Relevant research and modeling work has been published on soil covers that limit oxygen influx to tailings and their subsequent effect on oxidation of pyrite/pyrrhotite. For example, Nicholson et al. (1989) demonstrated that, in addition to a fine particle size of the cover material, maintenance of high moisture content is essential to minimize diffusion of oxygen. Furthermore, although erosion and stability are important considerations with respect to slope, one must also consider that the slopes are influx zones for oxygenated air entering the pitwall (Murray 1977). Buoyant air, heated by pyrite oxidation, tends to rise, drawing fresh air into the slopes. Accordingly, convective airflow may develop within the sloughed pitwall rock, maintaining oxidation despite burial of the acid-forming material. In order to inhibit continued oxidation, the slopes need low-permeability or oxygen-consumptive covers (Guo and Cravotta 1996).

Subaqueous disposal of reactive waste rocks by flooding of open pits or dumping into natural lakes or impoundments has been practiced by many mining operations, although such practices require a thorough understanding of local and regional hydrology and hydrogeology (Filipek et al. 1996). In Canada, the recommended method of preventing AMD is to dispose of sulfide waste rocks and tailings deposits subaqueously (Pederson et al. 1994). While this is feasible in areas of high precipitation, low temperature, and abundance of lakes, it cannot be considered in areas of low precipitation and lack of fresh water lakes. Disposal of sulfidic waste into sea water has also been practiced by mining operators in Norway (Sengupta 1994) and in Papua New Guinea; river disposal of tailings has been practiced for some time at the Pogera Gold and Ok Tedi copper mines in Papua New Guinea (Harries and Ritchie 1988).

Subaqueous disposal controls acid generation from sulfidic wastes by limiting the diffusion rate of oxygen through the water cover (Filipek et al. 1996). However, research over a 3-year period at the Noranda Technology Centre in Pointe Clare, Quebec, Canada, has shown that although water covers can reduce the rate of acid generation by 99.7%, the concentration of metals in the surface water can still exceed regulatory limits (Payant et al. 1995). The efficacy of subaqueous disposal and the environmental impact on the aquatic life are yet to be fully assessed.

Other barrier methods such as deliberate construction of a **hardpan** (using electrochemical methods), which consists of a cementitious iron oxyhydrate matrix, may also help control infiltration by water and, thus, reduce or prevent the generation of AMD (Ahmed 1995).

3.6 TREATMENT OF ACID MINE DRAINAGE

There are a number of overlapping approaches to the treatment of AMD, which are nominally categorized here as active, passive, and active–passive hybrid systems. In the recent past, research into the use of ion exchange resins and natural and synthetic zeolites to treat AMD has also shown promising results (Mondale et al. 1995, Ríos et al. 2008). The most common methods employed in the treatment of AMD are through chemical and biological processes. The chemical treatment processes include complexation, oxidation, and reduction principles. Although all of these methods show promising inhibitive properties in chemical sense, they are not all found to be practical under field conditions.

3.6.1　*In Situ* Neutralization

In sulfidic deposits, acid-generating minerals such as pyrite often occur in close association with acid-neutralizing minerals such as calcite ($CaCO_3$) and dolomite [$CaMg(CO_3)_2$] that normally occur as late stage mineralization. The acid produced from pyrite oxidation is neutralized, *in situ*, by dissolution of these basic minerals if in contact with the migrating AMD, resulting in precipitation of sulfate salts (Equations 3.12 and 3.13). These chemical reactions must also be examined to fully understand the processes occurring and to be able to predict the chemistry of solutions resulting from the combination of oxidation and neutralization processes (Morin et al. 1995).

$$CaCO_3 + H_2SO_4 \rightarrow CaSO_4 + H_2O + CO_2 \tag{3.12}$$

$$CaMg(CO_3)_2 + 2H_2SO_4 \rightarrow CaSO_4 + MgSO_4 + 2H_2O + CO_2 \tag{3.13}$$

Silicate minerals such as plagioclase feldspars also have the potential to neutralize acid under specific pH conditions (Mills 1997), by the dissolution reactions listed below (Equations 3.14 through 3.19). In acidic conditions, the silicate minerals rapidly decompose clay products as well as releasing significant amount of Al^{3+} and K^+ in the soil solution. The dissolution of $Fe(OH)_3$ is also considered an acid-consuming reaction (Equation 3.19).

Muscovite dissolution

$$KAl_2[AlSi_3O_{10}](OH)_2(s) + H^+ + 3/2H_2O \rightarrow K^+ + 3/2Al_2Si_2O_5(OH)_4 \tag{3.14}$$

Biotite dissolution

$$KMg_{1.5}Fe_{1.5}AlSi_3O_{10}(OH)_2 + 7H^+ + 1/2H_2O \rightarrow K^+ + 1.5Mg^{2+} +$$

$$1.5Fe^{2+} + H_4SiO_4 + 1/2Al_2Si_2O_5(OH)_4 \tag{3.15}$$

Albite dissolution

$$NaAlSi_3O_8 + H^+ + 9/2H_2O \rightarrow Na^+ + 2H_4SiO_4 + 1/2Al_2Si_2O_5(OH)_4 \tag{3.16}$$

Anorthite dissolution

$$CaAl_2Si_2O_8 + 2H^+ + H_2O \rightarrow Ca^{2+} + Al_2Si_2O_5(OH)_4 \tag{3.17}$$

K-feldspar dissolution

$$KAlSi_3O_8 + H^+ + 9/2H_2O \rightarrow K^+ + 2H_4SiO_4 + 1/2Al_2Si_2O_5(OH)_4 \tag{3.18}$$

Iron oxyhydroxide dissolution

$$Fe(OH)_3 + 3H^+ \rightarrow Fe^{3+} + 3H_2O \tag{3.19}$$

The *in situ* neutralization processes described above highlight the fact that mineralogy of the pyritic rocks is a key factor in defining the composition of the AMD and emphasizes the importance of mineralogical characterization in managing AMD. The amelioration of AMD in pyritic materials is primarily based on the neutralization of the acid as it occurs naturally *in situ* conditions.

3.6.2 Active Treatment

Active treatment systems basically involve neutralization of AMD with liming materials and precipitation of metals as hydroxides. There are numerous studies carried out on the variations of this technique but the ultimate process involves the addition of base to neutralize acid (Murdock et al. 1995, Taylor et al. 1997). Liming materials commonly used in neutralization techniques are hydrated lime ($Ca(OH)_2$), sodium bicarbonate ($NaHCO_3$), and sodium hydroxide ($NaOH$). Agricultural limestone and phosphate rocks have also been used to treat AMD.

In conventional lime treatment, there are five basic steps following collection of the AMD: (1) equalization (i.e., mixing) to minimize variations in water quality, (2) aeration to oxidize Fe^{2+} to the less-soluble Fe^{3+}, (3) neutralization to increase pH to precipitate metals as hydroxides, (4) sedimentation to separate water and solids, and (5) sludge disposal (Murdock et al. 1995, Taylor et al. 1997). Lime is normally introduced into the system as a 5%–20% (by weight) water-based slurry, although it is sometimes applied as a dry powder when the water volume to be treated is low. The principal reactions are summarized in Equations 3.20 through 3.22 where M represents dissolved metals (Fe, Zn, Cu, Zn, Mn, Al, etc.):

$$Ca(OH)_2 + H_2SO_4 \rightarrow Ca^{2+} + SO_4^{2-} + 2H_2O \tag{3.20}$$

$$Ca(OH)_2 + MSO_4 \rightarrow M(OH)_2 + Ca^{2+} + SO_4^{2-} \tag{3.21}$$

$$3Ca(OH)_2 + M_2(SO_4)_3 \rightarrow 2M(OH)_3 + 3CaSO_4 \tag{3.22}$$

Alternative liming materials such as fluidized bed boiler ash (FBA) generated from fossil fuel–fired boilers have been much investigated for use as a liming material in acid agricultural soils because of its high alkalinity. Because FBA contains both lime and gypsum, it has been shown to ameliorate both surface and subsurface acidity (Wang et al. 1994). Its use as AMD ameliorant in reclamation of coal mine spoils has been investigated by several workers (Dick et al. 1994, Stehouwer et al. 1995). Similar other by-products such as fluidized bed combustion ash (FBC), fluidized bed waste (FBW), and flue gas desulfurization sludge (FGD) have been investigated as possible ameliorants for AMD in coal mine waste (Bhumbla 1992). Since FBA is a waste product, its use in large quantities in reclamation of coal refuse serves as a viable alternative for its disposal.

Major drawbacks of the lime treatment system are the large volume of potentially toxic sludge produced (which must be physically removed) and a relatively high pH required to remove metals such as Mn. Some metal hydroxides such as $Fe(OH)_3$ and $Al(OH)_3$ may re-dissolve in the highly alkaline solutions required to complete metal precipitation, necessitating a multi-stage treatment to reduce all metals to acceptable concentrations. Sludges derived from the liming of AMD are chemically unstable and will partially re-dissolve if exposed to a sufficiently acidic environment (Kuyucak et al. 1995, Murdock et al. 1995).

To address some of these drawbacks, a number of refinements to standard liming treatments have been developed. Two such examples include the High Density Sludge™ (HDS) process and the patented Noranda Technology Centre (NTC) process, which are being applied increasingly within the mining industry. The HDS process produces a more compact and higher density sludge (Murdock et al. 1995) and is carried out in aerated reactors. Part of the settled sludge is recycled to the beginning of the process, where it is mixed with the lime slurry. The NTC process uses pH-controlled reactors in which sludge density and settling rates are improved relative to the HDS process (Kuyucak et al. 1995).

However, the benefit of such lime treatment is a proven technology with well documented and understood mechanisms of metal removal and acid neutralization. Despite the commercial availability of processes such as HDS and NTC, increasingly stringent legislation is likely to drive mine operators to look beyond the use of lime to avoid incurring growing disposal costs, and to avoid the possibility of future liability and litigation. One alternative to which serious consideration has been given is the use of passive treatment systems.

3.6.3 PASSIVE TREATMENT

Passive treatment systems encompass a number of discrete neutralization processes. These include anoxic ponds (APO), limestone ponds (LSP), anoxic limestone drains (ALD), alkalinity-producing system (APS), reverse alkalinity–producing system (RAPS), open limestone channels (OLC), and aerobic and anaerobic wetlands (Faulkner and Skousen 1984, Taylor et al. 1997). Porous reactive alkaline walls (i.e., permeable reactive barriers) have also been proposed as a means of treating subsurface AMD *in situ*. This approach has only been attempted on a limited scale, although the preliminary results appear promising in terms of both economic and technical performance (Blowes et al. 1995). Passive treatment systems have demonstrated substantial mitigation in AMD quality in some cases, while in other situations, less-dramatic results have been obtained (Brodie and Hedin 1994, Skousen 1995). High concentration of Fe and Al in the AMD often precipitate as hydroxide when in contact with the alkaline treatment systems, causing an "armoring effect" on the limestone, thereby reducing the generation of further alkalinity and impeding flow through the drain (Filipek et al. 1996).

The use of sulfate-reducing bacteria in open pits or underground workings has been suggested as a means of treating AMD *in situ* (Kuyucak and St.-Germain 1994). This approach might be suitable for low-load scenarios, where a suitable organic substrate for bacterial growth is locally and cheaply available. However, efficiency might be compromised by a single addition if the substrate is so deep that mass transfer is detrimentally affected. The operational temperature at depth is also an issue as this will influence bacterial activity.

Constructed or engineered aerobic and anaerobic wetlands are passive treatment systems, which attempt to duplicate natural systems and use chemical and biological processes to reduce dissolved metal concentrations and neutralize acidity (Taylor et al. 1997). Compared with conventional active chemical treatment by liming, passive methods generally require more land area, but use cheaper materials to support the chemical and biological processes, and require less operational attention and maintenance (Taylor et al. 1997). However, wetlands are not "walk-away" solutions but, rather, low-maintenance, low-energy systems designed to treat effluent AMD and are not suitable treatment options for *in situ* neutralization of AMD.

3.7 SUMMARY AND CONCLUSIONS

AMD from mines containing sulfidic ores and its preventive and control measures are a major environmental issue in the mining industry. The understanding of AMD, its prediction, and treatment are the subject of a substantial research effort by government, the mining industry, universities, and research establishments. Increasing worldwide awareness of the environmental impacts of mining activities has directed government agencies to lay down strict guidelines for reclamation and rehabilitation of mined lands.

Acid generation from pyrite oxidation is the principal cause of AMD. The understanding of the biochemistry of pyrite oxidation and the resulting by-products has led to significant advances in AMD mitigation technology. The nature of pyrite oxidation itself is *site specific* and dependent upon a number of factors including host rock mineralogy, weathering conditions, forms and distribution of pyrite, and local climatic conditions. AMD generation depends primarily on the ratio of acid-producing sulfide minerals to acid-neutralizing alkaline minerals. The presence or absence of acid-neutralizing material is generally regarded as the most critical factor in determining the intensity and migration of AMD in the mine waste rocks.

Predictive techniques employed in the assessment of AMD do not have standardized applications as yet mainly because acid generation in mine sites is *site specific*. Appropriate predictive testing along with waste characterization and scientific interpretation of the data are essential if proper AMD prevention and management practices are to be developed, disseminated, and sustained. Despite the uncertainties in acid–base accounting (ABA) screening criteria, the ABA method of evaluating acid-generating potential of sulfidic mine waste materials has become mandatory requirement of regulatory guidelines in the mining industry.

Liming, organic waste application, and, less commonly, bactericides have so far been the common amendment materials in the vegetative reclamation of mine sites. However, these amendments have had varying successes in alleviating low-pH conditions created by AMD mainly because of difficulties in assessing the lime requirement of sulfidic waste rocks. Because acid generation from pyrite oxidation is a continual process, the ABA method of estimating the liming requirement of sulfidic waste rock has been inadequate for providing long-term pH modification suitable for sustaining plant growth.

There is an apparent lack of kinetic data on comparative effectiveness of alkaline materials in *in situ* amelioration of AMD conditions in sulfidic waste rock materials. The *site-specific* characteristics of the AMD necessitate investigative evaluations of the ameliorating effectiveness of alkaline amendments in creating suitable plant growth media. Major emphasis has been placed on investigations regarding mechanisms and probable technologies (e.g., phosphate coatings, cover materials, collection and treatment with lime additions, and bactericides) for prevention and control of AMD primarily arising from coal refuse, mill tailings, and waste rock piles. Many of the procedures and techniques are site specific in nature and often expensive to implement in field-scale applications. Very little research has been undertaken to investigate the *in situ* characterization and mitigation of AMD conditions on active mine pitwalls although many of the waste rock characterization procedures and preventive measures can equally be applied to pitwall rock materials.

ACKNOWLEDGMENT

This research was partly supported by Australia Research Council Discovery-Projects (DP140100323).

REFERENCES

Ahmed, S. M. 1994. Surface chemical method of forming hardpan in pyrrhotite tailings and prevention of the acid mine drainage. *International Land Reclamation and Mine Drainage Conference and Third International Conference on the Abatement of Acidic Drainage*, Pittsburgh, PA, April 24–29, 1994, pp. 57–66.

Ahmed, S. M. 1995. Chemistry of pyrrhotite hardpan formation. In *Proceedings of Sudbury '95—Mining and the Environment*, May 28–June 1, 1995, Sudbury, Ontario, Canada, eds. T. P. Hynes and M. C. Blanchette, pp. 171–180. CANMET, Ottawa, Ontario, Canada.

Akcil, A., and S. Koldas. 2006. Acid Mine Drainage (AMD): Causes, treatment and case studies. *Journal of Cleaner Production*, 14(12–13):1139–1145.

American Society for Testing and Materials (ASTM). 1996. ASTM Designation: D 5744-96–Standard test method for accelerated weathering of solid materials using a modified humidity cell, ASTM, West Conshohocken, PA, 13pp.

Ayala-Parra, P., R. Sierra-Alvarez, and J. A. Field. 2016. Treatment of acid rock drainage using a sulfate-reducing bioreactor with zero-valent iron. *Journal of Hazardous Materials*, 308:97–105.

Bache, B. W. 1974. Soluble aluminium and calcium-aluminium exchange in relation to the pH of dilute calcium chloride suspension of acid soils. *Journal of Soil Science*, 25:320–332.

Backes, C. A., I. D. Pulford, and H. J. Duncan. 1986. Studies on the oxidation of pyrite in colliery soil. I. The oxidation pathway and inhibition of the ferrous-ferric oxidation. *Reclamation and Revegetation Research*, 4:279–291.

Bhumbla, D. K. 1992. Ameliorative effects of fly ashes, Sciences and Engineering, West Virginia University, Morgantown, WV (unpublished report), 11pp.

Blamey, F. P. C., D. G. Edwards, and C. J. Asher. 1983. Effects of aluminium, OH:Al and P:Al molar ratios and ionic strength on soybean root elongation in solution culture. *Soil Science*, 136:197–207.

Bloomfield, C., 1972. The oxidation of iron sulphides in soils in relation to the formation of acid sulfate soils, and of ochre deposits in field drains. *Journal of Soil Science*, 23:1–16.

Blowes, D. W., C. J. Ptacek, K. R. Waybrant, and J. G. Bain. 1995. In-situ treatment of mine drainage water using porous reactive walls. In *11th Annual General Meeting of BIOMINET: Biotechnology and the Mining Environment*, eds. L. Lortie, W. D. Gould, and S. Rajan, pp. 119–128. Ottawa, Ontario, Cananda, January 16, 1995.

Bolan, N. S., D. C. Adriano, and D. Curtin. 2003. Soil acidification and liming interactions with nutrient and heavy metal transformation and bioavailability. *Advances in Agronomy*, 78:216–272.

Bolan, N. S., B. J. Ko, C. W. N. Anderson et al. 2008. Manipulating bioavailability to manage remediation of metal contaminated soils. In *Chemical Bioavailability in Terrestrial Environment*, eds. Naidu, R. et al., pp. 657–678. Elsevier, Amsterdam, the Netherlands.

Bolan, N. S., A. Kunhikrishnan, R. Thangarajan et al. 2014. Remediation of heavy metal(loid)s contaminated soils—To mobilize or to immobilize? *Journal of Hazardous Materials*, 266:141–166.

Brady, K. B., E. F. Perry, R. L. Beam, D. C. Bisko, M. D. Gardner, and J. M. Tarantino. 1994. Evaluation of acid-base accounting to predict the quality of drainage at surface coal mines in Pennsylvania, USA. In *Proceedings of the International Land Reclamation and Mine Drainage Conference and Third International Conference on the Abatement of Acidic Drainage*, April 24–29, 1994. Pittsburgh, PA, USBM SP 06A-94, vol. 1, pp. 138–146.

Brodie, G. A. and R. S. Hedin. 1994. Constructed wetland for treating acidic drainage. In *International Land Reclamation and Mine Drainage Conference and the Third International Conference on the Abatement of Acidic Drainage*, April 25–29, 1994, Pittsburgh, PA.

Bruynesteyn, A. and R. P. Hackel. 1984. Evaluation of acid production potential of mining waste materials. *Minerals and the Environment*, 4(1):5–8.

Caruccio, F. T., L. R. Hossner, and G. Geidel. 1988. Pyritic materials: Acid drainage, soil acidity and liming. In *Reclamation of Surface-Mined Lands*, ed. L. R. Hossner, pp. 10–17. CRC Press, Boca Raton, FL.

Cravotta, C. A. 1996. Municipal sludge use in coal-mine reclamation and potential effects on the formation of acidic mine drainage. PhD dissertation, Pennsylvania State University, State College, PA.

Cravotta, C. A., K. B. Brady, M. W. Smith et al. 1990. Effectiveness of alkaline addition at surface mines in preventing or abating acid mine drainage: Part 1. Geochemical considerations. In *Proceedings of the 1990 Mining and Reclamation Conference and Exhibition*, Charleston, WV, eds. J. Skousen, J. Sencindiver, and D. Samuel, Vol. 1 West Virginia University, Morgantown, WV, pp. 221–226.

Dent, D. 1986. Chemical and physical processes in acid sulphate soils, acid sulphate soils: A baseline for research and development. International Institute for Land Reclamation and Improvement/ILRI Publication 39, Wageningen, the Netherlands.

Diao, Z., T. Shi, S. Wang et al. (2013). Silane-based coatings on the pyrite for remediation of acid mine drainage. *Water Research*, 47(13):4391–4402.

Dick, W. A., R. C. Stehouwer, J. H. Beeghly, J. M. Bigham, and R. Lal. 1994. Dry flue gas desulfurization by-products as amendment for reclamation of acid mine spoil. In *International Land Reclamation and Mine Drainage Conference and the Third International Conference on the Abatement of Acidic Drainage*, April 24–29, 1994, Pittsburgh, PA, pp. 129–138.

diPretoro, R. S. and H. W. Rauch. 1988. Use of acid-base accounts in premining prediction of acid drainage potential: A new approach from northern West Virginia. *U.S. Bureau of Mines IC*, 9183:1–10.

Doyle, F. M. 1990. Acid mine drainage from sulphide ore deposits. In *Sulphide Deposits—Their Origin and Processing*, eds. P.M.J. Gray, G.J. Bowyer, and J.F. Castle. The Institution of Mining and Metallurgy, London, U.K., pp. 301–310.

Erickson, P. M. and R. Hedin. 1988. Evaluation of overburden analytical methods as a means to predict post-mining coal mine drainage quality. *U.S. Bureau of Mines IC*, 9183:11–19.

Evangelou, V. P. and Y. L. Zhang. 1995. A review: Pyrite oxidation mechanisms and acid mine drainage prevention. *Environmental Science and Technology*, 25(2):141–199.

Faulkner, B. and J. G. Skousen. 1984. Treatment of acid mine drainage by passive treatment systems. In *Proceedings of the International Land Reclamation and Mine Drainage Conference*, U.S. Bureau of Mines SP 06A-94, Pittsburgh, PA, pp. 250–257.

Filipek, L., A. Kirk, and W. Schafer. 1996. Control technologies for ARD. In *Mining Environmental Management*, pp. 4–8.

Gray, N. F. 1997. Environmental impacts and remediation of acid mine drainage: A management problem. *Enviornmental Geology*, 30:62–71.

Gray, N. F. and M. Sullivan. 1995. The environmental impact of acid mine drainage, PhD dissertation. University of Dublin, Trinity College, Dublin, Ireland.

Guo, W. and C. A. Cravotta. 1996. Oxygen transport and pyrite oxidation in unsaturated coal-mine spoil. In *13th Annual Meeting of the American Society for Surface Mining and Reclamation*, May 18–23, 1996, American Society for Surface Mining and Reclamation, Princeton, WV, pp. 3–14.

Gurung, S. R., R. B. Stewart, P. Loganathan, and P. E. H. Gregg. 1996. Aluminium-organic matter-fluoride interaction during soil development in oxidised mine waste. *Soil Technology*, 9:273–279.

Haferburg, G. and E. Kothe. 2007. Microbes and metals: Interactions in the environment. *Journal of Basic Microbiology*, 47(6):453–467.

Harries, J. R. and A. I. M. Ritchie. 1988. Rehabilitation measures at the Rum Jungle Mine site. In *Environmental Management of Solid Waste—Dredged Material and Mine Tailings*, eds. Salomons, W., and U. Forstner. Springer-Verlag, New York.

Hedin, R. S. and P. M. Erickson. 1988. Relationship between the initial geochemistry and leachate chemistry of weathering overburden samples. In *Proceedings of the 1988 Annual Mine Drainage and Surface Mine Reclamation Conference*, April 17–22, 1988, American Society for Surface Mining and Reclamation and the U.S. Department of Interior, pp. 21–28. U.S. Bureau of Mines, Pittsburgh, PA.

Hedin, R. S., G. R. Watzlaf, and R. W. Nairn. 1994. Passive treatment of acid mine drainage with limestone. *Journal of Environmental Quality*, 23(6):1338–1345.

Helyar, K. L., M. K. Conyers, and D. N. Munns. 1993. Soil solution aluminium activity related to theoretical Al mineral solubilities in four Australian soils. *Journal of Soil Science*, 44:317–333.

Hem, J. D. 1963. Deposition and solution of manganese oxides. U.S. Geological Survey Water Supply Paper 1667-B, pp. B1–B42.

Herbert, R. B. 1995. Precipitation of Fe oxyhydroxides and jarosite from acidic groundwater. *GFF*, 117:81–85.

Herr, C., C. O'Neall, and N. F. Gray. 1996. Metal contamination from open-cast copper and sulphur mining in southeast Ireland. *Land Degradation and Development*, 7:161–174.

Humber, A. J. 1995. Separation of sulphide minerals from mill tailings. In *Proceedings of Sudbury '95—Mining and the Environment*, May 28–June 1, 1995, Sudbury, Ontario, Canada, eds. Hynes, T. P. and M. C. Blanchette, pp. 149–152. CANMET, Ottawa, Ontario, Canada.

Hutchison, I. P. G. and R. D. Ellison. 1989. Prediction of acid generation potential. In *Mine Waste Management: A Resource for Mining Industry Professionals, Regulators and Consulting Engineers*, eds. I. P. G. Hutchison and R. D. Ellison. Lewis Publishers, London, U.K., pp. 139–170.

Jambor, J. L. and D. W. Blowes (eds.). 1994. *Short Course Handbook on Environmental Geochemistry of Sulfide Mine-Wastes*, Mineralogical Association of Canada, Volume 22, May 1994.

Johnson, D. B. and K. B. Hallberg 2005. Acid mine drainage remediation options: A review. *Science of the Total Environment*, 338(1):3–14.

Kerth, M. and H. Wiggering. 1990. Colliery spoil weathering in the Ruhr—Problems and solutions. In *Proceedings of the Third International Conference on the Reclamation, Treatment and Utilisation of Coal Mining Waste*, September 3–7, 1990, Glasgow, Scotland, ed. Rainbow, A. K. M., pp. 417–424.

Klein, R., J. S. Tischler, M. Muhling, and M. Schlomann. 2014. Bioremediation of Mine Water. *Advances in Biochemical Engineering and Biotechnology*, 141:109–172.

Kuyucak, N., D. Lyew, P. St-Germain, and K. G. Wheeland. 1991. In-situ treatment of AMD in open pits. In *Proceedings of the Second International Conference on the Abatement of Acidic Drainage*, September 1991, Montreal, Quebec, Canada, pp. 335–353.

Kuyucak, N., S. Payant, and T. Sheremata. 1995. Improved lime neutralisation process. In *Proceedings of the Sudbury '95—Mining and the Environment*, May 28–June 1, 1995, Sudbury, Ontario, Canada, eds. T. P. Hynes and M. C. Blanchette. CANMET, Ottawa, Ontario, Canada, pp. 129–137.

Kuyucak, N. and P. St.-Germain. 1994. In situ treatment of acid mine drainage by sulphate reducing bacteria in open pits: Scale-up experiences. In *Proceedings of the International Land Reclamation and Mine Drainage Conference and the Third International Conference on the Abatement of Acidic Drainage*, April 24–29, 1994, Pittsburgh, PA, pp. 303–310.

Kwong, Y. T. J. 1995. Influence of galvanic sulphide oxidation on mine water chemistry. In *Proceedings of Sudbury '95—Mining and the Environment*, May 28–June 1, 1995, Sudbury, Ontario, Canada, eds. Hynes, T. P. and M. C. Blanchette, pp. 477–483. CANMET, Ottawa, Ontario, Canada.

Lamb, D., K. Venkatraman, N. S. Bolan, N. Ashwath, G. Choppala, and R. Naidu. 2014. Phytocapping: An alternative technology for the sustainable management of landfill sites. *Critical Reviews in Environmental Science and Technology*, 44:561–637.

Lapakko, K. 1988. Prediction of acid mine drainage from Duluth Complex mining wastes in northeastern Minnesota. In *Mine Drainage and Surface Mine Reclamation*, Pittsburgh, PA, pp. 180–190. USDI Bureau of Mines Information Circular 9183.

Lindsay, M. B. J., M. C. Moncur, J. G. Bain et al. 2015. Geochemical and mineralogical aspects of sulfide mine tailings. *Applied Geochemistry*, 57:157–177.

Luo, Y. and D. L. Rimmer. 1995. Zinc-copper interactions affecting plant growth on a metal-contaminated soil. *Environmental Pollution*, 88:79–83.

Mackinnon, I. D. R., H. R. A. Exelby, D. Page, and B. Singh. 1997. Kaolin amorphous derivatives for the treatment of acid mine drainage. In *Proceedings of the Third Australian Acid Mine Drainage Workshop*, July 15–18, 1997, Darwin, Australia, eds. R. W. McLean and L. C. Bell, pp. 195–210. Australian Centre for Mine Site Rehabilitation Centre, Brisbane, Queensland, Australia.

MEND. 1995. MINEWALL 2.0, Series of four reports: Literature review. User's guide. Application of MINEWALL to three minesites and programmer's notes and source code, plus one diskette, MEND, Ottawa, Ontario, Canada.

Miller, S. D. and J. Jeffrey. 1995. Advances in the prediction of acid generating mine waste materials. In *Proceedings of the Second Australian Acid Mine Drainage Workshop*, March 28–31, 1995, eds. N. J. Grundon and L. C. Bell, pp. 33–41. Australian Centre for Minesite Rehabilitation Research, Charters Towers, Queensland, Australia.

Mills, C. 1995. Technical review of the acid rock drainage (ARD) and metal leaching aspects of the metallurgical testwork, milling practices and tailings monitoring for the Huckleberry Project. Report to BC ministry of energy, mines and petroleum resources, Vancouver, British Columbia, Canada, 34pp.

Mills, C. 1996. ARD Web page. An introduction to acid rock drainage. http://www.enviromine.com/ard/ accessed (October 24, 2016).

Mills, C. 1997. An assessment of the results of acid base accounting (ABA) and mineralogical testwork on eight samples from the proposed Minto, Yukon Territory Minesite. Report to The Selkirk First Nation, Pelly Crossing, Yukon, Canada.

Mondale, K. D., R. M. Carland, and F. F. Aplan. 1995. The comparative ion exchange capacities of natural sedimentary and synthetic zeolites. *Minerals Engineering*, 8(4):535–548.

Morin, K. A. 1990. *Acid Drainage from Mine Walls: The Main Zone Pit at Equity Silver Mines*, Morwijk Enterprises Limited, Vancouver, British Columbia, Canada.

Morin, K. A. and N. M. Hutt. 1995. MINEWALL 2.0: A technique for predicting water chemistry in open-pit and underground mines. In *Proceedings of Sudbury '95—Mining and the Environment*, May 28–June 1, 1995, Sudbury, Ontario, Canada, eds. T. P. Hynes and M. C. Blanchette, pp. 1007–1016, CANMET, Ottawa, Ontario, Canada.

Morin, K. A., N. M. Hutt, and K. D. Ferguson. 1995. Measured rates of sulphide oxidation and acid neutralisation in kinetic tests: Statistical lessons from the database. In *Proceedings of Sudbury '95—Mining and the Environment*, May 28–June 1, 1995, Sudbury, Ontario, Canada, eds. T. P. Hynes and M. C. Blanchette, pp. 525–536. CANMET, Ottawa, Ontario, Canada.

Mueller, R. F., D. E. Sinkbeil, J. Pantano et al. 1996. Treatment of metal contaminated groundwater in passive systems: A demonstration study. In *Proceedings of the 1996 National Meeting of the American Society for Surface Mining and Reclamation*, May 19–25, 1996, Knoxville, TN, pp. 590–598.

Mulder, J., N. van Breemen, and H. C. Eijck. 1989. Depletion of soil aluminium by acid deposition and implication for acid neutralisation. *Nature*, 337:247–249.

Murdock, D. J., J. R. W. Fox, and J. G. Bensley. 1995. Treatment of acid mine drainage by the high density sludge process. In *Proceedings of Sudbury '95—Mining and the Environment*, May 28–June 1, 1995, Sudbury, Ontario, Canada, eds. T. P. Hynes and M. C. Blancjette, pp. 431–439. CANMET, Ottawa, Ontario, Canada.

Murray, D. R. 1977. Pit Slope Manual. Supplement 10-1, CANMET Report 77-31. Department of Energy, Mines and Resources, Canada, Ottawa, Ontario, 112pp.

MVTI. 1994. *The Bactericide Technology. ProMac Mine Acid Control System*. MVTechnologies, Inc. (MVTI), Akron, OH, pp. 1–17.

Nicholson, R. V., R. W. Gillham, J. A. Cherry, and E. J. Reardon. 1989. Reduction of acid generation in mine tailings through the use of moisture-retaining cover layers as oxygen barriers. *Canadian Geotechnical Journal*, 26:1–8.

Nordstrom, D. K. 1982. Aqueous pyrite oxidation and the consequent formation of secondary iron minerals. In *Acid Sulfate Weathering*, eds. Kittrick, J. A., D. S. Fanning, and L. R. Hossner, Soil Science Society of America Special Publication 10, pp. 37–56.

Nyavor, K., N. O. Egiebor, and P. M. Fedorak. 1996. Suppression of microbial pyrite oxidation by fatty acid amine treatment. *The Science of the Total Environment*, 182:75–83.

O'Hagan, M. and F. T. Caruccio. 1986. The effect of admixed limestone on rates of pyrite oxidation in low, medium, and high sulfur rocks. In *National Symposium on Mining, Hydrology, Sedimentology, and Reclamation*, University of Kentucky, Lexington, KY, pp. 3–10.

Orava, D. A. and R. C. Swider 1996. Inhibiting acid mine drainage throughout the mine life cycle. *CIM Bulletin*, 89:52–56.

Payant, S., L. C. St-Arnaud, and E. Yanful. 1995. Evaluation of techniques for preventing acidic rock drainage. *In Proceedings of the Sudbury '95—Mining and the Environment*, May 28–June 1, 1995, Sudbury, Ontario, Canada, eds. T. P. Hynes and M. C. Blanchette, pp. 485–494. CANMET, Ottawa, Ontario, Canada.

Pederson, T. F., J. J. McNee, B. Mueller, D. H. Flather, and C. A. Pelletier. 1994. Geochemistry of submerged tailings in Anderson Lake, Manitoba: Recent results. In *International Land Reclamation and Mine Drainage Conference and the Third International Conference on the Abatement of Acidic Drainage*, U.S. Department of Interior, Pittsburgh, PA.

Perry, M. and K. Brady. 1995. Influence of neutralisation potential on surface mine drainage quality in Pennsylvania. In *West Virginia Surface Mine Drainage Task Force Symposium*, April 4–5, 1995, Morgantown, WV. U.S. Bureau of Mining and Reclamation, Harrisburg, PA, pp. 1–15.

Pierce, W. G., N. Belzile, M. E. Wiseman, and K. Winterhalder. 1994. Composted organic waste as anaerobic reducing covers for long term abandonment of acid-generating tailing. In *Proceedings of the International Land Reclamation and Mine Drainage Conference and the Third International Conference on the Abatement of Acidic Drainage*, April 24–29, 1994, Pittsburgh, PA, pp. 148–157.

Pratt, A. R., H. W. Nesbitt, and J. R. Mycroft. 1996. The increase reactivity of pyrrhotite and magnetite phases in sulphide mine tailings. *Journal of Geochemical Exploration*, 56:1–11.

Price, W. A. 1997. *Guidelines and Recommended Methods for the Prediction of Metal Leaching and Acid Rock Drainage at Minesites in British Columbia, British Columbia Ministry of Employment and Investment*. Energy and Minerals Division, Smithers, BC, 143pp.

Ribet, I., C. J. Ptacek, D. Blowes, and J. L. Jambor. 1995. The potential for metal release by reductive dissolution of weathered mine tailings. *Journal of Contaminant Hydrology*, 17:239–273.

Ricthie, A. I. M. 1997. The performance of covers. In *Proceedings of the Third Australian Acid Mine Drainage Workshop*, July 15–18, 1997, eds. R. W. McLean, L. C. Bell, and Darwin, pp. 135–145. Australian Centre for Minesite Rehabilitation Research, Brisbane, Queensland, Australia.

Ríos, C. A., C. D. Williams, and C. L. Roberts. 2008. Removal of heavy metals from acid mine drainage (AMD) using coal fly ash, natural clinker and synthetic zeolites. *Journal of Hazardous Materials*, 156(1–3):23–35.

Ritcey, G. M. 1989. *Tailings Management—Problems and Solutions in the Mining Industry*. Process Metallurgy, 6. Elsevier, Amsterdam, the Netherlands, 970pp.

Ritchie, G. S. P. 1989. The chemical behaviour of aluminium, hydrogen and manganese in acid soils. In *Soil Acidity and Plant Growth*. Academic Press Australia, pp. 1–60.

Robertson, A. M. 1996. The importance of site characterisation for remediation of abandoned mine lands. In *Managing Environmental Problems at Inactive and Abandoned Metals Mine Sites*, pp. 8–13. USEPA Seminar Publication No. EPA/625/R-95/007, October 1996.

Robertson, J. D. and K. D. Ferguson. 1995. Predicting acid mine drainage. *Mining Environmental Management*, December 1995: 4–8.

Roman, R. J. and B. R. Benner. 1973. The dissolution of copper concentrates. *Minerals Science and Engineering*, 5:3–24.

Salcedo, I. H., B. G. Ellis, and R. E. Lucas. 1979. Studies in soil manganese II. Extractable manganese and plant uptake. *Soil Science Society of American Journal*, 43:138–141.

Scheetz, B. E., M. R. Silsbee, and J. Schuek. 1995. Field applications of cementitious grouts to address the formation of acid mine drainage. In *Proceedings of Sudbury '95—Mining and the Environment*, May 28–June 1, 1995, Sudbury, Ontario, Canada, eds. T. P. Hynes and M. C. Blanchette. CANMET, Ottawa, Ontario, Canada, pp. 935–944.

Schwertmann, U. and R. M. Taylor. 1989. Iron oxides. In *Minerals in Soil Environments*, eds. J. B. Dixon and S. B. Weed. Soil Science Society of America Book Series 1, Soil Science Society of America, Madison, WI, pp. 379–438.

Sengupta, M. 1994. *Acid Rock Drainage and Metal Migration, Environmental Impacts of Mining. Monitoring, Restoring, and Control.* Lewis Publishers, Boca Raton, FL, pp. 167–259.

Shelp, G., W. Chasworth, G. Spiers, and L. Liu. 1995. Cathodic protection of a weathering ore body. In *Proceedings of Sudbury '95—Mining and the Environment*, May 28–June 1, 1995, Sudbury, Ontario, Canada, eds. Hynes, T. P. and M. C. Blanchette, . CANMET, Ottawa, Ontario, Canada, pp. 1035–1042.

Sherlock, E. J., R. W. Lawrence, and R. Poulin. 1995. On the neutralisation of acid rock drainage by carbonate and silicate minerals. *Environmental Geology*, 25:43–54.

Skousen, J. G. 1995. Anoxic limestone drains for acid mine drainage treatment. *Green Lands*, 25(1):29–38.

Sobek, A. A., W. A. Schuller, J. R. Freeman, and R. M. Smith. 1978. Field and laboratory methods applicable to overburden and mine soils. USEPA Report No. 600/2-78-054, USEPA, Cincinnati, OH, 203pp.

Sopper, W. E. 1992. Reclamation of mine land using municipal sludge. *Advances in Soil Science*, 17: 351–430.

Stehouwer, R. C., P. Sutton, P. Fowler, and W. A. Dick. 1995. Minespoil amendment with dry flue gas desul-phurization by-products: Element solubility and mobility. *Journal of Environmental Quality*, 24:165–174.

Stichbury, M., G. Béchard, L. Lortie, and W. D. Gould. 1995. Use of inhibitors to prevent acid mine drainage. In *Proceedings of the Sudbury '95—Mining and the Environment*, May 28–June 1, 1995, Sudbury, Ontario, Canada, eds. T. P. Hynes and M. C. Blanchett. CANMET, Ottawa, Ontario, Canada.

Sullivan, P. J. and J. L. Yelton. 1988. An evaluation of trace element release associated with acid mine drainage. *Environmental Geological Science*, 12(3):181–186.

Tao, H. and L. Dongwei. 2014. Presentation on mechanisms and applications of chalcopyrite and pyrite biole-aching in biohydrometallurgy—A presentation. *Biotechnology Reports*, 4:107–119.

Taylor, G. F. and M. R. Thornber. 1995. The mechanisms of sulphide oxidation and gossan formation. In *Proceedings of the 17th International Geochemical Exploration Symposium*, May 1995, Townsville, Queensland, Australia, pp. 115–138.

Taylor, J. R., C. L. Waring, N. C. Murphy, and M. J. Leake. 1997. An overview of acid mine drainage control and treatment options, including recent advances. In *Proceedings of the Third Australian Acid Mine Drainage Workshop*, July 15–18, 1997, eds. R. W. McLean, L. C. Bell, and Darwin. Australian Centre for Minesite Rehabilitation Research, Brisbane, Queensland, Australia, pp. 147–159.

USEPA. 1994. Acid mine drainage prediction. US Environmental Protection Agency, Office of Solid Waste, Special Wastes Branch, Washington, DC. EPA 530-R-94-036, 37pp.

Vanberk, W. and F. Wisokzky. 1995. Sulfide oxidation in brown coal overburden and chemical modelling of reactions in aquifers influenced by sulfide oxidation. *Environmental Geology*, 26:192–196.

Wang, H. L., M. J. Hedley, and N. S. Bolan. 1994. Chemical properties of fluidised bed boiler ash relevant to its use as a liming material and fertiliser. *New Zealand Journal of Agricultural Research*, 39:249–256.

Watson, A. 1995. Practical engineering options to minimise AMD potential. In *Proceedings of the Second Australian Acid Mine Drainage Workshop*, March 28–31, 1995, Charters Towers, Queensland, Australia, eds. N. J. Grundon and L. C. Bell. Australian Centre for Minesite Rehabilitation Research, Brisbane, Queensland, Australia, pp. 53–65.

Wei, X. and F. A. Wolfe. 2013. Minerals and mine drainage. *Water Environment Research*, 85(10):1515–1547.

Wijesekara, H., N. S. Bolan, M. Vithanage et al. 2016. Utilization of biowaste for mine spoil rehabilitation. *Advances in Agronomy*, 138:97–173.

Wunderly, M. D., B. W. Blowes, and E. O. Frind. 1996. Sulfide mineral oxidation and subsequent reactive transport of oxidation products in the mine tailings impoundment-a numerical model. *Water Resources Research*, 32(10):3173–3187.

Wunderly, M. D., D. W. Blowes, E. O. Frind, C. J. Ptacek, and T. A. Al. 1995. A multicomponent reactive transport model incorporating kinetically controlled pyrite oxidation. In *Proceedings of Sudbury '95—Mining and the Environment*, May 28–June 1, 1995, Sudbury, Ontario, Canada, eds. T. P. Hynes and M. C. Blanchette, pp. 989–998. CANMET, Ottawa, Ontario, Canada.

Section II

Mine Site Rehabilitation Practices

4 Use of Biowaste for Mine Site Rehabilitation
A Meta-Analysis on Soil Carbon Dynamics

Hasintha Wijesekara, N.S. Bolan, Kim Colyvas,
Balaji Seshadri, Y.S. Ok, Yasser M. Awad, Yilu Xu,
Ramesh Thangavel, Aravind Surapaneni,
Christopher Saint, and Meththika Vithanage

CONTENTS

4.1 Introduction 59
4.2 Biowastes for Mine Site Rehabilitation: The Role of Organic Matter 62
4.3 Biosolids in Revegetation of Mined Lands: A Meta-Analysis for Quantification
     of Soil Organic Carbon 62
     4.3.1 Data Sources 63
     4.3.2 Methodology Used for Meta-Analysis 65
     4.3.3 Soil Carbon Dynamics 66
4.4 Challenges for Soil Carbon Sequestration in Mine Site Rehabilitation 68
4.5 Summary and Conclusions 68
Acknowledgments 69
Bibliography 69
References 71

4.1 INTRODUCTION

"Mining" refers to the excavation of economically important resources from terrestrial landmasses, thereby generating a large quantity of valuable precursors for commercial and industrial activities. Mineral products such as coal, aluminum, copper, iron, gold, and mineral sand are examples from the mining industry. Though mining advances global economic prosperity, this industry severely disturbs the land, water resources, and the environment (Figure 4.1). Mined waste materials such as tailings, subsoils, oxidized wastes, and fireclay are the main causes for land disturbance. Presence of potentially hazardous substances such as heavy metals in elevated concentrations in the mined waste materials has caused land contamination. Poor soil characteristics such as low-level organic matter and poor soil texture and structure have resulted in deterioration of the land, adversely affecting the establishment of plants and soil microbial flora and fauna (Boyer et al. 2011, Johnson 2003, Larney and Angers 2012, Sopper 1992). Disturbed mine sites are known to contaminate water resources

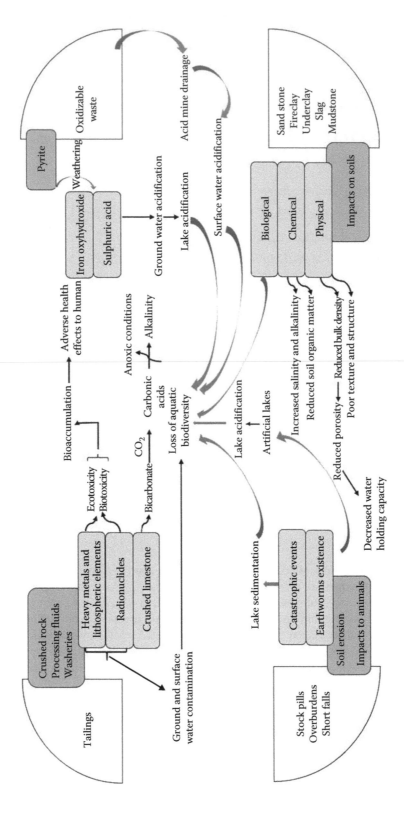

FIGURE 4.1 Possible pathways of the environmental impacts from various components of degraded mine sites.

in many countries, mainly from acid mine drainage (Bolan et al. 2003, Lindsay et al. 2015, Taylor et al. 1997). Therefore, these sites need to be rehabilitated to minimize potential environmental consequences, thereby enhancing their utilization. Revegetation of mine sites is one of the potential strategies that can be applied to improve these disturbed land masses. Here, infertile soil properties are improved by a series of processes such as land application of biowastes.

Globally, biowastes such as biosolids, municipal solid wastes (MSW), animal manures (e.g., poultry), and paper mill sludges are generated in large quantities. For example, in 2050, the potential generation of biosolids has been estimated to be 17.5×10^7 tons year^{-1} (Wijesekara et al. 2016a) and the estimated production of municipal solid waste is expected to be 2.2 billion tons year^{-1} in 2025 (Hoornweg and Bhada-Tata 2012). The expansion of intensive poultry and livestock production and paper mill industries has also resulted in generation of large quantities of biowastes. These biowastes are good sources of essential nutrients and organic matter in soils. It has been estimated that the potential generation rate of organic carbon and nitrogen from four types of biowastes in ten countries exceeds two billion and hundred million tons year^{-1}, respectively (Wijesekara et al. 2016b). Therefore, these biowastes are widely used to transform unproductive soils into productive and healthy ones, by improving their chemical, biological, and physical properties in degraded mine sites.

Among the biowastes, using biosolids for mine site rehabilitation has been identified as a sustainable strategy, because of their richness in organic matter and essential nutrients (Brown et al. 2014, Wijesekara et al. 2016a). This helps in enhancing the physical, chemical, and biological properties of degraded soils, thereby improving soil health. They, therefore, need greater attention. Biosolids are abundant resources available in all countries, and they increase day by day. Wijesekara et al. (2016a) compared and estimated the biosolids generation of eleven countries based on 50 g person^{-1} day^{-1} and found that China and India are top generators of biosolids (Figure 4.2). A high content of organic matter in biosolids is an important factor for converting degraded soils into healthy soils (Bolan et al. 2013).

This chapter provides an outlook to understand soil carbon dynamics in degraded mine lands as impacted by application of biosolids. For this purpose, a meta-analysis is presented in which the effect of biosolids on the soil organic carbon in a series of soils varying in textural class and type is determined. The soils include reclaimed mine soils. Finally, challenges of using biowastes for soil carbon sequestration in mine site rehabilitation, to ensure their sustainable utilization, are discussed.

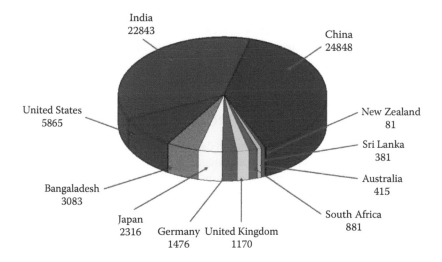

FIGURE 4.2 Projection of biosolids generation in eleven countries. Biosolids generation at 50 g person^{-1} per day was used for the estimation and numbers are denoted in Mg year^{-1}. (Adapted from Wijesekara, H. et al., *Adv. Agron.*, 138, 197, 2016b.)

4.2   BIOWASTES FOR MINE SITE REHABILITATION: THE ROLE OF ORGANIC MATTER

For decades, application of biowastes for mine spoil rehabilitation has been well understood and benefits for the degraded environment have been reported (Larney and Angers 2012, Wijesekara et al. 2016b). By utilizing biowastes, poor, degraded soil can be transformed to healthy soil, which can be validated with improvements in physical, chemical, and biological characteristics of the soil. The improvement in soil physical characteristics include lowering bulk density and temperature, increasing porosity (i.e., macroporosity, mesoporosity, and microporosity) and aggregation, improving hydraulic conductivity and water holding capacity, maintaining soil texture, and reducing erosion and sedimentation (Jones et al. 2012, Sopper 1992, Wijesekara et al. 2016b). In terms of chemical properties, biowaste helps in optimizing pH, electrical conductivity (i.e., increasing soluble salts such as chloride, sulphate, and sodium), cation exchange capacity, and content of nutrients (i.e., nitrate, nitrite, ammonium, phosphate, Ca, Mg, S, Mn, B), heavy metals, and organic matter (Nada et al. 2012, Sopper 1992). As discussed earlier, the biowaste induced increase in organic matter and the influence of physicochemical properties mentioned earlier enhance biological attributes of the soil, such as microbial biomass carbon (MBC), microbial enzymatic activities, and earthworm populations (Pepper et al. 2012, Wijesekara et al. 2016b). All these improved soil characteristics also lead to improved nutritional or growth responses of flora and fauna and groundwater quality in the surrounding environment (Larney and Angers 2012).

Among the fundamental building blocks of soils, soil organic matter is recognized as an important property that governs most of the soil characteristics. Soil organic matter refers to a heterogeneous mixture that comprises all living and nonliving organic material in the soils (Sanderman et al. 2009). Soil organic carbon is the carbon stored in soils, and it is a part of soil organic matter, which contains other elements such as hydrogen, oxygen, and nitrogen. Fundamentally, soil organic matter is generally composed of four major C groups, namely, alkyl, O-alkyl, aromatic, and carbonyl groups (Baldock et al. 1997). Four types or fractions of biologically significant soil organic carbon have been identified by the Commonwealth Scientific and Industrial Research Organisation (CSIRO), and they include crop residues, particulate organic carbon (i.e., relatively young carbon fraction of soil organic matter, which also represents a transitional stage in the humification process), humus, and recalcitrant organic carbon (Sanderman et al. 2009). These four types of soil organic carbons contain both primary energy sources for microorganisms and recalcitrant (pyrogenic) carbon compounds (Table 4.1). In addition to soil organic carbon, natural organic matter has a series of functions in soils (Table 4.2). Adding biowastes into soils or increasing exogenic carbon in soils can enhance soil organic matter or organic carbon content, and, thereby, soil quality (Li 2012). More importantly, increasing the soil organic carbon pool by adding biowastes is increasingly gaining attention towards soil carbon sequestration as a practical action for climate change mitigation.

4.3   BIOSOLIDS IN REVEGETATION OF MINED LANDS: A META-ANALYSIS FOR QUANTIFICATION OF SOIL ORGANIC CARBON

Low levels of organic matter are a common problem in degraded mine soils that affect the initial steps of remediation processes. Therefore, increasing organic matter in degraded mine soils is considered as the first step in the rehabilitation process (Castillejo and Castello 2010, Larney and Angers 2012). Biowastes, especially biosolids, as a good source of organic matter, have a major role in enhancing soil organic carbon in degraded soils. Over the past couple of decades, a number of studies have been reported on biosolids-induced increase in organic matter content, which has improved soil health. However, the experimental results from biosolids application related to soil organic matter changes are variable and dependent on factors such as the climatic region, the soil type, and properties of the biosolids. For instance, Silva et al. (2013) reported an unprecedented carbon accumulation in mined soils after application of biosolids at 100 dry Mg ha^{-1}, highlighting the role of organo-mineral associations in carbon stabilization in soils. In the above study, the authors demonstrated the effects of the age of the biosolids, the time after application

TABLE 4.1

Types or Fraction of Soil Organic Carbon (SOC), Their Characteristics, Sizes, and Key Functions in Soils

SOC Type or Fraction	Characteristics	Size	Degradability	Key Functions in Soils
Crop residues	Shoot and root residues found in or on the soil surface High carbohydrate carbon with lesser amounts of lignin and alkyl carbon	<2 mm	Readily broken down	Provide energy to soil biological processes
Particulate organic carbon	Individual pieces of plant debris; carbon that does not pass through a 0.45 μm filter	2 mm–0.05 mm	Broken down relatively quickly, but more slowly than crop residue	Important for soil structure formation, energy for biological processes, and provision of nutrients
Humus	Decomposed materials and dominated by molecules stuck to soil minerals Increased lignin and alkyl carbon	<0.05 mm	Great stability through hydrophobic interactions and hydrogen bonding	Plays a role in all key soil functions, provision of nutrients, that is, majority of available soil nitrogen is derived from humus
Recalcitrant organic carbon	Typically in the form of charcoal dominated by alkyl carbon	Charcoal <2 mm	Biologically stable, decompose very slowly	Behaves as carbon sink since unable to be decomposed by microorganisms

of the biosolids, and soil depth on carbon enrichment in soils. Contrastingly, a decreased soil organic carbon also has been reported after cessation of land application of biosolids (Li et al. 2013). Therefore, it is important to understand the factors affecting carbon accumulation in degraded mine soils and to estimate soil carbon storage to curb future challenges. For example, calculations of soil carbon storage are needed to see if it can mitigate global warming. Soil related factors, such as soil texture and depth of application of biosolids, are likely to influence carbon storage. So far, the role of soil texture on carbon dynamics has not been adequately investigated, in particular, for biosolids-treated, degraded mine soils. Therefore, the potential of carbon-storing capacity remains largely unclear for many soil textural classes.

The meta-analysis approach has been used as one of the best statistical approaches to compare quantitatively results derived from a range of studies based on different experimental variables (Borenstein et al. 2009). Several meta-analyses have been performed to estimate carbon sequestration in agricultural soils via management practices such as cultivation of cover crops (Lam et al. 2013, Poeplau and Don 2015), tillage (Luo et al. 2010), fertilization (Zhu et al. 2015), and precipitation (Guo and Gifford 2002). But none of the meta-analysis has been performed to estimate and understand the role of carbon storage in degraded mine soils followed by land application of biosolids. Therefore, a meta-analysis was conducted to quantify the soil organic carbon dynamics as affected by the application of biosolids on selected degraded mine soils and textural classes.

4.3.1 Data Sources

Articles were collected from Google Scholar, Web of Science, Science Direct, SCOPUS, and Springer Link. Search terms of "biosolids AND carbon sequestration" showed 65 papers in SCOPUS that had been published up to the cutoff date of November 8, 2016. Both laboratory (pot) and field experiments that had quantitative results were recorded to maximize the number of

TABLE 4.2

Summarized Information on Selected Natural Organic Matter's Origin, Their Characteristics and Functionality/Importance in Terms of Carbon Sequestration in Soils

Chemical Compound	Origin	Characteristics	Importance in Terms of Carbon Sequestration/Recalcitrant Nature/Monomer Complexity	References
Humic acid	Derived from organic matter metabolisms by microorganisms	Soluble under alkaline condition but not acidic condition (generally <2 pH)	High recalcitrant nature, main components of soil organic matter interacting with soil inorganic components	Lal, (2004); Sutton and Sposito (2005)
Fulvic acid	Derived from organic matter metabolisms by microorganisms	Soluble under all pH conditions	High recalcitrant nature, main components of soil organic matter interacting with soil inorganic components	Lal (2004); Sutton and Sposito (2005)
Humin	Derived from organic matter metabolisms by microorganisms	Insoluble fraction of humic substances under acidic or alkaline medium	High recalcitrant nature, main components of soil organic matter interacting with soil inorganic components	Lal (2004); Sutton and Sposito (2005)
Suberans	Periderm, the outer layer of plant stem, root, tuber or bark	Insoluble	ω-hydroxy acid and dicarboxylic acid fractions, saturated C16 and monounsaturated C18 acids	Kolattukudy (2001)
Murein or mucopeptide	Compound in bacteria cell wall	Insoluble	N-acetylglucosamine and N-acetylmuramic acid as alternating units for carbohydrate backbone and glucosamine, galactosamine, muramic acid, and diaminopimelic acid as amino sugar compounds	Kogel-Knabner (2006)
Cutin or chitin	Compound in fungi cell wall	Insoluble	10,16-dihydroxy C16 acid, 18-hydroxy-9,10 epoxy C18 acid, and 9,10,18-trihydroxy C18 acid as monomers	Kolattukudy (2001)
Glomalin	Produced by arbuscular mycorrhizal (AM) fungi	Insoluble	A glycoprotein	Wright et al. (2006)

studies. Therefore, studies that did not report quantitative results were excluded from the meta-analysis. Corresponding authors were contacted whenever additional information such as standard deviations, number of controls and treatments, soil textural classes, and clay content were needed. To obtain more data (i.e., unpublished data), thereby reducing bias, steps such as contacting lead researchers on the topic of land application of biosolids and use of the grey literature were carried out. However, the majority of data were collected from studies published in peer-reviewed scientific journals (The Bibliography provides studies used for the meta-analysis.). The WebPlotDigitizer version 3.10 was used to mine data points whenever data were provided only in graphic format (Rohatgi 2016).

4.3.2 Methodology Used for Meta-Analysis

The meta-analysis was conducted to evaluate the effects of land application of biosolids on organic carbon dynamics in selected soils. For example, influence of the application of biosolids on the carbon storage capacity of degraded mine soils was assessed using eight soil textural classes: clay, clay loam, loam, sand, sandy loam, silt loam, silty clay, and silty clay loam, and two types of soils (i.e., unspecified and reclaimed soils). This study compares soil organic carbon dynamics between controls (i.e., no biosolids application) and biosolids-treated soils.

Firstly, for each study, control and treatment means were recorded, or calculated where necessary. Mostly, soil organic carbon data were extracted from literature. Assuming that organic matter contains 58% carbon, the factor of 1.72 was used to convert organic matter to soil organic carbon (Granato et al. 2004). In some cases, no attempt was made to distinguish inorganic carbon and organic carbon. This assumption was reasonable, because the reported soils were slightly or moderately acidic in status (Brown et al. 2011). Standard deviations were used as the measure of variances. They were included, when present, or calculated from published measures of variance (i.e., standard deviation = square root of sample variance), standard error (i.e., standard deviation = standard error × square root of sample number), and confidence intervals (i.e., standard deviation = [confidence interval × square root of sample number]/t value) in each study. A series of sensitivity analyses were performed to exclude outlier studies. Due to the large variety of soil types reported, some grouping of similar types and textural class was needed to maximize the number of studies that could be used in the analysis. Therefore, extracted data were categorized into ten groups based on soil type (i.e., two groups) or texture (i.e., eight groups), as stated before. The references that did not have enough information to set into the eight textural classes were included in unspecified and reclaimed soil groups.

Standardization of the results in the literature was undertaken through calculation of the effect size. The effect size is a value that reflects the magnitude of the treatment effect or the strength of a relationship between two variables (Borenstein et al. 2009). More importantly, the effect size allows quantitative, statistical information to be pooled from a range of studies that report results based on different experimental variables. Therefore, the effect size allows strong, statistical comparisons to be made between effects, which were calculated in Comprehensive Meta-Analysis Version 3 statistical software (CMA 2016). Finally, the mean of effect sizes (i.e., difference in means of control versus treated soils), 95% confidence intervals (CI), and mixed effect model value were derived based on sample sizes, means, and standard deviations. By contrast, the fixed and mixed effect models make different assumptions about the nature of variability associated with studies, thereby leading to different approaches for assigning weights to studies. The mixed (or random effects) model introduces two sources of variation, within study variability and between study variability. The fixed model assumes there is only sampling error within study (Borenstein, et al. 2009). The fixed effect model is based on there being no underlying differences between studies. The mixed effect model allows for systematic differences between studies, due to a range of unknown factors, to be incorporated into an additional variability term for the model giving a more realistic assessment of variability. For this reason, the mixed model is preferred for meta-analysis and has been used in this study.

4.3.3   SOIL CARBON DYNAMICS

The meta-analysis included 226 independent pairwise studies that covered a range of soil textural classes and types across twenty countries. Overall, increased soil carbon contents were observed for all the soil classes and types, including reclaimed soils. Several factors such as biosolids type, their rate, and depth of land application, land use and management practices (i.e., organic matter self-assemblage by long periods of nondisturbance under no-tillage), and climatic conditions are found to affect the amount of soil carbon. However, this study mainly focused on the effect of soil properties such as soil texture and mineralogy on soil carbon dynamics.

Soil texture refers to the relative size distribution of the sand, silt, and clay-sized particles that make up the mineral fraction of the soil (Kettler et al. 2001). The rate of soil organic carbon sequestration is dependent on soil texture and, hence, soil texture plays an important role in carbon stabilization in soils (Lal 2004). The mineral components present in differently sized fractions associate with soil organic carbon (i.e., forming clay- or silt-sized organomineral complexes), where the majority of soil organic carbon can be found (Post and Kwon 2000). Although, clay-sized organomineral complexes often have greater soil organic carbon accumulation than silt-sized organomineral complexes, their loss is more rapid than in silt-sized organomineral complexes, which indicates stability of carbon bound to the silt fraction (Christensen 1996). Further, sandy or sandy-to-loam soils do not hold soil organic carbon for long periods due to their low protective capacity because of their low clay content (Chan et al. 2003). Many studies show the accumulation of soil organic carbon by mineral complexes. For example, a study of soil micro-aggregates revealed that organic matter is physically protected with clay and silt aggregates and, hence, resists microbial attacks on carbon, and this leads to its sequestration in soils (Skjemstad et al. 1993). Vogel et al. (2014) used advanced spectroscopic techniques such as NanoSIMS to show that fresh organic matter is preferentially attached to residual organic matter–containing clusters in clay particles.

In our meta-analysis, we found the following order for soil organic carbon accumulation: silty clay loam > silt loam > reclaimed soils > sand > sandy loam > unspecified soils ≥ clay ≥ clay loam ≥ loam ≥ silty clay (Figure 4.3). All the analyzed soil textures and types showed a significant mean effect of carbon accumulation compared to the controls (p < 0.05). Hereafter, detailed explanations for selected soil textural classes are considered to identify their specific mineralogical characteristics in relation to soil carbon accumulation.

Silty clay loam soils showed the highest mean effect, indicating a high potential for carbon sequestration. According to the major soil textural classes from the United States Department of Agriculture, a silty clay loam soil contains 27%–40% clay and less than 20% sand (USDA 1987). Therefore, our first observation, that is, a greater enrichment of carbon in silty clay loam soils, is likely due to associations of clay and organic carbon. To get a better understanding of this observation, individual studies performed in mine spoils or degraded mine lands were carefully investigated in relevant textural groups. As an example, a surface application (i.e., 0–15 cm) of biosolids at a rate of 100% (v/v) in to a brownfield with steel mill slag materials, which was covered with a 15 cm silt clay loam soil layer, caused a higher carbon accumulation in 4 years compared to a control (Brose et al. 2016). In this study, the silt clay loam soil was layered on the slag to regulate water movement and to simulate a heavy-textured B-horizon found in many soil profiles at the study area. Silt loam soils recorded the second greatest potential for carbon enrichment. A silt loam soil contains 50% or more silt and 12%–27% clay, or 50%–80% silt and less than 12% clay (USDA 1987). Gardner et al. (2010) reported the influence of application of biosolids to two mine tailings sites that originally consisted of predominantly a silt loam soil and a sandy soil. The authors showed that high application rates of biosolids, that is, 100 Mg ha⁻¹ and 150 Mg ha⁻¹, would be required to increase soil carbon to above 2% for silt loam and sandy tailing soils, respectively. In addition, compared to 0–15 cm soils, the authors reported a lower soil carbon accumulation in deeper soil increments (15–30 and 30–45 cm) at both sites. Further, in the deeper soil layers, a

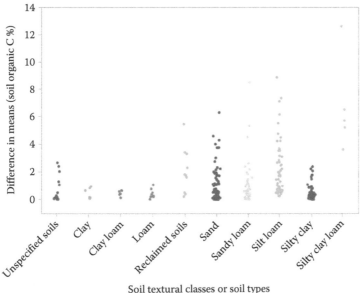

FIGURE 4.3 Effect of biosolids land application versus soil organic carbon dynamics. Y axis indicates the differences in means (% of soil organic carbon) between controls (i.e., non-biosolids application) and treatments (i.e., biosolids applied). X axis indicates eight soil textural classes and two soil types namely "unspecified soils" and "reclaimed soils."

higher accumulation of carbon was observed at the silt loam site than the sandy site, which indicated the effect of silt in soil carbon sequestration. By application of biosolids, the reclaimed soils changed their initial low soil carbon contents into considerably higher contents, thereby indicating the potential of biosolids to sequester soil carbon and increase soil health.

Reis et al. (2014) investigated carbon contents in different physical fractions of clay soils. They showed that organomineral complexes could be formed in clay soils, thereby increasing carbon accumulation in them. Across the investigated soils, silty clay loam and clay soils would be validated candidatures in relation to the soil carbon sequestration strategy, and need policy improvement at regional and global scale to secure economically (i.e., carbon credits) and environmentally sustainable soil carbon storage with application of biosolids. The meta-analysis showed that application of biosolids significantly increased ($p < 0.05$) soil carbon content with a narrow confidence interval (i.e., less variability) in sandy soils compared to clay soils. This observation suggests requirement of further statistical analysis, such as multiple-variable meta-analysis, to understand other effects (i.e., rate, and depth of biosolids land application) and validate obtained results. When all studies are considered, the effect size from the mixed model indicated 0.435 ($p < 0.05$) (Table 4.3).

Soil depth is another important factor that can be examined to quantify carbon accumulation in biosolids-treated soils. When the effect of depth was summarized in our study, biosolids significantly ($p < 0.05$) increased the organic carbon pool in the surface layer of soils (data not shown here). Therefore, surface incorporation of biosolids can be used as one of the main strategies to increase soil carbon sequestration. In addition to the effects of soil texture and depth, soil carbon sequestration has been reported to be significantly correlated with the application rate of biosolids (Tian et al. 2009). Generally, a higher application rate of biosolids causes a higher carbon accumulation in soils. However, this is not always true, because accelerated leaching and mineralization of organic matter can cause depletion of the carbon content of soils (Schwab et al. 2007, Toribio and Joan 2006). Therefore, detailed analyses are required to clarify carbon accumulation

TABLE 4.3

Summarized Data for the Meta-Analysis of Biosolids Land Application versus Soil Organic Carbon Difference between Means (Treatment–Control)

Soil Textural Classes or Soil Types	Number of Replications[a]	Mean	Standard Error	Lower Limit	Upper Limit	p Value
Clay loam	5	0.367	0.095	0.181	0.554	0.000
Clay	7	0.386	0.072	0.244	0.527	0.000
Loam	8	0.358	0.089	0.184	0.532	0.000
Silty clay	49	0.303	0.031	0.242	0.363	0.000
Unspecified soils	13	0.592	0.136	0.326	0.858	0.000
Reclaimed soils	10	1.855	0.392	1.086	2.624	0.000
Sand	65	1.201	0.165	0.879	1.524	0.000
Sandy loam	20	0.913	0.164	0.592	1.234	0.000
Silt loam	43	2.277	0.136	2.011	2.543	0.000
Silty clay loam	6	7.904	2.046	3.894	11.915	0.000
Overall	226	0.435	0.025	0.387	0.483	0.000

Note: The estimated mean effects for each soil type was derived analyzing across levels of subgroup within study and comparing effects at different levels of subgroup within study. The overall mean effect and confidence intervals for each texture or soil type group have been estimated with weights based on both within and between study variability using a mixed effects model.

[a] The replications are based on studies that generally had several application rates, observation periods, and soil depths.

in soil after application of biosolids. These analyses need to include studies of annual application rates, cumulative applications, and type of biosolids (i.e., fresh, dry, co-composted).

4.4   CHALLENGES FOR SOIL CARBON SEQUESTRATION IN MINE SITE REHABILITATION

A number of challenges are linked to soil carbon sequestration after land application of biowastes, in order to rehabilitate mine sites. Sites with biowastes added to degraded mined lands need to be monitored for groundwater and soil contamination, because there can be enhanced leaching of inorganic and organic nutrients, heavy metals, pathogens, and emerging contaminants after their application (Aguilar-Chavez et al. 2012, Gerba et al. 2002, Wang et al. 2008). Emissions of greenhouse gases and odorous gaseous from biowastes, during and after their application, have been identified, and they contribute to global warming and air pollution. Therefore, prevention steps should be aimed to minimize water, soil, and air pollution. To safeguard a healthy environment, regulations in relation to production of biowastes (i.e., co-composted and chemically stabilized biowastes) and land application (i.e., regulated annual and cumulative application rates for specific types of land use) have been developed and implemented in many countries. However, most regulations are focused only on minimizing pathogens and environmental contamination from heavy metals and major inorganic nutrients. Therefore, regulations need to be developed aimed at optimizing utilization of carbon content in biowastes.

4.5   SUMMARY AND CONCLUSIONS

A large quantity of biowastes is generated as a consequence of the increased human population. Because of the presence of high amounts of organic matter and nutrients, biowastes are used extensively as a soil amendment or fertilizer for land reclamation and revegetation. Biowastes enhance physical, chemical, and biological properties of degraded mine lands, thereby resulting in direct and indirect benefits. Soil carbon sequestration is one of the main benefits associated with land application of biowastes at degraded mine lands. Biosolids are one of the main types of biowastes that have a significant amount of organic matter. However, short- and long-term studies conducted after application of biosolids to degraded soils show contrasting results in terms of organic carbon in soil. Therefore, to evaluate the quantitative impact of application of biosolids on soil carbon dynamics in different soil textural classes and types, a meta-analysis was performed using data from the literature. Based on the meta-analysis, small but statistically significant effects were identified in all the analyzed soil textural classes and types, including reclaimed mine soils. Accumulation of soil organic carbon showed a clear order, as follows: silty clay loam > silt loam > reclaimed soils > sand > sandy loam > unspecified soils ≥ clay ≥ clay loam ≥ loam ≥ silty clay. Therefore, land application of biosolids can be used as a way to enhance soil carbon sequestration and develop soil health in soils with low fertility including degraded mine soils.

However, utilization of biowastes in degraded lands has a number of challenges. For example, there are concerns about soil contamination and air pollution. Therefore, further investigations are needed to understand the fundamental mechanisms and factors (i.e., soil types, influence of climatic conditions, stabilization and co-composting of biosolids, land management practices) on long-term carbon sequestration in soils.

ACKNOWLEDGMENTS

The authors would like to acknowledge the work carried out by the researchers whose published data were used for this meta-analysis and researchers who provided and clarified their data whenever necessary. This research was partly supported by Australian Research Council Discovery Projects (DP140100323). This chapter is one of the outcomes of the research project on "Carbon sequestration from land application of biosolids." We thank South East Water (Dr. Aravind Surapaneni), Western Water (William Rajendran), Gippsland Water (Mark Heffernan), City West Water (Sean Hanrahan), Yarra Valley Water (Andrew Schunke), and Transpacific Industries (Chris Hetherington) for supporting this research project.

BIBLIOGRAPHY

Albaladejo, J., J. Lopez, C. Boix-Fayos et al. 2008. Long-term effect of a single application of organic refuse on carbon sequestration and soil physical properties. *Journal of Environmental Quality* 37(6):2093–2099. doi: 10.2134/jeq2007.0653.

Alvarenga, P., P. Palma, A. P. Goncalves et al. 2008. Assessment of chemical, biochemical and ecotoxicological aspects in a mine soil amended with sludge of either urban or industrial origin. *Chemosphere* 72(11):1774–1781.

Asensio, V., F. A. Vega, and E. F. Covelo. 2014. Effect of soil reclamation process on soil C fractions. *Chemosphere* 95:511–518.

Brose, D. A., L. S. Hundal, O. O. Oladeji et al. 2016. Greening a steel mill slag brownfield with biosolids and sediments: a case study. *Journal of Environmental Quality* 45(1):53–61. doi: 10.2134/jeq2015.09.0456.

Brown, S., K. Kurtz, A. Bary et al. 2011. Quantifying benefits associated with land application of organic residuals in Washington state. *Environmental Science & Technology* 45(17):7451–7458. doi: 10.1021/es2010418.

Brown, S. L., I. Clausen, M. A. Chappell et al. 2012. High-Iron biosolids compost–induced changes in Lead and Arsenic speciation and bioaccessibility in co-contaminated soils. *Journal of Environmental Quality* 41(5):1612–1622. doi: 10.2134/jeq2011.0297.

Brown, S., M. Mahoney, and M. Sprenger. 2014. A comparison of the efficacy and ecosystem impact of residual-based and topsoil-based amendments for restoring historic mine tailings in the Tri-State mining district. *Science of the Total Environment* 485–486(0):624–632. doi: http://dx.doi.org/10.1016/j.scitotenv.2014.03.029.

Chowdhury, S., M. Farrell, and N. Bolan. 2014. Priming of soil organic carbon by malic acid addition is differentially affected by nutrient availability. *Soil Biology and Biochemistry* 77:158–169. doi: http://dx.doi.org/10.1016/j.soilbio.2014.06.027.

Coors, A., M. Edwards, P. Lorenz et al. 2016. Biosolids applied to agricultural land: Influence on structural and functional endpoints of soil fauna on a short- and long-term scale. *Science of the Total Environment* 562:312–326.

Eid, E. M., A. F. El-Bebany, S. A. Alrumman et al. 2017. Effects of different sewage sludge applications on heavy metal accumulation, growth and yield of spinach (*Spinacia oleracea* L.). *International Journal of Phytoremediation* 19:340–347.

Fernandez, J. M., M. A. Nieto, E. G. Lopez-de-Sa et al. 2014. Carbon dioxide emissions from semi-arid soils amended with biochar alone or combined with mineral and organic fertilizers. *Science of the Total Environment* 482–483:1–7.

Gardner, W. C., K. Broersma, A. Naeth et al. 2010. Influence of biosolids and fertilizer amendments on physical, chemical and microbiological properties of copper mine tailings. *Canadian Journal of Soil Science* 90(4):571–583. doi: 10.4141/cjss09067.

Granato, T. C., R. I. Pietz, G. J. Knafl et al. 2004. Trace element concentrations in soil, corn leaves, and grain after cessation of biosolids applications. *Journal of Environmental Quality* 33(6):2078–2089.

Harrison, R., D. Xue, C. Henry et al. 1994. Long-term effects of heavy applications of biosolids on organic matter and nutrient content of a coarse-textured forest soil. *Forest Ecology and Management* 66(1):165–177. doi: http://dx.doi.org/10.1016/0378-1127(94)90155-4.

He, Z. L., A. K. Alva, P. Yan et al. 2000. Nitrogen mineralization and transformation from composts and biosolids during field incubation in a sandy soil. *Soil Science* 165(2):161–169.

Ippolito, J. A., K. A. Barbarick, M. W. Paschke et al. 2010. Infrequent composted biosolids applications affect semi-arid grassland soils and vegetation. *Journal of Environmental Management* 91(5):1123–1130.

Kelly, G. 2008. *Application of Recycled Organics in Mine Site Rehabilitation.* Sydney, New South Wales, Australia: Forest Resources Branch, Science and Research, NSW Department of Primary Industries.

Kelly, J. J., E. Favila, L. S. Hundal et al. 2007. Assessment of soil microbial communities in surface applied mixtures of Illinois River sediments and biosolids. *Applied Soil Ecology* 36(2–3):176–183. doi: http://dx.doi.org/10.1016/j.apsoil.2007.01.006.

Lag-Brotos, A. J., I. Gomez, and J. Navarro-Pedreno. 2015. Sewage sludge use in bioenergy production. A case study of its effects on soil properties under *Cynara cardunculus* L. cultivation. *Spanish Journal of Agricultural Research* 13(1):1101. doi: http://dx.doi.org/10.5424/sjar/2015131-6145.

Lamb, D. T., S. Heading, N. Bolan et al. 2012. Use of biosolids for phytocapping of landfill soil. *Water, Air, & Soil Pollution* 223(5):2695–2705. doi: 10.1007/s11270-011-1060-x.

Larcheveque, M., C. Ballini, N. Korboulewsky et al. 2006. The use of compost in afforestation of Mediterranean areas: Effects on soil properties and young tree seedlings. *Science of the Total Environment* 369(1–3):220–230.

Lockwell, J., W. Guidi, and M. Labrecque. 2012. Soil carbon sequestration potential of willows in short-rotation coppice established on abandoned farm lands. *Plant and Soil* 360(1):299–318. doi: 10.1007/s11104-012-1251-2.

Madejon, E., P. Madejon, P. Burgos et al. 2009. Trace elements, pH and organic matter evolution in contaminated soils under assisted natural remediation: A 4-year field study. *Journal of Hazardous Materials* 162(2–3):931–938.

McIvor, K., C. Cogger, and S. Brown. 2012. Effects of biosolids based soil products on soil physical and chemical properties in urban gardens. *Compost Science & Utilization* 20(4):199–206. doi: 10.1080/1065657X.2012.10737049.

Noirot-Cosson, P. E., E. Vaudour, J. M. Gilliot et al. 2016. Modelling the long-term effect of urban waste compost applications on carbon and nitrogen dynamics in temperate cropland. *Soil Biology and Biochemistry* 94:138–153.

Ogut, M., and F. Er. 2015. Mineralizable carbon in biosolids/fly ash/sugar beet lime treated soil under field conditions. *Applied Soil Ecology* 91:27–36.

Oleszczuk, P. 2006. Persistence of polycyclic aromatic hydrocarbons (PAHs) in sewage sludge-amended soil. *Chemosphere* 65(9):1616–1626.

Oliver, I. W., A. Hass, G. Merrington et al. 2005. Copper availability in seven Israeli soils incubated with and without biosolids. *Journal of Environmental Quality* 34:508–513.

Ouimet, R., A. Pion, and M. Hebert. 2015. Long-term response of forest plantation productivity and soils to a single application of municipal biosolids. *Canadian Journal of Soil Science* 95(2):187–199. doi: 10.4141/cjss-2014-048.

Pawlett, M., L. K. Deeks, and R. Sakrabani. 2015. Nutrient potential of biosolids and urea derived organo-mineral fertilisers in a field scale experiment using ryegrass (*Lolium perenne* L.). *Field Crops Research* 175:56–63.

Perez-de-Mora, A., P. Burgos, E. Madejon et al. 2006. Microbial community structure and function in a soil contaminated by heavy metals: Effects of plant growth and different amendments. *Soil Biology and Biochemistry* 38(2):327–341.

Pitombo, L. M., J. B. do-Carmo, I. C. de-Maria et al. 2015. Carbon sequestration and greenhouse gases emissions in soil under sewage sludge residual effects. *Scientia Agricola* 72:147–156.

Quaye, A. K., and T. A. Volk. 2013. Biomass production and soil nutrients in organic and inorganic fertilized willow biomass production systems. *Biomass and Bioenergy* 57:113–125. doi: http://dx.doi.org/10.1016/j.biombioe.2013.08.002.

Ros, M., S. Klammer, B. Knapp et al. 2006. Long-term effects of compost amendment of soil on functional and structural diversity and microbial activity. *Soil Use and Management* 22(2):209–218. doi: 10.1111/j.1475-2743.2006.00027.x.

Sidhu, V., D. Sarkar, and R. Datta. 2016. Effects of biosolids and compost amendment on chemistry of soils contaminated with copper from mining activities. *Environmental Monitoring and Assessment* 188(3):1–9. doi: 10.1007/s10661-016-5185-7.

Silva, L. C. R., R. S. Correa, T. A. Doane et al. 2013. Unprecedented carbon accumulation in mined soils: the synergistic effect of resource input and plant species invasion. *Ecological Applications* 23(6):1345–1356. doi: 10.1890/12-1957.1.

Song, U., and E. J. Lee. 2010. Environmental and economical assessment of sewage sludge compost application on soil and plants in a landfill. *Resources, Conservation and Recycling* 54(12):1109–1116. doi: http://dx.doi.org/10.1016/j.resconrec.2010.03.005.

Spargo, J. T., M. M. Alley, R. F. Follett et al. 2008. Soil carbon sequestration with continuous no-till management of grain cropping systems in the Virginia coastal plain. *Soil and Tillage Research* 100(1–2):133–140. doi: http://dx.doi.org/10.1016/j.still.2008.05.010.

Thangarajan, R., S. Chowdhury, A. Kunhikrishnan et al. 2014. Interactions of soluble and solid organic amendments with priming effects induced by glucose. *Vadose Zone Journal* 13(7).1–8.

Torri, S., R. Alvarez, and R. Lavado. 2003. Mineralization of carbon from sewage sludge in three soils of the Argentine pampas. *Communications in Soil Science and Plant Analysis* 34(13–14):2035–2043. doi: 10.1081/CSS-120023235.

Trlica, A., and S. Brown. 2013. Greenhouse gas emissions and the interrelation of urban and forest sectors in reclaiming one hectare of land in the Pacific Northwest. *Environmental Science & Technology* 47(13):7250–7259. doi: 10.1021/es3033007.

Veeresh, H., S. Tripathy, D. Chaudhuri et al. 2003. Changes in physical and chemical properties of three soil types in India as a result of amendment with fly ash and sewage sludge. *Environmental Geology* 43(5):513–520. doi: 10.1007/s00254-002-0656-2.

Wijesekara, H. 2016. Biogeochemical mechanisms of biosolids application on carbon sequestration in soils. The University of Newcastle (unpublished data).

Zanuzzi, A., J. M. Arocena, J. M. van-Mourik et al. 2009. Amendments with organic and industrial wastes stimulate soil formation in mine tailings as revealed by micromorphology. *Geoderma* 154(1–2):69–75.

Zebarth, B. J., G. H. Neilsen, E. Hogue et al. 1999. Influence of organic waste amendments on selected soil physical and chemical properties. *Canadian Journal of Soil Science* 79(3):501–504.

REFERENCES

Aguilar-Chavez, A., M. Diaz-Rojas, M. D. R. Cardenas-Aquino et al. 2012. Greenhouse gas emissions from a wastewater sludge-amended soil cultivated with wheat (*Triticum* spp. L.) as affected by different application rates of charcoal. *Soil Biology and Biochemistry* 52:90–95. doi: http://dx.doi.org/10.1016/j.soilbio.2012.04.022.

Baldock, J. A., J. M. Oades, P. N. Nelson et al. 1997. Assessing the extent of decomposition of natural organic materials using solid-state 13C NMR spectroscopy. *Soil Research* 35(5):1061–1084. doi: http://dx.doi.org/10.1071/S97004.

Bolan, N. S., D. C. Adriano, and D. Curtin. 2003. Soil acidification and liming interactions with nutrient and heavy metal transformation and bioavailability. *Advances in Agronomy* 78:215–272.

Bolan, N. S., A. Kunhikrishnan, and R. Naidu. 2013. Carbon storage in a heavy clay soil landfill site after biosolid application. *Science of the Total Environment* 465(0):216–225. doi: http://dx.doi.org/10.1016/j.scitotenv.2012.12.093.

Borenstein, M., L. V. Hedges, J. P. T. Higgins et al. 2009. *Introduction to Meta-Analysis*. West Sussex, U.K.: Wiley.

Boyer, S., S. Wratten, M. Pizey et al. 2011. Impact of soil stockpiling and mining rehabilitation on earthworm communities. *Pedobiologia* 54S:S99–S102.

Brose, D. A., L. S. Hundal, O. O. Oladeji et al. 2016. Greening a steel mill slag brownfield with biosolids and sediments: A case study. *Journal of Environmental Quality* 45(1):53–61.

Brown, S., K. Kurtz, A. Bary et al. 2011. Quantifying benefits associated with land application of organic residuals in Washington State. *Environmental Science & Technology* 45(17):7451–7458. doi: 10.1021/es2010418.

Brown, S., M. Mahoney, and M. Sprenger. 2014. A comparison of the efficacy and ecosystem impact of residual-based and topsoil-based amendments for restoring historic mine tailings in the Tri-State mining district. *Science of the Total Environment* 485–486(0):624–632. doi: http://dx.doi.org/10.1016/j.scitotenv.2014.03.029.

Castillejo, J. M., and R. Castello. 2010. Influence of the application rate of an organic amendment (Municipal Solid Waste [MSW] Compost) on gypsum quarry rehabilitation in semiarid environments. *Arid Land Research and Management* 24(4):344–364. doi: 10.1080/15324982.2010.502920.

Chan, K. Y., D. P. Heenan, and H. B. So. 2003. Sequestration of carbon and changes in soil quality under conservation tillage on light-textured soils in Australia: A review. *Australian Journal of Experimental Agriculture* 43:325–334.

Christensen, B. T. 1996. Matching measurable soil organic matter fractions with conceptual pools in simulation models of carbon turnover: Revision of model structure. In *Evaluation of Soil Organic Matter Models: Using Existing Long-Term Datasets*, eds. D. S. Powlson, P. Smith, and J. U. Smith, pp. 143–159. Berlin, Germany: Springer.

CMA. 2016. Comprehensive meta-analysis version 3. https://www.Meta-Analysis.com. Accessed December 15, 2016.

Gardner, W. C., K. Broersma, A. Naeth et al. 2010. Influence of biosolids and fertilizer amendments on physical, chemical and microbiological properties of copper mine tailings. *Canadian Journal of Soil Science* 90(4):571–583. doi: 10.4141/cjss09067.

Gerba, C. P., I. L. Pepper, and L. F. Whitehead. 2002. A risk assessment of emerging pathogens of concern in the land application of biosolids. *Water Science and Technology* 46(10):225–230.

Granato, T. C., R. I. Pietz, G. J. Knafl et al. 2004. Trace element concentrations in soil, corn leaves, and grain after cessation of biosolids applications. *Journal of Environmental Quality* 33(6):2078–2089.

Guo, L. B., and R. M. Gifford. 2002. Soil carbon stocks and land use change: A meta analysis. *Global Change Biology* 8(4):345–360.

Hoornweg, D. and P. Bhada-Tata. 2012. *What a Waste: A Global Review of Solid Waste Management*. Washington, DC: Urban Development & Local Government Unit, World Bank. https://openknowledge.worldbank.org/handle/10986/17388. Accessed January 7, 2015.

Johnson, D. B. 2003. Chemical and microbiological characteristics of mineral spoils and drainage waters at abandoned coal and metal mines. *Water, Air, & Soil Pollution* 3:47–66.

Jones, B. E. H., R. J. Haynes, and I. R. Phillips. 2012. Addition of an organic amendment and/or residue mud to bauxite residue sand in order to improve its properties as a growth medium. *Journal of Environmental Management* 95(1):29–38.

Kettler, T. A., J. W. Doran, and T. L. Gilbert. 2001. Simplified method for soil particle-size determination to accompany soil-quality analyses. *Soil Science Society of America Journal* 65(3):849–852. doi: 10.2136/sssaj2001.653849x.

Kogel-Knabner, I. 2006. Chemical structure of organic N and organic P in soil. In *Nucleic Acids and Proteins in Soil*, eds. P. Nannipieri and K. Smalla, pp. 23–48. Heidelberg, Germany: Springer.

Kolattukudy, P. 2001. Polyesters in higher plants. In *Biopolyesters*. eds. W. Babel, and A. Steinbuchel, pp. 1–49. Berlin, Germany: Springer.

Lal, R. 2004. Soil carbon sequestration impacts on global climate change and food security. *Science* 304(5677):1623–1627. doi: 10.1126/science.1097396.

Lam, S. K., D. Chen, A. R. Mosier et al. 2013. The potential for carbon sequestration in Australian agricultural soils is technically and economically limited. *Scientific Reports* 3:2179. doi: 10.1038/srep02179.

Larney, F. J., and D. A. Angers. 2012. The role of organic amendments in soil reclamation: A review. *Canadian Journal of Soil Science* 92(1):19–38. doi: 10.4141/cjss2010-064.

Li, J. 2012. Effects of biosolids on carbon sequestration and nitrogen cycling. PhD dissertation, Virginia Polytechnic Institute and State University, Blacksburg, VA.

Li, J., G. K. Evanylo, K. Xia et al. 2013. Soil carbon characterization 10 to 15 years after organic residual application: Carbon (1s) K-edge near-edge x-ray absorption fine-structure spectroscopy study. *Soil Science* 178(9):453–464.

Lindsay, M. B. J., M. C. Moncur, J. G. Bain et al. 2015. Geochemical and mineralogical aspects of sulfide mine tailings. *Applied Geochemistry* 57:157–177.

Luo, Z., E. Wang, and O. J. Sun. 2010. Can no-tillage stimulate carbon sequestration in agricultural soils? A meta-analysis of paired experiments. *Agriculture, Ecosystems & Environment* 139(1–2):224–231. doi: http://dx.doi.org/10.1016/j.agee.2010.08.006.

Nada, W., O. Blumenstein, S. Claassens et al. 2012. Effect of wood compost on extreme soil characteristics in the Lusatian Lignite region. *Open Journal of Soil Science* 2:347–352.

Pepper, I. L., H. G. Zerzghi, S. A. Bengson et al. 2012. Bacterial populations within copper mine tailings: Long-term effects of amendment with Class A biosolids. *Journal of Applied Microbiology* 113(3):569–577. doi: 10.1111/j.1365-2672.2012.05374.x.

Poeplau, C., and A. Don. 2015. Carbon sequestration in agricultural soils via cultivation of cover crops - A meta-analysis. *Agriculture, Ecosystems & Environment* 200:33–41. doi: http://dx.doi.org/10.1016/j.agee.2014.10.024.

Post, W. M., and K. C. Kwon. 2000. Soil carbon sequestration and land-use change: Processes and potential. *Global Change Biology* 6(3):317–327. doi: 10.1046/j.1365-2486.2000.00308.x.

Reis, C. E. S. D., D. P. Dick, J. D. S. Caldas et al. 2014. Carbon sequestration in clay and silt fractions of Brazilian soils under conventional and no-tillage systems. *Scientia Agricola* 71:292–301.

Rohatgi, A. 2016. WebPlotDigitizer version 3.10. http://arohatgi.info/WebPlotDigitizer/. Accessed January 20, 2017.

Sanderman, J., R. Farquharson, and J. A. Baldock. 2009. Soil carbon sequestration potential: A review for Australian agriculture. In *A Report Prepared for Department of Climate Change and Energy Efficiency*. Adelaide, South Australia, Australia: Commonwealth Scientific and Industrial Research Organisation (CSIRO).

Schwab, P., D. Zhu, and M. K. Banks. 2007. Heavy metal leaching from mine tailings as affected by organic amendments. *Bioresource Technology* 98(15):2935–2941. doi: http://dx.doi.org/10.1016/j.biortech.2006.10.012.

Silva, L. C. R., R. S. Correa, T. A. Doane et al. 2013. Unprecedented carbon accumulation in mined soils: The synergistic effect of resource input and plant species invasion. *Ecological Applications* 23(6):1345–1356. doi: 10.1890/12-1957.1.

Skjemstad, J. O., L. J. Janik, M. J. Head et al. 1993. High energy ultraviolet photo-oxidation: A novel technique for studying physically protected organic matter in clay- and silt-sized aggregates. *Journal of Soil Science* 44(3):485–499. doi: 10.1111/j.1365-2389.1993.tb00471.x.

Sopper, W. E. 1992. Reclamation of mine land using municipal sludge. In *Soil Restoration*, eds. R. Lal and B. A. Stewart, pp. 351–431. New York: Springer.

Sutton, R., and G. Sposito. 2005. Molecular structure in soil humic substances: The new view. *Environmental Science & Technology* 39:9009–9015.

Taylor, J. R., C. L. Waring, N. C. Murphy et al. 1997. An overview of acid mine drainage control and treatment options, including recent advances. In *Proceedings of the Third Australian Acid Mine Drainage Workshop*, eds. R. W. McLean and L. C. Bell, Darwin, Northern Territory, Australia, July 15–18, 1997. Brisbane, Queensland, Australia: Australian Centre for Minesite Rehabilitation Research, pp. 147–159.

Tian, G., T. C. Granato, A. E. Cox et al. 2009. Soil carbon sequestration resulting from long-term application of biosolids for land reclamation. *Journal of Environmental Quality* 38(1):61–74. doi: 10.2134/jeq2007.0471.

Toribio, M., and R. Joan. 2006. Leaching of heavy metals (Cu, Ni and Zn) and organic matter after sewage sludge application to Mediterranean forest soils. *Science of the Total Environment* 363(1–3):11–21. doi: http://dx.doi.org/10.1016/j.scitotenv.2005.10.004.

USDA. 1987. *Soil Mechanics Level I, Module 3*. United States Department of Agriculture Textural Classification Study Guide, Washington, DC. https://www.nrcs.usda.gov/Internet/FSC_DOCUMENTS/stelprdb1044818.pdf.

Vogel, C., C. W. Mueller, C. Hoschen et al. 2014. Submicron structures provide preferential spots for carbon and nitrogen sequestration in soils. *Nature Communications* 5:2947 doi: 10.1038/ncomms3947.

Wang, L. K., N. K. Shammas, and G. Evanylo. 2008. Engineering and management of agricultural land application. In *Biosolids Engineering and Management*, ed. L. K. Wang, N. K. Shammas, and Y. T. Hung,
pp. 343–414. NJ: Totowa, NJ: Humana Press.

Wijesekara, H., N. S. Bolan, P. Kumarathilaka et al. 2016a. Biosolids enhance mine site rehabilitation and
revegitation. In *Environmental Materials and Waste: Resource Recovery and Pollution Prevention*, eds.
M. N. V. Prasad and K. Shih, pp. 45–74. London, U.K.: Elsevier Inc.

Wijesekara, H., N. S. Bolan, M. Vithanage et al. 2016b. Utilization of biowaste for mine spoil rehabilitation.
Advances in Agronomy 138:97–173.

Wright, S. F., K. A. Nichols, and W. F. Schmidt. 2006. Comparison of efficacy of three extractants to solubilize
glomalin on hyphae and in soil. *Chemosphere* 64:1219–1224.

Zhu, L., J. Li, B. Tao et al. 2015. Effect of different fertilization modes on soil organic carbon sequestration
in paddy fields in South China: A meta-analysis. *Ecological Indicators* 53:144–153. doi: http://dx.doi.
org/10.1016/j.ecolind.2015.01.038.

5 Rehabilitation of Biological Characteristics in Mine Tailings

Longbin Huang and Fang You

CONTENTS

5.1 Introduction 75
5.2 Organic Matter as Substrates for Biological Activities in Tailings 77
5.3 Importance of Organic Matter as Biological Substrates in Soil and Tailings 79
    5.3.1 Soil Organic Matter Decomposition and Coupled Nutrient Cycling Processes 79
    5.3.2 Mechanisms Involved in SOM Stabilization, Organo–Mineral Interactions,
        and Aggregation 79
5.4 Microbial Community Composition and Functions in Tailings 80
    5.4.1 Extremophiles-Dominant Microbial Communities in Tailings 81
    5.4.2 Characterization of Microbial Community Composition and Activities 82
5.5 Biogeochemical Changes in Tailings Induced by Ecological Engineering Practices 84
5.6 Introduction of Native Microbial Inoculum for Rehabilitating Microbial
    Communities in Engineered Tailing-Soil 86
5.7 Conclusions 87
References 88

5.1 INTRODUCTION

Mining and extraction of economic minerals inevitably cause land degradation/destruction and the loss of biological properties required for sustainable land use post mining. Australia is a leading producer of minerals, producing at least 19 minerals (in significant amounts), and the production of copper (Cu), black coal, lead (Pb), zinc (Zn), and iron (Fe) ores has increased markedly over the last decade (Mudd 2010; Sutton and Dick 1987). From the mid-1800s to 2008, cumulative Cu, Pb, and Zn production (kt) in Australia reached 20,473; 37,945; and 48,465, respectively, in which their production in Queensland accounted for 50%–60% (Mudd 2010). Intensive mining and processing activities generate vast volumes of tailings that have been deposited in the environment. Within mined landscapes, mine wastes (e.g., tailings) represent the extremity of negative environmental impacts on the land, regardless of mining methods (e.g., opencut or underground mining). As a result, the present chapter has focused on biological characteristics of metal mine tailings and ecological engineering inputs required to rehabilitate soil-like biological properties and conditions, which underpin soil formation in the tailings for sustainable phytostabilization with native plant communities. The biological properties discussed here have mainly focused on key aspects of microbial communities and rhizosphere biology, and the processes associated with the development of key physical and chemical conditions in the tailings, which are closely coupled with biological properties.

Tailings are mine wastes and mineral residues resulting from extraction of metals via various combinations of mining processing methods. Briefly, metal-rich ores are milled to a fine particle size and separated by flotation into metal-rich concentrates and by-products of low and uneconomic grade.

These by-products, referred to as mine tailings, are usually deposited in purposely constructed dams or storage facilities, either in the form of slurry and wet sedimentation or dry stacking (Lottermoser 2010). In Australia, the mining industry was estimated to produce 1750 million tons (Mt) of mine waste per year (Lottermoser 2010).

Tailings typically contain various hazardous chemicals and minerals in significant quantities, which are biologically toxic and nothing like natural soil and lack basic physical and chemical properties minimally required to support the colonization of even extremely tolerant plant species. These toxic constituents include heavy metals, metalloids, radioactive elements, acids, and bases, all of which pose long-term threats to both environmental and human health (Järup 2003). Finely grained tailing particles and dusts may contaminate the surrounding area through wind dispersion and/or water erosion. Furthermore, sulfide-rich tailings are potential sources of metals and acidity resulting from oxidation and mineral weathering (Bobos, Durães, and Noronha 2006). High concentrations of heavy metals and acid mine drainage have not only been found in mined sites (Aykol et al. 2003) but also in local streams and waterways (Fields 2003).

Phytostabilization (surface stabilization with plant cover) has been advocated as a sustainable and cost-effective solution across mined landscapes including those occupied by mine wastes (Mendez and Maier 2008). However, successful phytostabilization for nonpolluting and sustainable outcomes is challenging in tailings landscapes due to the extremely unfavorable conditions for biological activities (Huang et al. 2012). Base metal mine (e.g., Cu, Pb, Zn) tailings typically lack adequate quantity and quality of organic matter, in addition to other toxicity factors (Ye et al. 2000). Fresh tailings consist of crushed rock containing large proportions of silt and fine silt, and thus lacking proper physical structure for soil hydraulic functions and root penetration. Following sedimentation, tailings have a high bulk density ($>1.8 \times 10^3$ kg m^{-3}) and mechanical resistance; lack water-stable aggregates and macropores; and have very high levels of salts, heavy metals (and/or metalloids), and potential acidity (which cannot be immediately and sufficiently neutralized as in acid-sulfate soils) (Dold and Fontboté 2002; Huang et al. 2012; Lottermoser 2010). The inherent ecotoxicity from these contaminants can prevent natural colonization of plants in tailings for many decades (Ye et al. 2000). Conventional practices and amendments (e.g., fertilizer, ripping, organic amendments, mulching, and/or admixing with lime or gypsum) target the primary constraints in tailings related to plant establishment and growth (Ram and Masto 2010). However, to rehabilitate essential biological properties in the tailings for ecosystem establishment, a systematic approach of engineered pedogenesis (soil and soil horizons) of tailings into functional technosols is necessary through purposely oriented and stimulated processes to achieve nonpolluting and sustainable phytostabilization across tailings landscapes (Huang et al. 2012, 2014; Li and Huang 2015).

Based on the knowledge from natural ecosystems, soil formation is the consequence of alteration and transformation of parent minerals from long-term abiotic and biotic interactions (Jenny 1994). Decades to millennia are required to alter earth rock into parent materials for soil, with plant colonization greatly accelerating the soil formation at the latter phases (van Breemen and Buurman 2002). Continuous organic matter transformation (decomposition, humification, biochemical and carbon and nitrogen cycles), principally from plants incorporated into the soil, greatly modifies soil physicochemical properties and functions and biological characteristics. Particularly, soil microorganisms are the most active agents in stabilizing soil physical structure and facilitating soil functions.

In contrast, tailings' properties are mainly inherited from the mineralogical profile of the original ores and hosting rocks. Their geochemistry from mineral processing are normally characterized with abundant reactive minerals such as pyrite, chalcopyrite, and high specific surface area (due to their fine particle size). At best, these can be classified as engineered or novel parent minerals subject to rapid oxidation and dissolution when exposed to air and water. The biological characteristics in mine tailings are in sharp contrast to natural soil, in terms of biomass-derived substrates (organic matter) and biological structure (microbial communities) and functions (biogeochemical processes), due to the hostile physico-(geo)chemical conditions. To improve biological properties in tailings, it is a prerequisite to enhance microbial-mediated bioweathering of primary minerals for hydro-geochemical stabilization, which is the priming process to initiate soil formation in tailings and the development of soil microbial communities. The rehabilitation of biogeochemical properties and functions requires buildup and stabilization of organic carbon, recovery of soil-like microbial community and functions, and rehabilitation of organic matter decomposition, and nutrient cycling processes (Huang et al. 2014; Li and Huang 2015). In the following section, comparative analysis and discussions are made about organic matter decomposition and microbial community composition in soil and tailings. Ecological engineering approaches have been proposed and discussed about how to rehabilitate soil-like biological properties and characteristics in tailings for sustainable phytostabilization and ecosystem restoration.

5.2 ORGANIC MATTER AS SUBSTRATES FOR BIOLOGICAL ACTIVITIES IN TAILINGS

Soil biological activities evolve around soil organic matter (SOM), which is both the product and substrate of biological activities. SOM includes the organic fraction of soil, including plant, animal, and microbial residues, in fresh forms and forms at all stages of decomposition, and the relatively resistant soil humus (Nelson and Sommers 1982). In soil, SOM is comprised mainly from inputs of plant litter, dead roots, and root exudates (Tate et al. 2000), which play a critical role in maintaining soil fertility (Torn, Vitousek, and Trumbore 2005) and regulating chemical environment and stabilizing physical structure (Oades 1984). SOM represents the dominant source of plant-available nutrients (e.g., N, P, S) in natural ecosystems (Gardenas et al. 2011). Also, functional groups in the SOM are the foundation of organo–mineral interactions and aggregation, thus contributing to improving soil physical structure and hydraulic properties (Tisdall and Oades 1982). Although accounting for only approximately 5%, SOM is the critical substrate in many soil biogeochemical processes (e.g., SOM mineralization and nutrient mobilization, organo–mineral interactions, cation exchange and adsorption, and pollutant removal) (Rovira and Vallejo 2002).

Depending on its turnover time and decomposition rate, total organic carbon (OC) can be categorized into labile, slow, and stable OC pools (Jenkinson 1990). As summarized in Table 5.1, the labile OC pool turns over relatively rapidly, including molecules readily consumed by soil organisms and readily soluble organic molecules (Marriott and Wander 2006). The relatively slow pool of OC is normally represented by physically or chemically protected OC, which has slower turnover rates than the labile pool (Denef et al. 2001). The stable OC pool is mainly composed of humic polymers, intercalated OC in microaggregations, organo–mineral complexes, and charcoal (Jones et al. 2005), which collectively contribute to the chemical buffering capacity and soil physical structure and hydraulic properties (Brodowski et al. 2007).

Metal mine tailings are normally comprised of finely ground particles (ranging from fine silt to fine sand) and consist of residue minerals of high reactivity, with little OC content (Lottermoser 2010). The initial level of total organic carbon (TOC) in copper (Cu) tailings can be less than 0.1% (Ye et al. 2002). The extremely low levels of TOC deprive the tailings from the potential to develop basic physical, chemical, and biological properties and functions as "soil" for sustainable plant communities. As a result, rehabilitation of SOM and OC stabilization is critical to soil biological activities and soil formation in tailings, which is required for the development of geochemical stability, hydraulic properties and physical structure, and nutrient cycling and supplying capacity in tailing-soils (or technosols from tailings). Although TOC in the tailings can be rapidly increased by direct input of organic amendments, the formation and distribution of various OC functional pools in tailings may be a complex process. As a result, it is necessary to understand the characteristics of SOM and OC forms and distribution to evaluate effective options of ecological engineering to speed up soil development in tailings. However, until recently there was limited information available concerning OC pools or

TABLE 5.1

Turnover, Composition, Function, and Representative Fractions and Fractionation Methods of Organic Carbon (OC) Pools

OC Pools	Turnover Time	Percentage (%)	Composition	Functions	Representative Fractions	Fractionation Methods
Labile pool	Months to years	1–5	Molecules readily removed from the soil by living organisms. Readily soluble molecules	Early indicator of soil C dynamics. Energy source for microorganisms	Water soluble organic carbon (WSOC). Free OC in light fraction (<1.6 g · cm^{-3}). OC in the coarse fraction (>250 μm)	Water extraction. Density fractionation. Particle-size fractionation
Slow pool	Decades	60–85	Chemically recalcitrant but moderately decomposable materials in the form of macro-aggregate	Important source of mineralisable nutrients. Physical and chemical soil properties	OC in the aggregate (53–250 μm). Mineral-associated OC soluble in hydrofluoric acid (<1.6 g · cm^{-3})	Particle-size fractionation. HF demineralization
Stable pool	Millennia	10–40	Polymerized substrates. Microaggregates, organo-mineral complexes. Charcoal	Formation of stable aggregates. Soil structure and chemical buffering capacity	Silt + clay particles associated OC (<53 μm). Mineral-associated OC in density fractions in sodium polytungstate or other heavy liquids/solutions (1.6–2.0 g · cm^{-3}). OC resistant to oxidation unless under high temperature and pressure (e.g., charcoal)	Particle-size fractionation. Density fractionation. Dumas high-temperature combustion (above 450°C)

Note: HF, hydrofluoric acid.

fractions in tailings and, especially, those under field conditions. Our own research findings show that the amount of decomposed organic matter (e.g., plant litters, tree mulch, and woodchips) admixed in Cu and Pb–Zn tailings was small, due to the limited biomass, diversity, and functions of organotrophic microbial communities in the toxic tailings (Li et al. 2014, 2015; You 2015).

5.3 IMPORTANCE OF ORGANIC MATTER AS BIOLOGICAL SUBSTRATES IN SOIL AND TAILINGS

5.3.1 Soil Organic Matter Decomposition and Coupled Nutrient Cycling Processes

In natural ecosystems, plant-available nutrients are mainly from SOM decomposition and nutrient cycling. Mineralization of 1.5%–3.5% of the organic nitrogen (N) would provide sufficient mineral N for the growth of natural vegetation in most soils except for those with low SOM (Kemmitt et al. 2008). Also, ecophysiological requirements of plant species are closely linked to the SOM decomposition and nutrient cycling processes (Wardle et al. 2004). Specifically, a fertile and productive ecosystem is often colonized by fast-growing, short-lived plants with high-quality litter (high N content, low lignin) returning to soil, and the rhizosphere microbial community is likely to be bacteria dominated, which enable rapid decomposition rates to supply nutrients for plants at relatively higher rates. In an infertile and less productive ecosystem, plant species are slow growing and long lived with low-quality litter, coupled with rhizosphere microbial communities with more fungi enabling relatively slower litter decomposition and nutrient supply rates for the plants. More than 85% of litter decomposition in natural soil is mediated by microorganisms (Zhang et al. 2008). Decomposers (e.g., bacteria, fungi) quickly utilize newly added organic materials and start the decomposition process with the aid of enzymes, hydrolyzing complex molecules and polymers to smaller molecules. During decomposition, there is initially a large decrease of water-soluble organic matter in litter/organic amendments, followed by a decline in cellulose and hemicellulose and an increase of lignin (Sina 2003). Chemical composition analysis suggests that the degree of decomposition can be indicated by declines of carbohydrates, increases of relative proportion of alkyl C and carboxyl C, and the breakdown of lignin. In addition, the C:N ratio decreases as decomposition proceeds, as N is retained in microbial biomass or by-products while C is respired into CO_2 (Quideau et al. 2000).

Enhancing the formation and accumulation of OC pools in tailings is a critical aspect of rehabilitating biological properties in tailings into biogeochemically functional technosols. Therefore, ascertaining SOM decomposition and nutrient cycling processes in tailing-soil and their linkage with plant species are fundamental to the adoption of necessary and effective ecological engineering options for developing functional tailing-soil (or technosol). This gives rise to the necessary properties and functions that permit sustainable growth and development of target plant species and communities for tailings phytostabilization.

5.3.2 Mechanisms Involved in SOM Stabilization, Organo–Mineral Interactions, and Aggregation

The formation of stable SOM in soil is important to soil structure and biological functions. It is widely recognized that several biotic and abiotic processes are involved in SOM stabilization. These include (1) chemical recalcitrance of organic matter with complex structure retarding its degradation by microorganisms (e.g., lignin, wax) (Marschner et al. 2008); (2) interactions between organic molecules with inorganic (e.g., Ca^{2+}, Mg^{2+}, Fe^{3+}, Al^{3+}) or other organic substances (Kleber, Sollins, and Sutton 2007); and (3) physical protection by forming barriers between microbes and enzymes to organic substance (e.g., aggregates, clay-occluded OM) (Six et al. 2002).

Two typical forces between organic matter and mineral surfaces are observed, including a strong force (i.e., ligand exchange, polyvalent cation bridges) and a weak force (i.e., hydrophobic interaction, H-bonding and van der Waals forces) (Kogel-Knabner et al. 2008). Among them, ligand exchange and polyvalent cation bridges are the dominant mechanisms to form stable organo–mineral complexes in soil. For instance, in kaolinite and montmorillonite, contribution to absorption ability of different interactions are in the following order: cation bridging (40%) > ligand exchange (33%) > van der Waals force (22%) (Feng, Simpson, and Simpson 2005). The binding mechanisms depend strongly on the surface chemistry of SOM and clay minerals. Among functional groups of SOM, carboxylic C with negative charges has the highest potential to bind minerals, followed by phenolic and hydroxylic C (Kumar et al. 2007), while alkyl and aromatic C, which are relatively nonpolar or hydrophobic, are little involved in organo–mineral interactions.

In particular, organo–mineral interactions are the basis for aggregation and soil structure stability (Shang and Tiessen 1998). Plant rootlets, fungal hyphae, cell filaments, and secretions from living organisms interact with mineral components to form clumps and increase the size and stability of aggregates (Denef et al. 2001). Pores formed between and within aggregates allow gas exchange (large pores: 30–60 μm) and retention of water and nutrients (small pores: 0.2–60 μm) (Skopp 1981). Several studies have reported the significantly positive impacts of organic inputs on aggregation (Kong et al. 2011).

Despite a much improved understanding on the organo–mineral interactions and aggregation in soil (Kaiser and Guggenberger 2003), significant gaps remain in our current understanding about SOM stabilization processes in the mine tailings, which is anticipated to be distinct from those in natural soils depending on tailings mineralogy and forms of extraneous organic inputs. Organic amendments may stimulate organo–mineral interactions and aggregation in tailings (Yuan et al. 2016). Therefore, understanding mechanisms of SOM stabilization and their linkage to aggregation in tailings is important to the physical structure improvement and recovery of biogeochemical functions in the process of technosol formation, which are closely related to the mineralogy and geochemistry and local climatic conditions.

5.4 MICROBIAL COMMUNITY COMPOSITION AND FUNCTIONS IN TAILINGS

SOM dynamics driven by microorganisms is indicative of biogeochemical linkages between engineered soil and plant systems, which is a useful indicator for assessing the success of tailings rehabilitation (Harris 2003, 2009). As microorganisms rely on energy from SOM decomposition or mineral oxidization and nutrients (e.g., N and P) to grow (Torsvik and Øvreås 2002), forms of SOM, minerals, and levels of nutrients have marked impacts on microbial community composition and associated biogeochemical processes. Other environmental variables (e.g., moisture, temperature, pH, osmotic conditions) may also significantly influence composition and functions of microbial communities in the tailings. Some optimum conditions for microbial communities in soil have been summarized in Table 5.2 (Madigan et al. 2006). In spite of the recognized

TABLE 5.2

Environmental Factors and Optimum Conditions for a Microbial Community

Environmental Factor	Optimum Conditions
Available soil moisture	25%–85% water holding capacity
Oxygen	>10% air-filled pore space for aerobic degradation
Redox potential	Eh > 50 milli volts (mV)
Nutrients	C:N:P = 120:10:1 molar ratio
pH	6.5–8.0
Temperature	20°C–30°C
Heavy metals	≤700 ppm

importance of diverse and functioning microbial communities for tailings phytostabilization, the evolution of microbial communities in the engineered tailing-soil remains poorly understood in the trajectory of tailings weathering and soil development, due to inadequate investment into concerted and long-term research trials at mine sites.

5.4.1 Extremophiles-Dominant Microbial Communities in Tailings

Tailings represent an extremely and geochemically dynamic environment with a strong selection pressure on microbial colonizers. The extremity and intensity of physical and geochemical stresses are closely related to the shift of structure and composition and functions of microbial communities, due to the loss and/or gain of species/function diversity (Schimel et al. 2007). Tailings are typical of low diversity of microbial species and functions, with low levels of microbial biomass, microbial activities, and energy utilization efficiency. For instance, in Cu tailings, total cell counts are around 10^9 g^{-1} (Diaby et al. 2007), much lower than that (10^{12} g^{-1}) in typical and healthy soils (Or et al. 2007). Also, species richness is very low, with only a few species abundant in Pb–Zn tailings (Mendez et al. 2008).

Because of the lack of SOM, microbial communities in tailings mainly are comprised of microorganisms (i.e., extremophiles) that tolerate extreme geochemical conditions and rely on chemical energy from primary and secondary minerals in the tailings (Li et al. 2014; Schimel et al. 2007). Among them, the most intensively studied microorganisms are related to S and Fe oxidation (e.g., *T. ferrooxidans*) (Fortin et al. 2000). Some recent molecular examinations of microbial communities in base metal mine tailings (Kock and Schippers 2008) suggest that in addition to low microbial biomass and diversity, some essential groups (e.g., fungi) may be absent (Kock and Schippers 2008; Li et al. 2014) and result in a very limited capacity to decompose complex organic compounds (e.g., lignin) (Blanchette 1995). Furthermore, tailings are normally characterized with much lower microbial metabolic activities than soils (Chen et al. 2005). The microbial limitation in the tailings significantly lowers the rate of bioweathering of unstable minerals and development of functional technosols within foreseeable timeframes (e.g., years–decades). As a result, ecological engineering options are required to stimulate microbial-mediated processes involved in mineral weathering and hydrogeochemical stabilization, which forms the foundation for the development of soil-like microbial communities in the tailings-soils.

Many sulfidic metal mine tailings, such as Cu-, Pb–Zn-, and Ni mine tailings, are biologically toxic and not hospitable for natural colonization of soil microorganisms and plants, because of their acidic pH conditions, elevated soluble metal concentrations, and acute heavy metal toxicities (Ortega-Larrocea et al. 2010). In extremely acidic tailings, all phylotypes identified are closely related to S- and Fe-oxidizing bacteria with a high degree of phylotype dominance (e.g., *Leptospirillus ferriphilum*, *Sulfobacillus*, *Acidimicrobium ferrooxidans*) (González-Toril et al. 2003). The relative degree of microbial diversity and community complexity increases with pH rise to slightly acidic conditions in the tailings, where microorganisms with neutrophilic growth preference are colonizing (e.g., *thiosulfate*, ferrous iron and arsenite oxidizers), thus permitting the recolonization of microbial communities resembling those in local natural soil (Mendez and Maier 2008).

Sulphur-/Fe-oxidizing bacteria are the most intensively investigated microbes, which mediated biogeochemical processes in sulfidic tailings, such as pyrite weathering and acid mine drainage formation (Fortin et al. 1995; Mielke et al. 2003). However, the most important microbial activities in soil are decomposition and turnover of structurally complex and highly diverse organic matter from plants and associated nutrient cycling (Berg 2000; Moretto, Distel, and Didoné 2001). Yet, to date, knowledge concerning the evolution of soil-like microbial communities and associated SOM decomposition is limited, except for a few studies on the monitoring of SOM decomposition and nutrient cycling processes in tailings (Hulshof et al. 2003; Moynahan et al. 2002).

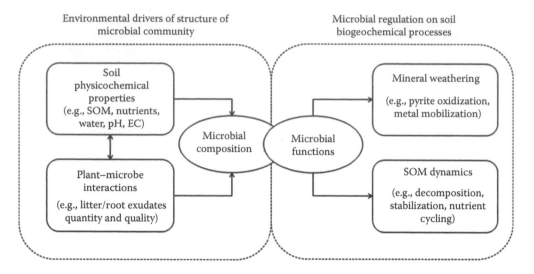

Environmental drivers of structure of microbial community

Microbial regulation on soil biogeochemical processes

FIGURE 5.1   A conceptual linkage between composition and functions of a microbial community.

Rehabilitating a soil-like habitat in tailings is required, in order to sustain diverse and functioning microbial communities, which underpin the development of biogeochemical functions and sustainable development of target plant communities. Microbial communities in tailings may be highly adapted to the changing physicochemical properties in the matrix and are stimulated by rhizosphere effects of plants introduced in the tailings subject to remediation by *in situ* engineering practices. Therefore, the associated biogeochemical processes may be greatly influenced, resulting in various levels of ecological changes (SOM content, metal and nutrient levels in its solution) in the context of soil development and formation (Figure 5.1). Detailed evaluation of biogeochemical dynamics in tailings in response to ecological engineering options is required in order to develop suitable options for stimulating recolonization and development of soil-like microbial communities and technosol with near-soil properties and conditions. In particular, detailed investigations need to be carried out to understand key environmental drivers of changes in microbial community composition and functions in the amended tailings, which form the basis for developing efficient ecological engineering strategies and options to engineer functional technosols for sustainable phytostabilization of the tailing landform in the mined landscape.

5.4.2   Characterization of Microbial Community Composition and Activities

Methods commonly applied to quantify microbial communities include direct enumeration, chloroform fumigation, and methods based on molecular technology (e.g., phospholipid fatty acid analysis (PLFA), 16S rRNA) at higher resolution (Table 5.3). Advancement in biotechnology has allowed the development of molecular techniques using biomarkers for measurements of the whole or selected parts of microbial communities, which have received significant development in the last decade (Chen et al. 2008; Li et al. 2014). Collectively this has enhanced knowledge of microbial communities in less known and complex soil ecosystems.

In general, methods based on molecular technology provide extensive information about taxa present in soil or tailings, but without providing the information about microbial species effectively and specifically facilitating biogeochemical processes in soil. It is common to combine microbial activities with microbial composition for microbial community characterization in complex systems. Depending on the biogeochemical processes concerned, various methods can be used to quantify microbial activities in soil or tailings (Table 5.4), such as microbial respiration, community-level

TABLE 5.3

Methods Applied to Quantify Microbial Community

Index	Methods Description	Advantages and Limitation	Reference
Microbial biomass	Fumigation extraction of organic C	Classical and reliable Both living and dead microorganisms will be extracted	Chotte, Ladd, and Amato (1998)
CFU	Dilution plating and culturing methods, less than 1% of the microorganism are culturable	Easy and economic operation Suitable for culturable microorganisms only	Vieira and Nahas (2005)
Phospholipid fatty acids (PLFAs) analysis	Phospholipid fatty acids have signature molecules presenting in all living cells. Specific fatty acid methyl esters are used as an accepted taxonomic discriminator for species identification.	Easily extractable molecules Reveal the presence and abundance of particular organisms Rely on chromatography of phospholipid component with medium resolution Not for species identification False signature under specific conditions	Frostegård, Tunlid, and Bååth (2011)
Nucleic acid techniques (16S rRNA)	DNA extraction, PCR amplification, and DGGE differentiation	High resolution Suitable for complex habitat Detect unknown species Multiple steps bias such as sample storage, extraction, and amplification Not suitable for large sample processing	Klindworth et al. (2012)
FISH	Fixed cell, 16S or 23S rRNA is hybridized with fluorescently labelled taxon-specific oligonucleotide probes, viewed with scanning confocal laser microscopy.	Direct identification and quantification of individual species and groups Not suitable to the nutrient-poor soils Familiar with sample for probe choice	Hill et al. (2000)

TABLE 5.4

Methods for Measurement of Microbial Activities

Index	Methods Description	Advantages and Limitation	Reference
Basal respiration	Carbon dioxide (CO_2) evolution	Low resolution Difficult to be separated from root respiration, OM decomposition	
Community-level physiological profiles	Based on BIOLOG system, quantify spectrophotometrically of color change	Suitable for culturable microorganisms Time lag bias resulting from microbial competition Does not match with real environment	Degens et al. (2001)
Enzyme activities	Substrate induced respiration, N mineralization, nitrification Potential denitrification activity, N-fixation	Specific processes	Alef and Nannipieri (1995)

physiological profiles based on the BIOLOG® system (Cookson et al. 2007; Garland 1997), or specific enzymatic activities (Burns 1982).

Respiration is commonly used as an integrated indicator to measure entire metabolic processes, based on CO_2 evolution with relatively low resolution. The BIOLOG approach combines both functional diversity and degradation rates and is suitable for culturable microorganisms, but sometimes results in bias from microbial competition (Degens et al. 2001). An enzyme assay provides information at a high resolution of biological processes related to biogeochemical functions such as C, N, and P mineralization. The activities of some of the enzymes measure entire metabolic processes (e.g., dehydrogenase), whereas others (e.g., invertase, cellulose, protease, urease, phosphatase) measure specific key processes involved in nutrient cycling, such as C, N, and P (Alef and Nannipieri 1995).

The above methods ranging from molecular to physiological analyses can be used to characterize microbial community composition and functions and greatly improve our understanding of the relationship among changes of microbial composition and functions and physicochemical changes induced by ecological engineering in tailings. The expected findings are critical to the development of ecological engineering options for *in situ* engineering of technosols from the tailings.

5.5 BIOGEOCHEMICAL CHANGES IN TAILINGS INDUCED BY ECOLOGICAL ENGINEERING PRACTICES

An organic amendment (OA) has commonly been used as one of the options to rehabilitate SOM and associated soil structure and functions in degraded/polluted soils as well as mine tailings (Huang et al. 2012; Ros et al. 2003; Tejada et al. 2006). Diverse microbial communities are often observed in tailings following OA (Pepper et al. 2012). The most commonly used OAs in tailings include crop residues and their compost; municipal and industrial wastes and manure; and uncomposted organic materials, such as sludge, plant residues, and biochar. All of them contain essential elements (e.g., C, N, P) with variable concentrations (Table 5.5), due to the differences in their origin and processing methods (Quilty and Cattle 2011). In general, compared to fertile OAs such as compost, manure, sludge, and plant residues, biochar is relatively low in nutrient quality, where low amounts of labile OC is readily available for microbial colonizers.

Physicochemical conditions in tailings can be greatly modified by OA. The major impacts of OAs on tailings properties and colonizing plants have been summarized in Table 5.6. The primary benefits of these OAs in tailings are the high OC content and plant nutrients. OA also improves nutrient retention capacity in tailings through significantly increasing CEC (Gardner et al. 2012;

TABLE 5.5
Physicochemical Properties of Typical Organic Amendments (OAs) Used for Tailings

Type of OAs	pH	EC (cm dm⁻¹)	CEC (cmol₊ kg⁻¹)	OC (%)	N (g kg⁻¹)	P (g kg⁻¹)
Compost	4.0–9.7	1.3–36	29–236.3	7.7–60.1	1.3–30.2	0.4–16.2
Manure	6.3–9.1	1.9–7.3	na	42.7–72.0	1.8–35.8	9.4–42.3
Sludge	4.8–7.8	0.27–16.0	18–33	28.1–48.4	6.8–65.0	5.2–48.6
Biochar	4.5–12.0	0.05–1.05	0.06–61.1	31–98	0.6–34.7	0.02–30.1
Plant residues	5.2–7.7	na	na	41.0–52.6	4.3–25.5	0.3–3.7

Data collected from Angın (2013); Baker, White, and Pierzynski (2011); Bolan, Baskaran, and Thiagarajan (1996); Chiu, Ye, and Wong (2006); Hoorens, Aerts, and Stroetenga (2003); Romero, Benítez, and Nogales (2005); Schwab, Zhu, and Banks (2007).

na, not available.

TABLE 5.6

Examples of Organic Amendments (OAs) Applied in Tailings and Impacts on Tailings and Plants

OA Types	Tailings Types	Duration	Impacts on Tailings	Impacts on Plants	Reference
Manure, compost, and biosolids	Pb/Zn/Cu	4 months	Increase N, P, K Decrease extractable Pb, Zn, Cu	Greater vegetation cover and dry weight yield	Chiu et al. (2006)
Compost, pig manure	Pb/Zn	1 year	Alter pH Increase TOC	Deeper root system Greater dry weight yield	Ye et al. (2000)
Domestic refuse	Pb/Zn	3 months	Decrease EC Increase TOC and macronutrients Decrease both total and extractable heavy metals	Greater vegetation cover and dry weight yield	Shu et al. (2002)
Biosolids	Cu	3 years	Decrease bulk density Increase water retention Increase EC, CEC, TOC and microbial activity	Increase biomass production; plant cover	Gardner et al. (2012)
Pine bark	Base metal	48 hours	Increase dissolved OC Increase water extractable metals	Not available	Munksgaard and Lottermoser (2010)
Mushroom compost	Pb/Zn	6 months	Improve physical and chemical status Decrease metal content	Increase dry weight yield	Jordan, Mullen, and Courtney (2008)
Paper sludge	Au	6 years	Slow decomposition	Not available	Cousins et al. (2009)
Biosolids	Cu	10 years	Not available	Noxious weed dominance	Borden and Black (2011)

Shu et al. 2002). Although biochar may not be beneficial to microorganisms in terms of nutrient supply (Fellet et al. 2011), it is able to retain nutrients because of its surface charge and area properties (Laird et al. 2010). Specifically, as intrinsic densities of organic materials are much lower than minerals, they are able to hold more water (Tester 1990). Thus, tailings receiving OA often have considerably lower bulk density and greater water holding capacity (WHC) compared with the tailings without OA (Brown, Enright, and Miller 2003).

Several studies suggest that OA immobilizes heavy metals through adsorption, complexation, reduction, and volatilization effects (Park et al. 2011). For instance, pig manure greatly decreased extractable concentrations of Pb, Zn, and Cd in tailings (Ye et al. 1999). However, the effects of OA on metal mobility, as reported in literature, are conflicting—possibly as a result from the interaction of several environmental variables (e.g., pH, CEC). Sludge, pine bark, and woodchips may even induce metal release in readily soluble forms when applied to tailings (Li et al. 2013). Besides, OA may increase EC in tailings due to elevated dissolution of minerals and salts, which may further exacerbate the levels of salinity in tailings at least in the short term, especially in semiarid areas (Chiu, Ye, and Wong 2006). Therefore, a comprehensive assessment is necessary for optimum utilization of OA in tailings to identify (1) OA-induced changes in physicochemical conditions in the tailings, (2) the underlying biogeochemical processes, (3) and potential impacts on physiological requirements of target plant species (see Wardle et al. 2004).

In the context of engineered pedogenesis of mine tailings, the selection of OA should focus on the effectiveness in stimulating the weathering of reactive minerals and speeding up hydro-geochemical stabilization as the first priority, which forms the basis for further improvement of biogeochemical capacity towards functional technosols. The perceived ecological linkages between soil and target plant subsystems must also be considered when selecting OA in relation to associated nutrient loads (such as N content), because soil fertility can influence the pattern of plant community development (Huang et al. 2012; Wardle et al. 2004).

5.6 INTRODUCTION OF NATIVE MICROBIAL INOCULUM FOR REHABILITATING MICROBIAL COMMUNITIES IN ENGINEERED TAILING-SOIL

Introducing native or cultured microbes through addition of soil, isolated microbes, and soil microbial extracts may speed up microbial community rehabilitation in tailings, which are also undergoing physicochemical changes induced by other amendments (e.g., OA) (Li et al. 2015; van de Voorde et al. 2012; You 2015). In previous studies, microbial inoculation was found to help establish diverse and functional microbial communities in tailings and improve plant growth (de-Bashan et al. 2010; Grandli et al. 2009; Li et al. 2015).

Inoculation of functional microbes such as plant growth promoting bacteria (PGPB) and mycorrhizal fungi has been found to be beneficial to root nutrient acquisition and plant tolerance of heavy metals (Alguacil et al. 2011; Ma et al. 2006). For instance, inoculating PGPB (e.g., N_2-fixing bacteria, phosphate and potassium solubilisers) enhanced plant growth in Pb–Zn tailings in an arid area and alleviated metal toxicity in plants (Grandlic et al. 2009; Wu et al. 2006). Many pot and field experiments have shown the potential of arbuscular mycorrhizal fungi (AMF) in facilitating plant establishment in tailings (Ma et al. 2006). In addition, mycorrhizal colonization of plant roots also helps reduce translocation of heavy metals to shoots by binding them to the cell walls of the fungal hyphae (Chen et al. 2007), but effectiveness of mycorrhizal colonization varies among the introduced fungal isolates (Orłowska et al. 2005).

Plant cover provides both intangible and tangible benefits in mine tailings landscapes, such as surface stabilization, land quality improvement, and increased biodiversity in tailings landscapes. Due to the presence of many physical and biogeochemical constraints, natural colonization of diverse native plants in tailings is largely unsuccessful and without significant physicochemical improvement, except for limited number of highly tolerant plant species (Huang et al. 2012; Shu et al. 2002). However, soil development in tailings may be stimulated by the colonization of tolerant pioneer plants if even only seasonally, because root turnover can provide inputs of organic matter and stimulate the colonization of rhizosphere microbial communities, in addition to root-induced improvement of physical structure in the tailings. For example, it is reported that initially low microbial diversity in nonvegetated tailings was rapidly increased after plant establishment and succession later (Alguacil et al. 2011). Even a relatively low plant cover is sufficient to stimulate microbial community recovery (Moynahan et al. 2002). It is worth noting that the establishment of pioneer plant species is far from the rehabilitation of target plant communities consisting of keystone species and expected species diversity, which might be achieved at a later stage during plant community succession.

The rhizosphere in the immediate vicinity of growing roots is known to promote microbial biomass, diversity, and activities, where roots produce organic exudates, including enzymes and organic compounds (Bais et al. 2006). Some root exudates (e.g., *isofalvones*) are able to attract beneficial bacteria (e.g., *Bradyrhizobium japonicum*), working as symbiotic signals to microorganisms in nutrient-poor soil (Dakora and Phillips 2002). For instance, legume plants routinely produce flavonoid molecules in root exudates to induce transcription of nodulation genes, leading to nodule formation and N_2 fixation, particularly in infertile soil (Dakora and Phillips 1996). Besides, root exudates stimulate the formation of mycorrhizal fungi association with roots, by inducing spore germination and/or hyphal growth in vesicular-arbuscular fungi, bringing about the overall improvement in the capacity of nutrient (particularly, P) and water acquisition (Gilbert et al. 2000). In addition, many low-weight organic compounds in root exudates may chelate cations to form organo–mineral complexes, thus reducing their availability to plants and microbes (Compant et al. 2005). As a result, the introduction of tolerant native plants as pioneer plants may greatly stimulate the development of rhizosphere microbial communities, in terms of biomass, diversity, and functions, which critically contribute to the development of functional technosols in the tailings.

Relative abundance of microorganisms and associated functions in the rhizosphere have been found to be species specific (Carrasco et al. 2010). Plants tend to actively select specific rhizosphere microorganisms to establish the habitat for themselves (Doornbos et al. 2012). Field-grown potato and wheat are associated with a distinct ascomycete community in the rhizosphere (Viebahn et al. 2005). The root system of *L. spartum* presents a higher cellulose content than that of *P. miliaceum*, favoring colonization and growth of fungi-producing extracellular cellulases (Carrasco et al. 2010). Symbiosis of AMF and hyphae with legumes (*Acacia* spp.) has been observed in many native acacia species in Australia (Herrera et al. 1993), while they are very low or absent in some species (e.g., *Ptilotus* spp., *Triodia* spp.) (Jasper et al. 1989). Much less is known about microbial communities in the rhizosphere of native plant species under field conditions, compared to those of crops.

Many metallophytes are recommended for tailings phytostabilization due to their tolerance of hostile habitat conditions, such as *Cynodon dactylon, Festuca rubra, Agrostis tenuis, Agrostis stolonifera, Typha latifolia*, and *Phragmites australis* (Archer and Caldwell 2004). However, due to the relatively high amount of metals in tailings, it is impossible to rely on hyperaccumulating plant species (e.g., accumulating Cd > 0.01%, Cu, Pb > 0.1% dry weight in plants) (van der Ent et al. 2013) to remove large quantities of metalloids and metals from metal mine tailings, even assuming these plants could grow adequate biomass. Gramineous grasses and legumes are generally the favorable options of pioneer plants for phytostabilization of tailing surface at the initial stage, due to their adaptation to nutrient deficiency and fast growing trait (Li 2006). Native plant species well adapted to local climatic conditions are preferred to be used as pioneer plants, without residual risks of weed proliferation and associated negative impacts on native plant species (Singh et al. 2002).

5.7 CONCLUSIONS

Biological characteristics in metal mine tailings are critical to soil formation and tailings rehabilitation. Engineered pedogenesis in metal mine tailings may be initiated and accelerated by microbially mediated weathering and transformation of minerals, which are considered to be fundamental indicators of pedogenesis, particularly those abundant in reactive primary minerals (e.g., pyrites) (Uzarowicz and Skiba 2011). Subsequently, soil-like structure and functions may be developed through coupled physical, chemical, and biological processes to stimulate soil formation in tailings (Remenant et al. 2009). Therefore, a comprehensive evaluation of pedogenesis in tailings requires clear characterization of mineralogical, hydrogeochemical, and biogeochemical processes, in response to various ecological engineering inputs (Huang et al. 2014).

When comparing the tailings with natural soils in terms of mineralogy, fertility, and microbial communities, the most significant differences are elevated metals of lithogenic origin, small amounts of clay minerals and organic matter, and stressed microbial communities dominated by extremophiles in base metal mine tailings (except for red mud) (Li and Huang 2015; Lottermoser 2010). Tailings are normally deficient in SOM due to the absence of plant cover or low productivity of plants. Soil organic matter levels in tailings tend to increase with the progress of vegetation establishment and development (Lorenz and Lal 2007; Ussiri et al. 2006). However, the efficiency of OC stabilization in the tailings seems to be low. For example, in Cu tailings rehabilitated more than 20 years with a dominant grass vegetation the levels of TOC were only 18% of those in soil samples from the reference site (Huang et al. 2011). Accumulation of organic matter is often reported as a primary pedogenetic process occurring in technosol formation (Hernández-Soriano et al. 2013). Different processes, such as solute (e.g., sulfates, carbonates) movement, aggregation involving minerals and organic materials, and changes in structure and hydrodynamics (Hartmann et al. 2010) have also been shown to occur during technosol formation. Ecological engineering inputs and direct/indirect interactions with natural climatic or environmental factors are expected to accelerate the processes of physical, chemical, and biological changes and influence technosol formation in the tailings.

Although advanced extraction technologies have extracted the majority of metals for profit purpose, the concentrations of heavy metals and metalloids (e.g., As, Cu, Mn, Pb, Cd, Zn) in tailings remain very high, which largely exceed the ecological investigation levels (Li and Huang 2015). There has been increasing evidence that phytostabilization of base metal mine tailings requires more knowledge about how to develop functional soil and root zones to support target plant communities, rather than simply relying on the unrealistic potential of hyperaccumulating plant species (Huang et al. 2012; Li and Huang 2015; Monserie et al. 2009). The pedogenesis on coal mining sites has been studied with pioneer plant cover, and the function of soil fauna and organic matter accumulation has been investigated (Hafeez et al. 2012; Novo et al. 2013). Yet, few studies have been conducted to understand the mineral transformation and translocation in the early stage of pedogenesis (Huot et al. 2014). It is apparent that the success of phytostabilization of tailings landscapes depends on the development of hydrogeochemically stable and biogeochemically functional tailing-soils (technosols) by utilizing effective ecological engineering options, in order to reach the expected economic and ecological sustainability necessary in the mining and mineral processing industry (Huang et al. 2014).

REFERENCES

Alef, K., and P. Nannipieri. 1995. *Methods in Applied Soil Microbiology and Biochemistry*. London, U.K.: Academic Press.

Alguacil, M.M., E. Torrecillas, J. Kohler, and A. Roldán. 2011. A molecular approach to ascertain the success of *"in situ"* AM fungi inoculation in the revegetation of a semiarid, degraded land. *Sci. Total Environ.* 409(15):2874–2880.

Angın, D. 2013. Effect of pyrolysis temperature and heating rate on biochar obtained from pyrolysis of safflower seed press cake. *Bioresour. Technol.* 128:593–597.

Archer, M.J.G., and R.A. Caldwell. 2004. Response of six Australian plant species to heavy metal contamination at an abandoned mine site. *Water Air Soil Pollut.* 157(1–4):257–267.

Aykol, A., M. Budakoglu, M. Kumral, A.H. Gultekin, M. Turhan, V. Esenli, F. Yavuz, and Y. Orgun. 2003. Heavy metal pollution and acid drainage from the abandoned Balya Pb–Zn sulfide Mine, NW Anatolia, Turkey. *Environ. Geol.* 45(2):198–208.

Bais, H.P., T.L. Weir, L.G. Perry, S. Gilroy, and J.M. Vivanco. 2006. The role of root exudates in rhizosphere interactions with plants and other organisms. *Annu. Rev. Plant Biol.* 57:233–266.

Baker, L.R., P.M. White, and G.M. Pierzynski. 2011. Changes in microbial properties after manure, lime, and bentonite application to a heavy metal-contaminated mine waste. *Appl. Soil Ecol.* 48(1):1–10.

Berg, B. 2000. Litter decomposition and organic matter turnover in northern forest soils. *Forest Ecol. Manag.* 133(1–2):13–22.

Blanchette, R.A. 1995. Degradation of the lignocellulose complex in wood. *Can. J. Bot.* 73(S1):999–1010.

Bobos, I., N. Durães, and F. Noronha. 2006. Mineralogy and geochemistry of mill tailings impounds from Algares (Aljustrel), Portugal: Implications for acid sulfate mine waters formation. *J. Geochem. Explor.* 88(1–3):1–5.

Bolan, N.S., S. Baskaran, and S. Thiagarajan. 1996. An evaluation of the methods of measurement of dissolved organic carbon in soils, manures, sludges, and stream water. *Commun. Soil Sci. Plant Anal.* 27(13–14):2723–2737.

Borden, R.K., and R. Black. 2011. Biosolids application and long-term noxious weed dominance in the Western United States. *Restor. Ecol.* 19(5):639–647.

Brodowski, S., W. Amelung, L. Haumaier, and W. Zech. 2007. Black carbon contribution to stable humus in German arable soils. *Geoderma* 139(1–2):220–228.

Brown, J., N.J. Enright, and B.P. Miller. 2003. Seed production and germination in two rare and three common co-occurring *Acacia* species from south-east Australia. *Aust. Ecol.* 28(3):271–280.

Burns, R.G. 1982. Enzyme activity in soil: Location and a possible role in microbial ecology. *Soil Biol. Biochem.* 14(5):423–427.

Carrasco, L., A. Gattinger, A. Fließbach, A. Roldán, M. Schloter, and F. Caravaca. 2010. Estimation by PLFA of microbial community structure associated with the rhizosphere of *Lygeum spartum* and *Piptatherum miliaceum* growing in semiarid mine tailings. *Microb. Ecol.* 60(2):265–271.

Chen, B.D., Y.G. Zhu, J. Duan, X.Y. Xiao, and S.E. Smith. 2007. Effects of the arbuscular mycorrhizal fungus *Glomus mosseae* on growth and metal uptake by four plant species in copper mine tailings. *Environ. Pollut.* 147(2):374–380.

Chen, G.L., M. Liao, and C. Huang. 2005. Effect of combined pollution by heavy metals on soil enzymatic activities in areas polluted by tailings from Pb–Zn–Ag mine. *J. Environ. Sci.* 17(4):637–640.

Chen, Y.Q., G.J. Ren, S.Q. An, Q.Y. Sun, C.H. Liu, and J.L. Shuang. 2008. Changes of bacterial community structure in copper mine tailings after colonization of reed (*Phragmites communis*). *Pedosphere* 18(6):731–740.

Chiu, K.K., Z.H. Ye, and M.H. Wong. 2006. Growth of *Vetiveria zizanioides* and *Phragmities australis* on Pb/Zn and Cu mine tailings amended with manure compost and sewage sludge: A greenhouse study. *Bioresour. Technol.* 97(1):158–170.

Chotte, J.L., J.N. Ladd, and M. Amato. 1998. Measurement of biomass C, N and ^{14}C of a soil at different water contents using a fumigation-extraction assay. *Soil Biol. Biochem.* 30(8–9):1221–1224.

Compant, S., B. Duffy, J. Nowak, C. Clément, and E.A. Barka. 2005. Use of plant growth-promoting bacteria for biocontrol of plant diseases: Principles, mechanisms of action, and future prospects. *Appl. Environ. Microb.* 71(9):4951–4959.

Cookson, W.R., M. Osman, P. Marschner et al. 2007. Controls on soil nitrogen cycling and microbial community composition across land use and incubation temperature. *Soil Biol. Biochem.* 39(3):744–756.

Cousins, C., G.H. Penner, B. Liu, P. Beckett, and G. Spiers. 2009. Organic matter degradation in paper sludge amendments over gold mine tailings. *Appl. Geochem.* 24(12):2293–2300.

Dakora, F.D., and D.A. Phillips. 1996. Diverse functions of isoflavonoids in legumes transcend anti-microbial definitions of phytoalexins. *Physiol. Mol. Plant Pathol.* 49(1):1–20.

Dakora, F.D., and D.A. Phillips. 2002. Root exudates as mediators of mineral acquisition in low-nutrient environments. *Plant Soil* 245(1):35–47.

de-Bashan, L.E., J.P. Hernandez, Y. Bashan, and R.M. Maier. 2010. *Bacillus pumilus* ES$_4$: Candidate plant growth-promoting bacterium to enhance establishment of plants in mine tailings. *Environ. Exp. Bot.* 69(3):343–352.

Degens, B.P., L.A. Schipper, G.P. Sparling, and L.C. Duncan. 2001. Is the microbial community in a soil with reduced catabolic diversity less resistant to stress or disturbance? *Soil Biol. Biochem.* 33(9):1143–1153.

Denef, K., J. Six, K. Paustian, and R. Merckx. 2001. Importance of macroaggregate dynamics in controlling soil carbon stabilization: Short-term effects of physical disturbance induced by dry-wet cycles. *Soil Biol. Biochem.* 33(15):2145–2153.

Diaby, N., B. Dold, H.R. Pfeifer, C. Holliger, D.B. Johnson, and K.B. Hallberg. 2007. Microbial communities in a porphyry copper tailings impoundment and their impact on the geochemical dynamics of the mine waste. *Environ. Microbiol.* 9(2):298–307.

Dold, B., and L. Fontboté. 2002. A mineralogical and geochemical study of element mobility in sulfide mine tailings of Fe oxide Cu–Au deposits from the Punta del Cobre belt, northern Chile. *Chem. Geol.* 189(3–4):135–163.

Doornbos, R.F., L.C. van Loon, and P.M. Bakker. 2012. Impact of root exudates and plant defense signaling on bacterial communities in the rhizosphere. A review. *Agron. Sustain. Dev.* 32(1):227–243.

Fellet, G., L. Marchiol, G.D. Vedove, and A. Peressotti. 2011. Application of biochar on mine tailings: Effects and perspectives for land reclamation. *Chemosphere* 83(9):1262–1267.

Feng, X., A.J. Simpson, and M.J. Simpson. 2005. Chemical and mineralogical controls on humic acid sorption to clay mineral surfaces. *Org. Geochem.* 36(11):1553–1566.

Fields, S. 2003. The earth's open wounds: Abandoned and orphaned mines. *Environ. Health Perspect.* 111(3):A154.

Fortin, D., B. Davis, G. Southam, and T.J. Beveridge. 1995. Biogeochemical phenomena induced by bacteria within sulfidic mine tailings. *J. Ind. Microbiol.* 14(2):178–185.

Fortin, D., M. Roy, J.P. Rioux, and P.J. Thibault. 2000. Occurrence of sulfate-reducing bacteria under a wide range of physico-chemical conditions in Au and Cu–Zn mine tailings. *FEMS Microbiol. Ecol.* 33(3):197–208.

Frostegård, Å., A. Tunlid, and E. Bååth. 2011. Use and misuse of PLFA measurements in soils. *Soil Biol. Biochem.* 43(8):1621–1625.

Gardenas, A.I., G.I. Agren, J.A. Bird et al. 2011. Knowledge gaps in soil carbon and nitrogen interactions—From molecular to global scale. *Soil Biol. Biochem.* 43 (4):702–717.

Gardner, W.C., M.A. Naeth, K. Broersma, D.S. Chanasyk, and A.M. Jobson. 2012. Influence of biosolids and fertilizer amendments on element concentrations and revegetation of copper mine tailings. *Can. J. Soil Sci.* 92(1):89–102.

Garland, J.L. 1997. Analysis and interpretation of community-level physiological profiles in microbial ecology. *FEMS Microbiol. Ecol.* 24(4):289–300.

Gilbert, G.A., J.D. Knight, C.P. Vance, and D.L. Allan. 2000. Proteoid root development of phosphorus deficient lupin is mimicked by auxin and phosphonate. *Ann. Bot.* 85(6):921–928.

González-Toril, E., E. Llobet-Brossa, E.O. Casamayor, R. Amann, and R. Amils. 2003. Microbial ecology of an extreme acidic environment, the Tinto River. *Appl. Environ. Microbiol.* 69(8):4853–4865.

Grandlic, C.J., M.W. Palmer, and R.M. Maier. 2009. Optimization of plant growth-promoting bacteria-assisted phytostabilization of mine tailings. *Soil Biol. Biochem.* 41(8):1734–1740.

Hafeez, F., F. Martin-Laurent, J. Béguet, D. Bru, J. Cortet, C. Schwartz, J.L. Morel, and L. Philippot. 2012. Taxonomic and functional characterization of microbial communities in Technosols constructed for remediation of a contaminated industrial wasteland. *J. Soil Sediments* 12(9):1396–1406.

Harris, J.A. 2003. Measurements of the soil microbial community for estimating the success of restoration. *Eur. J. Soil Sci.* 54(4):801–808.

Harris, J.A. 2009. Soil microbial communities and restoration ecology: Facilitators or followers? *Science* 325(5940):573–574.

Hartmann, P., H. Fleige, and R. Horn. 2010. Changes in soil physical properties of forest floor horizons due to long-term deposition of lignite fly ash. *J. Soil Sediments* 10(2):231–239.

Hernández-Soriano, M.C., A. Sevilla-Perea, B. Kerré, and M.D. Mingorance. 2013. Stability of organic matter in anthropic soils: A spectroscopic approach. In: *Soil Processes and Current Trends in Quality Assessment*, M.C. Hernandez-Soriano, ed. InTech, 444pp, 10.5772/55632. Available from: http://www.intechopen.com/books/. Accessed on June 29, 2016.

Herrera, M.A., C.P. Salamanca, and J.M. Barea. 1993. Inoculation of woody legumes with selected arbuscular mycorrhizal fungi and rhizobia to recover desertified Mediterranean ecosystems. *Appl. Environ. Microb.* 59(1):129–133.

Hill, G.T., N.A. Mitkowski, L. Aldrich-Wolfe, L.R. Emele, D.D. Jurkonie, A. Ficke, S. Maldonado-Ramirez, S.T. Lynch, and E.B. Nelson. 2000. Methods for assessing the composition and diversity of soil microbial communities. *Appl. Soil Ecol.* 15(1):25–36.

Hoorens, B., R. Aerts, and M. Stroetenga. 2003. Does initial litter chemistry explain litter mixture effects on decomposition? *Oecologia* 137(4):578–586.

Huang, L., T. Baumgartl, and D. Mulligan. 2012. Is rhizosphere remediation sufficient for sustainable revegetation of mine tailings? *Ann. Bot.* 110(2):223–238.

Huang, L., T. Baumgartl, L. Zhou, and D. Mulligan. 2014. The new paradigm for phytostabilising mine wastes—Ecologically engineered pedogenesis and functional root zones. Paper read at *Life-of-Mine 2014 Conference*, Brisbane, Queensland, Australia, July 16–18, 2014.

Huang, L., F. Tang, Y. Song, C. Wan, S. Wang, W. Liu, and W. Shu. 2011. Biodiversity, abundance, and activity of nitrogen-fixing bacteria during primary succession on a copper mine tailings. *FEMS Microbiol. Ecol.* 78(3):439–450.

Hulshof, A.H.M., D.W. Blowes, C.J. Ptacek, and W.D. Gould. 2003. Microbial and nutrient investigations into the use of in situ layers for treatment of tailings effluent. *Environ. Sci. Technol.* 37(21):5027–5033.

Huot, H., M.O. Simonnot, F. Watteau, P. Marion, J. Yvon, P. De Donato, and J.L. Morel. 2014. Early transformation and transfer processes in a Technosol developing on iron industry deposits. *Eur. J. Soil Sci.* 65(4):470–484.

Järup, L. 2003. Hazards of heavy metal contamination. *Br. Med. Bull.* 68(1):167–182.

Jasper, D.A., L.K. Abbott, and A.D. Robson. 1989. Acacias respond to additions of phosphorus and to inoculation with VA mycorrhizal fungi in soils stockpiled during mineral sand mining. *Plant Soil* 115(1):99–108.

Jenkinson, D.S. 1990. The turnover of organic-carbon and nitrogen in soil. *Philos. Trans. R. Soc. Lond. Ser. B Biol. Sci.* 329(1255):361–368.

Jenny, H. ed. 1994. *Factors of Soil Formation : A System of Quantitative Pedology*. Toronto, Ontario, Canada: General Publishing Company, Ltd.

Jones, D.L., S.J. Kemmitt, D. Wright, S.P. Cuttle, R. Bol, and A.C. Edwards. 2005. Rapid intrinsic rates of amino acid biodegradation in soils are unaffected by agricultural management strategy. *Soil Biol. Biochem.* 37(7):1267–1275.

Jordan, S.N., G.J. Mullen, and R.G. Courtney. 2008. Utilization of spent mushroom compost for the revegetation of lead-zinc tailings: Effects on physico-chemical properties of tailings and growth of *Lolium perenne*. *Bioresour. Technol.* 99(17):8125–8129.

Kaiser, K., and G. Guggenberger. 2003. Mineral surfaces and soil organic matter. *Eur. J. Soil Sci.* 54(2):219–236.

Kemmitt, S.J., C.V. Lanyon, I.S. Waite, Q. Wen, T.M. Addiscott, N.R.A. Bird, A.G. O'Donnell, and P.C. Brookes. 2008. Mineralization of native soil organic matter is not regulated by the size, activity or composition of the soil microbial biomass—A new perspective. *Soil Biol. Biochem.* 40(1):61–73.

Kleber, M., P. Sollins, and R. Sutton. 2007. A conceptual model of organo mineral interactions in soils: Self-assembly of organic molecular fragments into zonal structures on mineral surfaces. *Biogeochemistry* 85(1):9–24.

Klindworth, A., E. Pruesse, T. Schweer, J. Peplies, C. Quast, M. Horn, and F.O. Glöckner. 2012. Evaluation of general 16S ribosomal RNA gene PCR primers for classical and next-generation sequencing-based diversity studies. *Nucl. Acids Res.* 41(1):1–11.

Kock, D., and A. Schippers. 2008. Quantitative microbial community analysis of three different sulfidic mine tailing dumps generating acid mine drainage. *Appl. Environ. Microb.* 74(16):5211–5219.

Kogel-Knabner, I., K. Ekschmitt, H. Flessa, G. Guggenberger, E. Matzner, B. Marschner, and M. von Lützow. 2008. An integrative approach of organic matter stabilization in temperate soils: Linking chemistry, physics, and biology. *J. Plant Nutr. Soil Sci.* 171(1):5–13.

Kong, A.Y.Y., K.M. Scow, A.L. Córdova-Kreylos, W.E. Holmes, and J. Six. 2011. Microbial community composition and carbon cycling within soil microenvironments of conventional, low-input, and organic cropping systems. *Soil Biol. Biochem.* 43(1):20–30.

Kumar, P.P., A.G. Kalinichev, and R.J. Kirkpatrick. 2007. Molecular dynamics simulation of the energetics and structure of layered double hydroxides intercalated with carboxylic acids. *J. Phys. Chem. C* 111(36):13517–13523.

Laird, D., P. Fleming, B. Wang, R. Horton, and D. Karlen. 2010. Biochar impact on nutrient leaching from a Midwestern agricultural soil. *Geoderma* 158(3):436–442.

Li, M.S. 2006. Ecological restoration of mineland with particular reference to the metalliferous mine wasteland in China: A review of research and practice. *Sci. Total Environ.* 357(1–3):38–53.

Li, X., and L. Huang. 2015. Toward a new paradigm for tailings phytostabilization—Nature of the substrates, amendment options, and anthropogenic pedogenesis. *Crit. Rev. Environ. Sci. Technol.* 45(8):813–839.

Li, X., L. Huang, P.L. Bond, Y. Lu, and S. Vink. 2014. Bacterial diversity in response to direct revegetation in the Pb–Zn–Cu tailings under subtropical and semi-arid conditions. *Ecol. Eng.* 68:233–240.

Li, X., F. You, P.L. Bond, and L. Huang. 2015. Establishing microbial diversity and functions in weathered and neutral Cu–Pb–Zn tailings with native soil addition. *Geoderma* 247–248:108–116.

Li, X., F. You, L. Huang, E. Strounina, and M. Edraki. 2013. Dynamics in leachate chemistry of Cu–Au tailings in response to biochar and woodchip amendments: A column leaching study. *Environ. Sci. Eur.* 25(1):32.

Lorenz, K., and R. Lal. 2007. Stabilization of organic carbon in chemically separated pools in reclaimed coal mine soils in Ohio. *Geoderma* 141(3–4):294–301.

Lottermoser, B.G. 2010. *Mine Wastes: Characterization, Treatment and Environmental Impacts*. Berlin, Germany: Springer.

Ma, Y., N.M. Dickinson, and M.H. Wong. 2006. Beneficial effects of earthworms and arbuscular mycorrhizal fungi on establishment of leguminous trees on Pb/Zn mine tailings. *Soil Biol. Biochem.* 38(6):1403–1412.

Madigan, M.T., J.M. Martinko, and J. Parker. 2006. *Brock Biology of Microorganisms*, 11th edn. Upper Saddle River, NJ: Prentice Hall.

Marriott, E.E., and M.M. Wander. 2006. Total and labile soil organic matter in organic and conventional farming systems. *Soil Sci. Soc. Am. J.* 70(3):950–959.

Marschner, B., S. Brodowski, A. Dreves et al. 2008. How relevant is recalcitrance for the stabilization of organic matter in soils? *J. Plant Nutr. Soil Sci.* 171(1):91–110.

Mendez, M., and R. Maier. 2008. Phytoremediation of mine tailings in temperate and arid environments. *Rev. Environ. Sci. Biotechnol.* 7(1):47–59.

Mendez, M.O., J.W. Neilson, and R.M. Maier. 2008. Characterization of a bacterial community in an abandoned semiarid lead-zinc mine tailing site. *Appl. Environ. Microb.* 74(12):3899–3907.

Mielke, R.E., D.L. Pace, T. Porter, and G. Southam. 2003. A critical stage in the formation of acid mine drainage: Colonization of pyrite by *Acidithiobacillus ferrooxidans* under pH-neutral conditions. *Geobiology* 1(1):81–90.

Monserie, M., F. Watteau, G. Villemin, S. Ouvrard, and J. Morel. 2009. Technosol genesis: Identification of organo-mineral associations in a young Technosol derived from coking plant waste materials. *J. Soil Sediments* 9(6):537–546.

Moretto, A.S., R.A. Distel, and N.G. Didoné. 2001. Decomposition and nutrient dynamic of leaf litter and roots from palatable and unpalatable grasses in a semi-arid grassland. *Appl. Soil Ecol.* 18(1):31–37.

Moynahan, O.S., C.A. Zabinski, and J.E. Gannon. 2002. Microbial community structure and carbon-utilization diversity in a mine tailings revegetation study. *Restor. Ecol.* 10(1):77–87.

Mudd, G.M. 2010. The environmental sustainability of mining in Australia: Key mega-trends and looming constraints. *Resour. Policy* 35(2):98–115.

Munksgaard, N.C., and B.G. Lottermoser. 2010. Effects of wood bark and fertilizer amendment on trace element mobility in mine soils, Broken Hill, Australia: Implications for mined land reclamation. *J. Environ. Qual.* 39(6):2054–2062.

Nelson, D.W., and L.E. Sommers. 1982. Total carbon, organic carbon, and organic matter. In: *Methods of Soil Analysis. Part 2. Chemical and Microbiological Properties*, A.L. Page, ed. Madison, WI: American Society of Agronomy.

Novo, L.A.B., E.F. Covelo, and L. González. 2013. Phytoremediation of amended copper mine tailings with *Brassica juncea. Int. J. Min. Reclam. Environ.* 27(3):215–226.

Oades, J.M. 1984. Soil organic matter and structural stability: Mechanisms and implications for management. *Plant Soil* 76(1–3):319–337.

Or, D., B.F. Smets, J.M. Wraith, A. Dechesne, and S.P. Friedman. 2007. Physical constraints affecting bacterial habitats and activity in unsaturated porous media—A review. *Adv. Water Resour.* 30(6–7):1505–1527.

Orłowska, E., P. Ryszka, A. Jurkiewicz, and K. Turnau. 2005. Effectiveness of arbuscular mycorrhizal fungal (AMF) strains in colonisation of plants involved in phytostabilisation of zinc wastes. *Geoderma* 129(1–2):92–98.

Ortega-Larrocea, M.P., B. Xoconostle-Cázares, I.E. Maldonado-Mendoza, R. Carrillo-González, J. Hernández-Hernández, M.D. Garduño, M. López-Meyer, L. Gómez-Flores, and M.C.A. González-Chávez. 2010. Plant and fungal biodiversity from metal mine wastes under remediation at Zimapan, Hidalgo, Mexico. *Environ. Pollut.* 158(5):1922–1931.

Park, J.H., D. Lamb, P. Paneerselvam, G. Choppala, N. Bolan, and J. Chung. 2011. Role of organic amendments on enhanced bioremediation of heavy metal(loid) contaminated soils. *J. Hazard. Mater.* 185(2–3):549–574.

Pepper, I.L., H.G. Zerzghi, S.A. Bengson, B.C. Iker, M.J. Banerjee, and J.P. Brooks. 2012. Bacterial populations within copper mine tailings: Long-term effects of amendment with Class A biosolids. *J. Appl. Microbiol.* 113(3):569–577.

Quideau, S.A., M.A. Anderson, R.C. Graham, O.A. Chadwick, and S.E. Trumbore. 2000. Soil organic matter processes: Characterization by ^{13}C NMR and ^{14}C measurements. *Forest Ecol. Manag.* 138(1–3):19–27.

Quilty, J.R., and S.R. Cattle. 2011. Use and understanding of organic amendments in Australian agriculture: A review. *Soil Res.* 49(1):1.

Ram, L.C., and R.E. Masto. 2010. An appraisal of the potential use of fly ash for reclaiming coal mine spoil. *J. Environ. Manag.* 91:603–617.

Remenant, B., G.L. Grundmann, and L. Jocteur-Monrozier. 2009. From the micro-scale to the habitat: Assessment of soil bacterial community structure as shown by soil structure directed sampling. *Soil Biol. Biochem.* 41(1):29–36.

Romero, E., E. Benítez, and R. Nogales. 2005. Suitability of wastes from olive-oil industry for initial reclamation of a Pb/Zn mine tailing. *Water Air Soil Pollut.* 165(1–4):153–165.

Ros, M., M.T. Hernandez, and C. García. 2003. Soil microbial activity after restoration of a semiarid soil by organic amendments. *Soil Biol. Biochem.* 35(3):463–469.

Rovira, P., and V.R. Vallejo. 2002. Labile and recalcitrant pools of carbon and nitrogen in organic matter decomposing at different depths in soil: An acid hydrolysis approach. *Geoderma* 107(1–2):109–141.

Schimel, J., T.C. Balser, and M. Wallenstein. 2007. Microbial stress response physiology and its implications for ecosystem function. *Ecology* 88(6):1386–1394.

Schwab, P., D. Zhu, and M.K. Banks. 2007. Heavy metal leaching from mine tailings as affected by organic amendments. *Bioresour. Technol.* 98(15):2935–2941.

Shang, C., and H. Tiessen. 1998. Organic matter stabilization in two semiarid tropical soils: Size, density, and magnetic separations. *Soil Sci. Soc. Am. J.* 62(5):1247–1257.

Shu, W.S., H.P. Xia, Z.Q. Zhang, C.Y. Lan, and M.H. Wong. 2002. Use of Vetiver and three other grasses for revegetation of Pb/Zn mine tailings: Field experiment. *Int. J. Phytoremediat.* 4(1):47–57.

Sina, M.A. 2003. *The Ecology of Soil Decomposition*. Oxon, U.K.: CABI Publishing.

Singh, A.N., A.S. Raghubanshi, and J.S. Singh. 2002. Plantations as a tool for mine spoil restoration. *Curr. Sci. India* 82(12):1436–1441.

Six, J., P. Callewaert, S. Lenders, S. De Gryze, S.J. Morris, E.G. Gregorich, E.A. Paul, and K. Paustian. 2002. Measuring and understanding carbon storage in afforested soils by physical fractionation. *Soil Sci. Soc. Am. J.* 66(6):1981–1987.

Skopp, J. 1981. Comment on "Micro-, meso-, and macroporosity of soil". *Soil Sci. Soc. Am. J.* 45(6):1246.

Sutton, P., and W.A. Dick. 1987. Reclamation of acidic mined lands in humid areas. *Adv. Agron.* 41:377–405.

Tate, K.R., N.A. Scott, D.J. Ross, A. Parshotam, and J.J. Claydon. 2000. Plant effects on soil carbon storage and turnover in a montane beech (*Nothofagus*) forest and adjacent tussock grassland in New Zealand. *Soil Res.* 38(3):685–697.

Tejada, M., C. Garcia, J.L. Gonzalez, and M.T. Hernandez. 2006. Use of organic amendment as a strategy for saline soil remediation: Influence on the physical, chemical and biological properties of soil. *Soil Biol. Biochem.* 38(6):1413–1421.

Tester, C.F. 1990. Organic amendment effects on physical and chemical properties of a sandy soil. *Soil Sci. Soc. Am. J.* 54(3):827–831.

Tisdall, J.M., and J.M. Oades. 1982. Organic matter and water-stable aggregates in soils. *J. Soil Sci.* 33(2):141–163.

Torn, M.S., P.M. Vitousek, and S.E. Trumbore. 2005. The influence of nutrient availability on soil organic matter turnover estimated by incubations and radiocarbon modeling. *Ecosystems* 8(4):352–372.

Torsvik, V., and L. Øvreås. 2002. Microbial diversity and function in soil: From genes to ecosystems. *Curr. Opin. Microbiol.* 5(3):240–245.

Ussiri, D.A.N., R. Lal, and P.A. Jacinthe. 2006. Soil properties and carbon sequestration of afforested pastures in reclaimed minesoils of Ohio. *Soil Sci. Soc. Am. J.* 70(5):1797–1806.

Uzarowicz, L., and S. Skiba. 2011. Technogenic soils developed on mine spoils containing iron sulphides: Mineral transformations as an indicator of pedogenesis. *Geoderma* 163(1–2):95–108.

van Breemen, N., and P. Buurman. 2002. *Soil Formation*. Dordrecht, the Netherlands: Kluwer Academic Press.

van de Voorde, T.F.J., W.H. van der Putten, and T.M. Bezemer. 2012. Soil inoculation method determines the strength of plant–soil interactions. *Soil Biol. Biochem.* 55:1–6.

van der Ent, A., A.J.M. Baker, R.D. Reeves, A.J. Pollard, and H. Schat. 2013. Hyperaccumulators of metal and metalloid trace elements: Facts and fiction. *Plant Soil* 362(1–2):319–334.

Viebahn, M., C. Veenman, K. Wernars, L.C. van Loon, E. Smit, and P.A.H.M. Bakker. 2005. Assessment of differences in ascomycete communities in the rhizosphere of field-grown wheat and potato. *FEMS Microbiol. Ecol.* 53(2):245–253.

Vieira, F.C.S., and E. Nahas. 2005. Comparison of microbial numbers in soils by using various culture media and temperatures. *Microbiol. Res.* 160(2):197–202.

Wardle, D.A., R.D. Bardgett, J.N. Klironomos, H. Setala, W.H. van der Putten, and D.H. Wall. 2004. Ecological linkages between aboveground and belowground biota. *Science* 304(5677):1629–1633.

Wu, S.C., K.C. Cheung, Y.M. Luo, and M.H. Wong. 2006. Effects of inoculation of plant growth-promoting rhizobacteria on metal uptake by *Brassica juncea*. *Environ. Pollut.* 140(1):124–135.

Ye, Z.H., W.S. Shu, Z.Q. Zhang, C.Y. Lan, and M.H. Wong. 2002. Evaluation of major constraints to revegetation of lead/zinc mine tailings using bioassay techniques. *Chemosphere* 47(10):1103–1111.

Ye, Z.H., J.W.C. Wong, M.H. Wong, A.J.M. Baker, W.S. Shu, and C.Y. Lan. 2000. Revegetation of Pb/Zn mine tailings, Guangdong province, China. *Restor. Ecol.* 8(1):87–92.

Ye, Z.H., J.W.C. Wong, M.H. Wong, C.Y. Lan, and A.J.M. Baker. 1999. Lime and pig manure as ameliorants for revegetating lead/zinc mine tailings: A greenhouse study. *Bioresour. Technol.* 69(1):35–43.

You, F. 2015. Rehabilitation of organic carbon and microbial community structure and functions in Cu–Pb–Zn mine tailings for in situ engineering technosols. PhD thesis, Sustainable Minerals Institute, The University of Queensland, St. Lucia, Queensland, Australia.

Yuan, M., Z.P. Xu, T. Baumgartl, and L. Huang. 2016. Organic amendment and plant growth improved aggregation in Cu/Pb–Zn tailings. *Soil Sci. Soc. Am. J.* 80(1):27–37.

Zhang, W.J., H.A. Xiao, C.L. Tong et al. 2008. Estimating organic carbon storage in temperate wetland profiles in Northeast China. *Geoderma* 146(1–2):311–316.

6 Nanoscale Materials for Mine Site Remediation

Tapan Adhikari and Rajarathnam Dharmarajan

CONTENTS

6.1 Introduction 95
6.2 Conventional Remediation Techniques 97
    6.2.1 Physical Remediation 97
    6.2.2 Chemical Remediation 97
        6.2.2.1 Chemical Leaching 97
        6.2.2.2 Chemical Fixation 97
        6.2.2.3 Electrokinetic Remediation 98
        6.2.2.4 Vitrification Technique 98
    6.2.3 Biological Remediation 98
        6.2.3.1 Phytoremediation 98
        6.2.3.2 Microbial Remediation 99
    6.2.4 Animal Remediation 99
6.3 Nanoscale Materials and Processing 99
6.4 Nanoscale Materials and Their Applications for Remediation 101
    6.4.1 Nano-Remediation of Organic Contaminants 101
        6.4.1.1 Nanometal and Metal Oxide Particles 101
        6.4.1.2 Carbon Nanotubes 102
    6.4.2 Nano-Remediation of Inorganic Contaminants 102
        6.4.2.1 Nano Zero-Valent Iron and Nano Iron Oxide
                  Particles (nZVI (Fe^0)/nFeO) 103
        6.4.2.2 Zeolites 104
        6.4.2.3 Phosphate-Based Nanoparticles 104
        6.4.2.4 Nano Iron Sulfide Particles 104
        6.4.2.5 Carbon Nanotubes 105
6.5 Advantages and Disadvantages of Nanoscale Materials 105
6.6 Conclusions 106
References 106

6.1 INTRODUCTION

In the era of global competition, mineral exploitation has been significantly increased resulting in pressure on the environment in the form of massive deforestation, soil pollution, and erosion. Despite global economic importance, mineral industries have adversely affected the ecosystems across the world. The impact of mine waste in soil depends on its type and composition, commodity being mined, type of ore, and technologies used to process the ore. Mining types and activities are several, which include surface mining, underground mining, openpit mining, in situ mining, pillar mining, slope mining, block caving, and quarrying. And thus mine waste materials vary in their physical and chemical composition and potential for soil contamination. The different types of mine waste materials are overburden, waste rock, tailings, slags, mine water, sludge, and gaseous wastes. Overburden includes the soil and rock that are removed to gain access to the ore deposits at openpit mines. It is usually dumped on the surface at mine sites where it will not hinder further expansion of the mining operation. Waste rock contains minerals in concentrations considered too low to be extracted at a profit. It is often stored in heaps on the mine site. Tailings are finely ground rock and mineral waste products of mineral processing operations. They also contain leftover processing chemicals, and usually are deposited in the form of water-based slurry into tailings ponds. Slags are nonmetallic by-products from metal smelting. Mine water is produced in a number of ways at mine sites and varies in its quality and potential for environmental contamination. Sludge is produced at active water treatment plants used at some mine sites and consists of the solids that have been removed from the water as well as any chemicals. Gaseous wastes are produced during high-temperature chemical processing such as smelting, and consist of particulate matter and oxides of sulfur.

The aftermath of mining work thus leads to generous mining wastes that significantly change the land forms on the surface. These waste products interrupt the aesthetic value of the land and different components of soil, such as soil structure and horizons, microbial populations, and nutrient cycling phenomena, which are vital for supporting a healthy soil ecosystem. Subsequently they create an environmentally challenging situation to society, and hence it is always preferred that the original conditions are reestablished. Mine wasteland space includes the exposed uncovered area, slack soil heaps, ravage rock and overload surfaces, subsided land areas, and other despoiled land by mining services, among which the ravage rocks often create extremely unfavorable conditions against reestablishment back to pristine conditions. The overburden waste nearby mining areas has several undesirable factors, namely, high bioavailable toxic heavy metals, enhanced sand content, moisture scarcity, augmented compaction, and reduced organic matter content. Adverse effects of mine dumps to ecosystems are many and include soil erosion, soil/water pollution, geoenvironmental calamities, disturbed biodiversity, and subsequent destruction of socioeconomic conditions.

In metal mining industries, more than 90% of ore is rejected as tailings after recovery of the target metal, and they are stored in tailings impoundments. These tailings often contain substantial quantities of other metals whose amounts are well below the cut-off grade. These metals cannot be recovered by conventional metallurgical process and remain in the tailings as waste. In the mining industry, ore-processing activities for extraction of metal generate mill tailings that are the primary component of mine waste and a major source of soil and water pollution. Tailings consist of mostly coarse (silt or sand) sized particles; they lack major nutrients and are devoid of organic matter. Under these situations, mine tailings are biotoxic due to their low pH and highly toxic heavy metal contents. As a result, most of the tailings disposal sites are converted into barren land, and devoid of heterotrophic microbial communities. However, this type of situation also aggravates metal leaching phenomena during the rainy season that ultimately causes acid mine drainage (AMD), which badly affects the surrounding surface and ground water. AMD is a widespread, acknowledged environmental problem that causes adverse impacts on soil ecology through acidification, elevated levels of iron, sulfate, and soluble toxic metals.

Being a nonrenewable resource, the high grade minerals that have been exploited in the past by mining sectors cannot be replenished. Hence, during the last decade the focus has shifted towards exploitation of low-grade minerals, along with geological surveys to identify more reserves. Owing to diversity in composition of minerals and their low metal contents, waste disposal management subsequently turns out to be a tedious task. As the mineral industries keep a hawk's eye on the day-to-day production, recovery, reserve of ore, and waste disposal, the estimation of an adverse impact on the soil environment and its remediation are also increasingly needed through the exploitation of the latest available technology. To comprehensively and effectively address the adverse impacts of mine spoil, a holistic approach should be adopted for restoration of original soil conditions in tailing embankments and waste dumping sites. Given such drastic conditions, mining industries are placing much emphasis on establishing appropriate remediation techniques and measures to mitigate the adverse impacts from mining activities to the environment, besides saving costs.

Remediation is a process to restore the ecological integrity of the spoilt and derelict mine sites. It includes management of all types of physical, chemical, and biological properties of soil, such as soil texture, soil aggregation, water-holding capacity, compaction, land slope, soil color, soil pH, soil fertility, microbial diversity, and various soil nutrient cycles that make the degraded soil productive. Remediation techniques for mine tailings or mine spoil soil have grown and evolved over the past years through the adoption of new technologies, thus enhancing the existing remediation processes. However, current active remediation technologies are tedious, demanding cumbersome procedures, and are often costly affairs due to the need of routine maintenance. But with the advent of nanotechnology, the remediation process based on a nanotechnology platform is considered as more plausible, which may be explored for the decontamination of mine spoil soil. Nanotechnology is the creation, synthesis, study, manipulation, design, and application of devices, functional materials, and systems through control of matter at the nanometric scale (1–100 nm). At atomic and molecular levels, fabricated nano materials have unique physicochemical properties that can be exploited for multiple technological applications including mine-site remediation.

6.2 CONVENTIONAL REMEDIATION TECHNIQUES

6.2.1 Physical Remediation

Thermal desorption and soil replacement methods are the two physical remediation techniques that are mainly followed. Thermal desorption is based on the contaminants' volatility, and heating of the contaminated soil using steam, microwave, or infrared radiation to volatilize the pollutants is carried out. These volatilized pollutants (e.g., Hg, As) are then collected using vacuum negative pressure or carrier gas, thereby achieving their removal. This method can be adopted at different temperatures in the range 90°C–560°C, depending on the nature of the soil and the types of toxic elements present. This technique is simple and can be carried out through mobile devices. In the soil replacement method, clean soil is used to replace, or partly replace, the contaminated soil with the aim of diluting the pollutant concentration and to increase the soil environmental capacity. This method is suitable for contaminated soil within a small area. Moreover, the replaced soil should be ameliorated plausibly, as otherwise it will cause secondary pollution. This method is different from soil spading, where the contaminants are deeply placed in the soil to allow natural debasing. The soil replacement method is large in working volume, expensive, and is only suitable for soil within a small area that is polluted severely.

6.2.2 Chemical Remediation

6.2.2.1 Chemical Leaching

Chemical leaching is a method to wash the polluted soil using fresh water, reagents, and other fluids or gas that can leach the contaminants from the soil. Through various mechanisms, such as ion exchange, precipitation, adsorption, and chelation, the heavy metals and other pollutants in the soil can be transferred from soil to liquid phase, which are eventually recovered from the leachate. Alkali metal salts are used to extract toxic elements such as As, and ethylenediaminetetraacetic acid (EDTA) is used to remove organic pollutants like polychlorinated biphenyls (PCBs) (Ehsan et al. 2007). However, it has to be noted that employing EDTA is expensive and also that it is a poorly biodegradable substance.

6.2.2.2 Chemical Fixation

In the chemical fixation method, soil-conditioning reagents or materials are added into the contaminated soil to make the heavy metals insoluble or hardly movable, thereby decreasing the migratory ability of heavy metals to reach water, plants, and other environmental media. The remediation materials include clays, metallic oxides, and biomaterials. Hodson and Valsami-Jones (2000) investigated the capability of bone meal additions to immobilize pollutant metals in soils, and reported that bioavailability of metals was reduced through the formation of metal phosphates. Batch experiments and subsequent extraction of metals from controls and bone meal amended soils using 0.01 M $CaCl_2$ and diethylenetriaminepentaacetic acid (DTPA) depicted that bone meal additions decreased the availability of the metals in the soil solution. There was also a report on the remediation of contaminated soil by attapulgite clay (Fan et al. 2007), which showed that adding a moderate amount of attapulgite clay could reduce Cd concentrations to 46%, with the soil quality and productivity of the crops not being affected. Zhang et al. (2009) evaluated the chemical fixation efficiency of phosphate rock, furfural, and weathered coal on contaminated soil. The results showed that all three conditioning agents could reduce the concentration of heavy metals such as Cu, Zn, Pb, and Cd to a certain extent. The chemical fixation could remediate the soil with a low concentration of contaminants; however, the bioavailability of fixed heavy metals may be changed with the environmental conditions changing. In addition, the use of conditioning agents could change the soil structure, which may impose some adverse effects on the microbes in the soil.

6.2.2.3 Electrokinetic Remediation

Electrokinetic remediation is a new remediation method based on the application of voltage from the two sides of soil thereby forming an electric field gradient. With the induced potential gradient, pollutants will be carried out to the poles depending on the types, by means of electromigration, electro-osmosis, or electrophoresis, with subsequent processing for their removal. This method is suitable for low-permeable soil and has the advantages of easy installation and operation with low operating cost and without destroying the original natural environment. However, the main drawback of direct electrokinetic remediation is that the pH value of the soil system cannot be controlled, and also the efficacy of the treatment is low when compared to other processes.

6.2.2.4 Vitrification Technique

This method involves heating of soil at a temperature range of 1400°C–2000°C with subsequent refreezing. After cooling, the melt will take the form of a vitreous rock which captures and immobilizes the heavy metals and isolates them from the environment. At such high temperatures, organic contaminants will volatilize or decompose instantly. The steam and the pyrolysis products produced are collected by off-gas treatment systems. It was reported that the strength of the formed vitreous structure is 10 times higher than that of a concrete structure. In this application, for ex situ remediation, the energy can be supplied in several ways, including burning fossil fuel, direct electrode heating, through an arc, by plasma, and by applying microwaves. For in situ remediation, the heating can be done through electrodes inserted into the contaminated soil. Through this technology, high efficiency in removing heavy metals can be achieved; however, the process is complicated and requires a lot of energy for melting, thus making it less cost effective, and it has only a limited application.

6.2.3 Biological Remediation

6.2.3.1 Phytoremediation

Phytoremediation uses living green plants to fix or absorb contaminants, thereby mitigating the risks arising from them. This approach encompasses three main methods, namely, *phytostabilization*—fixing heavy metals by plants through adsorption and precipitation, and thus reducing their migration and bioavailability to roots; *phytovolatilization*—transferring heavy metals into the volatile state or adsorbing the metals and transferring them into gaseous matter using special compounds secreted by roots; and *phytoextraction*—absorbing heavy metals using tolerant and accumulating plants followed by transfer and storage in aboveground parts. Knowledge of absorption characterization of different plants and screening of plant species with high uptake capabilities are the key requirements in this technique (Ferro et al. 1999). The U.S. Department of Energy regulations for pollutant uptake plants are as follows: (1) They should have high accumulating efficiency under a low contaminant concentration. (2) They should be able to accumulate many different kinds of heavy metals. (3) They should have a fast growth rate with large biomass. (4) They should have pest and disease resistance.

6.2.3.2 Microbial Remediation

Microorganisms cannot degrade and destroy heavy metals directly, but they can affect the migration and transformation through changing their physical and chemical characterizations. The remediation mechanisms include extracellular complexation, precipitation, redox reaction, and intracellular accumulation. The effective and simple technique for extraction of valuable metals from low-grade ores and mineral concentrates is microbial leaching. This technology is considered to be strong and a potential option for remediation of mining sites, decontamination of sewage sludge, and remediation of soils and sediments. Lamber and Weidensaul (1991) reported that at low soil-pH levels the presence of mycorrhizae substantially increased the uptake of Cu and Zn by shoots. Abdel-Aziz et al. (1997) investigated vesicular-arbuscular (VA) mycorrhizae as a biological agent in ameliorating the toxicity of heavy metals in soils for the growth of faba bean. Inoculation with VA mycorrhizae induced a significant increase in parameters such as bean growth, nodule number and weight, and phosphorous and nitrogen contents as compared with the noninoculated treatments. It was found in the sewage sludge treated soil, where heavy metals were present in high concentrations, that inoculation with VA mycorrhizae reduced the concentration of heavy metals.

However, biological remediation is vulnerable to various external parameters such as temperatures, oxygen, moisture, and the pH value. It also has some limitation in such applications because certain microorganisms can only be used for selective contaminants. Also, other microbes causing secondary pollution might be produced.

6.2.4 Animal Remediation

Animal remediation has been tried to some extent. For example, lower animals have been used to absorb or degrade heavy metals, thereby removing and inhibiting their toxicity. Treatments using combinations of mulches of earthworms and straw enhanced plant Cu concentration (Jones and Leyval 1997). Kou et al. (2008) investigated Pb accumulation by earthworms by spiking soil with different concentrations of Pb. The results confirmed that earthworms could accumulate Pb efficiently with the accretion amount rising with the increase of Pb concentration.

6.3 NANOSCALE MATERIALS AND PROCESSING

Nanoscale materials contain one or more components that have at least one dimension in the range of 1–100 nm, and they include nanoparticles, nanofibers, nanotubes, nanocomposite materials, and nanostructured surfaces (Ramsden 2016). Nanoparticles (NP) are defined as a subset of nanoscale materials with single particles and with a diameter < 100 nm. Agglomerates of NP can be larger than 100 nm in diameter, but will be included in the "nanoscale" category because they may break down from weak mechanical forces or in solvents. Nanofibers and nanotubes are subclasses of nanoparticles, which have two dimensions. Nanocomposite materials are fabricated from two or more different nanoscale materials or one nanoscale material that is united with bulk type materials with typical properties for a particular application. Nanostructured surfaces include fullerenes and carbon nanotubes, and they subsist as hollow spheres, namely, buckyballs, ellipsoids, or tubes, which are made up of mainly carbon. In nanoscale dimensions, properties of materials can be different and unique for two main reasons:

1. Nanoscale materials possess a larger surface area than the same mass of material prepared in a bulk form. Although the materials are inert in their larger form, the nanoscale forms are more chemically reactive and show enhanced electrical and magnetic properties.
2. Below 50 nm, the laws of classical physics give way to quantum effects, provoking optical, electrical, and magnetic behaviors different from those predicted for the same material at a larger scale.

These effects can impart materials with beneficial physicochemical properties such as exceptional electrical conduction or resistance, a high capacity for storing or transferring heat, enhanced catalytic properties, and increased redox capabilities. Even the noble metals, gold or platinum, which are usually chemically inert, are able to catalyze chemical reactions at nanoscales. Because of their unique nature and properties, nanoscale materials are increasingly employed for soil remediation with pronounced success (US EPA 2007, 2008). The available nanoscale materials include natural, incidental, and engineered/manufactured groups. Naturally occurring nanoscale materials, like clays, organic matter, and iron oxides in soil play a vital role in biogeochemical processes (Klaine et al. 2008). The sources of incidental nanoscale materials are atmospheric emissions, industrial effluent streams, agricultural operations, fuel combustion, and weathering (Klaine et al. 2008). Engineered or manufactured nanoscale materials are designed with specific properties and may be released into the environment through industrial or environmental applications.

A variety of NPs are commercially available in the form of dry powder or liquid dispersions. The commercially important NPs mostly are metal/metalloid oxides such as silicon (SiO_2), titanium

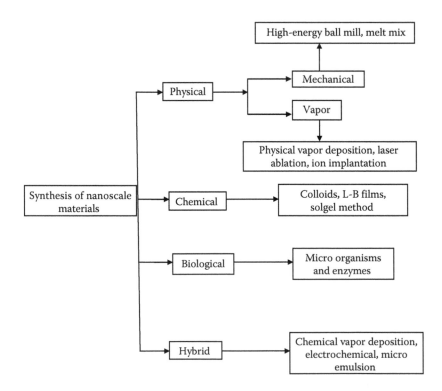

FIGURE 6.1 Different methods for the synthesis of nanoscale materials. *Note:* L-B, Langmuir-Blodgett films.

(TiO$_2$), aluminum (Al$_2$O$_3$), iron (Fe$_3$O$_4$), zinc (ZnO), copper (Cu$_2$O$_3$) and zirconium (ZnO$_2$). Also of increasing importance are mined oxide nanoparticles, such as indium-tin oxide (In$_2$O$_3$SnO$_2$), antimony-tin oxides (Sb$_2$O$_3$–SnO$_2$), and barium titanate (BaTiO$_3$). Besides, a number of pure metal NPs of gold (Au), silver (Ag), platinum (Pt), and cobalt (Co) are increasingly available for multiple applications. Organic nanoscale materials are also commercially available, which include buckyballs, carbon nanotubes, and graphene. Several methods are available for the synthesis of NPs either with single species or with multiple species as nanocomposites as depicted in Figure 6.1, and numerous literature references are available on this topic, including their physicochemical characterizations.

6.4 NANOSCALE MATERIALS AND THEIR APPLICATIONS FOR REMEDIATION

6.4.1 NANO-REMEDIATION OF ORGANIC CONTAMINANTS

In mine-affected soil and proximal areas, occurrences of organic contaminants are less pronounced than their inorganic counterparts, and their availability is mainly through anthropogenic activities and, to a smaller extent, natural sources. Organic pollutants differ in charge and solubility, but mostly are nonpolar in nature. Organic compounds such as hydrocarbons are poorly soluble in water and sorb readily to hydrophobic soil particles like soil organic matter. Typical organic contaminants that are of most concern include polycyclic aromatic hydrocarbons (PAHs), nitroaromatic compounds (e.g., aniline, trinitrotoluene), polychlorinated biphenyls (PCBs), chlorinated phenols (e.g., 2,4,6-trichlorophenol, pentachlorophenol), and dioxins. In particular PAHs and nitroaromatic compounds are potent toxic compounds, among which some are carcinogenic. They bind to other organic matter and have reduced mobility and higher resistance to biodegradation. Aromatic organic compounds are often demarcated by the number of rings they hold, which may range from one to five (Anderson et al. 1974). Lighter, mono-aromatic (one ring) compounds include benzene, toluene, ethylbenzene, and xylenes (Huguenin et al. 1996). Aromatics with two or more rings are termed polycyclic aromatic hydrocarbons (PAHs) (Anderson et al. 1974). Coal ash and slag, which can be recognized by the presence of off-white grains in soil, gray heterogeneous soil, or (coal slag) bubbly, vesicular pebble-sized grains typically contain significant concentrations of PAHs such as benzo(b)fluoranthene, benzo(a)anthracene, benzo(a)pyrene, indeno(cd)pyrene, benzo(k)fluoranthene, anthracene, and phenanthrene. These PAHs are known human carcinogens, and the acceptable limits of concentrations in soil are typically around 1 mg kg^{-1}.

Most of the conventional remediation methods to mitigate organic contaminants in soil have some disadvantages. Some of these methods generate daughter compounds that are more noxious to the environment than their precursors. Currently the impacts of nanotechnology are increasingly obvious in all areas of science and technology, including the field of soil ecology. Development and deployment of nanotechnology for remediation of organic contaminated soil are already in place. Treatment and remediation using nanotechnology have seemingly experienced the most growth in recent years.

6.4.1.1 Nanometal and Metal Oxide Particles

Nanometal particles have been used in mitigating organic contaminants for the past few decades. They represent a viable, commercially available nanotechnology for organic contaminant remediation. Most common and successful examples are nano zero-valent iron (nZVI, Fe^0) particles and emulsified zero-valent iron nanoparticles (EZVI). These materials, through redox reactions, effectively remove several potent organic contaminants such as tetrachloroethene (TCE), vinyl chloride (VC), cis-1, 2-dichloroethylene (c-DCE), polychlorinated biphenyls (PCBs), 1-1-1-tetrachloroethane (TCA), nitroaromatic- and halogenated aromatics (Tratnyek et al. 2003). Generally the metallic iron particles act as efficient reducing agents and react mostly through redox (Fenton-type) reaction mechanisms, capable of changing numerous persistent organic contaminants to gentler compounds. Naturally elemental iron stabilizes compounds by acquiring the valence states of either +2 or +3. On the contrary, Fe^0, through its excess of electrons, has an overall charge of "zero," thus making it readily more reactive. Its main advantage is its low cost and its main drawback is the formation of iron oxides and other corrosive by-products on its surface. However, in addition to its high-ameliorating effectiveness, the technology has the advantage of using an ecofriendly material. Also, because of its small size, it can be utilized in the subsurface soil layer.

Other nanoscale level metals that can reduce organic contaminants are zero-valent zinc (Zn^0) and palladium (Pd) particles. Zn^0 with a reduction potential of -0.763 V has a greater reducing power than Fe^0, which has a reduction potential of -0.44 V, and Zn^0 has been employed to eliminate halogenated hydrocarbons with the resultant products containing fewer halogen atoms. The reaction pathways for such reactions include dihalo-elimination (elimination of two halo atoms, which are of two types: α elimination—from the same carbon atom, and β elimination—from two different carbon atoms) and hydrogenolysis (substitution of halogen atom by hydrogen) (Arnold et al. 1999). Nano particles of Pd, when doped onto iron or gold nano particles (thus forming bimetallic nano particles), more efficiently degrade organic compounds than when these metal particles are engaged alone. This is attributed to the creation of high surface activity by Pd acting as catalyst (Lien and Zhang 2007). Given its success, it has been demonstrated that Pd particles can be encapsulated with alginate, polyurethane, or polyacrylamide nanoparticles for such applications in batch reactors (Hennebel et al. 2009).

Recently, researchers (Kamat et al. 2002, Kamat 2012) used titanium oxide-(TiO_2) and zinc oxide-(ZnO) based nanostructured materials for extensive remediation of different types of organic compounds. They are a group of photocatalysts, and a free radical formation (OH$\cdot$ and O_2^-) is the basis for their degradation reactions. The whole remediation process in such a case is dictated by the nature of nanostructured materials as well as doping of any transition metal, which may further govern the decontamination rate.

These remediating nanoparticles can be delivered to the contaminated mine site water beds by direct injection by gravity flow (Lien and Zhang 2007). However, the complete environmental risk assessment of releasing nanoparticles into environmental water is not available yet. Little is known about their ecotoxicity, bioaccumulation potential, and persistency in the environment (Grieger et al. 2010).

6.4.1.2 Carbon Nanotubes

Carbon nanotubes (CNTs) are categorized into two main classes based on the structures: single-walled carbon nanotubes (SWCNT) and multiwalled carbon nanotubes (MWCNT). Because CNTs possess a large surface area, a tubular structure, and have a nonpolar property, they may be used as a promising adsorbent material for nonpolar organic contaminants such as PAHs, trihalomethanes, dioxin, DDT and its metabolites, and herbicides. The purified CNTs have two to three times more adsorption capacities for organic contaminants in comparison to activated carbon. In addition, municipal sludge and other solid wastes after mixing with small amounts of CNTs could be safely utilized for soil application to sustain soil quality and reduce waste-disposal expenses. Aggregation and precipitation phenomena hinder the direct application of CNTs in soil solution due to their tremendous hydrophobicity. To overcome these shortfalls, the surface structure of CNTs has been modified by introducing hydrophilic functional groups and also through the addition of surfactants, polymers, and natural organic matter.

6.4.2 Nano-Remediation of Inorganic Contaminants

Soil contamination by heavy metals is a widespread problem around the world, where excessive concentration of heavy metals such as Pb, Zn, Cr, Cu, Cd, Hg, and As can be found in mining sites. One of the largest problems associated with the persistence of heavy metals is the potential for bio-accumulation and biomagnification causing heavier exposure for some organisms than is present in the environment alone. Through precipitation of their compounds or by ion exchange into soils and muds, heavy metal pollutants can be localized and laid dormant. Unlike organic pollutants, heavy metals do not decay and thus pose a different kind of challenge for remediation. Mining and milling of metal ores coupled with industries cause wide distribution of metal contaminants in soil. During mining, tailings (heavier and larger particles settled at the bottom of the flotation cell during mining) are directly discharged into natural depressions including onsite wetlands, resulting in elevated concentrations. Extensive Pb and Zn ore mining, and smelting in particular, have resulted in contamination of soil that poses a risk to human and ecological health. Many reclamation methods used for these sites are lengthy and expensive and may not restore soil productivity fully. For the removal of inorganic contaminants, different types of nanoscale materials are being utilized and are gaining a gradual momentum for such applications.

6.4.2.1 Nano Zero-Valent Iron and Nano Iron Oxide Particles (nZVI (Fe^0)/nFeO)

Nano zero-valent iron particles, discussed earlier, can be also used to decontaminate heavy metals from soil due to their reduction potential (E^0, Fe^{2+}/Fe^0) of -0.44 V and hence, acting as a strong reducing agent. Hypothetically, nZVI can immobilize reductively some toxic heavy metals with E^0 much more positive than -0.44 V. Good examples are CrO$_4^{2-}$/Cr^{3+} (E^0 = +1.56 V) and Cr$_2$O$_7^{2-}$/Cr^{3+} (E^0 = +1.36 V). Both CrO$_4^{2-}$ and Cr$_2$O$_7^{2-}$ species are generally more soluble and more noxious than their low-valent counterparts (Cr^{3+}) in the soil ecosystem (Kotas and Stasicka 2000). Through redox reactions, nZVI converts the oxidation state of metal ions from a higher state to a lower state, thus reducing the solubility, mobility, and toxicity of those heavy metals. Studies have shown that nZVI is able to reduce higher-valent Cr^{+6} to low-valent Cr^{+3} in soils. Franco et al. (2009) noted that 97.5% of Cr^{+6} in a polluted soil could be reduced to Cr^{+3} by nZVI, which drastically decreased the chromium (VI) toxicity in the affected soil with the mine tailings. Selenium species, namely, SeO$_4^{-2}$ (or Se^{6+}) and SeO$_3^{-2}$ or (Se$_4^+$) are more soluble and mobile in the soil environment and more toxic than the low-valent species like Se^0 and Se^{-2}. The toxicity and solubility of Se are significantly reduced in mine-spoiled soil by the application of nZVI, which transforms the high-valent species to the low-valent ones. nZVI is able to remove up to 0.1 mole Se/mole Fe from dissolved Se^{6+}. The compound, FeSe, in the solid phase as the reduced Se$_2^-$ species, which was transformed from S^{6+}, was identified by X-ray absorption near edge structure (XANES) spectroscopy and X-ray absorption fine structure (EXAFS) spectroscopy. nZVI is also capable of reducing some other toxic elements in soil such as Hg^{+2}, Ni^{+2}, Ag$^+$, Cd^{+2}, As^{+3}, and As^{+5}. Adsorption of the ions on the nZVI particle shells that consist of a layer of iron oxidation products, and reduction of metal ions to zero-valent metals on the nZVI surfaces, are the possible explanation for this decontamination phenomenon.

However, because of rapid nanoparticle agglomeration and interactions with surfaces of the ambient porous media, the mobility of bare nZVI is limited to a few centimeters in the subsurface environment. The stability and mobility of an nZVI suspension are required to degrade the contaminants in situ, once they are injected underground. But there is no concrete proof on the mobility of such products in the field. But, in case of stabilized nZVI, it may travel 1 m from the source of injection (O'Carroll et al. 2013). Moreover, dissolved oxygen oxidizes nZVI very quickly resulting in the precipitated formation of maghemite and magnetite. These imply that risks of nZVI occurrence in the soil environment and further contact of organisms with nZVI are not important, at least when considering the current state of nZVI technology (Reinsch et al. 2010, O'Carroll et al. 2013). There is no report of a field study in which nZVI particles have been applied for amelioration of polluted soil, which is completely different from other remediation applications like groundwater decontamination process. For reclamation of mine soil and vegetation establishment, a thin soil surface layer is generally important for plant root establishment. The suspension of nanoparticles should be generally applied over the targeted land surface. The size of the nanoparticles should be set in such a way so that the particles will be preferably fixed within the polluted surface layer only after the whole targeted soil has been saturated and treated with the particles, thus decreasing the risk of leakage of nanoscale materials and preventing secondary contamination to adjacent water bodies. From this point of view, nZVI and other nanoparticles with exceedingly high mobility should not be used for surface soil amelioration. Applications with other nanoparticles based on zinc, manganese, and aluminum cause phytotoxicity. Besides, these particles tend to be reduced to lower-valent states and eventually lose their adsorption capacities.

Naturally occurring Fe oxide nanocrystals are capable of sorbing both inorganic and organic contaminants through surface complexation as well as surface precipitation. Various engineered iron oxide nanoparticles with high adsorption capacity for pollutants and their eco-friendly features are being fabricated and utilized for in situ soil remediation processes. Direct application of nano-Fe oxide solutions to polluted sites removes the labile fractions of toxic heavy metals from soil solution through sequestering or adsorption phenomena, which decrease the bioavailability and mobility of these toxic ions in soil. Further, application of industrial sludge rich in iron oxides to polluted soils is a common practice for arresting heavy metal availability. The interaction of nFeOs with mine soils could successfully reduce the bioavailability of soil-bound toxic heavy metals. Besides, nFeO, other nanometal oxides, such as titanium oxide (TiO_2), cobalt oxide (Co_3O_4), and manganese oxide (Mn_3O_4), have been tried in acidic soils. They are relatively inert and responsive to the reduced environment, as in waterlogged soils or wetlands. However, these nanoparticles generate more reactive oxygen species (ROS) than their respective salt solutions, which may have an impact on plant organisms.

6.4.2.2   Zeolites

Toxic heavy metals in polluted soils and sediments can be immobilized by using natural and synthesized zeolites. Hence, the chance of releasing these toxic ions in soils, along with their impact on plants, can be reduced. Significant reductions (42%–72%) of the labile and easily available fractions of these heavy metals occurred when mine soils contaminated by Zn, Pb, Cu, and Cd were treated with synthesized zeolites at rates of 0.5%–5% by weight (Edwards et al. 1999). The metals were immobilized due to increased soil pH caused by the zeolites. Application of Na-type zeolites tends to release Na^+ to the soil solution, which badly affects plant growth, albeit the bad effects of the heavy metals can be mitigated. Hence Ca-type zeolites are advocated for heavy metal remediation at the sites where building of new plantations is intended.

6.4.2.3   Phosphate-Based Nanoparticles

Generally, application of phosphate-based nanoparticles decontaminates heavy metal-polluted soils by transforming the heavy metals to highly stable and insoluble phosphate compounds. This is especially applicable for stabilizing Pb compounds. For example, the solubility products of common lead compounds in soils, namely, anglesite ($PbSO_4$), cerussite ($PbCO_3$), galena (PbS), and litharge (PbO) were measured as $10^{-7.7}$, $10^{-12.8}$, $10^{-27.5}$, and $10^{+12.9}$, respectively, in comparison to Pb phosphate compounds, such as pyromorphites [$(Pb_5(PO_4)_3X, X = F^-, Cl^-, Br^-,$ and OH^-], which have solubility products less than 10^{-71} (Ruby et al. 1994). Lead phosphates are less soluble than other Pb phases found in soils. Phosphate amendments transform less stable Pb species to a more stable species, and the process is thermodynamically favored, and it subsequently reduces the leachability and bioavailability of Pb. However, the application of solid phosphate for such applications is hindered by the large size of the particles which limits the phosphate mobility and delivery, thereby preventing phosphate from reaching heavy metals in subsoil layers. To overcome this problem nanosized iron phosphate (vivianite) particles have been synthesized and applied for heavy metal immobilization (Liu and Zhao 2007).

6.4.2.4   Nano Iron Sulfide Particles

Sulfide-based nano particles are utilized particularly to reduce the contamination of Hg and As in soils and sediments. Sulfide (S^{-2}) ligands as well as coordination surfaces provide the basis of the adsorption mechanism. Sulfide (S^{2-}) sorbs the heavy metals in the reduced environment like in sediments and water-logged soils. It forms highly insoluble metal sulfides. Hence, in this case, the heavy metals are all bound as metal-sulfides and, consequently, soluble metals in the pore water are reduced. Moreover, sulfide (S_2^-) removes Hg from aqueous solution and represses the formation of the noxious methyl-mercury (CH_3Hg) in the soil ecosystem. FeS nanoparticles are potential amendments for in situ remediation of different heavy metals including Hg in soils. But the drawback of using FeS nanoparticles is that under aerobic conditions, FeS nanoparticles are not stable, so chances of release of the sorbed metal ions occur and it may create toxicity problems for plants and microorganisms in soils. Acid mine drainage (AMD) is a severe environmental issue in almost all derelict mining sites. Acidity is present in the drainage effluent, and iron sulfide minerals are exposed upon the mining operations that, upon oxidation, develop acidity in the soil. They lead to barren land. Hence addition of FeS to soils, as advocated by previous investigations (Singer and Stumm 1970, España et al. 2005), may aggravate AMD and soil acidity problems at mining sites. Caution must be observed in selecting suitable ameliorating agents for remediation of contaminated soils. Stable immobilizing agents like iron oxide nanoparticles or phosphate-based nanoparticles might be suggested for ameliorating AMD-polluted soils.

6.4.2.5 Carbon Nanotubes

Because of their nonpolar property, raw CNTs are not suitable for sorption of polar metal ions. But when the CNTs' surface is chemically modified and a large amount of oxygen-containing polar functional groups (e.g., –COOH, –OH or –C=O) are created, sorption capacity increases significantly. This is attributed to the presence of functional groups that create negative charges on carbon surfaces; simultaneously, oxygen atoms in functional groups readily donate a single pair of electrons to metal ions, which increases the cation adsorption capacity of CNTs. Nitric acid treated multi-walled carbon nanotubes (MWCNTs) were used successfully for the sorption of various toxic heavy metal ions, like Pb^{2+} (97.08 mg g^{-1}), Cu^{2+} (24.49 mg g^{-1}), and Cd^{2+} (10.86 mg g^{-1}) from aqueous solutions (Li et al. 2003).

6.5 ADVANTAGES AND DISADVANTAGES OF NANOSCALE MATERIALS

Nanoscale materials have quickly gained popularity as an option in the field for mine site remediation. Among them, metal oxide nanoparticles are continuously being developed for the prevention of pollution. Nanotechnology has several advantages, which include the following:

- It provides environmental benefits, including cleaner, more efficient industrial processes and a better ability to identify and get rid of pollution by improving soil and water quality.
- It offers high precision manufacturing by controlling the amount of waste, enhanced efficiency of industrial plants, and minimizing environmental damages.

Although new nanoscale materials are progressively being developed with time, the shapes and sizes of the materials mainly govern their toxicity. Even a minor change in a chemical functional group will affect the degree of toxicity of the nanoscale material. This warrants a detailed and in-depth analysis of nanoscale materials. With the continuous advent of state-of-the-art technological development, these studies are being done.

Moreover, at all stages of nanotechnology—from its manufacturing, storage, transport, application, recycling, and disposal—a full risk assessment is necessary to investigate the adverse impacts to the environment. Without assessing their toxicity, handling of nanoscale materials is always a risk, and the toxicity depends upon their specific type. For example, soluble nanoparticles are difficult to separate from waste if not properly handled, and a dedicated method should be available. It will be a major environmental concern if such a waste product is disposed inappropriately. Another issue with nanoscale materials is that they might not be detectable after release into the environment, which in turn can create difficulties if remediation is needed. Thus relevant assessment methods need to be developed and put in place to detect nanoscale particles in the environment. These methods need to determine accurately the shape and surface area of the particles, the main factors that define their toxic properties. These structure–function relationships will assist in estimating the toxicity levels enabling full risk assessments for any newly developed nanoscale material during manufacture or application. Such assessments should also take into consideration their toxicological hazard, the probability of exposure, and the environmental and biological fate, transport, transformation, and persistence. Further, whenever the use of a scarce nanoscale material is prompted, an effective strategy of recycling and recovery should be in place.

6.6 CONCLUSIONS

The potential and promise of nanotechnology are growing in different sectors with excellent socio-economic benefits. However, while selected nanotechnology uses have reached the commercialization process, nanotechnology studies related to mine site remediation still remain largely at the bench scale. In general, a lack of assembled information exists on how these individual nanotechnologies might eventually be implemented, the type of sites where they would be appropriate, and the general feasibility for up-scaling. Selection of the right nanoscale materials for the relevant mine site is of prime importance that will guide the future direction of nano-remediation. Although there are many hurdles in implementing this technology, the world forum of scientists has tried to adopt and accept it after continuous refining. Even though nanoscale materials may decontaminate polluted soil, it may have an adverse impact on the environment in other ways. The effects of nanoscale materials and their chemistry in soil and water are still unknown. Its adverse impacts on ecology spanning the soil–plant–animal continuum need to be studied. Thus before using these materials commercially, the ethics of engineering and biosafety protocols need to be defined and developed. To address the ill-effects of nanoscale particles on the environment, "eco-nano–toxicology," a new branch of science, has emerged that aims at studying the movement of nanoscale materials through the biosphere. Also, to solve this issue, international bodies such as the USEPA, the OECD, and NanoImpactNet (EU) aim to set up regulations and legislation to ascertain that nanoparticles themselves will not become a source of pollution. In-depth research on the potential risks and exposure routes of nanoscale materials is required in order to achieve sustainable and safe development. This should be augmented further through the evaluation of full life-cycle assessments of nanoscale materials in remediation applications for a better tomorrow.

REFERENCES

Abdel-Aziz, R. A., S. M. A. Radwan, and M. S. Dahdoh. 1997. Reducing the heavy metals toxicity in sludge amended soil using VA mycorrhizae. *Egypt J Microbiol* 32:217–234.

Anderson, J. W., J. M. Neff, and B. A. Cox. 1974. Characteristics of dispersions and water soluble extracts of crude and refined oils and their toxicity to estuarine crustaceans and fish. *Mar Biol* 27:75–88.

Arnold, W. A., W. P. Ball, and A. L. Roberts. 1999. Polychlorinated ethane reaction with zero-valent zinc: Pathways and rate control. *J Contam Hydrol* 40:183–200.

Edwards, R., I. Rebedea, N. W. Lepp, and A. J. Lovell. 1999. An investigation into the mechanism by which synthetic zeolites reduce labile metal concentrations in soils. *Environ Geochem Health* 21:157–173.

Ehsan, S., S. O. Prasher, and W. D. Marshall. 2007. Simultaneous mobilization of heavy metals and polychlorinated biphenyl (PCB) compounds from soil with cyclodextrin and EDTA in admixture. *Chemosphere* 68:150–158.

España, J. S., E. L. Pamo, E. Santofimia, O. Aduvire, J. Reyes, and D. Barettino. 2005. Acid mine drainage in the Iberian Pyrite Belt (Odiel river watershed, Huelva, SW Spain): Geochemistry, mineralogy and environmental implications. *Appl Geochem* 20:1320–1356.

Fan, D. F., S. S. Huang, and Q. L. Liao. 2007. Restoring experiment on cadmium polluted vegetable lands with attapulgite of varied dose. *Jiangsu Geol* 31:323–328.

Ferro, A. M., S. A. Rock, J. Kennedy, J. J. Herrick, and D. L. Turner. 1999. Phytoremediation of soils contaminated with wood preservatives: Greenhouse and field evaluations. *Int J Phytoremediation* 1:289–306.

Franco, V., L. M. Da Silva, and W. F. Jardim. 2009. Reduction of hexavalent chromium in soil and ground water using zerovalent iron under batch and semi-batch conditions. *Water Air Soil Poll* 197:49–60.

Grieger, K. D., A. Fjordbøge, N. B. Hartmann, E. Eriksson, P. L. Bjerg, and A. Baun. 2010. Environmental benefits and risks of zero-valent iron nanoparticles (nZVI) for in situ remediation: Risk mitigation or trade-off? *J Contam Hydrol* 118:165–183.

Hennebel, T., P. Verhagen, H. Simoen, B. De Gusseme, S. E. Vlaeminck, N. Boon, and W. Verstraete. 2009. Remediation of trichloroethylene by bio-precipitated and encapsulated palladium nanoparticles in a fixed bed reactor. *Chemosphere* 76:1221–1225.

Hodson, M. E., and E. Valsami-Jones. 2000. Bonemeal additions as a remediation treatment for metal contaminated soil. *Environ Sci Technol* 34:3501–3507.

Huguenin, M. T., D. H. Haury, and J. C. Weiss. 1996. *Injury Assessment: Guidance Document for Natural Resources and Services under the Oil Pollution Act of 1990*. Silver Spring, MD: Damage Assessment and Restoration Program, National Oceanic and Atmospheric Administration.

Jones, E. J., and C. Leyval. 1997. Uptake of [109]Cd by roots and hyphae of a *Glomus* with high and low concentration of cadmium. *New Phytol* 135:353–360.

Kamat, P. V. 2012. TiO_2 nanostructures: Recent physical chemistry advances. *J Phys Chem C* 116:11849–11851.

Kamat, P. V., R. Huehn, and R. Nicolaescu. 2002. A "sense and shoot" approach for photocatalytic degradation of organic contaminants in water. *J Phys Chem B* 106:788–794.

Klaine, S. J., P. J. J. Alvarez, G. E. Batley, and T. E. Fernandes. 2008. Nano particles in the environment: Behavior, fate, bioavailability, and effects. *Environ Toxicol Chem* 27:1825–1851.

Kotas, J., and Z. Stasicka. 2000. Chromium occurrence in the environment and methods of its speciation. *Environ Pollut* 107:263–283.

Kou, Y. G., X. Y. Fu, and P. Q. Hou. 2008. The study of lead accumulation of earthworm in lead polluted soil. *Environ Sci Manage* 33:62–64.

Lamber, D. H., and T. Weidensaul. 1991. Element uptake by mycorrhizal soybean from sewage-treated soil. *Soil Sci Soc Am J* 55:393–398.

Li, Y. H., J. Ding, and Z. Luan. 2003. Competitive adsorption of Pb^{2+}, Cu^{2+} and Cd^{2+} ions from aqueous solutions by multiwalled carbon nanotubes. *Carbon* 41:2787–2792.

Lien, H.-L., and W.-X. Zhang. 2007. Nanoscale Pd/Fe bimetallic particles: Catalytic effects of palladium on hydrodechlorination. *Appl Catal B* 77:110–116.

Liu, R., and D. Zhao. 2007. Reducing leachability and bioaccessibility of lead in soils using a new class of stabilized iron phosphate nanoparticles. *Water Res* 41:2491–2502.

O'Carroll, D., B. Sleep, M. Krol, H. Boparai, and C. Kocur. 2013. Nanoscale zero valent iron and bimetallic particles for contaminated site remediation. *Adv Water Resour* 51:104–122.

Ramsden, J. J. 2016. *Nanotechnology: An Introduction*, 2nd cdn. Amsterdam, the Netherlands: Elsevier.

Reinsch, B. C., B. Forsberg, R. L. Penn, C. S. Kim, and G. V. Lowry. 2010. Chemical transformations during aging of zerovalent iron nanoparticles in the presence of common groundwater dissolved constituents. *Environ Sci Technol* 44:3455–3461.

Ruby, M. V., A. Davis, and A. Nicholson. 1994. In situ formation of lead phosphates in soils as a method to immobilize lead. *Environ Sci Technol* 28:646–654.

Singer, P. C., and W. Stumm. 1970. Acidic mine drainage: The rate-determining step. *Science* 167:1121–1123.

Tratnyek, P. G., M. M. Scherer, T. L. Johnson, and L. J. Matheson. 2003. Permeable reactive barriers of iron and other zero-valent metals. In *Chemical Degradation Methods for Wastes and Pollutants, Environmental and Industrial Applications*, ed. M. A. Tarr, pp. 371–421. New York: Marcel Dekker.

US Environmental Protection Agency. 2008. Nanotechnology for site remediation fact sheet. Solid Waste and Emergency Response. https://clu-in.org/download/remed/542-f-08-009.pdf. Accessed May 16, 2017.

US Environmental Protection Agency, Science Policy Council. 2007. Nanotechnology white paper. https://www.epa.gov/sites/production/files/2015-01/documents/nanotechnology_whitepaper.pdf. Accessed May 16, 2017.

Zhang, L. J., Y. Zhang, and D. H. Liu. 2009. Remediation of soils contaminated by heavy metals with different amelioration materials. *Soils* 41:420–424.

Section III

Post Mine Site Land-Use Practices

7 Profitable Beef Cattle Production on Rehabilitated Mine Lands

Dee Murdoch and Rajasekar Karunanithi

CONTENTS

7.1 Introduction.. 111
7.2 Rehabilitation of Mine Lands 112
    7.2.1 Infrastructure... 112
    7.2.2 Land Form Establishment.. 112
    7.2.3 Growing Media.. 112
    7.2.4 Revegetation 113
    7.2.5 Maintenance Revegetation Works.. 113
    7.2.6 Determining Stocking Rates 114
7.3 Beef Cattle Production on Rehabilitated Mine Lands: Case Studies 115
    7.3.1 Pasture and Grazing Research, Queensland, Australia........................ 115
        7.3.1.1 Sustainable Grazing on Rehabilitated Lands in the
                Bowen Basin—Stages 1 and 2 115
        7.3.1.2 Outcomes of the Project... 115
    7.3.2 Pasture and Grazing Trials, Hunter Valley, Australia.......................... 116
        7.3.2.1 Pasture Establishment Using Biosolids and Compost Treatments 116
        7.3.2.2 Assessment of Sustainable Beef Cattle Production 118
7.4 Summary and Conclusions 121
References... 121

7.1 INTRODUCTION

The Australian beef cattle industry is one of the most efficient and ranks third largest in beef export in the world, contributing 4% of beef supply. As on 2013, the meat value produced from beef cattle, in Australia is estimated to be $12.3 billion (Fastfacts, 2013). Beef cattle production ranges from intensive farms on fertile lands to extensive range lands. With the increase in human population and increase in affordability of meat-based food, the demand for beef cattle is also increasing.

The main goal of the mine rehabilitation plan is to ensure that while the site is left safe, stable, and nonpolluting, the landscape can be put to optimal use post mining and closure. However, in the face of evolving shareholder expectations and the current international economic climate, there is an increasing pressure to ensure that land use, both during and after mining operations, is more profitable and sustainable. Hence, the goal of many rehabilitation programs is to reinstate the pre-mining capability of grazing land, with mined lands being revegetated with pasture species and areas of trees over grass to provide enhanced habitat for both native animals and domesticated stock. The focus on the earthworks and rehabilitation program across these mining operations is to provide stable landforms compatible with the surrounding landscape that will allow optimal post-mining land use in terms of current social and economic constraints (Mattiske, 2016).

7.2 REHABILITATION OF MINE LANDS

To achieve the objective of sustainable beef cattle farming, a wide range of land management activities need to be undertaken. These are best classified in the context of the following phases of the rehabilitation program:

- Decommissioning of infrastructure
- Landform establishment
- Growing media
- Revegetation
- Maintenance

The activities undertaken at each phase where the post-mining land use is cattle production are described next.

7.2.1 INFRASTRUCTURE

All infrastructure including roads, not required post mining, will be removed and rehabilitated. Any proposed and/or existing cut and cover tunnels that are located under public roads may be partially filled, allowing safe post-mining access for cattle and farm vehicles.

7.2.2 LAND FORM ESTABLISHMENT

Reshaping principally involves recontouring overburden emplacement areas into the designed shape for final rehabilitation. Post placement the shaping of overburden will usually be undertaken using bulldozers. Ideally, reshaping will result in a stable landform with slopes and drainage patterns that blend in with the surrounding natural topography. Slope stability is integral to rehabilitation design, and slopes in excess of 18° (approximately 35%) are not favored due to the safety issues associated with accessing the site for farming equipment and for the purpose of cattle movement. However, slopes steeper than 18° may be necessary in some locations to ensure rehabilitation merges seamlessly with adjacent undisturbed land.

Once bulk reshaping is completed, the landform is deep-ripped and the final trim/rock raking is undertaken. The ripping loosens up any near surface strata within the landform that have been compacted during placement, aiding root penetration during vegetation establishment. The final trim smooths out any washouts and gullies, rough edges, temporary access tracks, local steep slopes, and prepares the surface for revegetation. Rock-raking is the final stage of reshaping and removes or buries exposed surface rock greater than 200 mm in diameter. Rocks are either buried within the spoil structure or may be left in groups on the surface as fauna habitat. This raking is usually done along the contour, leaving a cultivated surface that minimizes the risk of erosion until vegetation can be established.

To supply a drinking water point for cattle, appropriate water storage dams will be incorporated into the landscape. These may also serve the purpose of sediment control features during the initial phase of the rehabilitation program. In addition, these dams will be revegetated with plant species (e.g., pasture species and emergent reeds) suitable to ensure stability of the dam wall and batters. These areas may also provide potential localized habitat for native fauna with varying water depths, island refuges, large woody debris, and/or rock stockpiles retained in situ.

7.2.3 GROWING MEDIA

While it is recognized that the growing media may in part comprise shallow soils and the potential for the presence of sodic subsoils, the mine land management program will generally ensure that there are no rocky outcrops. Proactive management of erosion (wind and water) will be carried out, and cropping for agriculture will not be included in the program. These measures together with the use of soil ameliorants including though not limited to biosolids, lime, and gypsum, will enable the land to return to the desired post-mining land and soil capability classes.

Soil/spoil ameliorants will be spread and integrated into the surface layer to address issues such as soil sodicity and assist with soil structural properties, while also providing a growing media nutrient regime suitable for the establishment of pasture species. Once the material has been top-dressed, the surface will be contour disc- or chisel-ploughed to integrate the topdressing material. The area will then be contour cultivated to create seed entrapments and microclimates prior to sowing.

7.2.4 Revegetation

The main revegetation steps may include

- Species selection
- Sowing rates and species proportions
- Tube stock densities—in the case of tree/shade species
- Consideration of habitat augmentation
- Seed pretreatment requirements
- Seed spreading and planting techniques
- Soil amelioration and fertilizer requirements
- Use of temporary cover crops to assist soil stabilization
- Protection from vertebrate pest species, domesticated stock, and unauthorized access

The area will be sown and fertilized with the selected pasture and/or tree seed mixes shortly after spreading the topsoil to avoid loss in activity of preexisting microflora and to mitigate the exposure of the soil surface to the forces of wind and rain erosion. Fertilizer is not usually required where biosolids have been applied. Tube-stock may be planted in areas providing visual screens and habitat/refuge corridors.

A typical species list sown for post-mining in the Hunter Valley of NSW, Australia, in approximate kilograms per hectare, for the establishment of pastures for a post-mining grazing land use includes:

Rhodes grass (1 kg/ha), Couch grass (2 kg/ha), Rye grass (4 kg/ha), Clover (6 kg/ha) varieties, Haifa white clover (2 kg/ha), Woolly Pod Vetch (4 kg/ha), green panic (5 kg/ha), *Sirosa phularis* (4 kg/ha), Sephi Barrel Medic (4 kg/ha), Lucerne (4 kg/ha), and Kikuyu (1 kg/ha).

These species would be adapted to local conditions with varying proportions of tropical and temperate grass and legume species, ensuring year round growth of productive pastures.

The focus of this mix is to establish a vegetation cover that ensures surface stability, reduction in the risk of soil erosion while also providing a plant community suitable of sustaining beef cattle. During this phase of the rehabilitation program, cattle may be introduced, under a carefully managed program, to the rehabilitated lands with a purpose of enhancing nutrient cycling via consumption of grown feed and production of manure and the trampling and incorporation of plant material (green and dead) into the surface soil layer (Commonwealth of Australia, 2006).

7.2.5 Maintenance Revegetation Works

As with all successful grazing-based systems, maintenance works are required in terms of fertilizer and vegetation enhancement to ensure successful growth of cattle. Maintenance works to be implemented at this phase of the rehabilitation program may include

- Soil sampling for the purpose of defining fertilizer and seeding regimes
- Application of defined fertilizer—in terms of rates and mix
- Oversowing of pasture with legumes

Beef cattle production levels are primarily determined by pasture intake. There are a range of parameters that influence this, with the primary factors being

- Pasture quantity
- Pasture quality (digestibility)
- Species composition

Pasture quantity: Usually expressed in terms of kilogram of dry matter (DM) per hectare, this refers to the total mass of pasture if cut at ground level and includes both green and dead material. Pasture components that determine herbage mass are height, density, and water content. The critical herbage mass for cattle is 700–2900 kg DM/ha/head. Out of these, plant height has the most significant impact on how much livestock will consume.

Pasture digestibility and species composition: There are numerous factors influencing pasture intake, with digestibility and the proportion of legume the most practical measures. Digestibility is generally the most useful measure for pasture quality, and refers to the proportion of feed an animal can use to satisfy nutritional requirements. Digestibility is positively related to the energy and protein content of the pasture. Digestibility is also related to the speed of digestion. The higher the digestibility, the quicker it will pass through the animal, which in turn allows higher intake and hence higher growth rates (Hart et al., 2009).

Herbage mass and digestibility interact with each other to determine the amount of pasture that will be consumed. Once digestibility declines below 65% for lactating stock and 55% for dry stock, livestock will experience unsatisfactory performance and weight loss no matter how much herbage is available. Other factors influencing digestibility include species composition, plant parts, growth stage, and species preference. The species composition refers to the proportion of species present, with percentage of legumes being the most important. Legumes usually have higher digestibility and protein levels at the same stage of maturity, and livestock tend to consume a greater quantity of legume in preference to grass. Leaf material has a higher digestibility than stems. Pasture management that maintains a high proportion of leaf will have higher digestibility and livestock performance. The digestibility of pasture declines as it matures, especially following the onset of flowering. Livestock will selectively graze preferred species, which can result in higher performance than predicted from digestibility measurements. Selective grazing also has the potential to reduce or eliminate preferred species from a pasture, especially under continuous grazing, due to limited opportunity for them to recover following grazing or a stress period (e.g., drought).

7.2.6 DETERMINING STOCKING RATES

The stocking rate depends on a number of factors including nature of grazing animal (sheep vs cattle), age of cattle (heifer vs lactating cattle), and pasture production capacity of the grazing paddock (Maczkowiack et al., 2012). For example, the current stocking rate for beef cattle in the Hunter Valley region of New South Wales, Australia is 5 dry sheep equivalent (D.S.E) per hectare on the areas of existing pasture, though, with good seasonal conditions and increased fertilizer application and seed sowings, this could increase to 7 D.S.E. per hectare for unmined lands.

As cattle graze, they defoliate a pasture through direct consumption and through trampling as they move around. If grazing pressure is managed through the assessment of plant cover and animal health and weight gains, then the pasture is able to recover with no loss in long-term yields or ground cover. Light grazing may even stimulate the pasture and improve productivity as compared to an ungrazed pasture. However, in areas where grazing practices are not managed correctly the pasture is unable to maintain regrowth. This may then result in a decline in ground cover, plants contract in size (above- and below-ground) and become less vigorous and the growing media is exposed to erosion and the associated loss of nutrients. With time, changes in species composition can also occur, with weeds and less palatable grasses becoming more prevalent. The key consideration from the land managers' perspective is the maximum long-term sustainable carrying capacity, which is determined by the broad quality of the pasture environment (Lankester, 2013).

7.3   BEEF CATTLE PRODUCTION ON REHABILITATED MINE LANDS: CASE STUDIES

7.3.1   PASTURE AND GRAZING RESEARCH, QUEENSLAND, AUSTRALIA

7.3.1.1   Sustainable Grazing on Rehabilitated Lands in the Bowen Basin—Stages 1 and 2

The following is a summary of the research undertaken by the team from Centre for Mined Land Rehabilitation, University of Queensland. This 2-year project officially commenced in February 2000, and consisted of three grazing trials established at Goonyella Riverside, Norwich Park, and Blackwater mines on sites that varied in soil type and age since establishment (3 to >25 years). At Goonyella Riverside, a short-term but intensive grazing experiment was designed to rapidly create a range of carefully controlled degradation states. Cattle were rotated through small paddocks (up to 0.8 ha in size) in two periods of four weeks to simulate a total of eight different grazing pressures. Available forage, cover, and species composition were monitored throughout the trial period.

The trials at Norwich Park and Blackwater mines were set up under continuous grazing regimes testing three grazing pressures in paddocks up to 35 ha in size. Both trials were designed to validate and expand on the conclusions drawn from the intensive short-term grazing experiment at Goonyella Riverside mine, and to provide demonstration of practical management outcomes. Again, available forage, cover, and species composition were monitored before and following the introduction of cattle in February 2000 (Blackwater) and May 2000 (Norwich Park).

Under the 2-year time horizon, assessment of the three trials can provide preliminary indicators of grazing capacity. A study was undertaken to develop a model to predict long-term sustainable stocking rates. The model predicted grass production using the concept of rainfall-use efficiency, or the ability of a pasture to produce forage per unit of rainfall. With additional long-term rainfall records and some assumptions, "safe" stocking rates can be calculated. Predictions for each site were compared with likely suitable grazing pressures observed in the trials. The study involved estimation of peak seasonal pasture growth in ungrazed plots on all three sites, with results linked to measured site and soil quality parameters.

7.3.1.2   Outcomes of the Project

Stage 1 demonstrated that pasture-based rehabilitation on opencut coal mines in central Queensland has the capacity to support cattle grazing. Estimated sustainable stocking rates on pasture rehabilitation sites at Blackwater, Norwich Park, and Goonyella Riverside mines were 2.7, 2.2, and 5.9 ha/head, respectively. Estimates were based on a model of pasture productivity that predicted the amount of feed available for cattle consumption, yet left sufficient cover for erosion control. The predicted sustainable stocking rates were in agreement with preliminary recommended rates from the three grazing trials and were comparable with stocking rates reported for improved pastures on unmined land in the region.

The second stage of the project aimed to build confidence in determining whether grazing was likely to be a sustainable land use option in the rehabilitated landscapes of the Bowen Basin following opencut coal mining. This project has defined the biophysical parameters that allow land managers to make an informed, scientific decision about which sections or parcels of land within the rehabilitated landscapes should be appropriate for supporting a managed grazing regime. Sustainable stocking rates will vary with pasture productivity, which is in turn linked to factors affecting the retention of soil moisture. At the higher end of site productivity such as the Norwich Park trial site, sustainable stocking of at least 3 ha/head could be achieved, and this is equivalent to rates on the better improved pastures on unmined lands in the region.

7.3.2 Pasture and Grazing Trials, Hunter Valley, Australia

The following trials were undertaken to assess the sustainability and profitability of grazing on mined lands in the Hunter Valley of New South Wales, Australia.

7.3.2.1 Pasture Establishment Using Biosolids and Compost Treatments

The purpose of this trial was to assess the quality and quantity of pasture produced on rehabilitation sites for beef cattle grazing and to confirm that rehabilitation practices are capable of achieving future land use targets. Four treatments were established, including a control, biosolids, and two compost treatments. Quadrats were assessed across each treatment for species diversity, herbage mass, and forage quality. Data were collected in March 2015, approximately 18 months following sowing, and compared to results taken 12 months after sowing in September 2014. The pasture mix used was based on the species as listed in Table 7.1, all of which have been historically used as part of the revegetation program.

The pasture productivity and dry matter production increased significantly in the control and biosolids treatments, but was recorded as being markedly lower in the compost treatments (Tables 7.2 and 7.3). Plant mortality due to shading in the biosolids treatment is expected to have underestimated pasture and leaf mass in this treatment (Table 7.4). Weeds were suppressed on all treatments by subtropical grasses, but remained substantially higher in the control. The weed content was inversely proportional to nutrient inputs, and the productivity of subtropical grasses. Pasture species diversity remained strongest in the Control. Rhodes grass and Green Panic continued the trend of suppressing weeds and other pasture species with prostrate growth habits. Composite samples of pasture (excluding weeds) were prepared for forage quality analysis. This included a composite sample of "total pasture" (including leaf and stalk) and a composite sample of pasture leaf only (green and dry). These results are shown in Tables 7.5 and 7.6.

Based on pasture production and digestibility, the minimum herbage mass required to maintain a satisfactory level of cattle production was estimated (Table 7.7). The biosolids treatment offers the highest potential stocking rate at around 1 dry cow/ha in the autumn, which is equivalent to 7.3 DSE/ha. In contrast, potential stocking rates on the biosolids treatment in the early spring (September 2014) were equivalent to 2 dry cows/ha or 14.6 DSE/ha. These results compare favorably to unimproved native and naturalized pastures in the Hunter valley, which average 2–3 DSE/ha on equivalent class 4 agricultural land.

TABLE 7.1
Pasture Species Mixture for Vegetation

Pasture Species	Quantity (kg/ha)
Rhodes Grass	1
Couch	5
Rye	4
Sub. Clover—Seaton Park	3
Sub. Clover—Clare	3
Haifa White Clover	2
Woolly Pod Vetch	4
Green Panic	5
Sirosa Phalaris	4
Sephi Barrel Medic	4
Lucerne (summer)	4
Kikuyu	1

TABLE 7.2
Total Herbage Mass (Grams) and Pasture Height (March 2015)

Pasture Type	Control	Biosolids	Compost 1	Compost 2
Grass pasture (g)	2376	2111	1295	1098
Legume pasture (g)	0	0	0	0
Weed (g)	216	12	81	29
Total herbage mass (g)	2592	2123	1376	1127
Average pasture height (cm)	43	49	47	42

TABLE 7.3
Total Dry Matter Pasture Available (March 2015)

Category	Control	Biosolids	Compost 1	Compost 2
Pasture available (g)	2376	2111	1295	1098
% Dry matter	47.1	52.8	41	47.5
Kg of dry matter	1.12	1.11	0.53	0.52
DM equivalent (kg/ha)	7460.6	7430.7	3539.7	3477.0

Note: Total pasture available refers to the amount of total pasture potentially available, including grass and legume minus the mass of weed.

TABLE 7.4
Pasture Leaf Available (March 2015)

Category	Control	Biosolids	Compost 1	Compost 2
Total pasture available (g)	2376	2111	1295	1098
% Leaf	27.4	28.5	52.1	59.0
Total leaf available (g)	651.0	601.6	674.7	647.8
% Dry matter	62.8	67.5	53.8	58.2
kg dry matter	0.41	0.41	0.36	0.38
DM equivalent (kg/ha)	2725.6	2707.4	2419.9	2513.5

TABLE 7.5
Total Pasture Quality (Leaf + Stalk)

Category	Control	Biosolids	Compost 1	Compost 2
% Dry matter digestibility	50.0	53.0	49.0	50.0
Crude protein (% dry matter)	4.1	4.7	1.0	2.3
Metabolizable energy (MJ/kg DM)	6.9	7.5	6.7	6.9

TABLE 7.6
Pasture Leaf Quality

Category	Control	Biosolids	Compost 1	Compost 2
% Dry matter digestibility	53.0	54.0	51.0	51.0
Crude protein (% dry matter)	7.5	7.5	3.7	3.7
Metabolizable energy (MJ/kg DM)	7.4	7.7	7.2	7.2

TABLE 7.7
Minimum Herbage Mass (kg Green DM/ha) to Maintain Satisfactory Production Levels in Cattle

	Pasture Digestibility		
Cattle Class	75%	68%	60%
Dry cow	700	1100	2600
Pregnant cow (7–8 months not lactating)	900	1700	ns
Lactating cow (calf 2 months)	1100	2200	ns
Growing stock (% of potential growth)			
30 (0.39 kg/day)	600	1100	2900
50 (0.61 kg/day)	800	1600	ns
70 (0.85 kg/day)	1200	2600	ns
90 (1.12 kg/day)	2200	ns	ns

Source: Bell, A., Pasture assessment and livestock production—PRIMEFACT 323, NSW Department of Primary Industries, Orange, New South Wales, Australia, December 2006.

Notes: Predicted growth rates in brackets are based on a weaned 13-month old steer of approximately 320 kg from a cow with standard reference weight of 500 kg.

ns—not suitable, at this digestibility no matter how much pasture is available, dry or pregnant stock are unlikely to maintain weight, lactating stock are likely to experience unacceptable weight loss, and growing stock will not be achieving targeted weight gain.

The biosolids treatment continues to show higher forage quality and higher potential stocking rates than the other treatments. Indicative pasture productivity and stocking rates are approximately 2–3 times higher than native pasture of equivalent land class. The compost treatments had significantly lower digestibility and protein levels, and reflect the lower pasture productivity observed in these treatments. Reasons for this result are unclear and further soil testing is required. Grazing is likely to complement long-term objectives of improving pasture species diversity and forage quality.

7.3.2.2 Assessment of Sustainable Beef Cattle Production

The objective of this study was to ascertain if rehabilitated mine land supports cattle grazing on a sustainable basis. The trial was established on rehabilitated mined land ranging from 3 to 10 years since the establishment of the pasture and adjoining unmined grazing land. The results of the trial reported here cover the period December 2012 to June 2014. Steers for the trial came from similar breeding to reduce variation between cattle. Cattle were randomly allocated to rehabilitated or natural pasture. The stocking rate equivalent of 1 steer/2.4 ha was selected, this being as suggested by local agronomists, but also being slightly above the district average for year round stocking.

TABLE 7.8
Stock Type, Pasture, and Soils

Category	Rehabilitated	Natural (Unmined)
Stock type	30 steers/70 ha (1 steer/2.4 ha)	30 steers/70 ha (1 steer/2.4 ha)
Pasture	Rhodes grass dominant, Couch, Kikuyu, medic	Redgrass wiregrass dominant, Paspalum, weeping grass
Soils	Neutral-alkaline	Mildly acidic
	Phosphorus (Colwell) 5–60 g/kg	Phosphorus (Colwell) 6–14 g/kg
	CEC 10–19 meq/100 g of soil	CEC 2.5–7 meq/100 g of soil
	Salinity (EC) 0.4–1.6 Se	Salinity (EC) 0.3–0.7 dS/m
	Sodicity (ESP) 0.5%–6.3%	Sodicity (ESP) 1.6%–3.5%
	Organic carbon 0.8%–4.6%	Organic carbon 1.7%–2.5%

Baseline monitoring parameters included cattle weights, blood testing from a sub-sample of 10 cattle in each treatment—protein, P, Se, Cu, Ca, Mg, Mn—to determine mineral nutrition status of the cattle (at entry and end of trial), pasture ground cover, species, and feed value. Ongoing monitoring included monitoring of pastures and cattle weights every 2–4 months to coincide with cattle rotations, regular inspections to assess suitability of stocking rates, pasture availability/condition, cattle condition, and health and carcass feedback from point of sale.

Details on the pasture and stock are shown in Table 7.8. The following results were defined in terms of the pasture quality:

- Quality of pasture was relatively low throughout most of trial due to high levels of rank low quality pasture and poor seasonal conditions limiting fresh growth.
- Little pasture growth occurred during drought periods. Cattle weight loss occurred during winter 2013 (April–August) and low cattle growth rates occurred spring 2013 (August–November 2013). This is a period when pasture quality and quantity declined significantly.
- Higher cattle growth rates were achieved after rain in November 2013, when fresh pasture growth supported cattle growth rates of 0.9 kg/hd/day (both rehabilitation and natural pastures).
- Good autumn rain produced good pasture response when quality of the pasture increased cattle growth rates during the period from March 2014 to May 2014 (1.2 kg/hd/day Rehabilitated, 1.0 kg/hd/day Natural).
- Average ground cover levels have been maintained above the 70% level throughout the trial. This has minimized the potential for erosion. Figure 7.1a and b show beef cattle grazing on rehabilitated land, Hunter Valley, Australia.

In context of cattle weight the following results were obtained:

- On entry, there was no significant weight difference between natural pasture and rehabilitation steers.
- Results after approximately 18 months showed rehabilitation cattle grew quicker and the overall weight difference was +79 kg in rehabilitation cattle over natural pasture cattle.
- At each of the subsequent weighing (7 over 533 days) cattle grazing on the rehabilitated pastures were significantly heavier than the cattle grazing on the natural pastures.
- Period of highest weight gain was recorded between late March 2014 and end of April 2014, which corresponded with fresh feed availability stimulated after rain.

(a)

(b)

FIGURE 7.1 Beef cattle grazing on rehabilitated land: (a) Hunter Valley, Australia and (b) Bowen Basin, Queensland, Australia.

In context of carcass comparison, the following results were obtained:

- All the cattle were removed from the trial and sold in June 2014 because they were in a marketable weight and condition.
- All cattle were slaughtered on June 3, 2014 at Meat Standards Australia accredited abattoir (Hunter Valley Quality Meats, Scone, NSW).
- Results showed that the extra weight and condition of the rehabilitation cattle related to $220 per head or $6588 overall more for the carcasses.
- Nearly 25% greater return was obtained from cattle grazing rehabilitated pasture than cattle grazing natural pastures.

7.4 SUMMARY AND CONCLUSIONS

Beef cattle production on mine lands can be undertaken after completing rehabilitation processes such as land form establishment, growing media placement and revegetation. The role of tropical and temperate, exotic and native pasture species and organic amendments to restore the rehabilitated mine land for sustainable environment and profitable beef cattle production is worth exploring. However, the case studies have indicated that tropical and temperate pastures which are synonymous with rehabilitated post mined lands have a feed quality and productivity capable of supporting a beef cattle farming system.

REFERENCES

Bell, A. December 2006. Pasture assessment and livestock production—PRIMEFACT 323, NSW Department of Primary Industries, Orange, New South Wales, Australia.

Commonwealth of Australia. 2006. Mine rehabilitation: Leading practice sustainable development program for the mining industry, Department of Industry, Tourism and Resources, Canberra, Australian Capital Territory, Australia.

Fastfacts. 2013. Meat and Livestock Australia, Canberra, Australian Capital Territory, Australia.

Hart, K. J., P. G. Martin, P. A. Foley, D. A. Kenny, and T. M. Boland. 2009. Effect of sward dry matter digestibility on methane production, ruminal fermentation, and microbial populations of zero-grazed beef cattle. *J. Anim. Sci.*, 87(10): 3342–3350.

Lankester, A. 2013. Conceptual and operational understanding of learning for sustainability: A case study of the beef industry in north-eastern Australia. *J. Environ. Manage.* 119: 182–193.

Maczkowiack, R. I., C. S. Smith, G. J. Slaughter, D. R. Mulligan, and D. C. Cameron. 2012. Grazing as a post-mining land use: A conceptual model of the risk factors. *Agric. Syst.* 109: 76–89.

Mattiske, A. 2016. Mine rehabilitation in the Australia mineral industry—An industrial report, Minerals council of Australia, Canberra, Australian Capital Territory, Australia.

8 Restoring Forests on Surface Coal Mines in Appalachia
A Regional Reforestation Approach with Global Application

Christopher D. Barton, Kenton Sena,
Teagan Dolan, Patrick Angel, and Carl Zipper

CONTENTS

8.1 Overview 124
8.2 Surface Coal Mining and Reclamation 124
    8.2.1 Surface Mining 124
    8.2.2 Pre-SMCRA Reforestation 126
    8.2.3 Post-SMCRA Reforestation 127
8.3 Forestry Reclamation Approach 129
    8.3.1 FRA Step 1: Create a Suitable Rooting Medium for Good Tree Growth
        That Is No Less Than 4 ft (1.2 m) Deep and Comprises Topsoil,
        Weathered Sandstone, and/or the Best Available Material 129
    8.3.2 FRA Step 2: Loosely Grade the Topsoil or Topsoil Substitutes Established
        in Step One to Create a Noncompacted Soil Growth Medium 130
    8.3.3 FRA Step 3: Use Less Competitive Ground Covers That Are Compatible
        with Growing Trees 130
    8.3.4 FRA Step 4: Plant Two Types of Trees—Early Successional Species
        for Wildlife and Soil Stability and Commercially Valuable Crop Trees 131
    8.3.5 FRA Step 5: Use Proper Tree Planting Techniques 131
8.4 Testing the FRA: Case Studies 131
    8.4.1 The Bent Mountain Project: Examining Spoil Chemical Influences
        on Reforestation Success 131
        8.4.1.1 Soil 132
        8.4.1.2 Tree Survival and Growth 132
        8.4.1.3 Understory Recruitment 133
        8.4.1.4 Hydrology and Water Chemistry 133
    8.4.2 The Starfire Project: Examining Spoil Physical Influences
        on Reforestion Success 135
        8.4.2.1 Survival 136
        8.4.2.2 Growth 136
8.5 The Appalachian Regional Reforestation Initiative 138
8.6 Impacts 141
8.7 Global Application 142
References 142

8.1 OVERVIEW

The Appalachian region of the United States is a land of contrast—people have suffered from poverty for decades, but the region abounds in natural resources. Appalachian forests support some of the greatest biological diversity in the world's temperate region, but extraction of its abundant coal reserves has impacted the landscape. Surface mining poses a significant threat to the region via forest loss and fragmentation. Since the implementation of the federal Surface Mining Control and Reclamation Act (SMCRA) of 1977 (Public Law 95-87), more than 600,000 ha have been mined for coal, and efforts to reforest these areas had historically resulted in high seedling mortality, slow growth, and poor production. Research showed that highly compacted soils with inappropriate chemical characteristics and intense competition from ground cover were the biggest impediments to the establishment of productive forests on surface mines in the eastern United States (Ashby et al. 1978; Burger and Torbert 1997; Graves et al. 2000). Other obstacles for reforesting surface mines included lack of careful selection of a rooting medium for tree roots, selection of tree species that were not suited to site conditions, and improper tree planting techniques. Realizing the significance of these problems, enhanced efforts to address reforestation shortcomings were examined by regulatory, mining, and research groups alike to ensure that forests are restored to the region. Drawing on the recommendations generated by surface mine reclamation research over the past 80 years, a five-step system to reforest coal-mined land called the Forestry Reclamation Approach (FRA) was developed. The steps in the FRA are to (1) create a suitable rooting medium for good tree growth that is no less than 1.2 m deep and comprises topsoil, weathered sandstone, and/or the best available material; (2) loosely grade the topsoil or topsoil substitutes placed on the surface to create a noncompacted growth medium; (3) use native and noncompetitive ground covers that are compatible with growing trees; (4) plant two types of trees—early succession species for wildlife and soil stability and commercially valuable crop trees; and (5) use proper tree planting techniques (Burger et al. 2005). Today, the FRA has been applied by many coal mining firms in the United States, millions of trees have been planted, ecosystem services have been returned to thousands of mined acres and the native Appalachian forest is returning (Zipper et al. 2011b). Successful reestablishment of the hardwood forest ecosystem that once dominated these sites, made possible by FRA, will provide a renewable, sustainable, multiuse resource that will create economic opportunities while enhancing the local and global environment. Although climate, geology, and soils will ultimately dictate agronomic prescriptions for global reforestation projects, the FRA provides a framework for developing a successful reforestation program that may be applied to other mining regions of the world.

8.2 SURFACE COAL MINING AND RECLAMATION

8.2.1 Surface Mining

The Appalachian coalfield is one of the most important coal-producing regions in the United States. The basin covers 63,000 miles^2 (160,000 km^2) and extends from western Pennsylvania to north Alabama (Figure 8.1). At one time, two-thirds of U.S. coal was mined in the Appalachian basin using both underground and surface mining techniques (Ruppert 2001). The three major surface mining methods for coal in Appalachia are area, contour, and mountaintop removal mining. These methods involve the same basic steps: clearing the land of trees and other vegetation, removing the topsoil and rock over the coal, mining the coal, and reclaiming the land (Figure 8.2). Mechanized equipment such as bulldozers, scrapers, power shovels, bucket-wheel excavators, draglines, and similar machinery strip away or push aside the overlying soil and rock called overburden. The excavated overburden is initially placed in piles or long parallel ridges composed of an agglomeration of substrate, soil, and weathered and unweathered rocks called spoil.

The method of coal surface mining used depends on the local topography. The area mining method is commonly used in the flat to moderately rolling terrain found principally in the western

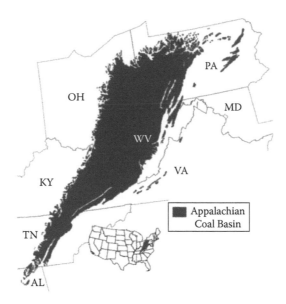

FIGURE 8.1 Location of the Appalachian coalfield within eastern United States.

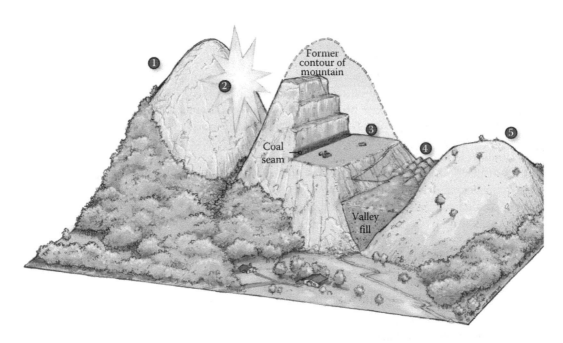

FIGURE 8.2 The steps associated with mountaintop surface mining in the mountainous areas of Appalachia. (1) *Site preparation*: involves removing vegetation from the area to be mined and development of erosion control structures. (2) *Overburden removal*: involves moving soil and rock above the coal seam using equipment or explosives. (3) *Mining of the coal*: involves isolating the coal seam, extracting the coal and removal off site. (4) *Creation of valley fills*: excess spoil is trucked or pushed in valleys. (5) *Reclamation*: spoil is graded to the "approximate original contour" of the pre-mined condition and revegetated. Area mining and contour mining are other forms of surface mining that occur in Appalachia that do not require removal of the entire mountaintop, but still produce excess spoil. (From Ware, C., Reclaiming mountains: A special report, *Lexington Herald-Leader*, Lexington, KY, 2009, http://www.kentucky.com/news/local/coal/article44006004.html, accessed August 1, 2007. With permission.)

and mid-western United States. A large dragline is often used to dig a trench through the overburden to expose the coal seam. After the coal is removed, a second pit is dug parallel to the first, and the overburden is deposited in the first trench. The process is repeated over the entire site. Until laws were passed in the 1950s and 1960s that required grading the spoil to resemble the terrain that existed prior to mining, partial grading or "strike-off reclamation" was required on area mines. This requirement consisted of making one or two bulldozer passes down the length of each parallel ridge, pushing spoil into the parallel valleys on both sides. When viewed from the air, these loosely graded parallel valleys and ridges of rubble, known as spoil banks, resembled the rough surface of a giant washboard.

The contour mining method is used in the hilly or mountainous terrain of the Appalachian coal fields where coal seams are exposed in outcrops on hillsides and mountainsides. Contour mines are usually large, but extend in great length and narrow width. First, an excavation is made in the mountainside above a coal seam and the coal is exposed as the overburden is removed. This creates a working bench and a vertical wall of rock and soil called a highwall. It is possible to make multiple cuts and mine more than one seam, which enlarges the disturbance. Mining extends into the hill to the point where the overburden is too thick to make further exposure of the coal cost-effective. Before the State and Federal laws regulating contour mining were passed, the overburden was usually blasted and/or pushed onto the downslope, or hillside below the working area of the mine, to expose the coal seam. The final result was an almost level bench, following the contour around the mountains, with a steep outer slope consisting of very loose and highly erodible spoil and the exposed highwall. Under current laws, spoil is retained on the bench to eliminate the highwall. But all of the spoil that is generated during mining will usually not fit back onto the bench because it swells when it is displaced from its natural state. Therefore, disposal areas are needed for the excess. The types of disposal sites characteristic of contour mines are head-of-hollow fills, valley fills, and durable rock fills. Each serves the same purpose of providing a site for disposal of excess spoil material.

Mountaintop removal mining is also used in the hilly or mountainous terrain of the Appalachian coal fields, where it is economically feasible to remove the whole top of a mountain to gain access to a coal seam or several coal seams. Mountaintop removal mining is actually an adaptation of area mining and contour mining and involves several steps. First, disposal sites are prepared to hold any excess spoil. Next, an excavation is made in the mountainside above the coal seam just as in the contour mining method. Overburden from this cut is placed in the excess spoil disposal sites. A second cut is made parallel to the first. Then overburden from the succeeding cuts is placed in the cut previously made, similar to the area mining method. The mountaintop is progressively reduced by a series of parallel cuts and the entire hilltop is removed all the way down to the coal seams. Excess spoil is permanently stored in much larger head-of-hollow fills, valley fills, or durable rock fills than what is usually found on contour mining operations. The spoil that remains on the mountaintop is backfilled and then graded to create a huge plateau of flat or gently rolling reclaimed land.

8.2.2 Pre-SMCRA Reforestation

Prior to SMCRA, reclamation efforts on surface mines in the United States varied widely. Economics usually took precedence over adequate reclamation, and many sites were left in a disturbed state. Ruined, denuded landscapes created many environmental issues—spoil piles were subjected to accelerated erosion, rockslides, and landslides; streams became clogged with sediment and acid mine drainage; groundwater was permanently impaired; dangerous highwalls were left exposed; and topsoil was buried or allowed to wash away. Even after 50–60 years, some of these sites remain destabilized.

As these practices continued, outrage and concern by the public grew and mining companies frequently responded by implementing tree planting projects (DenUyl 1955). By the 1940s, many states had passed laws that required coal operators to revegetate mined land, which was generally achieved through afforestation/reforestation efforts (Mickalitis and Kutz 1949; Potter et al. 1951;

FIGURE 8.3 Examples of 55-year-old black walnut trees planted on pre-SMCRA spoil banks in Pike County, IN. (Courtesy of Ron Rathfon.)

Evilsizer 1980; Plass and Powell 1988). Trees were planted at the rate of 1500–2000 trees ha^{-1} in loosely graded spoil banks of the Midwestern United States and in the loose spoil on the downslopes of Appalachia. Forests on these early mines, established decades ago, are now maturing to valuable, productive, and diverse forests (Pope 1989; Ashby 1991). Southern Illinois University researchers have reported that mine sites subjected to the loosely graded strike-off reclamation of the 1950s and 1960s resulted in areas with the highest site indexes in Illinois for yellow-poplar (*Liriodendron tulipifera*), white oak (*Quercus alba*), and black walnut (*Juglans nigra*) (Ashby et al. 1978) (Figure 8.3). Many Midwestern United States and Appalachian sites mined and reforested under pre-SMCRA conditions have repeatedly been shown to support healthy and productive forests across a wide range of environmental conditions (Plass 1982; Andrews 1992).

8.2.3 Post-SMCRA Reforestation

Passage of the Surface Mining Control and Reclamation Act in 1977 challenged the coal industry and regulators to adopt a different mindset regarding reclamation than what existed prior to the law. It required surface mine operators to restore the approximate original contour (AOC) of disturbed land, to eliminate highwalls by backfilling and grading them, to replant trees and grass, and to prevent the pollution and sedimentation of streams. Working in tandem with the AOC and highwall elimination stipulations was the prohibition of placing spoil on the downslope in the mountainous areas of the coal fields. In the decades prior to SMCRA, gravity transport of spoil was common as it was pushed or cast upon the downslope by surface mine operators. Although the spoil placement that resulted from this practice created a loose rooting medium that was conducive to the growth of trees, these areas had a propensity for being unstable. Landslides, massive slumping of spoil, serious erosion and sedimentation, and the endangerment of coal field citizens were commonplace in many areas on pre-SMCRA surface mines. SMCRA improved the landforms created by surface mining by increasing stability, which enhanced human safety and reduced sedimentation in the Appalachian

FIGURE 8.4   An introduced elk species thrives on the expansive grasslands in Appalachia that were created after the passage of SMCRA. The site pictured located in Perry County, KY, was entirely forested prior to being mined for coal. (Courtesy of John Cox.)

region; however, SMCRA's implementation has not been accompanied by widespread replacement of forests disturbed by mining (Burger and Torbert 1992).

A desire to prevent land instability and sedimentation associated with pre-SMCRA surface mining led to widespread adoption of the practice of intensive grading and soil compaction with mine equipment and establishment of aggressive ground covers (Angel et al. 2005). As a consequence, forest reclamation took a backseat to establishment of herbaceous ground covers and forest fragmentation increased within the region (Wickham et al. 2007; Townsend et al. 2009). Early efforts by mine operators to reforest under SMCRA proved problematic, in part because these efforts were conducted without the benefit of scientific knowledge that is available today; as a result, mine operators and regulators came to believe that post-mining land uses such as hay and pasture land were easier and cheaper to achieve than forests. As a result, where majestic forests had once stood, now lay several hundred thousand acres of sparse grasslands (Angel et al. 2005; Wickham et al. 2007, 2013) (Figure 8.4).

The problems associated with establishing trees on surface mines in the 1980s and 1990s did not go unnoticed by reforestation researchers. Numerous studies reported that reestablishing productive forests on surface mines reclaimed after the enactment of SMCRA were extremely difficult, with the biggest impediment being excessive compaction (Ashby et al. 1978; Davidson et al. 1983; Ashby 1991; Burger and Torbert 1997; Burger 1999; Graves et al. 2000). According to Boyce (1999), early interpretation of SMCRA in the field "...pitted quick ground cover and water quality against reforestation, and water quality prevailed." When reforestation efforts failed on pre-SMCRA surface mines, it was usually the result of soil chemistry rather than soil physical conditions. Conversely, on post-SMCRA mine sites the opposite was often demonstrated (Davidson et al. 1983; Conrad et al. 2002; Sweigard et al. 2007a).

Early studies in Illinois and Virginia reported similar results after comparing forest growth on surface mines reclaimed after the implementation of SMCRA. Ashby and Kolar (1998) tested tree productivity on Illinois sites that were undisturbed, sites that were mined and left ungraded, and sites that were mined, graded, and topsoiled. After 13 years, tree height and stem diameter were greater on the ungraded mine sites than all other sites, including the undisturbed site. Tree heights on the sites that were mined, graded, and topsoiled were the lowest. Torbert et al. (2000) reported 11-year test planting results for three pine species planted on a pre-SMCRA and a post-SMCRA reclaimed mined site in Virginia. Trees on the pre-SMCRA mined site were planted on the bench created by conventional contour mining. On the post-SMCRA site, trees were planted on the portion returned to AOC. The heights and diameters of all three pine species were greater at the pre-SMCRA mined site than at the post-SMCRA mined site.

Davidson et al. (1983) reported short-term findings by Reising who observed in 1982 that only 21% of the forested areas being surface mined were being returned to forests. Conversion of native forests to unmanaged hay and pasture lands and wildlife plantings became the norm (Plass and Powell 1988; Andrews 1992; Torbert et al. 1995). Unfortunately, grazing to support the reclaimed hay and pasture lands was not widely performed and the sites often were abandoned from these designated uses. Over time, thick covers of highly competitive and aggressive exotic grasses, legumes, forage, and shrub species—that were often not native to the region—colonized the reclaimed areas (Zipper et al. 2011a; Franklin et al. 2012; Wickham et al. 2013). Such species have the propensity to arrest natural succession and the development of a more diverse biological community for many decades (Ashby 1987; Groninger et al. 2007; Franklin et al. 2012).

8.3   FORESTRY RECLAMATION APPROACH

Research has demonstrated that forested vegetation can be established on surface coal mines, and that it can be both productive and diverse. When properly prepared for forest vegetation, mine soils can provide a deep rooting medium, be rich in geologically derived nutrients, such as calcium, magnesium, and potassium, and produce forests with similar growth characteristics as that of pre-mined native forest (Zipper et al. 2013). Rodrigue et al. (2002) reported forest growth on 12 of 14 selected older coal mine sites in the eastern and midwestern United States that was similar to local unmined forests. Casselman et al. (2007) measured a 50-year "site index" (expected average height after 50 years) of 32.3 m for a 26-year-old white pine stand growing on an uncompacted mine site in Virginia, considerably greater than the average site index for the southern Appalachians. The 50-year site index of yellow poplar growing on 40–50-year-old Tennessee mine sites with uncompacted soils averaged 32.3 m, which was greater than the 26.5 m regional average (Franklin and Frouz 2007). Cotton et al. (2012) showed that 10-year-old yellow poplar and white oak growing on loose dumped Kentucky spoils exhibited similar stem diameters to that of regenerating nonmined stands of the same age.

It is clear that both the productivity and the diversity of forest vegetation on surface coal mines are influenced by reclamation practices. Several studies (Torbert et al. 1990; Angel et al. 2008; Emerson et al. 2009; Wilson-Kokes et al. 2013; Zipper et al. 2013; Sena et al. 2014) have reported that mine-soil properties and practices used to construct them greatly influence growth of planted trees.

Based on these observations, reclamation scientists have developed a method of reclaiming mined land called the "Forestry Reclamation Approach" (FRA) that can be used to produce site conditions that are favorable to forest vegetation (see Burger et al. 2005, 2010). The FRA is presented as a series of "five steps," all of which are necessary for successful mine reforestation.

Native Appalachian forests host numerous plant species, and it would be impossible to establish the full forest community on the mine site by seeding and planting. The FRA is intended to establish site conditions suitable for colonization by native vegetation whose seeds are carried by birds and wind, while planting tree species with large seeds, such as oaks that are important to the native forest, but do not volunteer quickly into reclaimed areas.

8.3.1   FRA Step 1: Create a Suitable Rooting Medium for Good Tree Growth That Is No Less Than 4 ft (1.2 m) Deep and Comprises Topsoil, Weathered Sandstone, and/or the Best Available Material

The properties of spoil materials used to create a mine soil will influence both the survival and growth of planted trees and the "natural succession" of native vegetation that is essential to ecosystem development (Groninger et al. 2007). The "best available" growth medium will depend on the local conditions. While topsoil is a valuable resource for reclamation and is recommended for use when available—its availability being limited and selected—alternate growth media are often used to support productive forestland (Burger and Zipper 2011).

During mining operations, all highly alkaline materials with excessive soluble salts and all highly acidic or toxic material should be covered with 4–6 ft of a suitable rooting medium that will support trees. Growth media with low to moderate levels of soluble salts, equilibrium pH of 5.0–7.0, low pyretic sulfur content, and textures conducive to proper drainage are preferred (Burger et al. 2005). Native hardwood diversity and productivity will be best on soils with a sandy loam texture, where the pH is between 5 and 7. These types of soils can be formed from overburden materials comprising weathered brown or unweathered gray sandstone, especially if these materials are mixed with natural soils. Shale may be used in combination with sandstone; however, high concentrations of shale and other fine-grained spoil materials should be avoided. Many times these materials have higher pH values, which encourages heavy ground cover but inhibits tree growth. On remining sites, topsoil/sandstone may not be available in sufficient quantities. In these cases, a combination of spoil materials will be required to create the best available growth medium.

8.3.2   FRA Step 2: Loosely Grade the Topsoil or Topsoil Substitutes Established in Step One to Create a Noncompacted Soil Growth Medium

Because soil compaction hinders survival and growth of planted trees, minimizing soil compaction is essential for effective mine reforestation. A loose mine soil with a rough surface and favorable chemical properties is best for forest reestablishment. On the other hand, excessive soil density will depress the survival and growth of planted trees (Torbert and Burger 1994; Sweigard et al. 2007a), recruitment of nonplanted forest species (Angel et al. 2008; Sena et al. 2015), and the economic value of timber that eventually grows on the mine site (Burger and Zipper 2011).

A loose soil suitable for reforestation can be achieved by limiting reshaping and grading of surface materials to the minimum needed for the desired land configuration (Sweigard et al. 2007a). In area mining, haul trucks are used to dump (end-dump) the growth medium in a tight arrangement, and final grading is accomplished with one or two light passes with a dozer to strike off the tops of the dump piles. Likewise, in a dragline operation, the growth medium is placed in piles and a dozer lightly grades the area leaving a rough, noncompacted growth medium. In steep slope mining areas, the majority of the backfill is placed and compacted as usual, but the final 4–6 ft of growth medium is dumped and lightly graded to achieve the required final grade. This low compaction technique will actually reduce erosion, provide enhanced water infiltration and restore the hydrologic balance, and allow trees to achieve good root penetration. Placing soils to create loose, rough, uneven surfaces will actually reduce erosion on many mine sites because this condition allows more water infiltration, leaving less water runoff to erode and carry soil. The increased water infiltration aids tree growth, and the loose soil aids rooting by planted trees. The rough soil surface also aids plant recruitment by providing microsites to hold and germinate seeds carried to the mine site by wind, birds, and other wildlife.

8.3.3   FRA Step 3: Use Less Competitive Ground Covers That Are Compatible with Growing Trees

Ground cover vegetation used in reforestation requires a balance between soil coverage and competition for the light, water, and space required by trees. Fast growing grasses and legumes that have been used traditionally in coal mine reclamation, such as fescue (*Festuca arundinacea*), sericea lespedeza (*Lespedeza cuneata*), and certain clovers (e.g., *Melilotus officinalis*), are too aggressive and competitive for use with trees. Alternatively short-statured, bunch-forming grasses, such as rye (*Secale cereale*), red top (*Agrostis palustris*), perennial ryegrass (*Lolium perenne*), and orchardgrass

(*Dactylis glomerata*) are less competitive and can be established on mine sites easily. Ground-cover legumes convert nitrogen, an essential plant nutrient, from atmospheric to plant-available forms, and legume species such as birdsfoot trefoil (*Lotus corniculatus*) and Ladino clover (*Trifolium repens*) are less competitive than those used traditionally in reclamation. These slower growing, less competitive ground covers can be used without causing excessive erosion because the loosely placed mine soils recommended in Step 2 allow greater rates of water infiltration. To further discourage heavy ground-cover growth, applied fertilizers are low in nitrogen, but include sufficient phosphorus and potassium to support optimal tree growth (Burger et al. 2005).

The tree-compatible ground cover is typically sparse over the first growing season. The sparse ground cover minimizes competition with the planted seedlings allowing recruitment by non-seeded native plants that will initiate natural succession. If the mine soils have been prepared in accord with FRA recommendations, seeded and volunteer herbaceous plant cover will commonly approach 100% by the third growing season.

8.3.4 FRA Step 4: Plant Two Types of Trees—Early Successional Species for Wildlife and Soil Stability and Commercially Valuable Crop Trees

Crop trees are long-lived species that can produce saleable forest products. Northern red oak (*Quercus rubra*), white oak, green ash (*Fraxinus pennsylvanica*), black cherry (*Prunus serotina*), sugar maple (*Acer saccharum*), yellow-poplar, and other native hardwood species are commonly planted as crop trees on mines that are using the FRA.

Early successional "nurse trees" are also planted to enhance the soil's organic matter and nitrogen contents, and to attract seed-carrying wildlife. Eastern redbud (*Cercis canadensis*), hawthorn (*Crataegus mollis*), and dogwood (*Cornus florida*) are commonly planted as nurse trees. Leguminous nurse species, such as redbud convert atmospheric nitrogen to plant available forms, thus aiding accumulation of this essential nutrient in the emerging forest system. Tree species that produce edible fruits and seeds at a relatively young age, such as dogwood, are often planted to attract wildlife that carries seed from unplanted species into the mine site.

8.3.5 FRA Step 5: Use Proper Tree Planting Techniques

In order for the mine site to be reforested successfully, tree seedlings must be planted properly (Davis et al. 2010). Mine sites are often planted by commercial contractors, as their planting crews can care for seedlings properly prior to planting and get large mine sites planted quickly. Proper tree planting requires a planting hole that is deep enough to accommodate the seedling's root system, and use of FRA Steps 1 and 2 to prepare the soil material will leave the soil loose, making it easier for tree planters to open an adequate hole.

Additional information on each of the FRA's five steps can be found at: http://arri.osmre.gov.

8.4 TESTING THE FRA: CASE STUDIES

8.4.1 The Bent Mountain Project: Examining Spoil Chemical Influences on Reforestation Success

In order to investigate more thoroughly the influence of mine spoil composition on reforestation outcomes, a series of 0.4-ha spoil test plots were established on Bent Mountain, in Pike County, Kentucky, USA (37° 35.88′ N, 82° 24.31′ W). Three spoil types (weathered brown sandstone, unweathered gray sandstone, and mixed sandstones and shale), representing the dominant spoil types found in the region, were end-dumped in duplicate (two plots per spoil type, for a total of six plots) and strike-off graded with one pass of a light dozer (as described in FRA #2 above). Plots were planted in early 2005 with four native hardwood species: red oak (*Quercus rubra*), white oak (*Quercus alba*), green ash (*Fraxinus pennsylvanica*), and yellow poplar (*Liriodendron tulipifera*). No soil amendments were applied, and no groundcover was seeded. In order to investigate spoil type influence on hydrology and water chemistry, plots were constructed on a graded base layer, designed so that interflow would drain out one side via drainage tile. A pit was constructed at each tile outflow. Interflow from each plot passed through a monitoring station and was drained via holes drilled to an underlying deep mine. This design permitted investigation of water flow and chemistry patterns without contaminating adjacent plots. (See Angel et al. 2008, for a detailed description of plot construction.)

8.4.1.1 Soil

In Appalachia, development of healthy and productive soils is critical to reforestation. Because native soil is frequently not retained during surface mining operations, mine soils used for surface mine reclamation are typically composed of crushed overburden, and are characterized by low organic matter content and high conductivity (Zipper et al. 2013). Differences among spoil types in soil physical and chemical properties were observed at the beginning of the Bent Mountain experiment. While soil pH was slightly acidic in BROWN, both GRAY and MIXED were slightly alkaline. Also, soil moisture content was consistently higher in BROWN than GRAY or MIXED (Angel et al. 2008). These differences were still evident in 2013 where soil pH in BROWN was 6.0, and pH in GRAY and MIXED was consistently >8.0. Similarly, clay content was higher on BROWN than GRAY and MIXED resulting in higher soil water holding capacity in the BROWN. While soil nitrogen and phosphorus did not differ among treatments, we observed that nitrogen-fixing species were common in BROWN and uncommon on GRAY and MIXED (Sena et al. 2015).

8.4.1.2 Tree Survival and Growth

Survival of all planted species was statistically similar on all plots in 2013 (64% on GRAY, 68% on MIXED, and 86% on BROWN), although white oak was apparently declining on GRAY and MIXED. Green ash survival was >90% on all spoil types through 2013, indicating that it is an excellent candidate for mine reclamation even on soils unfavorable for other species (Sena et al. 2015).

While tree survival was similar across spoil types, tree growth on BROWN rapidly outpaced growth on GRAY and MIXED (Table 8.1). By 2013, average tree volume for planted species on BROWN was 50× higher than GRAY, and 6× higher than MIXED (Sena et al. 2015).

TABLE 8.1

Tree Height, Volume, and Diameter Means and Standard Errors for all Planted Species by Spoil Type and Year

Year	Spoil	Height (cm)	Diameter (mm)	Volume (cm³)
2005	BROWN	36.8a ± 0.64	5.7a ± 0.1	16.1a ± 1.1
	GRAY	36.3a ± 0.68	5.7a ± 0.1	16.2a ± 0.95
	MIXED	37.0a ± 0.28	5.6a ± 0	15.2a ± 0.45
2006	BROWN	51.5 ± 4.18	10.6 ± 0.65	88.6 ± 21.05
	GRAY	37.7 ± 1.26	7.25 ± 0.15	28.6 ± 2.05
	MIXED	41.7 ± 0.71	7.85 ± 0.25	37.9 ± 1.7
2007	BROWN	66.0 ± 5.70	13.8 ± 0.75	237.6 ± 76.6
	GRAY	34.6 ± 2.39	7.6 ± 0.6	34.8 ± 6.9
	MIXED	44.9 ± 0.78	9.35 ± 0.25	85.7 ± 10.3
2013	BROWN	312.6a ± 7.80	48.7a ± 0.6	12,220a ± 292
	GRAY	50.5c ± 13.6	11.3c ± 2.8	236.7c ± 115
	MIXED	114.3b ± 3.3	22.0b ± 1.3	1,837b ± 277

Note: Means with the same letter are not statistically different (p = 0.05) among S types within year.

8.4.1.3   Understory Recruitment

In this study, groundcover species were not seeded during plot establishment. Seeded groundcover, although considered beneficial for erosion control (especially on steep slopes), can negatively impact reforestation outcomes by competing with planted trees for scarce nutrients and water. In the second growing season after plot establishment, natural vegetative recruitment resulted in 42% groundcover in BROWN, compared to 1% on GRAY and 2.6% on MIXED. This disparity was still obvious in 2013, when ground cover on BROWN was nearly 100%, while only 10% on GRAY and 20% on MIXED (Angel et al. 2008; Sena et al. 2015).

Soil development on reclaimed surface mined sites is critical for reforestation success. Soil pH is a major determinant of soil chemical function, especially controlling mineral and nutrient availability. Soil pH of native Appalachian forests is typically fairly acidic (4.8–5.6), and the native Appalachian trees we planted (especially white oak) are well-adapted to conditions in that range. Thus, pH was likely a dominant factor contributing to soil suitability on BROWN and unsuitability on GRAY and MIXED. In addition, moisture availability tends to be a critical factor on surface mined sites. Because mine soils typically have low fine particle fractions and little to no organic matter, moisture retention can be very poor, and soils can become xeric during hot, dry periods. The clay fraction was higher in BROWN than GRAY or MIXED, contributing to higher soil field capacity.

Vegetative recruitment and tree growth were also likely a major factor in mid- to long-term ecosystem development on BROWN. Soil organic matter is crucial for moisture retention and microbiome development. Ground cover on BROWN rapidly outpaced ground cover on GRAY or MIXED, indicating that soil on BROWN was more favorable for vegetative recruitment and growth. While ground cover can be detrimental to planted tree growth if competition is high, slow recruitment of ground cover over the first few growing seasons (rather than ground cover seeding at tree planting) likely permitted tree establishment and reduced competition. In addition, recruited ground cover contributed to the development of a thick litter layer on BROWN (little to no litter was observed on GRAY or MIXED), as well as improved nutrient availability through nitrogen fixation.

A major concern on reforested surface mine sites is colonization by competitive invasive species, especially lespedeza (*Lespedeza cuneata*), princess tree (*Paulownia tomentosa*), and autumn olive (*Elaeagnus umbellata*). We observed rapid and heavy colonization by these species on BROWN, with lespedeza representing 43% of total vegetative groundcover. However, planted tree growth (as well as volunteer tree growth) on BROWN was rapid, and we observed canopy closure on BROWN plots in 2013. In plot areas where canopy closure was taking place, understory community composition was starkly different from open-canopy areas. Notably, lespedeza was shaded out and locally extirpated from shaded areas, and native shade-tolerant species were recruited, including *Heuchera* spp., *Hydrangea* spp., and *Polystichum acrostichoides* (Sena et al. 2015). Clearly, BROWN plots are much more rapidly progressing toward native forest conditions than GRAY or MIXED.

8.4.1.4   Hydrology and Water Chemistry

Water drained from reclaimed surface mined sites is often poor, even decades after reclamation. Poor water quality contributes to negative ecological consequences, with observed impacts for aquatic insects and vertebrates (e.g., Pond et al. 2008, Price et al. 2015). Water quality in Appalachia is particularly important because the Appalachian region is a global hotspot for salamander, crayfish, and mussel diversity. Thus, an important outcome for surface mine reclamation should be improved water quality. In addition to water quality, water flow patterns are known to be significantly disrupted by surface mining. Surface mine sites typically exhibit low infiltration and high runoff rates and are anecdotally associated with increased flood incidence (e.g., Negley and Eshleman, 2006).

The Bent Mountain project represented one of the first attempts to investigate water quality and hydrology issues under FRA reclamation practices.

Water chemistry of interflow drained from these sites was initially different across spoil types. For example, electrical conductivity (EC), an important generic water quality metric used in assessing risk thresholds for aquatic macroinvertebrate communities in the region, was much higher in GRAY and MIXED than in BROWN. In general, during the first 3 years, water chemistry parameters indicated sharp downward trends, suggesting rapid weathering of minerals from the crushed overburden mine soils (Agouridis et al. 2012). However, by 2013, no significant treatment effect was observed for any water chemistry parameter. In addition, regression slopes for all parameters for all spoil types were similar to zero, suggesting that the rapid weathering period lasted only the first few years after establishment (Sena et al. 2014). Most importantly, by the 9th growing season after establishment, average EC of water discharged from all plots was similar to the 500 μS cm^{-1} threshold proposed by Pond et al. (2008). These data suggest that, although spoil types differ widely in suitability for tree growth and vegetative recruitment, few differences exist in chemical water quality of interflow. However, these data support application of FRA methods. While conventionally reclaimed surface mines have been documented discharging water with EC > 2000 μS cm^{-1} as long as 10 years after reclamation (Lindberg et al. 2011), our plots rapidly (~9 years) approximated target EC levels (300–500 μS cm^{-1}, Figure 8.5).

Hydrology on conventionally reclaimed surface mine sites is frequently disrupted by heavy compaction and consequently low infiltration. FRA plots are constructed using minimal compaction techniques, which can improve infiltration and tree growth. We monitored flow patterns of interflow discharged from these plots, and we observed distinct seasonal variation among spoil types. While discharge was similar across spoil types during the dormant season, discharge from BROWN was significantly lower than GRAY or MIXED during the growing season. We conclude that a greater percentage of incident precipitation was being evapotranspired on BROWN than on GRAY or MIXED, resulting in lower volumes of discharge during the growing season (Sena et al. 2014).

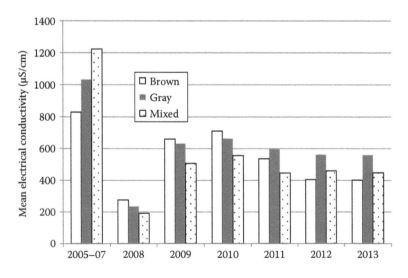

FIGURE 8.5 Electrical conductivity of flow-weighted samples, separated into sampling groups (2005–2007) and (2008–2013).

8.4.2 THE STARFIRE PROJECT: EXAMINING SPOIL PHYSICAL INFLUENCES ON REFORESTATION SUCCESS

Excessive compaction of spoil material in the backfilling and grading process is often considered one of the biggest impediments to the establishment of productive forests on surface mine lands. A pilot study was implemented at the Starfire Surface Mine near Hazard, Kentucky, USA, in 1996 to demonstrate the impact of compaction on reforestation (thereby evaluating FRA # 2). The objective of this project was to examine reforestation outcomes of low-compaction reclamation techniques that may provide a better environment and growing medium for trees than the conventional reclamation techniques practiced under the requirements of SMCRA. In this study, three levels of spoil compaction were examined: conventionally compacted spoil (control), lightly graded (end-dumped and struck-off) spoil, and uncompacted (end-dumped) spoil.

Starfire is located in eastern Perry County and western Knott County, Kentucky (37° 24′ N, 83° 08′ W) in the Cumberland Plateau physiographic region, which is predominately forested. Starfire has operated as a mountaintop-removal operation since the early 1980s. Multiple coal seams were mined, and overburden was removed using both dragline and truck/shovel operations. The thickness of the topsoil covering the site prior to mining was relatively thin; thus, a mixture of nontoxic, nonacidic shale stratum and brown weathered sandstone in the overburden at the mine was used as a soil substitute material during reclamation.

In 1996 and 1997, nine 1-ha reclamation cells were developed at the mine to represent three subsurface compaction treatments. The cells were constructed on top of mined land that had been reclaimed to hay and pastureland in the late 1980s. Three compacted cells were constructed using normally accepted spoil handling techniques that resulted in a smooth graded surface (control). The remaining six cells comprised new spoil material from the mining operation. Large earth moving trucks were used to end-dump the spoil material in consecutive piles that tightly abutted each other until the entire cell was filled. Three of these end-dumped cells were graded with a small bulldozer (D-8), which struck-off or leveled the tops of the consecutive spoil piles (strike-off). The final three end-dumped cells were not graded and represent the uncompacted treatment (end-dumped). Surface amendments (wheat straw/manure mulch and wood mulch) were individually applied to one of the three plots per compaction treatment at a rate of 85 m^3 ha^{-1}.

The resulting microtopography of the nine cells varied from extremely smooth to extremely rough. The three cells containing the compacted spoil were smooth and relatively level, with no boulders and very little surface variation. The three cells containing the lightly graded (strike-off) spoil were relatively flat with small undulations near the bases of where the spoil piles were loose-dumped. These light compaction cells have more surface variation than the compacted cells. The three cells containing the end-dumped, uncompacted treatment exhibited the highest surface variation. These cells were extremely rough and were characterized by unleveled spoil piles with high tops and low depressions with large boulders interspersed.

Six bare-root tree species, including eastern white pine (*Pinus strobus*), white ash (*Fraxinus americana*), black walnut (*Juglans nigra*), yellow poplar (*Liriodendron tulipifera*), white oak (*Quercus alba*), and northern red oak (*Quercus rubra*) were planted on the site. Each tree species was randomly allotted to three plots (three replications) within each reclamation cell. Tree seedlings were planted on 1.8 × 1.8 m spacing, providing 121 trees in each growth plot. All reclamation cells were seeded with a mixture of grasses and legumes, including annual rye (*Secale cereale*), perennial rye (*Lolium perenne*), orchard grass (*Dactyis glomerata*), birdsfoot trefoil (*Lotus corniculatus*), and Appalow lespedeza (*Serecia lespedeza*, var. *Appalow*) at the following rates of application: 33.61 kg ha^{-1} for the annual rye and 5.61 kg ha^{-1} for each of the other species.

Annual monitoring of seedling survival, height, and diameter was conducted through 2005 and periodically thereafter. Even-aged naturally regenerated stands were also sampled across a chronosequence of five age classes (5, 10, 20, 40, and 80 years old) for white oak (WO) and yellow-poplar (YP) to develop reference growth curves (height and diameter) for these species on unmined sites (see Cotton et al. 2012 for further information on methodology). We compared the growth data from the reforested sites to those from the reference stands to determine if growth trajectories were similar.

8.4.2.1 Survival

The use of uncompacted and strike-off methods have resulted in significantly greater survival and growth for planted species than that of traditional reclamation methods. In 2005, survival for all species was 72% for end-dumped, 54% for strike-off, and 23% for conventional (control) treatments. Individual species response showed a similar trend as that exhibited by combined species with the exception of white ash, which showed similar survival (80%–82%) across treatments (Angel et al. 2006). Yellow poplar and white oak both showed significant survival differences between treatments after 17 years of growth (Table 8.2). In both cases, end-dump survival was excellent (>80%) and conventional was poor (<30%), with strike-off being intermediate. Variation in compaction in these treatments was confirmed by annual bulk density assessments from 1998 to 2001 (Conrad et al. 2002). Bulk density was highest in the conventional treatment and lowest in the end-dumped treatment. The addition of surface amendments showed a meager benefit, but results varied by species and by treatment (Angel et al. 2006). A later survey of two species, white oak and yellow-poplar, in 2014 showed that survival rates remained similar to those observed in 2005.

8.4.2.2 Growth

Deviation of naturally regenerating growth (diameter and height) on nonmined land is depicted with the black dashed lines in Figures 8.6 and 8.7. Mean diameter and height for the two species planted in our treatments on the mine spoil with straw/manure mulch surface amendment are depicted by the gray line (linear regression). Yellow poplar and white oak exhibited greater initial height growth by subsurface treatment with the addition of the straw/manure compost (Angel et al. 2006; Cotton et al. 2012), however, that difference was not observed in 2014. The figures show that height and diameter growth are influenced by compaction and reclamation method. Species differences are also observable. Seedlings growing in the mine spoil exhibited better diameter growth than height growth, which may be attributable to the open grown conditions and wide initial plant spacing of 3 m, which minimizes competition between seedlings during early growing years (Cotton et al. 2012). At Year 17, height and diameter growth of yellow poplar was higher in strike-off treatments than that observed in the end-dump treatments. This trend is the opposite of that observed in the first 10 years of growth (Angel et al. 2006) possibly due to differences in survival. The fewer surviving trees on strike-off have less hardwood competition for resources, which resulted in more growth.

TABLE 8.2
Yellow Poplar and White Oak Percent Survival for Starfire Mine Reforestation Cells, 1997–2005 and 2014

Method	Species	1997	1998	1999	2000	2001	2002	2003	2004	2005	2014
End-dump	YP	93	86	77	64	83	82	79	80	77[a]	83[a]
	WO	88	69	87	70	83	80	84	81	80[1]	84[1]
Strike-off	YP	94	63	61	51	59	57	54	52	52[b]	51[b]
	WO	94	55	78	66	68	71	70	69	69[2]	70[2]
Conventional	YP	59	50	30	9	15	15	9	11	10[c]	11[c]
	WO	49	25	49	25	27	27	24	21	25[3]	28[3]

Note: Treatments for 2005 followed by the same letter (for YP), or number (for WO) are not significantly different ($\alpha \geq 0.05$). YP, yellow poplar; WO, white oak.

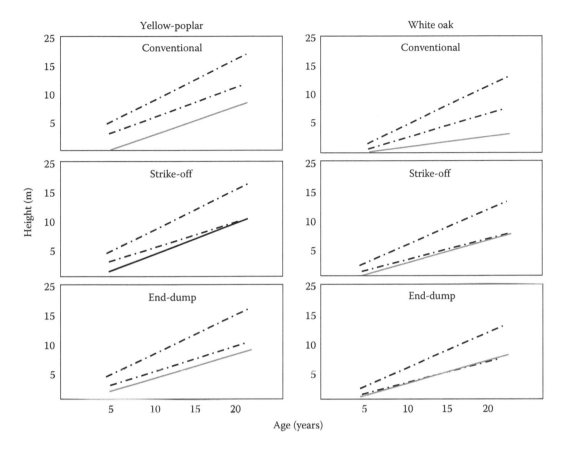

FIGURE 8.6 Chronosequence of height comparing deviation of tree heights from nonmined reference stands ±1 SD (black dashed lines) to mean tree heights from reforestation plots with a straw/manure mulch surface amendment (gray solid line). At Year 17 (2014), when last measured, tree heights were approaching the low side of the reference curve for strike-off and end-dump treatments, but remain shorter than one would expect in a naturally regenerating site. Tree height in conventional is well below what would be expected in a natural stand. The forward regression of height data past Year 17 suggests that height growth in these young stands is less than would be expected by natural regeneration, regardless of subsurface treatment.

Thus, end-dump appears to foster good conditions for stand initiation, but strike-off creates more favorable conditions for stem exclusion. Conventional reclamation holds little promise for the reestablishment of native forests on mined lands.

These data suggest that FRA techniques (low compaction) set reclaimed sites on a recovery trajectory that can be similar to the native forest (Figure 8.8). Low compaction spoil placement techniques improved tree growth and survival and also accelerated natural succession compared to conventional reclamation techniques (Groninger et al. 2007). At Starfire, canopy closure on strike-off and end-dump treatments was achieved by Year 10 for most species. This change in microclimate and forest floor attributes likely contributed to colonization of 27 tree species that were not planted at the site. In addition, the number of naturally colonized individuals was approximately 10× greater for the end-dump treatment than the conventional treatments (477 stems per acre versus 46 stems per acre). The uncompacted spoil in the end-dumped treatment not only provided sites for seed germination, but were also more favorable for small mammal colonization than compacted spoil (Larkin et al. 2008). These data suggest that forest restoration was accelerated using low compaction FRA techniques.

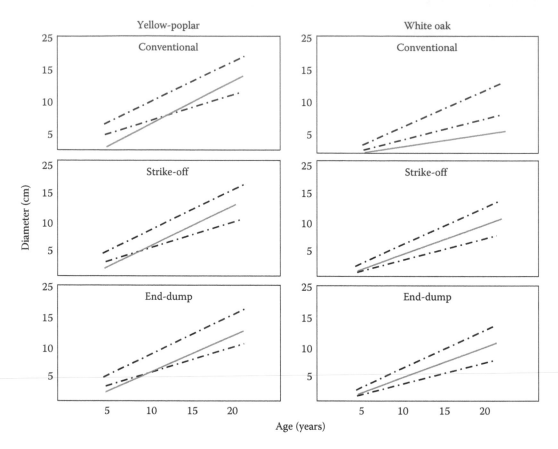

FIGURE 8.7 Chronosequence of diameter comparing deviation of tree diameter from reference stands ±1 SD (black dashed lines) to mean tree diameters from reforestation plots with a straw/manure mulch surface amendment (gray solid line). Diameters at Year 17, when last measured, fall within the expected range as those found in reference stands on strike-off and end-dump treatments. Conventional yellow poplar diameter growth appears to be good at Year 17, but the small sampling size (11% survival) may skew the data (the few that survived were on plot edges and grew well). Unlike height growth, the forward regression of diameter data past Year 17 suggests that diameter growth in these young stands is similar to would be expected by natural regeneration in strike-off and end-dump treatments.

8.5   THE APPALACHIAN REGIONAL REFORESTATION INITIATIVE

In response to years of research demonstrating that productive forests can be established on surface coal mines in Appalachia using the FRA, the U.S. Office of Surface Mining Reclamation and Enforcement and state regulatory authorities in Appalachia created the Appalachian Regional Reforestation Initiative (ARRI) to promote establishment of healthy, productive forest habitat on active and abandoned mine lands in the eastern U.S. coal fields. ARRI was a mechanism for the mining regulatory authorities to communicate best management practices for establishing forests on post–mining lands with the mining industry, stakeholders, and even those who regulate mining and reclamation practices. ARRI's goals are to plant more high-value hardwood trees on surface mines, increase the survival rates and growth rates of those trees, and expedite the establishment of forest habitat through natural succession. The ARRI Core Team was created to coordinate activities by landowners, the coal industry, university researchers, watershed and conservation groups, and government agencies that have an interest in converting reclaimed mined lands to productive forestland.

The ARRI Science Team was established to ensure that the recommended reforestation methods are based on scientific principles and research findings, and to conduct scientific research on forestry reclamation practices. See the following website for more information about the reforestation initiative in Appalachia: http://arri.osmre.gov/

Focused efforts by ARRI and its partners to promote FRA have been embraced by many in the coal industry within Appalachia, and significant changes in reclamation practices have resulted. Since ARRI's inception in 2004, approximately 95 million trees have been planted on over 56,000 ha. ARRI is "forward looking," diligently working to educate and train active mining industry and

(a)

(b)

FIGURE 8.8 Images from the strike-off reclamation cells showing growth of seedlings in 2001 (a), 2003 (b). Silvicultural practices (thinning) are being considered to improve stand conditions in these cells. (*Continued*)

(c)

(d)

FIGURE 8.8 (*Continued*)   Images from the strike-off reclamation cells showing growth of seedlings in 2006 (c), and 2013 (d). Silvicultural practices (thinning) are being considered to improve stand conditions in these cells.

regulatory personnel about the FRA in order to reclaim new surface mine disturbances to forests from this point forward.

ARRI is also "looking backward" at the estimated 600,000 ha of nonforested, unused post-bond release mined lands that could be available for reforestation in the eastern United States (Zipper et al. 2013). The reforestation guidelines for unused mined land developed by ARRI scientists (Sweigard et al. 2007b; Burger and Zipper 2011; Burger et al. 2013) have been applied with success for restoring unused mined land to native forests. Through a not for profit organization called Green Forests Work (GFW), the practice of returning forests to previously mined land has grown substantially. The GFW program is an economic development plan for Appalachia that focuses on restoring ecosystem services on mine-scarred lands and creating jobs in the process. Successful reestablishment of the hardwood forests that once dominated these lands will provide a renewable, sustainable multiuse resource that will create economic opportunities while enhancing the local and global environment. The jobs include everything from nursery jobs, equipment operators, tree planters, forest managers, and wildlife biologists to those that may manage these sites for timber, recreation, renewable energy, and climate change mitigation. More importantly, the success exhibited by these partnerships has resulted in formation of more partnerships. From 2009 to 2016, GFW and ARRI have partnered with state and federal agencies, watershed groups, coal operators, conservation groups, environmental organizations, faith-based groups, and numerous universities, colleges, and high schools to coordinate 259 tree planting projects/events throughout Appalachia. These events involved 726 partner organizations and 14,277 volunteers and participants. The result has been the planting of over 1.8 million trees on about 3000 acres of previously reclaimed mine sites where reforestation was not attempted, or where the results were undesirable. These projects can empower local communities to participate in beautification and enhancement of land that is initially often abandoned. In addition, millions of U.S. dollars have been invested in a region of the United States that is economically distressed.

Forests are a renewable resource. By recreating forests where no forests currently exist, the economic opportunities provided by this program will not only provide for the Appalachian people today but will put those lands on a trajectory that will ensure that a forest is available for use by future Appalachian citizens. The Appalachian forest is one of the most beautiful in the world, is one of the region's most valuable assets, and has played an integral part in the rich cultural heritage of the mountain people. As support for the program grows, GFW can proceed in developing a skilled green workforce to restore, protect, and manage this natural resource that is so vital to the region's current and future prosperity. For more information on GFW see www.greenforestswork.org.

8.6 IMPACTS

Through the FRA, we are not only restoring forests but we are also restoring an "environmental infrastructure"—ecosystem processes and services provided by native forests—to coal-mined Appalachian landscapes. The high-value commercial forests resulting from the FRA will contribute to the timber industry's continued viability as a major employer throughout Appalachia over the long term. Use of woody biomass for energy production has gained much attention recently, not only for its potential to serve as an inexpensive and domestic supply of power, but also for the possible environmental and rural development benefits it offers. Woody biomass can be used to generate electric power, and new and emerging technologies allow conversion of biomass materials to liquid fuels (Williams et al. 2009). Research has demonstrated the potential of reforested coal mine sites to support woody biomass crops with average growth rates far exceeding those of native hardwoods, with potential for harvest as energy crops on rotations as short as 10–15 years (Stringer and Carpenter 1986; Fields-Johnson et al. 2008; Brinks et al. 2011). Establishment of biomass plantations on favorably configured and located coal mine sites will support management and harvest jobs, and can potentially create additional jobs by meeting capacity for feedstock production that will attract cellulosic biorefinery investors to the Appalachian region.

If carbon-emission restrictions become law, it is likely that coal-burning electric-power producers will be called upon to offset carbon emissions to the atmosphere. Planting trees on productive post-mined soils is a way to produce a measurable carbon sink (Maharaj et al. 2007; Amichev et al. 2008). Forests growing on good quality mine sites can sequester 3–5 times more carbon than the grasslands that were established through the original reclamation (Burger and Zipper 2011). Reforestation of older mine sites can sequester carbon that can be used to offset U.S. carbon emissions from fossil fuels.

Society at large also benefits from restoration of productive forests, as these constitute an "environmental infrastructure" that produces ecosystem services of tangible value to global and Appalachian communities. For example, forested landscapes help to maintain clean water in rivers and streams, an ecosystem service that is of significant value throughout Appalachia and beyond. This service is of special value in those Appalachian streams that, although affected by coal mining, continue to support exceptional biodiversity in the form of rare, threatened, or endangered aquatic species, including fish and mollusks. The reestablishment of forests will also aid in restoring watershed protection services to streams draining Appalachian landscapes, such as reduction of the peak flows that can cause flooding and the maintenance of water flows during dry weather periods (Sena et al. 2015). Flooding causes millions of dollars in damages in coal-mined areas of the Appalachian region each year; restoration of native forests within extensively mined but nonreforested watersheds would reduce the intensity and frequency of flooding impacts downstream. As demonstrated by Taylor et al. (2009a,b), spoil materials prepared for reclamation on experimental plots as per FRA have infiltration characteristics similar to unmined forest lands. Targeted reforestation can also reduce forest fragmentation that has been caused by mining (Wickham et al. 2007), as needed to restore habitat for wildlife species that depend on large expanses of unbroken forest, including rapidly declining bird species such as the Cerulean Warbler (*Dendroica cerulea*). Reforestation with native species will also improve landscape aesthetics, thus enhancing the capacity of communities in coal-mined areas to serve as tourist destinations and to support tourism-related businesses and jobs.

8.7 GLOBAL APPLICATION

Although this discussion entails a restoration program specific to the Appalachian region of the United States, the FRA was formulated using basic principles of soil science, forestry, ecology, engineering, and agronomy that are universal and applicable to multiple ecosystems. Climate, geology, and soils will ultimately dictate which species are best suited for a particular ecoregion, but the techniques described in the FRA as they pertain to the determination of spoil suitability, the minimization of soil physical and chemical stressors, the reduction of herbaceous competition, the planting of early and late successional tree species, and the use of good planting methods can be applied to surface mined areas throughout the globe. Moreover, ARRI and GFW can be used as models for demonstrating how partnerships are developed and successfully implemented. Following ARRI and GFW, these steps can be emulated to encourage global reforestation of surface mined lands: (1) through research and discovery, develop improved techniques for restoring ecosystems impacted by surface mining; (2) through outreach, advance knowledge by demonstrating improved reforestation techniques to all partners; and (3) through scholarship and engagement, develop leaders who will carry the work forward to ensure that forested ecosystem services are returned for use by future generations.

REFERENCES

Agouridis, C. T., P. N. Angel, T. J. Taylor, C. D. Barton, R. C. Warner, X. Yu, and C. Wood. 2012. Water quality characteristics of discharge from reforested loose-dumped mine spoil in eastern Kentucky. *Journal of Environmental Quality* 41:454–468.

Amichev, B. Y., J. A. Burger, and J. A. Rodrigue. 2008. Carbon sequestration by forests and soils on mined land in the Midwestern and Appalachian coalfields of the U.S. *Forest Ecology and Management* 256:1949–1959.

Andrews, J. A. 1992. Soil productivity model to assess forest site quality on reclaimed surface mines. Masters thesis. Virginia Polytechnic Institute and State University, Blacksburg, VA.

Angel, P. N., C. D. Barton, R. Warner, C. Agouridis, T. Taylor, and S. Hall. 2008. Tree growth, natural regeneration, and hydrologic characteristics of three loose-graded surface mine spoil types in Kentucky. Pp. 28–65, in: *Proceedings of the 25th National Meeting of the American Society of Mining and Reclamation (ASMR)*, ASMR, Lexington, KY.

Angel, P. N., V. M. Davis, J. A. Burger, D. Graves, and C. E. Zipper. 2005. The Appalachian regional reforestation initiative. Appalachian Regional Reforestation Initiative, US Office of Surface Mining, Pittsburgh, PA. Forest Reclamation Advisory No. 1.

Angel, P. N., D. H. Graves, C. D. Barton, R. C. Warner, P. W. Conrad, R. J. Sweigard, and C. A. Agouridis. 2006. Surface mine reforestation research: Evaluation of tree response to low compaction reclamation techniques. Pp. 45–58, in: *Proceedings of the Seventh ICARD and ASMR Conference*, St. Louis, MO, March 26–30, 2006.

Ashby, W. C. 1987. Forests. Chapter 7. Pp. 89–107, in: W. R. Jordan, and M. E. Gillph (ed.), *Restoration Ecology: A Synthetic Approach to Ecological Research*. Cambridge University Press, Cambridge, U.K.

Ashby, W. C. 1991. Factors limiting tree growth in southern Illinois under SMCRA. Pp. 287–293, in: *Proceedings of the Seventh National Meeting of the ASMR*, Durango, CO, May 14–17, 1991.

Ashby, W. C., C. Kolar, M. L. Guerke, C. F. Pursell, and J. Ashby. 1978. Our reclamation future with trees. Southern Illinois University at Carbondale, Coal Extraction and Utilization Research Center, Carbondale, IL, 100pp.

Ashby, W. C., and C. A. Kolar. 1998. Thirteen-year hardwood tree performance on a Midwest surface mine. Pp. 124–133, in: D. Throgmorton, J. Nawrot, J. Mead, J. Galetovic, and W. Joseph (eds.), *Mining-Gateway to the Future. Proceedings of the 25th Anniversary and 15th Annual Meeting of the American Society for Surface Mining and Reclamation*, St. Louis, MO, May 17–22, 1988.

Boyce, S. 1999. Office of Surface Mining (OSM) revegetation team survey results. Pp. 31–35, in: K. C. Vories and D. Throgmorton (ed.), *Proceedings of the Enhancement of Reforestation at Surface Coal Mines: Tech. Interactive Forum*, Fort Mitchell, KY, March 23–24, 1999.

Brinks, J. S., J. M. Lhotka, C. D. Barton, R. C. Warner, and C. T. Agouridis. 2011. Effects of fertilization and irrigation on American sycamore and black locust planted on a reclaimed surface mine in Appalachia. *Forest Ecology and Management* 261:640–648.

Burger, J. A. 1999. Academic research perspective on experiences, trends, constraints and needs related to reforestation of mined land. Pp. 63–74, in: K. C. Vories and D. Throgmorton (eds.), *Proceedings of the Enhancement of Reforestation at Surface Coal Mines: Technical Interactive Forum*, Fort Mitchell, KY, March 23–24, 1999.

Burger, J. A., D. Graves, P. N. Angel, V. M. Davis, and C. E. Zipper. 2005. The forestry reclamation approach. Appalachian Regional Reforestation Initiative, US Office of Surface Mining, Pittsburgh, PA. Forest Reclamation Advisory Number 2.

Burger, J. A., and J. L. Torbert. 1992. Restoring forests on surface-mined land. Virginia Cooperative Extension Publication No. 460-123. Blacksburg, VA. 16 p.

Burger, J. A., and J. L. Torbert. 1997. Restoring forests on surface-mined land. Virginia Cooperative Extension. Virginia Polytechnic Institute and State University, Blacksburg, VA. Available online at http://www.ext. vt.cdu/pubs/mincs/460-123/460-123.html (verified March 1, 2008).

Burger, J. A., and C. E. Zipper. 2011. How to restore forests on surface-mined land. Virginia Cooperative Extension Publication 460-123. Virginia Tech, Blacksburg, VA.

Burger, J. A., C. E. Zipper, P. N. Angel, N. Hall, J. G. Skousen, C. D. Barton, and S. Eggerud. 2013. Establishing native trees on legacy surface mines. U.S. Office of Surface Mining. Forest Reclamation Advisory Number 11, U.S. Office of Surface Mining. Pittsburgh, PA, 10pp. http://arri.osmre.gov/fra.htm. Accessed September 28, 2016.

Burger, J. A., C. E. Zipper, and J. G. Skousen. 2010. Establishing groundcover for forested postmining land uses. Virginia Cooperative Extension Publication 460-124, U.S. Office of Surface Mining, Pittsburgh, PA.

Casselman, C. N., T. R. Fox, and J. A. Burger. 2007. Thinning response of a white pine stand on a reclaimed surface mine in southwest Virginia. *Northern Journal of Applied Forestry* 24:9–13.

Conrad, P. W., R. J. Sweigard, D. H. Graves, J. M. Ringe, and M. H. Pelkki. 2002. Impacts of spoil conditions on reforestation of surface mine land. *Mining Engineering* 54:39–47.

Cotton, C., C. Barton, J. Lhotka, P. Angel, and D. Graves. 2012. Evaluating reforestation success on a surface mine in Eastern Kentucky. Pp. 16–23, in: L. E. Riley, D. L. Haase, and J. R. Pinto (eds.), *National Proceedings of the Forest and Conservation Nursery Associations—2011, Proceedings of the RMRS-P-65*, Huntington, WV, July 25–28, 2011. USDA Forest Service, Rocky Mountain Research Station, Fort Collins, CO.

Davidson, W. H., R. J. Hutnik, and D. E. Parr. 1983. Reforestation of mined land in the northeastern and north-central U.S. *Northern Journal of Applied Forestry* 1:7–11.

Davis, V., J. Franklin, C. E. Zipper, and P. N. Angel. 2010. Planting hardwood tree seedlings on reclaimed land in Appalachia. Appalachian Regional Reforestation Initiative, US Office of Surface Mining. Forest Reclamation Advisory Number 7, U.S. Office of Surface Mining, Pittsburgh, PA.

DenUyl, D. 1955. Hardwood tree planting experiments strip mine spoil banks of Indiana. Purdue University, Agricultural Experiment Station Bulletin No. 619. Lafayette, IN, 16pp.

Emerson, P., J. G. Skousen, and P. Ziemkiewicz. 2009. Survival and growth of hardwoods in brown versus gray sandstone on a surface mine in West Virginia. *Journal of Environmental Quality* 38:1821–1829.

Evilsizer, B. 1980. Reclamation with trees in Illinois. Pp. 15–16, in: *Proceedings for the Symposium on Trees for Reclamation in the Eastern US*, Lexington, KY, October 27–29, 1980. USDA Forest Service GTR-61.

Fields-Johnson, C., C. E. Zipper, D. Evans, T. R. Fox, and J. A. Burger. 2008. Fourth-year tree response to three levels of silvicultural input on mined lands. In: *Proceedings of the 2008 National Meeting of the American Society of Mining and Reclamation*, Richmond, VA, June 14–19, 2008.

Franklin, J. A., and J. Frouz. 2007. Restoration of soil function on coal mine sites in eastern Tennessee 50 years after mining. In: *Proceedings of Ecological Restoration in a Changing World. Ecological Society of America and Society for Ecological Restoration Joint Conference*, San Jose, CA, August 5–10, 2007.

Franklin, J. A., C. E. Zipper, J. A. Burger, J. G. Skousen, and D. F. Jacobs. 2012. Influence of herbaceous ground cover on forest restoration of eastern US coal surface mines. *New Forests* 43:905–924.

Graves, D. H., J. M. Ringe, M. H. Pelkki, R. J. Sweigard, and R. Warner. 2000. High value tree reclamation research. Pp. 413–421, in: R. K. Singhal and A. K. Mehrotra (eds.), *Environmental Issues and Management of Waste in Energy and Mineral Production*. Balkema, Rotterdam, the Netherlands.

Groninger, J., J. G. Skousen, P. N. Angel, C. B. Barton, J. A. Burger, and C. E. Zipper. 2007. Mine Reclamation Practices to Enhance Forest Development through Natural Succession. Appalachian Regional Reforestation Initiative, US Office of Surface Mining. Forest Reclamation Advisory Number 5.

Larkin, J. L., D. S. Maehr, J. J. Krupa, J. J. Cox, K. Alexy , D. E. Unger, and C. D. Barton. 2008. Small mammal response to vegetation and spoil conditions on a reclaimed surface mine in eastern Kentucky. *Southeastern Naturalist* 7:401–412.

Lindberg, T. T., E. S. Bernhardt, R. Bier, A. M. Helton, R. B. Merola, A. Vengosh, and R. T. Di Giulio. 2011. Cumulative impacts of mountaintop mining on an Appalachian watershed. *Proceedings of the National Academy of Science of the United States of America* 108:20929–20934.

Maharaj, S., C. D. Barton, T. A. D. Karathanasis, H. D. Rowe, and S. M. Rimmer. 2007. Distinguishing "new" from "old" organic carbon in reclaimed coal mine sites using thermogravimetry: II Field validation. *Soil Science* 172:302–312.

Mickalitis, A. B., and D. B. Kutz. 1949. Experiments and observations on planting areas "stripped" for coal in Pennsylvania. *Pennsylvania Forests and Waters* 3:62–66, 70.

Negley, T. L., and K. N. Eshleman. 2006. Comparison of stormflow responses of surface-mined and forested watersheds in the Appalachian Mountains, USA. *Hydrological Processes* 20:3467–3483.

Plass, W. T. 1982. The impact of surface mining on the commercial forests of the United States. pp. 1–7, in: C. A. Kolar, and W. C. Ashby (eds.), *Post-Mining Productivity with Trees*, Department of Botany, Southern Illinois University, Carbondale, IL, March 31–April 2, 1982.

Plass, W. T., and J. L. Powell. 1988. Trees and shrubs. p. 176–198, in: L. R. Hossner (ed.), *Reclamation of Surface Mined Lands*, Vol. 2. CRC Press, Boca Raton, FL.

Pond, G. J., M. E. Passmore, F. A. Borsuk, L. Reynolds, and C. J. Rose. 2008. Downstream effects of mountaintop coal mining: comparing biological conditions using family- and genus-level macroinvertebrate bioassessment tools. *Journal of the North American Benthological Society* 27:717–737.

Pope, P. E. 1989. Reforestation of minelands in the Illinois Coal Basin. Purdue University, Agricultural Experiment Station Bulletin No. 565. Department of Forestry and Natural Resources. West Lafayette, IN, 12pp.

Potter, S. H., S. Weitzman, and G. R. Trimble. 1951. Reforestation of strip-mined lands in West Virginia. USDA Forest Service Northeastern Forest Experiment Station, Radnor, PA, Paper No. 43, 28pp.

Price, S. J., B. L. Muncy, S. J. Bonner, A. N. Drayer, and C. D. Barton. 2015. Effects of mountaintop removal mining and valley filling on the occupancy and abundance of stream salamanders. *Journal of Applied Ecology* 53:459–468.

Rodrigue, J. A., J. A. Burger, and R. G. Oderwald. 2002. Forest productivity and commercial value of pre-law reclaimed mined land in the eastern United States. *Northern Journal of Applied Forestry* 19:106–114.

Ruppert, L. F. 2001. Chapter A—Executive summary—Coal resource assessment of selected coal beds and zones in northern and central Appalachian Basin coal regions. US Geological Survey Professional Paper 1625-C. US Geological Survey, Reston, VA.

Sena, K., C. Barton, P. Angel, C. Agouridis, S. Hall, and R. Warner. 2015. Influence of spoil type on afforestation success and natural vegetative recolonization on a surface coal mine in Appalachia, United States. *Restoration Ecology* 23:131–138.

Sena, K., C. Barton, P. Angel, C. Agouridis, and R. Warner. 2014. Influence of spoil type on chemistry and hydrology of interflow on a surface coal mine in the eastern US coalfield. *Water, Air, and Soil Pollution* 225:1–14.

Stringer, J. W., and S. B. Carpenter. 1986. Energy yield of black locust biomass fuel. *Forest Science* 32(4):1049–1057.

Sweigard, R., J. A. Burger, D. Graves, C. E. Zipper, C. D. Barton, J. G. Skousen, and P. N. Angel. 2007a. Loosening compacted soils on mined sites. Appalachian Regional Reforestation Initiative, Forest Reclamation Advisory Number 4, U.S. Office of Surface Mining, Pittsburgh, PA.

Sweigard, R., J. A. Burger, C. E. Zipper, J. G. Skousen, C. D. Barton, and P. N. Angel. 2007b. Low compaction grading to enhance reforestation success on coal surface mines. Appalachian Regional Reforestation Initiative, US Office of Surface Mining. Forest Reclamation Advisory Number 3, U.S. Office of Surface Mining, Pittsburgh, PA.

Taylor, T. J., C. T. Agouridis, R. C. Warner, and C. D. Barton. 2009a. Runoff curve numbers for loose-dumped spoil in the Cumberland Plateau of eastern Kentucky. *International Journal of Mining, Reclamation and Environment* 23:103–120.

Taylor, T. J., C. T. Agouridis, R. C. Warner, C. D. Barton, and P. Angel. 2009b. Hydrologic characteristics of loose-dumped spoil in the Cumberland Plateau of eastern Kentucky. *Hydrological Processes* 23:3372–3381.

Torbert, J. L., and J. A. Burger. 1994. Influence of grading intensity on ground cover establishment, erosion, and tree establishment on steep slopes. Pp. 226–231, in: *Proceedings of the 1994 Annual Meeting of the ASMR*, Pittsburgh, PA, April 24–29, 1994.

Torbert, J. L., J. A. Burger, and W. L. Daniels. 1990. Pine growth variation associated with overburden rock type on a reclaimed surface mine in Virginia. *Journal of Environmental Quality* 19:88–92.

Torbert, J. L., J. A. Burger, and T. Probert. 1995. Evaluation of techniques to improve white pine establishment on an Appalachian minesoil. *Journal of Environmental Quality* 24(5):869–873.

Torbert, J. L., S. H. Schoenholtz, J. A. Burger, and R. E. Kreh. 2000. Growth of three pine species on pre- and post-SMCRA land in Virginia. *Northern Journal of Applied Forestry* 17:95–99.

Townsend, P. A., D. Helmers, C. Kingdon, B. McNeil, K. M. de Beurs, and K. N. Eshleman. 2009. Changes in the extent of surface mining and reclamation in the Central Appalachians detected using a 1976–2006 Landsat time series. *Remote Sensing of Environment* 113:62–72.

Ware, C., Reclaiming mountains: A special report, *Lexington Herald-Leader*, Lexington, KY, 2009.

Wickham, J., P. B. Wood, M. C. Nicholson, W. Jenkins, D. Druckenbrod, G. W. Suter, M. P. Strager, C. Mazzarella, W. Galloway, and J. Amos. 2013. The overlooked terrestrial impacts of mountaintop mining. *BioScience* 63:335–348.

Wickham, J. D., K. Riitters, T. Wade, M. Coan, and C. Homer. 2007. The effect of Appalachian mountaintop mining on interior forest. *Landscape Ecology* 22:179–187.

Williams, P. R., D. Inman, A. Aden, and G. A. Heath. 2009. Environmental and sustainability factors associated with next-generation biofuels in the U.S.: What do we really know? *Environmental Science and Technology* 43:4763–4775.

Wilson-Kokes, L., P. Emerson, C. DeLong, C. Thomas, and J. Skousen. 2013. Hardwood tree growth after eight years on brown and gray mine soils in West Virginia. *Journal of Environmental Quality* 42:1353–1362.

Zipper, C. E., J. A. Burger, C. D. Barton, and J. G. Skousen. 2013. Rebuilding soils on mined land for native forests in Appalachia. *Soil Science Society of America Journal* 77:337–349.

Zipper, C. E., J. A. Burger, J. M. McGrath, J. A. Rodrigue, and G. I. Holtzman. 2011a. Forest restoration potentials of coal-mined lands in the eastern United States. *Journal of Environmental Quality* 40:1567–1577.

Zipper, C. E., J. A. Burger, J. G. Skousen, P. N. Angel, C. D. Barton, V. Davis, and J. A. Franklin. 2011b. Restoring forests and associated ecosystem services on Appalachian coal surface mines. *Environmental Management* 47:751–765.

9 Recreating a Headwater Stream System on a Valley Fill in Appalachia, USA

Carmen T. Agouridis, Christopher D. Barton, and Richard C. Warner

CONTENTS

9.1 Introduction...148
9.2 Project Site..149
9.3 Pre-Restoration Conditions..150
    9.3.1 Water Quality..150
    9.3.2 Soils...153
    9.3.3 Vegetation...153
    9.3.4 Habitat...153
9.4 Design Components...153
    9.4.1 Reference Reaches..153
        9.4.1.1 Geomorphology..153
        9.4.1.2 Water Quality and Habitat..154
    9.4.2 Valley Reconfiguration and Hydrologic Modifications....................155
    9.4.3 Stream Dimension, Pattern, and Profile...156
        9.4.3.1 Main Channel..156
        9.4.3.2 Ephemeral Channels...157
        9.4.3.3 Bioreactor-Wetland Treatment System...............................158
        9.4.3.4 Forested Zones..159
        9.4.3.5 Erosion Control...159
9.5 Monitoring...159
    9.5.1 Geomorphology...159
    9.5.2 Hydrology..160
    9.5.3 Water Quality..160
    9.5.4 Vegetation...161
        9.5.4.1 Forested Zones..161
        9.5.4.2 Ground Cover..161
    9.5.5 Habitat...161
9.6 Results...162
    9.6.1 Geomorphology...162
    9.6.2 Hydrology..163
    9.6.3 Water Quality..164
    9.6.4 Vegetation...166
        9.6.4.1 Riparian Forest Census...166
        9.6.4.2 Upland Forest Census...166
        9.6.4.3 Ground Cover..168
    9.6.5 Habitat...168

9.7   Lessons Learned 168
        9.7.1   Isolation of Problematic Spoils.. 169
        9.7.2   Groundwater Connection.. 169
        9.7.3   Colonization of Macroinvertebrates 170
        9.7.4   Invasive Species... 170
        9.7.5   Browse 170
References.. 170

9.1   INTRODUCTION

Between 60% and 80% of the cumulative channel length in mountainous areas, such as eastern Kentucky, comprises headwater streams (Shreve 1969). These headwater streams are vital components of the landscape and to the overall function of aquatic ecosystems. Headwater streams serve as the primary pathways for water, sediment, and organic matter transport to higher-order stream systems (May and Gresswell 2003). Gomi et al. (2002) noted that headwater regions receive and mediate the majority of surface runoff in mountainous watersheds, thereby serving as regulators of flow intensity (e.g., flooding) for lower gradient regions. Additionally, headwater streams support large populations of macroinvertebrates, amphibians, and fish thus effectively increasing biodiversity as well as providing habitat for rare and endangered species (Petranka and Murray 2001; Lowe and Likens 2005; Meyer et al. 2007; Clark et al. 2008; Brannon and Purvis 2008).

Surface mining, and in particular mountaintop mining with its associated valley fills, poses a significant environmental risk to aquatic and terrestrial life through the direct conversion of headwater streams to uplands and the associated degradation of downstream receiving waters. See Chapter 8 for a description of this mining practice. Among the potential consequences to downstream waters are increased sedimentation, elevated specific conductance, and substantial changes in pH (Pond et al. 2008; U.S. EPA 2011). Sedimentation homogenizes the streambed and important habitat features such as riffles and pools become embedded with fine sediments (Barbour et al. 1999; Burdon et al. 2013). Organic matter processing is reduced through the burial of woody debris and leaf litter and the redistribution of aquatic biota associated with these habitat features (Cummins et al. 1989; Sponseller and Benfield 2001). With regard to water chemistry, one of the largest impacts of mountaintop mining and valley fills involves dramatic increases in specific conductance or total dissolved solids (Pond 2004; Hartman et al. 2005; Pond et al. 2008). The average specific conductance among reference streams in Kentucky's Eastern Coal Fields (ECF) was 63 μS cm^{-1}, whereas the average of streams draining valley fills was 1096 μS cm^{-1} (Pond 2004). Biota in forested headwater streams of the ECF are accustomed to an aquatic environment with low dissolved solids. Pond et al. (2008) and Green et al. (2000) found significantly fewer taxa and lower percentage of individuals belonging to the insect orders Ephemeroptera, Plecoptera, and Trichoptera (EPT taxa) when specific conductance was greater than 500 μS cm^{-1}. From a food web perspective, these effects trickle through the food chain thus impacting a multitude of species that utilized the habitat prior to mining. Moreover, habitat features such as leaf litter and woody debris, rock outcrops, vernal pools, and canopy cover, which are key components for some species, such as woodland salamanders, are completely destroyed by mining and the subsequent conventional reclamation techniques that yield grasslands where forests once stood (Ford et al. 2002; Zipper et al. 2011).

In light of the importance of headwater streams, development of practical stream restoration techniques for mined lands is crucial in order to restore impacted headwater stream systems. Stream restoration efforts are predominately focused on waterways in agricultural and urban settings (Bernhardt et al. 2005) with a relatively limited number of projects on mined lands (Palmer and Hondula 2014). Often employing natural channel design (NCD) techniques, stream restoration practitioners utilize reference reaches to design new channels with the appropriate dimensions, patterns, and profiles (Rosgen 1998; Hey 2006). The crux of the NCD approach is the ability of the designer to locate appropriate reference reaches, ones with similar hydrogeological characteristics (e.g., watershed area, valley type, geology, soils, and climate) to the impacted reach. Mined lands, particularly valley fills, present a unique challenge to stream restoration designers in that the valley configuration is notably altered from its prior natural state, the geologic age of the material in the structure itself is relatively young as buried strata are now exposed meaning the weathering process is new, and the water table is now located tens to hundreds of meters below the reclaimed ground surface.

Furthermore, modifications to vegetative communities (e.g., the conversion of forests to grasslands) result in altered hydrology, water chemistry, and habitat (Bosch and Hewlett 1982; Zipper et al. 2011). Efforts to restore forested communities on mined lands in Appalachia have resulted in the Forestry Reclamation Approach (FRA), which consisted of five steps: (1) selecting the best available growth medium, (2) minimizing compaction during placement of the rooting medium, (3) selecting appropriate tree species based on approved post-mining land use and site specific characteristics, (4) using tree compatible ground cover, and (5) using proper tree planting techniques (Burger et al. 2005; Zipper et al. 2011; See Chapter 8). Research surrounding the FRA has documented successful regrowth of hardwood forests (Torbert and Burger 1994; Graves et al. 2000; Sena et al. 2015) while replicating pre-mining hydrologic conditions (Taylor et al. 2009a,b; Sena et al. 2014). With regard to water chemistry, findings indicate that following an initial flush in fines during the first few years, a rapid decrease followed by stabilization occurs, with many water quality constituents, including specific conductance, following the first few years of spoil placement (Agouridis et al. 2012; Sena et al. 2014).

The ability to successfully restore or recreate streams on mined lands is linked to the restoration of the streams' watershed, including uplands and riparian areas, and not just the stream channel. In an effort to replace lost headwater stream form and function, improve water quality, and enhance aquatic and terrestrial habitat, researchers at the University of Kentucky recreated a headwater stream system on a valley fill utilizing NCD techniques and the FRA. The project resulted in the creation and enhancement of 1450 m of ephemeral, intermittent, and perennial streams, 0.1 ha of vernal pools, and 16.2 ha of hardwood forest. Following construction, the project was monitored with regard to geomorphology, hydrology, vegetation, water quality, and habitat.

9.2   PROJECT SITE

The project was located at the University of Kentucky's Robinson Forest, which is an over 6000 ha experimental forest located in southeastern Kentucky. Robinson Forest comprises eight discontinuous properties, with the main block composed of over 4100 ha. Research, extension, and educational activities conducted at the forest are largely performed on the main block. The forest sits within the rugged eastern section of the Cumberland Plateau, and its landscape consists of long, rectilinear side slopes cut into a horizontally bedded substrate of sandstone, shale, siltstone, and coal (Smalley 1984). The vegetation is typical of the mixed mesophytic forest region and ranges from xeric oak-pine-dominated stands to rich mesic cove hardwoods (Braun 1950; Carpenter and Rumsey 1976). The climate is humid and temperate with an average annual rainfall of 123 cm and mean monthly temperatures ranging from about 2°C in January to about 24°C in July (U.S. DC 2002).

The University of Kentucky acquired Robinson Forest in 1923 after extensive logging. As such, the majority of the forest consists of 90+-year-old second growth. During the early to mid-1990s, a portion of Robinson Forest was mined for coal, resulting in the creation of several valley fills. In the valley that was once known as Guy Cove, over 1.8 million m³ of spoil was placed over the top of a headwater stream system (Figure 9.1). The Guy Cove valley fill consisted of a crown, which was designed to slope away from the face toward the back of the fill at a slope of 1%; a face, which was constructed of three 15.2 m tall benches, each with a 3% back slope and a 2:1 (H:V) drop; two groin drains (rock-lined) along both sides of the face; and an underdrain (approximately 5 m × 2.5 m in size) (Figure 9.2). The presence of ponded water (approximately 0.2 ha) at the back of the crown indicated that the underdrain was experiencing some level of clogging. Valley side slopes along the crown were regular and steep (2:1 H:V) with ephemeral gullies interspersed throughout. A sediment-laden pond was located at the toe of the valley fill (Figure 9.3). Because a lesser amount of coal was present

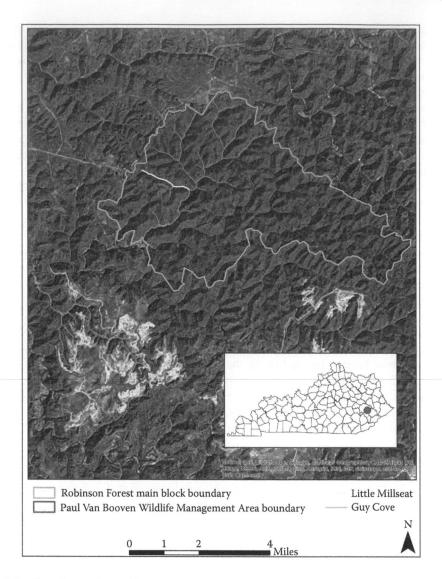

FIGURE 9.1 Guy Cove is located in eastern Kentucky on a mined section of the University of Kentucky's Robinson Forest.

than anticipated (e.g., the coal seam ended before reaching the valley ridgetop), Guy Cove was not completely surface mined. Instead, a 9 ha section at the most upgradient portion (ridgetop) of the valley experienced some clearing and grubbing (e.g., removal of trees) only. A small spring-fed channel flows nearly all year round (average baseflow of 55 m^3 day^{-1}) from this unmined section. This streamflow was directed into the underdrain (Figure 9.4) at the interface with the mined section. The Guy Cove watershed is 44 ha in size, and prior to restoration, had a drainage density of 0.0034 m m^{-2}.

9.3 PRE-RESTORATION CONDITIONS

9.3.1 WATER QUALITY

Movement of water through the unconsolidated fill has resulted in poor water quality for the watershed and downstream environments (Table 9.1). Examination of water samples collected on a monthly basis starting in June 2004 indicated that specific conductance, chloride, sulfate, calcium,

FIGURE 9.2   The Guy Cove valley fill prior to creation of a headwater stream system. *Note:* forested section in the top center of the picture is an area that was not mined and where a small stream channel remained relatively intact.

FIGURE 9.3   Prior to restoration, a 0.02 ha sediment-laden pond was located at the toe of the Guy Cove valley fill.

magnesium, potassium, and sodium concentrations in Guy Cove were well above that observed from a nearby unmined reference watershed, Little Millseat (LMS) (Mastin et al. 2011). The concentrations observed reflect a problem that is prevalent throughout the mined portion of the EKC. Of particular concern was specific conductance and manganese levels, which were both above discharge water quality standards for active mine sites (Code of Federal Regulations 1996). Evidence of erosion and

FIGURE 9.4 A spring-fed channel is located in the 9 ha unmined section of Guy Cove. This channel serves as a continual source of surface water for the created main stream located on the crown of the valley fill.

TABLE 9.1

Average Pre-Construction Water Quality Characteristics from a Nonmined Reference Reach (LMS), an Upgradient Stream That Drained into the Fill (Seep), and a Stream Channel below the Valley Fill (Guy Cove)

Site[a]	pH	EC (µS cm^{-1})	Cl (mg L^{-1})	SO$_4$ (mg L^{-1})	NO$_3$ (mg L^{-1})	Ca (mg L^{-1})	Fe (mg L^{-1})	Mn (mg L^{-1})
LMS (30 year)[b]	6.5	46	0.6	10	0.13	2.3	na[d]	na
Seep[c]	7.9	478	1.3	225	0.09	47	0.3	4.0
Guy Cove[c]	7.0	1723	2.3	1293	0.01	137	0.7	21.4

[a] LMS, Little Millseat reference watershed; Seep, natural seepage from unmined upgradient area; Guy Cove, downstream of the fill and underdrain.
[b] Average from weekly samples collected over a 30-year period.
[c] Samples collected monthly from June 2004 to June 2008.
[d] na, Not analyzed.

deposition was common throughout Guy Cove. Fine grain sands had accumulated in low gradient areas and had resulted in a notable level of embeddedness in receiving waters. An iron rich flocculent was observable in waters emanating from the underdrain, which indicated the presence and oxidation of pyrite (FeS_2). Low pH values were typically observed where iron precipitation was present; however, buffering from high dissolved solid concentrations and subsequent alkalinity production had maintained a neutral pH and limited trace element mobility from the overburden. Analyses were below detection for arsenic (As), selenium (Se), lead (Pb), mercury (Hg), cadmium (Cd), and chromium (Cr) in water samples obtained from Guy Cove.

9.3.2 SOILS

Prior to mining, the project site was composed of soils from the Dekalb-Marowbone-Lantham-Cloverlick-Shelocata-Cutshin, Shelocta-Gilpin-Hazelton, and Shelocta-Gilpin-Kimper complex mapping units (Hayes 1991). Soils in the Dekalb-Marowbone-Lantham complex are moderately deep, well-drained, and formed from sandstone, shale, and siltstone. These soils are located on the upper third section of the hillslopes that are typically xeric and contain rock outcroppings. Soils in the Cloverlick-Shelocata-Cutshin complex are very deep, well-drained soils formed in colluvium on steep slopes. They occupy the north- and east-facing coves and cool slopes. The soils in the Shelocta-Gilpin-Hazleton complex consist of steep, well-drained, soils formed in colluvium and residuum. These soils occur on the warmer slopes with south and east aspects.

Following mining, only a sparse amount of ridgetop soils from the Dekalb-Marowbone-Lantham complex was remaining, primarily from the Marowbone soil series. The rest of the site is composed of shallow to very deep spoils with varying degrees of weathering and soil development. A soil pit was dug in 2004 and described by U.S. Natural Resource Conservation Service's Soil Survey staff. As with most surface mines in the area, the soil was classified as an Udorthent in the Fairpoint soil series (loamy-skeletal, mixed, active, nonacid, mesic Typic Udorthents).

9.3.3 VEGETATION

The regenerated forest located in the 9 ha unmined section of Guy Cove comprised yellow poplar (*Liriodendron tulipifera*), white oak (*Quercus alba*), sweet birch (*Betula lenta*), American beech (*Fagus grandifolia*), red maple (*Acer rubrum*), black walnut (*Juglans nigra*), and American sycamore (*Platanus occidentalis*) (Maupin 2012; Blackburn-Lynch 2015). Along the crown, face, and on 50% of the side slopes adjacent to the crown, vegetation was predominately herbaceous, consisting of ground cover species that were seeded during the initial reclamation (Figure 9.2).

9.3.4 HABITAT

Rapid Bioassessment Protocols (RBP) following Barbour et al. (1999) were conducted in Guy Cove, on the stream located in the unmined, upgradient section and on the stream located below the toe of the valley fill. The RBP scores were 120 (suboptimal) and 100 (marginal), respectively. Lack of epifaunal substrate, velocity/depth variations (e.g., riffles, runs, pools, and glides), bank stability, vegetation, and riparian width along with increased sediment deposition and embeddedness were the most significant habitat impairments.

9.4 DESIGN COMPONENTS

9.4.1 REFERENCE REACHES

9.4.1.1 Geomorphology

Natural channel designs use reference reaches from stream reaches within the same valley types, flow regimes, and sediment regimes. While preference is typically given to reaches located upstream/downstream of project sites or in nearby watersheds, reference reaches that are from similar but spatially distant locations can be utilized as geomorphic references as long as they exhibit similar characteristics to those of the impacted stream (Rosgen 1998; Hey 2006; Johnson and Fecko 2008; Faustini et al. 2009; Blackburn-Lynch 2015). While preference was given to streams in eastern Kentucky, reference geomorphologic data were also obtained from streams in the anthracite coal mining region of Pennsylvania as well as the mountain and Piedmont regions of North Carolina. Use of reaches from outside Kentucky was dictated by the extensive change to watershed geomorphology after the mining. Given that it was cost prohibitive to remove and redistribute the entire fill, we chose to examine other geomorphic types with lower gradient streams for portions of the design. These reference reach data were used to design the main channel located along the crown of the valley fill and not the ephemeral channels.

9.4.1.2 Water Quality and Habitat

The LMS watershed, which is located in Robinson Forest, was identified as the primary water quality and habitat reference watershed for the restoration project (Figure 9.5). The watershed is 74.5 ha in size with a drainage density of 0.0038 m m^{-2} (Cherry 2006). The stream has a well-developed

FIGURE 9.5 The Little Millseat (LMS) watershed in Robinson Forest served as a water quality and habitat reference watershed for the project.

and undisturbed riparian zone with a fully shaded canopy. Trees in the watershed are mature (>90 years old) and past land use is well documented. Bank erosion is low and embeddedness is optimal for this ecotype. Both water quality and quantity have been monitored at LMS since 1971; water quality is high. Macroinvertebrate sampling (February and April) commenced in 2004. The timing of sampling is important to obtain winter emergent taxa (e.g., winter stonefly families Capniidae and Taeniopterygidae) and in spring prior to emergence of spring-emergent taxa (e.g., most Ephemeroptera). Sampling points were positioned in riffles.

9.4.2 Valley Reconfiguration and Hydrologic Modifications

Valley fills result in unnatural topography, particularly with respect to valley widths and slopes. Intermittent and perennial stream reaches in LMS have valley slopes of about 12% and 6%, respectively (Villines et al. 2015). Lower slopes, such as the one present on the crown prior to construction, encourage sediment deposition in mined lands. As this project was a retrofit of a valley fill, economic limits were present regarding the amount spoil we could move to reshape the width and slope of Guy Cove's valley. The design incorporated the use of multiple valley slopes along the crown starting with 1% at the most upgradient section of the project transitioning to a maximum slope of 5.5% before decreasing to a slope of 3% prior to transitioning to the face of the valley fill. The new valley slope resulted in the elimination of the uppermost bench along the face of the valley fill. To achieve a cut-fill balance within Guy Cove, all excavated spoil was used to create a narrower valley (Figure 9.6). Spoil was backfilled along the lower portion of the valley side walls and compacted, resulting in an average final grade of 3:1 (H:V). Following compaction, additional spoil was placed along the side slopes in accordance with the FRA (Burger et al. 2005). Along these slopes, spoil was dumped in piles that closely abutted each other with an approximate height of 2.5 m. Using an excavator, the tops of the piles were struck-off (see Chapter 8) to level the surface while minimizing mechanical compaction. The resulting surface was rocky and loosely compacted. Approximately 115,000 m³ of spoil were moved to reconfigure the valley.

FIGURE 9.6   A narrower valley was created by compacting spoil along the valley side walls and then placing additional spoil on top of the compacted layer in accordance with the FRA. Final side sloped were typically 3:1 (H:V).

FIGURE 9.7 One groin drain was removed while the other groin drain along with the benches were modified to create a single flow path along the face of the valley fill. Woody debris was also added to the remaining groin drain to increase habitat quality.

To promote the surface expression of waters (e.g., reduce infiltration into the fill), the crown was compacted during valley reconfiguration. Loaded dump trucks were repeatedly driven over the crown to increase compaction. Spoil was placed at the interface of the unmined and mined sections (i.e., entrance to the underdrain) and compacted in a similar manner. Along the face of the valley fill, the right groin drain, when facing downvalley, was removed and the benches were regraded to improve the ability of water to flow from the benches into the left groin drain (Figure 9.7). Large boulders along the left groin drain, which were blocking overland flow from the benches, were removed. Woody debris was also added to improve habitat conditions. These modifications were done so that the main channel on the crown of the valley fill would flow into a single groin drain (e.g., single channel).

9.4.3 Stream Dimension, Pattern, and Profile

9.4.3.1 Main Channel

The main channel consists of a 760 m section along the crown, nearly 120 m section along the face, and a nearly 140 m section below the toe of the valley fill. For the sections along the crown and below the toe, the stream was created. The typical design bankfull dimensions for the riffle cross sections along the main channel, excluding the face of the valley fill, were as follows: area = 1.4–1.5 m², width = 2.4–2.7 m, and mean depth = 0.2 m. Typical design bankfull dimensions for a pool cross section, excluding the face of the fill, were as follows: area = 1.8–3.4 m², width = 2.7–3.4 m, and mean depth = 0.2–0.4 m. As previously noted, for the portion of the main channel on the crown, slopes ranged between 1% and 5.5%, while values for sinuosity (distance along stream between two points ÷ straight line distance between the two points) ranged between 1.0 and 1.2 along the crown. Along the face, the slope was 25.5% with a sinuosity of 1.0. For the portion of the channel below the toe of the valley fill, slopes ranged between 2% and 4% with a sinuosity of 1.02.

Instream structures within the main channel predominately consisted of cross vanes, steps, and constructed riffles. Due to the remoteness of the project site coupled with the prevalence of a nearby forest, the cross vanes and steps were constructed of logs and anchored in place using boulders. Because of the long distance to a rock quarry, it was cost prohibitive to haul in rock for instream structures or riffles. As such, a Powerscreen (Dungannon, Ireland, U.K.) was used to sort spoil into four size classes of which the largest class (large gravel to cobble size material) was used to construct the riffles in the main channel.

9.4.3.2   Ephemeral Channels

Ephemeral channels convey water from the side slopes to the main channel during runoff-producing rain events. While ephemeral channels transport water for short periods of time, these streams are important to headwater ecosystems (McDonough et al. 2011). Four ephemeral channels (totaling 435 m in length) were constructed with their placement guided by the development of preexisting gullies (i.e., the presence of gullies at a particular location indicated overland flow had channelized at that point, hence a preferential flow path existed). Ephemeral channels were designed in the field largely on the basis of local slopes and available construction material (e.g., rock and logs). Due to the steepness of the ephemeral channels (typically >20%), these streams predominately exhibited a step-pool morphology (Figure 9.8).

FIGURE 9.8   Ephemeral channels were constructed using logs and rocks to create step-pool features.

9.4.3.3   Bioreactor-Wetland Treatment System

The exposure and oxidation of iron sulfide materials from surface coal mining activities result in the formation of acid drainage (AD), which adversely impacts approximately 20,000 km of streams and rivers in the United States (Kleinmann 1989). Pyrite oxidation and subsequent AD formation are a complex process involving hydrolysis, redox, and microbial reactions (Nordstrom 1982). The general stoichiometry can be described by the reaction:

$$FeS_{2(s)} + 3.75O_2 + 3.5H_2O \rightarrow Fe(OH)_{3(s)} + 2H_2SO_4$$

where iron sulfide and other mixed-metal sulfides decompose upon exposure to water and air producing sulfuric acid and insoluble ferric iron hydroxide from hydrolysis (Bigham et al. 1992). This ferric iron precipitate and the associated acidity are considered the principle causes for the degradation of water bodies receiving AD and for the endangerment of aquatic habitat that resides within. Many streams receiving AD exhibit pH ranges from 2 to 4. However, neutral mine drainage is common in areas that contain appreciable amounts of carbonate materials (as is the case at Guy Cove) (Agouridis et al. 2012).

In recent years, the development of low-cost passive treatment technologies has allowed for the utilization of natural chemical and biological processes to clean contaminated mine waters without the expense or potential hazards associated with chemical additions. Constructed wetlands, anoxic limestone drains (ALDs), and successive alkalinity producing systems (SAPS) are examples of such technologies that offer a potential solution to the AD problem (Barton and Karathanasis 1999). To address the water quality problem in Guy Cove, a passive treatment system was designed that incorporated the use of two bioreactors (similar to a SAPS design) and an artificial wetland (Figure 9.9). The bioreactor-wetland treatment system was constructed below the toe of the valley fill and adjacent to the main channel—a berm separated the two. Effluent from the wetland was directed back

FIGURE 9.9   A bioreactor-wetland treatment system (left side of figure) was used to treat a portion of the waters emanating from the underdrain. Effluent from the wetland was directed into the main channel (right side of figure).

into the main channel. The reader is referred to Mastin et al. (2011) for additional details regarding this treatment system. The bioreactor-wetland treatment system was designed to improve the quality of a portion of the waters emanating from the toe of the fill by (1) reducing redox conditions in the seepage to promote the precipitation of metal sulfides in the bioreactors and (2) filtering precipitated metal sulfides and hydroxides, complexation of organo-metal compounds, and surface exchange of cations in the wetland substrates. The wetland was planted with a mixture of wetland and bottom-land forested and sedge species such as buttonbush (*Cephalanthus occidentalis*), river birch (*Betula nigra*), bald cypress (*Taxodium distichum*), black gum (*Nyssa sylvatica*), silky dogwood (*Cornus amomum*), fox sedge (*Carex vulpinoidea*), Frank's sedge (*Carex frankii*), and tussock sedge (*Carex stricta*).

9.4.3.4   Forested Zones

Reforestation of the watershed was performed separately for riparian (adjacent to the created stream) and upland areas. The riparian corridor was divided into two zones paralleling the main channel. The inner zone (Zone 1), extending 5 m on either side of the stream, was densely stocked (1 m × 1 m centers) with shrubs and trees adapted to occasional flooding and frequent disturbance. One-year-old bare-root American sycamore, American beech, green ash (*Fraxinus pennsylvanica* var. *subintegerrima*), swamp chestnut oak (*Quercus michauxii*), and dogwood (*Cornus* sp.) seedlings, all common in riparian zones of mixed mesophytic forests, were used in this first riparian zone. A second zone (Zone 2) extending 10 m beyond the inner streambank zone was planted with additional trees and shrubs species (2 m × 2 m centers) that are accustomed to occasional flooding and saturated soils. Additional species included white oak, silver maple (*Acer saccharinum*), and river birch. Black willow (*Salix nigra*) stakes were densely planted (0.5 m × 0.5 m centers) in areas that were susceptible to bank erosion (intersections of ephemeral and main channels).

The remaining watershed to which FRA was applied (Zone 3) was planted with mixed meso-phytic hardwoods typical of eastern Kentucky and the Cumberland Plateau. Yellow poplar, white oak, northern red oak (*Quercus rubra*), chestnut oak (*Quercus prinus*), white ash (*Fraxinus americana*), sugar maple (*Acer saccharum*), eastern white pine (*Pinus strobus*), black locust (*Robinia pseudoacacia*), dogwood, and redbud (*Cercis canadensis*) seedlings were planted on 3 m × 3 m centers and mixed throughout the entire midslope and ridgetop regions.

In all three zones combined, over 30,000 seedlings were planted.

9.4.3.5   Erosion Control

Temporary (winter wheat and annual ryegrass) and permanent (orchard grass) seeding of ground cover for erosion control was performed along all streams excluding areas with FRA placed spoil. Seed was applied to a 6 m buffer on each side of the main channel and a 3 m buffer on each side of ephemeral channels at an approximate ratio of 4:2:1 winter wheat (*Triticum* sp.), ryegrass (*Lolium* sp.), orchard grass (*Dactylis* sp.), respectively. All seeded areas were top-dressed with approximately 112 kg ha^{-1} of 19–19–19 fertilizer and covered with erosion control blanket (coir along the streambanks, jute/straw floodplain).

9.5   MONITORING

9.5.1   Geomorphology

Permanent cross sections were installed throughout the main channel (n = 40) at fairly regular intervals encompassing all slope changes, as well as each of the four ephemeral channels (n = 18) following construction. At each cross section, two steel posts (1.2 m in length) were installed above bankfull elevation on each side of the bank. All cross sections were surveyed annually using techniques outlined by Harrelson et al. (1994). Following installation, locations of all bank pins as well as the longitudinal profiles of the main and ephemeral channels were surveyed using a Sokkia (Olathe, Kansas) total station equipped with a Carlson (Maysville, Kentucky) data logger.

Due to the length of the main channel, one representative reach and two representative riffle pebble counts were collected from five subreaches along the main channel, excluding the reach traversing the face of the fill. Representative reach data from all five subreaches were combined to determine the overall bed material composition. Additionally, the subreaches from the crown of the fill were combined to determine the overall bed material composition for that portion of the main channel. For the section of the channel along the face of the fill where only enhancement occurred, visual assessments were done. All pebble counts were performed annually in accordance to methods outlined in Harrelson et al. (1994). Due to the large number of step-pool structures present in the ephemeral channels, only visual assessments were performed.

9.5.2 Hydrology

Streamflow was monitored only on the main channel immediately upstream of the project site (GC01), at the crest or transition point from the crown to the face of the valley fill (GC02), and just upstream of the confluence with the wetland outlet (GC03). At these points, streamflow was measured using trapezoidal flumes and In-Situ Level Troll 500 (5 lb per square inch—gauged) pressure transducers (Fort Collins, Colorado).

Falling Rock (FR), located in Robinson Forest, is a perennial stream that was used as the hydrologic reference watershed for the restoration project. The watershed is 92.4 ha in size with a drainage density of 0.0040 m m^{-2} (Cherry 2006). Similar to LMS, the watershed has undergone extensive hydrologic and water quality monitoring. Falling Rock was used as a hydrologic reference reach in lieu of LMS because of extensive equipment failure during the post-restoration period at the latter watershed. Streamflow at FR was measured using a 3:1 side-sloped, broad-crested combination weir. Wharton Branch (WB) is a perennial stream located within 1 km of the project site. It is a mined headwater stream emanating from a valley fill, has a drainage of 44 ha, and is dominated by grasses, exotic shrubs, and conifers. Streamflow at WB was measured using an H-flume.

A weather station with a tipping bucket rain gage linked to a Campbell Scientific CR10X data logger (Logan, Utah) was used to record rainfall data (Cherry 2006). The weather station was located centrally to the project site, FR and WB.

Storm hydrographs (12 in 2010, 17 in 2011, 11 in 2012, and 14 in 2013) were analyzed to determine discharge volume, peak discharge, discharge duration, peak time, lag time, and response time. Discharge volumes and peak discharges were normalized by the drainage areas of the respective sites. Curve numbers ($\lambda = 0.05$ and 0.2) were computed using procedures outlined by the SCS (1972) and Hawkins et al. (2002). Baseflow volumes were computed monthly for each monitored site (e.g., GC01, GC02, GC03, FR, and WB) and normalized by the drainage area of the respective monitored sites. The number of days that surface water was present in the stream at each monitored site was determined using state height data, and in cases of low or no flow, Plant Cams (Windscapes, Calera, AL) were used to confirm flow presence. For additional details on the hydrologic assessment, the reader is referred to Blackburn-Lynch (2015).

9.5.3 Water Quality

Water samples were collected from GC01, GC02, GC03, LMS, FR, and WB on a biweekly basis from 2004 to 2013. Water samples were also collected from the bioreactor-wetland treatment systems (bioreactors 1 and 2, wetland influent and effluent) on a biweekly basis from 2009 to 2013. Sampling, preservation, and analytic protocols were performed in accordance to standard procedures (Greenberg et al. 1992). Water quality parameters such as pH and specific conductance (EC) were monitored in the field using a Yellow Springs Instrument (Yellow Springs, Ohio) 556 multiprobe monitor. Water samples were collected in plastic bottles and immediately stored in ice chests.

Field collected water samples were transferred to a refrigerator at 4°C as soon as feasible. All samples remained refrigerated until they were processed. Samples were filtered through 0.45 μm syringe filters prior to analysis. Sulfate was analyzed on a Dionex ion chromatograph system 2500 (ISC 2500). The setup contained an AS50 auto sampler, IS25 chromatograph, LC25 chromato-oven, and an EG50 eluent generator. Nitrate was analyzed with a Braun+Luebbe Auto Analyzer 3 using the colorimetric procedure. An ICP-OES (Varian Vista-Pro CCD Simultaneous, Varian Instruments, Palo Alto, CA) was used to measure dissolved iron (Fe) and manganese (Mn). Laboratory control standards and blanks were run for accuracy, and duplicates were run for precision.

9.5.4 Vegetation

9.5.4.1 Forested Zones

9.5.4.1.1 Riparian Forest Census

Sample plots were established in the riparian areas along the portion of main channel on the valley fill crown that encompassed both Zone 1 and Zone 2 plantings. Plots were designed in a manner to both adequately sample each species planted and to account for differences with respect to variations in topographic gradients associated with the main channel (e.g., differing valley slopes). Ten plots measuring 20 m × 10 m were established parallel to the stream channel (five plots on each side of the stream). Seedling growth (height and diameter) and survival were measured annually within the monitoring plots. Notes were taken on seedling health, vigor, growth form, and level of herbivory. Each tree within the plot was marked for identification purposes using an aluminum tree tag. Plot corners were permanently marked with rebar, and GPS coordinates were obtained to delineate the plot boundaries on a map.

9.5.4.1.2 Upland Forest Census

Sample plots were also established in the mixed species upland planting areas. Plots were designed in a manner to both adequately sample each species planted and to account for differences with respect to microtopographic variability among planting units. As such, multiple (8) 20 m × 20 m plots were established within planting units (four plots on each side of the stream in the FRA placed spoil). Seedlings were measured and plots marked following the same procedures used for the riparian forest census described above.

9.5.4.2 Ground Cover

Vegetative (woody plants <1 m and all herbs) covers of seeded and naturally colonized species were examined in two, 1 m² permanent ground-layer plots per seedling plot (riparian and upland). Vegetative cover in these plots was estimated by cover class. Cover classes of I–IV were assigned as follows: Class 1 = 0%–25% cover, Class 2 = 25%–50% cover, Class 3 = 50%–75% cover, and Class 4 ≥ 75% cover.

9.5.5 Habitat

Habitat was assessed using two metrics: annual RBP's for Kentucky's high gradient streams (KDOW 2011) and macroinvertebrate sampling. Benthic macroinvertebrates were sampled in the main channel for five periods between 2009 and 2012. Samples were collected in the upper 400 m of the restored reach to ensure the presence of water throughout the year (refer to Section 9.5.2 regarding days of surface flow at GC02). During each monitoring period, four to seven benthic samples were collected from riffles using a Wildco (Yulee, Florida) surber bottom sampler (250 μm mesh, 0.09 m² sample area) and preserved in the field with 90% alcohol. In the lab, each sample was rinsed with tap water through nested sieves (1 mm and 250 μm) for sorting. Organisms retained by the 1 mm sieve were collected under a tabletop magnifying glass, while those retained in the 250 μm sieve were subsampled (1/8 to 1/1) with a Folsom plankton splitter (Wildco, Yulee, Florida) and picked under a dissecting microscope. Aquatic insects were identified to families using Merritt et al. (2008) and Hilsenhoff (1995), while other invertebrates were identified to the following groups using Thorp and Covich (2010): oligochaetes to single taxa, and crustaceans and molluscs to classes. Densities (# m^{-2}) of total macroinvertebrates, insects, EPT (Ephemeroptera, Plecoptera, and Trichoptera), and % EPT of macroinvertebrates based on individual samples were computed. Results from all macroinvertebrates collected on each sampling date were combined to determine the family-based macroinvertebrate taxa richness, insect taxa richness, EPT richness, and Shannon-Weiner diversity index (H').

9.6 RESULTS

9.6.1 GEOMORPHOLOGY

For Year 1 following construction, average riffle bankfull cross section dimensions on the crown of the fill were as follows: area = 0.4 m^2, width = 2.3 m, and mean depth = 0.2 m. Below the toe of the fill, these riffle bankfull dimensions were as follows: area = 0.6 m^2, width = 2.3 m, and mean depth = 0.2 m. For pools (crown only), the bankfull dimension values were as follows: area = 0.8 m^2, width = 2.2 m, and mean depth = 0.4 m. In Year 5 following construction, average riffle bankfull cross section dimensions on the crown of the fill were as follows: area = 0.4 m^2, width = 2.4 m, and mean depth = 0.2 m. Below the toe of the fill, these riffle bankfull dimensions are as follows: area = 0.5 m^2, width = 2.7 m, and mean depth = 0.2 m. For pools (crown only), the bankfull dimension values are as follows: area = 0.7 m^2, width = 2.5 m, and mean depth = 0.3 m. A comparison of cross section data between Years 1 and 5 indicates that the main channel is laterally and vertically stable (Figure 9.10). In locations where the vegetation is well established, a greater level of bank formation is occurring, particularly in the lower reaches along the crown and in the reach below the toe of the fill. Streamside vegetation appears to be trapping sediments and in many instances reducing the width-to-depth ratio and decreasing the bank height ratio. Similar results with regard to lateral and vertical stability were found with the ephemeral channels. Comparisons of Year 1 and Year 5 longitudinal profiles for the main and ephemeral channels yielded similar results to those found with the permanent cross sections.

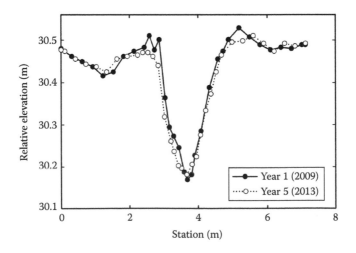

FIGURE 9.10 Typical riffle cross section on the main channel. A comparison of Years 1 and 5 shows the channel is vertically and laterally stabile.

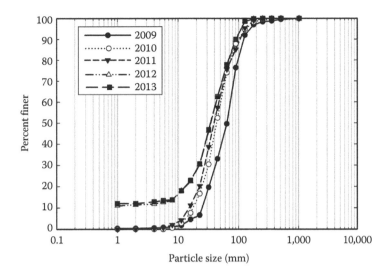

FIGURE 9.11   Between Years 1 and 5, the D_{50} of the main channel decreased slightly but continues to classify as gravel.

With regard to bed material, the main channel, excluding the face of the fill, had median bed material sizes (D_{50}) of 64 mm for Year 1 and 35 mm for Year 5 (Figure 9.11). As compared to Year 1 data, Year 5 bed material data showed a shift toward a finer particle size, though the D_{50} in all instances would classify the material as gravel. Weathering of spoil material is thought to play a role as finer material is flushed from the watershed.

9.6.2   HYDROLOGY

Results indicated that FR, GC03, and WB had the largest storm discharge volumes and peak discharges (both parameters were normalized for drainage area), while GC01 and GC02 had the smallest. Runoff curve numbers ($\lambda = 0.05$ and 0.2) were largest for GC03 and WB followed by GC01, GC02, and FR (see Taylor et al. 2009a for more information on runoff curve numbers). Discharge durations were longest for FR and GC01 and shortest for GC02, GC03, and WB. Lag time and peak time were longest for FR and GC01 and shorter for GC02, GC03, and WB. No differences were noted with respect to response time. Both GC03 and WB tended to respond more quickly to storm events than FR which agrees with Phillips (2004) who found that valley fills can increase the speed at which runoff reaches streams. The similarity between GC03 and WB with respect to storm hydrograph characteristics indicates that restoration efforts exerted little influence on the hydrology at the toe of the valley fill. As storm discharge from GC02 comprised about 15% of the volume at GC03, a significant amount of discharge continues to come from the underdrain potentially in the form of groundwater.

While storm discharge volumes at GC01 and GC02 were statistically similar, GC01 exhibited larger peak discharges and longer durations, peak times, and lag times. These differences between GC01 and GC02 were attributed to the low slope and heavily vegetated section of the main channel immediately down-gradient of GC01 (Sand-Jensen and Mebus 1996), the FRA surrounding the main channel (Taylor et al. 2009b), and most importantly the lack of groundwater connection at GC02 as the main channel is approximately 40 m above original ground surface at this location. The lack of connection between the main channel and the groundwater table means GC02 lacks a sustained form of subsurface flow. This lack of groundwater connection is evident when examining the number of days of flow. Low flows from GC01 (average of 55 m³ day⁻¹) were not sufficient, particularly during drier conditions characteristic of summer and fall months, to sustain baseflow

TABLE 9.2

Average Number of Days per Month Surface Waters Were Present at Each Monitoring Site for the Post-Construction Monitoring Period (2010–2013)

	Monitoring Site[a]				
	FR	GC01	GC02	GC03	WB
Days per month	28.7a[b]	24.5a	12.9b	28.7a	28.2a

[a] FR, Falling Rock; GC01, immediately upgradient of project site; GC02, interface between crown and face of valley fill; GC03, downstream end of project reach; and WB, Wharton Branch (perennial).

[b] Statistically significant differences within row noted by differing letter ($p \leq 0.05$).

conditions at GC02 (Table 9.2). As vegetation along the main channel continues to grow and evapotranspire at increased rates, it is anticipated that baseflow volumes at GC02 will continue to decrease (Sena et al. 2014). We speculate that as the forest matures (and evapotranspiration increases) and surface water continues to move through the watershed via the created stream, infiltration to the underdrain will diminish over time and perhaps flow volumes at GC03 will also decrease.

9.6.3 WATER QUALITY

Results from the post-construction monitoring period are presented in Table 9.3. Water qualities in the constructed stream section located on the crown of the fill are better than those observed during the pre-restoration sampling period. One parameter of note is the EC level at GC02, which dropped from a mean concentration of 852 μS cm^{-1} in 2010 to a mean concentration of 471 μS cm^{-1} in 2013. This period of initially elevated EC levels followed by a decrease as finer material is flushed through the system mirrored the trend seen by Agouridis et al. (2012) for FRA test plots. Concentrations of sulfate and manganese exhibited a similar decreasing trend over the 4-year post-construction monitoring period at GC02. Concentrations of dissolved ions on the crown were lower than observed exiting the underdrain. Groundwater discharging from the underdrain continues to be a source of contamination to downstream reaches. Unfortunately, little is known about the hydrology of these underdrains and the long-term ramifications they may have on the environment.

Since water with high EC has been implicated as a potential driver of macroinvertebrate assemblage shifts (Pond et al. 2008), developing reclamation techniques that mitigate high EC concentrations is a priority. The creation of the stream on top of the valley fill allowed for freshwater, primarily from precipitation, to move through the watershed without extensive comingling with spoil buried in the fill. This limited contact time between water and readily dissolvable minerals and dissolved solid concentrations was reduced. Sulfate has been identified in previous studies (Kennedy et al. 2003; Lindberg et al. 2011; Hopkins et al. 2013) as a major contributor to the elevated EC and associated macroinvertebrate toxicity. On the crown, sulfate was reduced from 871 to 161 mg L^{-1} during the post-construction period (81.5% reduction), which was encouraging. If the system remains at these levels, or continues to improve, water quality conditions should be sufficient to support sensitive macroinvertebrate species common to the region (Pond et al. 2008).

Flow through the bioreactor-wetland treatment system varied seasonally with the percentage of discharge treated by the system typically ranging between 10% and 35% of the total available flow (stream and underdrain combined) (Table 9.4). Although the bioreactors did not perform as well as bench-scale laboratory experiments predicted (Edwards et al. 2009;

TABLE 9.3

Average Post-Construction Water Quality Characteristics from Guy Cove, a Nonmined Reference Stream (LMS) and a Nearby Traditionally Constructed and Reclaimed Valley Fill (WB)[a]

Sample ID[b]	2010	2011	2012	2013
EC μS cm^{-1}				
LMS	52	43	45	42
GC01	479	457	526	447
GC02	852	827	598	471
GC03	1949	1724	1841	1598
WB	2343	2126	2397	1979
SO$_4$ mg L^{-1}				
LMS	9	15	8	7
GC01	525	417	134	71
GC02	871	815	309	161
GC03	1470	1780	1310	715
WB	1884	2465	1826	960
Mn mg L^{-1}				
LMS	0.3	0.3	0.2	na[c]
GC01	0.1	0.2	0.1	0.1
GC02	3.2	5.3	0.4	0.2
GC03	6.2	3.6	4.9	1.9
WB	7.4	15.5	11.4	8.0
pH				
LMS	5.8	6.6	6.2	5.5
GC01	7.5	7.8	7.2	7.4
GC02	6.0	6.3	6.3	6.9
GC03	6.2	6.6	6.2	6.6
WB	6.1	6.6	6.3	6.6

[a] Construction and planting activities occurred from July 2008 to July 2009.

[b] LMS, Little Millseat reference watershed; GC01, confluence of existing upgradient stream and beginning of created stream; GC02, interface between crown and face of valley fill; GC03, downstream end of project reach; WB, reference impacted (valley fill) watershed.

[c] na, Not analyzed.

Karathanasis et al. 2010), the bioreactor-wetland treatment system contributed to a reduction in EC, sulfate, and manganese concentrations by 86%, 16% and 7%, respectively. By Year 5, the bioreactors were adding alkalinity to the mine drainage and have become limited in their treatment efficiency. Bioreactor 1 has become a source of Fe to Bioreactor 2 indicating that the exchange capacity of the substrate has been exhausted. Sulfate and manganese concentrations leaving Bioreactor 2 showed a 22% reduction over the influent water. However, levels increase for both sulfate and manganese in the treatment wetland. These findings could indicate that the wetland has become a source of these constituents, but the most likely culprit is due to inclusion of water into the wetland that bypasses the bioreactors and does not receive treatment. Periodic replacement (recharge) of the filter media inside the bioreactors will be required to improve treatment efficiencies.

TABLE 9.4

Average Post-Construction Water Quality Characteristics from the Wetland and Bioreactor Treatment System at Guy Cove

Sample ID	2010	2011	2012	2013
EC μS cm^{-1}				
Influent	2040	1817	2037	1760
Effluent	1854	1319	1904	1664
SO$_4$ mg L^{-1}				
Influent	1510	1911	1572	847
Effluent	1418	1481	1398	793
Mn mg L^{-1}				
Influent	9.5	6.7	7.2	4.9
Effluent	8.2	4.8	5.6	4.1
Fe mg L^{-1}				
Influent	7.4	na[a]	8.9	1.9
Effluent	2.3	na	3.1	0.3
pH				
Influent	5.5	5.7	5.8	5.5
Effluent	6.4	6.9	6.1	6.4

[a] na, Not analyzed.

9.6.4 VEGETATION

After 5 years, vegetation establishment at Guy Cove has progressed despite several setbacks with regard to plant loss and damage from animal browse and in some cases human activities (one plot in riparian Zone 1 was lost). In 2013, browse was observed on 44% of trees in riparian sample plots and on 11% of those in the upland plots. Although browse appears to have a significant effect on height growth, survival of the planted seedlings does not appear to have suffered. Browse was initially attributed primarily to elk, but excrement from deer, cattle, and horses was observed on the site. By 2013, much of the vegetation established for erosion control was replaced by other naturally seeded species.

9.6.4.1 Riparian Forest Census

Table 9.5 contains height growth data for Year 1 and height growth and survival data for Year 5 Zones 1 and 2 plots. For all species, survival was 75% and ranged from 42% for silver maple to 100% for five species. Within the plots, tree density was calculated at 2805 trees ha^{-1}. Height growth over the 5-year period was observed, but damage from browse resulted in unusual growth patterns for the planted species (not linear as would be expected). Black locust, green ash, silver maple, and river birch exhibited extensive browse (>80%). Natural colonization of species was observed in the study plots. European alder (*Alnus glutinosa*) and Virginia pine (*Pinus virginiana*) were the most commonly observed.

9.6.4.2 Upland Forest Census

Sample plots were resurveyed in the mixed species upland planting areas (Zone 3) at the project site in Year 5 (Table 9.6). Survival was greater than 100% for sycamore and white pine due to natural colonization (population increased by over 100 specimens in the sample plots). Excluding sycamore

TABLE 9.5

Year 1 (Height) and 5 (Height, Survival, and Browse) Data for Primary Riparian Species Planted in Zones 1 and 2[a]

Species	Year 1 Height (m)	Year 5 Height (m)	Survival (%)	Browse (%)
Green ash	0.42	0.60	54	87
Beech	0.52	0.47	91	0
Sycamore	0.37	1.04	77	1
Swamp white oak	0.57	0.83	90	48
Silver maple	0.71	0.56	42	89
River birch	0.62	0.83	96	80
Black locust	0.36	1.70	100	83
Chestnut oak	0.61	1.14	100	0
N red oak	0.52	0.89	78	21
White oak[b]	0.53	1.64	100	0
White pine[b]	0.31	1.28	100	0
Yellow poplar[b]	0.55	0.45	100	50

[a] Year 1 = 2009; Year 5 = 2013.
[b] Year 5: Low number of survey trees in plot for these species (n < 5).

TABLE 9.6

Year 1 (Height) and 5 (Height, Survival, and Browse) Data for Primary Species Planted in Low Compacted Spoil Located in Upland Zone 3

Species	Year 1 Height (m)	Year 5 Height (m)	Survival (%)	Browse (%)
White ash	0.54	0.95	66	66
White oak	0.56	1.39	89	2
Chestnut oak	0.67	0.93	100	0
Sycamore	0.53	2.00	>100[a]	0
N red oak	0.57	0.81	86	28
Sugar maple	0.28	0.42	43	66
Yellow poplar	0.40	1.63	100	0
Black locust	0.67	2.72	85	33
White pine	0.35	1.57	>100[a]	0
Red bud	0.42	0.65	20	0

[a] 2013 populations greater than 2009 population likely due to natural colonization.

and white pine, overall survival was ≥85% for five of the other eight observed species. Within the plots, tree density was calculated at 1539 trees per ha. Height growth was improved over that observed in the riparian zones and browse was much lower for most species. The reduction in browse may be attributable to the hummocky nature of the FRA which is less hospitable to large herbivores. The following naturally colonized species were observed in the upland sample plots: quaking aspen (*Populus tremuloides*), sweet birch (*Betula lenta*), river birch, red maple (*Acer rubrum*), black gum,

European alder, Virginia pine, sourwood (*Oxydendrum arboreum*), tree of heaven (*Ailanthus altissima*), royal paulownia (*Paulownia tomentosa*), and autumn olive (*Elaeagnus umbellate*).

9.6.4.3 Ground Cover

Mean cover class was 2.03 and 2.68 for the riparian and upland plots, respectively. A native broomsedge (*Andropogon virginicus*) was the most common groundcover species found throughout the site. Broomsedge is commonly found in open areas such as abandoned fields, overgrazed pastures, cutover timber sites, and rights of way, and it grows on a wide variety of soils, preferring loose, sandy, moist sites with low fertility. On infertile soils, broomsedge is a long-lived competitor that has allelopathic chemicals which can have adverse effects on other plants trying to colonize (Rice 1972).

9.6.5 Habitat

Following construction of the stream on the crown of the valley fill, RBP values increased from 147 (flatter portion) and 139 (steeper portion) to 152 and 162, respectively, in Year 5. For three of the ephemeral streams, RBP values increased (100 to 114, 82 to 107, and 91 to 110), while values decreased on one (103 to 77). Along the face of the valley fill, RBP scores increased from 114 to 157, while at the toe of the valley fill, RBP values increased from 110 to 133. Based on the RBP scores and specific conductance values from Year 5 as well as the Ecological Integrity Index Calculator developed by the Louisville District Corps of Engineers, it is estimated that this project produced nearly 2200 Ecological Integrity Units (EIU). In the EKC, mitigation credit is assigned based on EIUs, which are units based on the length of stream created and/or restored and the integrity of the biotic communities, as inferred through specific conductance, and the quality of habitat assessed via RBP in relation to reference biologic stream conditions (U.S. ACE 2002).

Densities of total macroinvertebrates and insects increased between 2009 and 2012. Mean macroinvertebrate density was 639 m^{-2} on the first sampling date in 2009 and steadily increased reaching >25,000 m^{-2} on the last sampling date in 2012. The mean density of insects was 511 m^{-2} in 2009 and increased to >19,000 m^{-2} by 2012. Insects accounted for more than half (range 51%–81%) of the total macroinvertebrate density. The density of EPT taxa increased from 92 m^{-2} in 2009 to 3493 m^{-2} in 2012. EPT taxa accounted for 10%–61% of total macroinvertebrates sampled. A total of 22 macroinvertebrate taxa were collected in 2009 with total richness increasing to 33 taxa by 2012. Diversity index (H′) was 1.77 in 2009 and increased to 2.10 in 2011 before decreasing to 1.75 in 2012. The majority of the macroinvertebrates were composed of relatively small taxa such as small crustaceans, oligochaetes, chironomids, gastropods, and small-bodied stoneflies and mayflies. Oligochaete worms and chironomids were the dominant taxa during early sampling dates, but the presence of small crustaceans and gastropods became prominent during the later dates. Early stages of stoneflies (Nemouridae) and mayflies (Ameletidae) were collected with high density on several sampling dates.

9.7 LESSONS LEARNED

Following nearly 4 years of planning, the University of Kentucky in a first of its kind project utilized NCD techniques and knowledge gained by research on the FRA to create a headwater stream system on a valley fill at Guy Cove in 2008. This project was proof-of-concept in that it sought to answer questions on "how to" retrofit an existing valley fill in an effort to restore lost stream and watershed functions. During a 6-month period, 1450 m of headwater streams was created along with 0.1 ha of vernal pools and 16.2 ha of hardwood forest. A bioreactor-wetland treatment system was constructed to treat a portion of the waters emanating from the underdrain. Results from the monitoring of the geomorphology, hydrology, water quality, vegetation, and habitat at the project site (in addition to reference and impacted sites) over a 5-year post-construction period point to a number of successes. The created streams are vertically and laterally stable and have withstood large storm events (e.g., 139 mm, 50 year 24 hour return interval event in 2009 about 5 months post construction).

Tree growth and survival are quite good for most species, particularly in the uplands where the FRA was used. On the portion of the main channel located on the crown, both water quality and habitat metrics have shown signs of improvement and the bioreactor-wetland treatment system improved water quality for a portion of the underdrain discharge. Notably, the 760 m created stream on the crown did not exist after fill construction was completed and there was no surficial connection between the small headwater stream in the unmined portion of the watershed and the stream reach below the fill. This project provided an opportunity to put back what was lost, reconnect the streams, and, with improved water quality, restore lost habitat for wildlife.

Designing and implementing such a novel and extensive restoration/creation project is not without lessons learned. These lessons fall into five main categories: isolation of problematic spoils, groundwater connection, colonization of macroinvertebrates, invasive species, and browse. The authors recommend designers to carefully consider these points when developing future projects that seek to recreate headwater stream systems on mined lands.

9.7.1 ISOLATION OF PROBLEMATIC SPOILS

The potential for sandstones to produce poor water quality is not uniform. Some sandstones produce high EC or even selenium (Se) leachate, while others do not (Vesper et al. 2008). Warner and Agouridis (2010) recommend isolating such problematic spoils (e.g., source reduction) through the use of a low permeability barrier. Low permeability barriers are commonly used in landfills (Goldman et al. 1988) and in coal mining to isolate acid-producing spoils (Skousen et al. 2000). Research by da Rosa et al. (2013) suggests brown weathered sandstones are a viable material for constructing such barriers in the EKC. Warner and Agouridis (2010) also recommend the selection of low-reactive durable rock and the minimization/removal of fines when constructing underdrains. Construction of valley fills using these techniques produced discharge with EC levels similar to pre-mining conditions (e.g., 180 μS cm^{-1} pre-mining vs. 128 μS cm^{-1} 3 months post mining, unpublished data).

9.7.2 GROUNDWATER CONNECTION

Intermittent streams are seasonal in nature, flowing only when the local water table is elevated above the streambed (Villines et al. 2015). In the case of valley fills, the local water table is several meters below the ground surface. Therefore, the need exists to create a localized perched aquifer to sustain flows (e.g., baseflow) to any intermittent and perennial designed streams (Eberle and Razem 1985). While this project benefited from a spring upgradient of GC01, flows from this source (average baseflow of 55 m^3 day^{-1}) were insufficient to ensure baseflow existed throughout the main channel on the crown (i.e., to GC02). Lack of additional subsurface flow inputs in addition to other factors such as evapotranspiration from aquatic and riparian vegetation and increased pore space in the unconsolidated overburden resulted in GC02 exhibiting significantly fewer days of flow particularly during the summer and fall months (Hawkins 1998). Blackburn-Lynch (2015) used the assumed balance method to model baseflow conditions on the main channel on the crown at Guy Cove and found that infiltration into the surrounding soils (approximately 1–2 m depth) at a rate of 0.5 L s^{-1} accounted for 80%–85% of the reduction in baseflow. During the construction process, precautions were taken to reduce infiltration by compacting the final ground surface using loaded dump trucks. Following compaction, the stream was excavated. Future efforts should focus on creating a perched aquifer, and in essence a shallow hyporheic zone, by first compacting a subsurface layer using knowledge gained by da Rosa et al. (2013) and then placing additional spoil overtop before excavating the channel. This depth of spoil should be sufficient to ensure the created perched aquifer is not breached during the construction process. As noted in Section 9.7.1, care should be taken to use only nonproblematic spoils when constructing the perched aquifer and overlying hyporheic zone and channel. Additional consideration should be given to identifying localized seeps and springs along the side slopes, and if the quality of the emanating waters is good, to diverting the water to the channel via subsurface drains.

9.7.3 Colonization of Macroinvertebrates

The colonization of macroinvertebrates to restored streams depends in part on the source population in the surrounding landscape (Bilton et al. 2001; Petersen et al. 2004; Sundermann et al. 2011) and water chemistry. The constructed stream system at Guy Cove benefited from the presence of a diverse macroinvertebrate community, including several EPT taxa, in the reach of stream immediately upgradient of the project site. We attribute the collection of many macroinvertebrate taxa shortly after the completion of construction to this upgradient source. For project sites lacking such a connected source, consideration should be given to artificially colonizing the site by utilizing rocks from a donor site when constructing riffles. Also, on mined lands devoid of woody material, designers should seek opportunities to incorporate logs, twigs, leaves, and the like into instream structures (e.g., vanes, riffles, snags) and floodplain features (e.g., vernal pools). A forested riparian area will serve as a source of such carbon inputs but not for several years post construction. Designers should plan for this lack of input during the initial phases of the project.

9.7.4 Invasive Species

Mined lands are rife with invasive species such as autumn olive (*Elaeagnus umbellate*), sericea lespedeza (*Lespedeza cuneate*), Japanese stilt grass (*Microstegium vimineum*), Kentucky 31 fescue (*Festuca arundinacea*), and multiflora rose (*Rosa multiflora*). Many of these species were intentionally introduced during the mine reclamation process due to their ability to grow in poor soils that are often droughty and lacking in nutrients (Brothers 1990). Restoration efforts on mined lands should give consideration to invasive species management. In some instances, treatment using herbicides may be needed while, in other instances, continued growth of the hardwood forest will lead to canopy closure and thus shading and eventual dieback of invasive species.

9.7.5 Browse

Herbivory is an important consideration when designing plans to establish vegetation, particularly in mined lands. Consideration should be given to both wild and domestic animals. The project site is located centrally to the Kentucky Department of Fish and Wildlife Resources Elk Program elk reintroduction point (current estimated elk population is 10,000 animals) (Crank et al. 2014) and is within a few kilometers of Robinson Forest. Additionally, the grass and legume species often used to reclaim mined lands are sometimes used to graze cattle (Ditsch et al. 2006). Since the Great Recession, the presence of free-roaming cattle and horses on mined lands seems to have increased (Estep 2009; Kenning 2015). Use of the FRA, with its hummocky surface, seems to help reduce browse by these grazing animals. However, browse in riparian areas which have smoother ground surfaces and are close to the stream (i.e., water source for animals) is higher. Consideration should be given to protecting vegetation planted in riparian areas from browse whether it is through the use of tree shelters (Stange and Shea 2006) or even overplanting.

REFERENCES

Agouridis, C.T., P.N. Angel, T.J. Taylor, C.D. Barton, R.C. Warner, X. Yu, and C. Wood. 2012. Water quality characteristics of discharge from reforested loose-dump mine spoil in eastern Kentucky. *J Environ Qual* 41: 454–468.

Barbour, M.E., J. Gerritsen, B.D. Snyder, and J.B. Stribling. 1999. *Rapid Bioassessment Protocols for Use in Streams and Wadeable Rivers: Periphyton, Benthic Macroinvertebrates, and Fish*, 2nd ed. EPA 841-B-99-002. Washington, DC: U.S. Environmental Protection Agency; Office of Water.

Barton, C.D. and A.D. Karathanasis. 1999. Renovation of a failed constructed wetland treating acid mine drainage. *Environ Geol* 39: 39–50.

Bernhardt, E.S., M. Palmer, J.D. Allan, G. Alexander, K. Barnas, S. Brooks, and E. Sudduth. 2005. Synthesizing U.S. river restoration efforts. *Science* 308: 636–637.

Bigham, J.M., U. Schwertmann, and L. Carlson. 1992. Mineralogy of precipitates formed by the biogeochemical oxidation of Fe(II) in mine drainage. In *Biomineralization Processes of Iron and Manganese—Modern and Ancient Environments*, eds. H.G.W. Skinner and R.W. Fitzpatrick, pp. 219–232. Catena Supplement 21. Cremlingen-Destedt, Germany: Catena-Verlag.

Bilton, D.T., J.R. Freeland, and B. Okamura. 2001. Dispersal in freshwater invertebrates. *Annu Rev Ecol Evol* 32: 159–181.

Blackburn-Lynch, W. 2015. Development of techniques for assessing and restoring streams on surface mined lands. PhD dissertation, University of Kentucky, Lexington, KY.

Bosch, J.M. and J.D. Hewlett. 1982. A review of catchment experiments to determine the effect of vegetation changes on water yield and evapotranspiration. *J Hydrol* 55: 3–23.

Brannon, M.P. and B.A. Purvis. 2008. Effects of sedimentation on the diversity of salamanders in a southern Appalachian headwater stream. *J N C Acad Sci* 124: 18–22.

Braun, E.L. 1950. *Deciduous Forests of Eastern North America*. Philadelphia, PA: The Blakiston Company.

Brothers, T.S. 1990. Surface-mine grasslands. *Geogr Rev* 80: 209–225.

Burdon, F.J., A.R. McIntosh, and J.S. Harding. 2013. Habitat loss drives threshold response of benthic invertebrate communities to deposited sediment in agricultural streams. *Ecol Appl* 23: 1036–1047.

Burger, J., D. Graves, P. Angel, V. Davis, and C. Zipper. 2005. The forestry reclamation approach. Forestry Reclamation Advisory Number 2. Appalachian Regional Reforestation Initiative, Office of Surface Mining, Pittsburgh, PA.

Carpenter, S.B. and R.L. Rumsey. 1976. Trees and shrubs of Robinson Forest Breathitt County, Kentucky. *Castanea* 41: 277–282.

Cherry, M.A. 2006. Hydrochemical characterization of ten headwater catchments in Eastern Kentucky. MS thesis, University of Kentucky, Lexington, KY.

Clark, A., R. MacNally, N. Bond, and P.S. Lake. 2008. Macroinvertebrate diversity in headwater streams: A review. *Freshw Biol* 53: 1707–1721.

Code of Federal Regulations. 1996. Part 434. Coal mining point source category BPT, BAT, BCT limitations and new source performance standards. Office of Federal Register, Washington, DC.

Crank, D., G. Jenkins, W. Bowling, and J. Hast. 2014. 2014–2015 Kentucky Department of Fish and Wildlife Resources Elk report. http://fw.ky.gov/Hunt/Documents/1415ElkReport.pdf. Accessed June 25, 2016.

Cummins, K.W., M.A. Wilzbach, D.M. Gates, J.B. Perry, and W.B. Taliaferro. 1989. Shredders and riparian vegetation. *Bioscience* 39: 24–30.

da Rosa, M., C.T. Agouridis, and R.C. Warner. 2013. Potential use of weathered sandstones to construct a low permeability barrier to isolate problematic coal mine spoils. *J Am Soc Min Recl* 2: 49–67.

Ditsch, D.C., C.D. Teutsch, M. Collins, W. Dee Whittier, J. Rockett, C.E. Zipper, and J.T. Johns. 2006. ID-157: Managing livestock forage for beef cattle production on reclaimed surface-mined lands. University of Kentucky Cooperative Extension Service, Lexington, KY.

Eberle, M. and A.C. Razem. 1985. Effects of surface coal mining and reclamation on ground water in small watersheds in the Allegheny Plateau, Ohio. Water-Resources Investigations Report 85-4205. U.S. Geological Survey, Columbus, OH.

Edwards, J.D., C.D. Barton, and A.D. Karathanasis. 2009. Evaluating the use of a bioreactor mesocosm for manganese removal in alkaline mine drainage. *Water Air Soil Pollut* 203: 267–275.

Estep, B. 2009. Replacing coal with cattle. *Lexington Herald-Leader*. http://www.kentucky.com/news/local/watchdog/article44007057.html (accessed June 25, 2016).

Faustini, J.M., P.R. Kaufmann, and A.T. Herlihy. 2009. Downstream variation in bankfull width of wadeable streams across the conterminous United States. *Geomorphology* 108: 292–311.

Ford, W.M., B.R. Chapman, M.A. Menzel, and R.H. Odom. 2002. Stand age and habitat influences on salamanders in Appalachian cove hardwood forests. *Forest Ecol Manag* 155: 131–141.

Goldman, L.J., L.I. Greenfield, A.S. Damle, and G.L. Kingsbury. 1988. Design, construction, and evaluation of clay liners for waste management facilities. EPA/530/SW-86/007F. U.S. Environmental Protection Agency, Office of Solid Waste and Emergency Response, Washington, DC.

Gomi, T., R.C. Sidle, and J.S. Richardson. 2002. Understanding processes and downstream linkages of headwater systems. *Bioscience* 52: 906–916.

Graves, D.H., J.M. Ringe, M.H. Pelkki, R.J. Sweigard, and R.C. Warner. 2000. High value tree reclamation research. In *Environmental Issues and Management of Waste in Energy and Mineral Production*, eds. R.K. Singhal and A.K. Mehrotra, pp. 413–421. Rotterdam, the Netherlands: Balkema.

Green, J., M. Passmore, and H. Childers. 2000. *A Survey of the Condition of Streams in the Primary Region of Mountaintop Mining/Valley Fill Coal Mining*. Appendix, Mountaintop Mining/Valley Fill Programmatic Environmental Impact Statement. Wheeling, WV: U.S. Environmental Protection Agency, Region III.

Greenberg, A.E., L.S. Clesceri, and A.D. Eaton. 1992. *Standard Methods for the Examination of Water and Wastewater*. Washington, DC: American Public Health Association.

Harrelson, C.C., C. Rawlins, and J. Potyondy. 1994. Stream channel reference sites: An illustrated guide to field techniques. USDA Forest Service Rocky Mountain Forest and Range Experiment Station General Technical Report RM245, Fort Collins, CO.

Hartman, K.J., M.D. Kaller, J.W. Howell, and J.A. Sweka. 2005. How much do valley fills influence headwater streams? *Hydrobiologia* 531: 91–102.

Hawkins, J.W. 1998. Hydrogeologic characteristics of surface mine spoil. In *Coal Mine Drainage Prediction and Pollution Prevention in Pennsylvania*, eds. Brady, K.B.C., M.W. Smith, and J. Schueck, pp. 3.1–3.11. Harrisburg, PA: Pennsylvania Department of Environmental Protection.

Hawkins, R., D.E. Woodward, and R. Jiang. 2002. Investigation of runoff curve number abstraction ratio. *Proceedings of the USDA-NRCS Hydraulic Engineering Workshop*, Las Vegas, NV, July 28–August 1.

Hayes, R.A. 1991. Soil survey of Breathitt County, Kentucky. U.S. Department of Agriculture, Natural Resources Conservation Service, Lexington, KY.

Hey, R.D. 2006. Fluvial geomorphological methodology for natural stable channel design. *J Am Water Resour Assoc* 42: 357–374.

Hopkins, R.L., B.M. Altier, D. Haselman, A.D. Merry, and J.J. White. 2013. Exploring the legacy effects of surface coal mining on stream chemistry. *Hydrobiologia* 713: 87–95.

Hilsenhoff, W.L. 1995. Aquatic insects of Wisconsin. Keys to Wisconsin genera and notes on biology, habitat, distribution and species. Natural History Museums Council Publication No. 3. University of Wisconsin, Madison, WI.

Johnson, P.A. and B.J. Fecko. 2008. Regional channel geometry equations: A statistical comparison for physiographic provinces in the eastern U.S. *River Res Appl* 24: 823–834.

Karathanasis, A.D., J.D. Edwards, and C.D. Barton. 2010. Manganese and sulfate removal from a synthetic mine drainage through pilot scale bioreactor batch experiments. *Mine Water Environ* 29: 144–153.

KDOW (Kentucky Division of Water). 2011. Methods for assessing habitat in wadeable waters. Kentucky Department for Environmental Protection, Frankfort, KY.

Kennedy, A.J., D.S. Cherry, and R.J. Currie. 2003. Field and laboratory assessment of a coal processing effluent in the Leading Creek Watershed, Meigs County, Ohio. *Arch Environ Contam Toxicol* 44: 324–331.

Kenning, C. 2015. Free-roaming horses growing problem in E. KY. *Louisville Courier-Journal*. http://www.courier-journal.com/story/news/local/2015/02/06/free-roaming-horse-herds-growing-problem-eastern-kentucky/23003733/ (accessed June 25, 2016).

Kleinmann, R.L.P. 1989. Acid mine drainage: U.S. Bureau of Mines, research and developments, control methods for both coal and metal mines. *Eng Min J* 190: 16i–16n.

Lindberg, T.T., E.S. Bernhardt, R. Bier, A.M. Helton, R.B. Merola, A. Vengosh, and R.T. Di Giulio. 2011. Cumulative impacts of mountaintop mining on an Appalachian watershed. *Proc Natl Acad Sci USA* 108: 20929–20934.

Lowe, W. and G.E. Likens. 2005. Moving headwater streams to the head of the class. *Bioscience* 55: 196–197.

Mastin, C.B., J.D. Edwards, C.D. Barton, A.D. Karathanasis, C.T. Agouridis, and R.C. Warner. 2011. Development and deployment of a bioreactor for the removal of sulfate and manganese from circumneutral coal mine drainage. In *Bioreactors: Design, Properties and Applications*, eds. P.G. Antolli and Z. Liu, pp. 125–144. New York: Nova Science Publishers, Inc.

Maupin, T.P. 2012. Assessment of conductivity sensor performance for monitoring mined land discharged waters and an evaluation of the hydrologic performance of the Guy Cove Stream Restoration Project. MS thesis, University of Kentucky Lexington, KY.

May, C.L., and R.E. Gresswell. 2003. Processes and rates of sediment and wood accumulation in the headwater streams of the Oregon Coast Range, U.S.A. *Earth Surf Process Landf* 28: 409–424.

McDonough, O.T., J.D. Hosen, and M.A. Palmer. 2011. Temporary streams: The hydrology, geography, and ecology of non-perennially flowing waters. In *River Ecosystems: Dynamics, Management and Conservation*, eds. H.S. Elliot and L.E. Martin. New York: Nova Science Publishers, Inc.

Merritt, R.W., K.W. Cummins, and M.B. Berg. 2008. *An Introduction to the Aquatic Insects of North America*, 4th ed. Dubuque, IA: Kendall/Hunt Publishing Company.

Meyer, J.L., D.L. Strayer, J.B. Wallace, S.L. Eggert, G.S. Helfman, and N.E. Leonard. 2007. The contribution of headwater streams to biodiversity in river networks. *J Am Water Resour Assoc* 43: 86–103.

Nordstrom, D. 1982. Aqueous pyrite oxidation and the consequent formation of secondary iron minerals. In *Acid Sulfate Weathering*, eds. Kittrick, J.A., D.S. Fanning, and L.R. Hosner, pp. 37–62. Special Publication 10. Madison, WI: Soil Science Society of America.

Palmer, M.A. and K.L. Hondula. 2014. Restoration as mitigation: Analysis of stream mitigation for coal mining impacts in southern Appalachia. *Environ Sci Technol* 48: 10552–10560.

Petersen, I., Z. Masters, A.G. Hildrew, and S.J. Ormerod. 2004. Dispersal of adult aquatic insects in catchments of differing land use. *J Appl Ecol* 41: 934–950.

Petranka, J.W. and S.S. Murray. 2001. Effectiveness of removal sampling for determining salamander density and biomass: A case study in an Appalachian streamside community. *J Herpetol* 35: 36–44.

Phillips, J.D. 2004. Impact of surface mine valley fills on headwater floods in eastern Kentucky. *Environ Geol* 45(3): 367–380.

Pond, G.J. 2004. Effects of surface mining and residential land use on headwater stream biotic integrity in the Eastern Kentucky Coalfield region. Kentucky Department for Environmental Protection, Division of Water, Water Quality Branch, Frankfurt, KY.

Pond, G.J., M.E. Passmore, F.A. Borsuk, L. Reynolds, and C.J. Rose. 2008. Downstream effects of mountaintop coal mining: Comparing biological conditions using family- and genus-level macroinvertebrate bioassessment tools. *J N Am Benthol Soc* 27: 717–737.

Rice, E.L. 1972. Allelopathic effects of *Andropogon virginicus* and its persistence in old fields. *Am J Bot* 59: 752–755.

Rosgen, D.L. 1998. The reference reach—A blueprint for natural channel design. *Proceedings of the Wetland Engineering and River Restoration Conference*, ASCE, CDROM, Reston, VA.

Sand-Jensen, K. and J.R. Mebus. 1996. Fine-scale patterns of water velocity within macrophyte patches in streams. *Oikos* 76: 169–180.

SCS (Soil Conservation Service). 1972. *National Engineering Handbook. Section 4: Hydrology*. Washington, DC: U.S. Department of Agriculture.

Sena, K., C. Barton, P. Angel, C. Agouridis, and R. Warner. 2014. Influence of spoil type on chemistry and hydrology of interflow on a surface coal mine in the Eastern US Coalfield. *Water Air Soil Pollut* 225: 2171.

Sena, K., C. Barton, S. Hall, P. Angel, C. Agouridis, and R. Warner. 2015. Influence of spoil type on afforestation success and natural vegetative recolonization on a surface coal mine in Appalachia. *Restor Ecol* 23: 131–138.

Shreve, R.L. 1969. Stream lengths and basin areas in topographically random channel networks. *J Geol* 77: 397–414.

Skousen, J.G., A. Sexstone, and P.F. Ziemkiewicz. 2000. Acid mine drainage control and treatment. In *Reclamation of Drastically Disturbed Lands*, ed. F. Schaller, pp. 131–168. Madison, WI: American Society of Agronomy and American Society for Surface Mining and Reclamation.

Smalley, G.W. 1984. Classification and evaluation of forest sites on the Cumberland Mountains. General Technical Report GTP SO-50. USDA Forest Service Southern Forest Experiment Station, New Orleans, LA.

Sponseller, R.A. and E.F. Benfield. 2001. Relationships between land use, spatial scale and stream macroinvertebrate communities. *Freshw Biol* 46: 1409–1424.

Stange, E.E. and K.L. Shea. 2006. Effect of deer browsing, fabric mats, and tree shelters on *Quercurs rubra* seedlings. *Restor Ecol* 6: 29–34.

Sundermann, A., S. Stoll, and P. Haase. 2011. River restoration success depends on species pool of the immediate surroundings. *Ecol Appl* 21: 1962–1971.

Taylor, T.J., C.T. Agouridis, R.C. Warner, and C.D. Barton. 2009a. Runoff curve numbers for loose-dumped spoil in the Cumberland Plateau of eastern Kentucky. *Int J Min Reclam Environ* 23: 103–120.

Taylor, T.J., C.T. Agouridis, R.C. Warner, C.D. Barton, and P.N. Angel. 2009b. Hydrologic characteristics of Appalachian loose-dumped spoil in the Cumberland Plateau of eastern Kentucky. *Hydrol Process* 23: 3372–3381.

Thorp, J.H. and A.P. Covich. 2010. *Ecology and Classification of North American Freshwater Invertebrates*. London, U.K.: Academic Press.

Torbert, J.L. and J.A. Burger. 1994. Influence of grading intensity on ground cover establishment, erosion, and tree establishment on steep slopes. *Proceedings of the International Land Reclamation and Mine Drainage Conference and the Third International Conference on the Abatement of Acid Drainage*, Pittsburgh, CA.

U.S. Army Corps of Engineers. 2002. Executive summary: Stream assessment protocol for headwater streams in the Eastern Kentucky Coalfield Region. http://www.lrl.usace.army.mil/Missions/Regulatory/Mitigation/Eastern-Kentucky/ (accessed June 28, 2016).

U.S. DC (Department of Commerce). 2002. Climatography of the United States, No. 81 (15) monthly station normal of temperature, precipitation, and heating and cooling degree 1971–2000. Published by United States of Commerce, Asheville, NC.

U.S. EPA (Environmental Protection Agency). 2011. The effects of mountaintop mines and valley fills on aquatic ecosystems of the central Appalachian Coalfields. EPA/600/R-09/138F. Office of Research and Development, National Center for Environmental Assessment, Washington, DC.

Vesper, D.J., M. Roy, and C.J. Rhoads. 2008. Selenium distribution and mode of occurrence in the Kanawha Formation, southern West Virginia, USA. *Int J Coal Geol* 73: 237–249.

Villines, J.A., C.T. Agouridis, R.C. Warner, and C.D. Barton. 2015. Using GIS to delineate headwater stream origins in the Appalachian Coalfields of Kentucky. *J Am Water Resour Assoc* 51: 1667–1687.

Warner, R.C. and C.T. Agouridis. 2010. Enhanced environmental protection through new valley fill design techniques. Presented at the *23rd Annual Kentucky Professional Engineers in Mining Seminar*, Lexington, KY, August 20.

Zipper, C., J. Burger, J. Skousen, P. Angel, C. Barton, V. Davis, and J. Franklin. 2011. Restoring forests and associated ecosystem services on Appalachian coal surface mines. *Environ Manag* 47: 751–765.

10 Key Issues in Mine Closure Planning for Pit Lakes

Jerry A. Vandenberg and Cherie D. McCullough

CONTENTS

10.1   Introduction 175
10.2   Key Issues 175
      10.2.1   Determining Closure Objectives and Developing Closure Criteria 175
      10.2.2   Anticipating and Meeting Stakeholder and Regulator Expectations 177
      10.2.3   Predicting and Managing Water Balances 177
      10.2.4   Understanding Long-Term Vertical Mixing Regimes 179
      10.2.5   Identifying Contaminants of Concern 180
      10.2.6   Mitigating Acid and Metalliferous Drainage (AMD) 181
      10.2.7   Subaqueous Disposal of Liquid and Solid Mine Waste 182
      10.2.8   Health and Safety Issues 183
      10.2.9   Historical Reliability of Model Predictions 184
10.3   Conclusions 185
References 185

10.1   INTRODUCTION

Pit lakes can form when surface mines close and open pits fill with water, either through passive groundwater recharge, surface water diversion, or active pumping, or a combination of all of these. They often display poor water quality through Acid Mine Drainage, often also known as the more neutral and alkaline drainage inclusive term of Acid and Metalliferous Drainage (AMD). Historically, the success in relinquishing mines with pit lakes has varied tremendously: there are many well-known examples of legacy sites abandoned and still requiring rehabilitation and others requiring perpetual treatment. Conversely, there are many examples of other pit lakes internationally that have achieved various beneficial end uses (McCullough and Lund 2006, McCullough and Schultze 2015). Although case studies are often limited, mining companies contemplating new open pit mines nevertheless have a number of examples in both success and failure from which to draw lessons learned that can be used in future mine closure planning, for example, Castendyk (2011).

This chapter discusses key issues that should be addressed in the mine planning process to increase the likelihood of successful rehabilitation and closure of mine operations with pit lakes. Examples of issues and potential management strategies to address them are given with reference to previous experiences in North and South America, Australasia, and Asia.

10.2   KEY ISSUES

10.2.1   Determining Closure Objectives and Developing Closure Criteria

Pit lakes may demonstrate a range of risks and also a range of demonstrated end uses (McCullough and Lund 2006). Closure objectives and criteria applied to pit lakes are therefore site specific and dependent on stakeholder expectations, including final acceptance by the responsible regulatory agency (Jones and McCullough 2011) (Figure 10.1). In most jurisdictions, there are also no

FIGURE 10.1 There are lots of potential targets available; which one is appropriate will come down to site-specific context and stakeholder expectations.

guidelines specific to pit lake water quality, and so other guidelines—usually derived for protection of natural waterbodies—are typically applied by default. If pit lake water concentrations are below applicable generic water quality guidelines, then water quality would most likely be deemed acceptable, but this is rarely the case. More likely, site-specific objectives will need to be developed by the proponent of each pit lake. Site-specific objectives can be derived based on effects thresholds, technological limits, background concentrations, or combinations thereof.

Where discharge is expected, pit lakes may be expected to be managed as closed-circuit waterbodies until they achieve water quality that will not cause adverse effects to identified downstream end uses such as aquatic life. Then at that time they can be reconnected to the receiving environment. If water quality in the pit lakes is not adequate by the time the lakes fill, active treatment may be required, as well as water diversions around the pit lake. In poorly planned pit lakes, water treatment and/or water diversions may need to occur in perpetuity—with resulting high residual risk of failure at some point in the future, for example, McCullough et al. (2012, 2013).

There are three nested layers that can be used to define and gauge success in pit lake closure:

1. *End use*—Will the pit lake and associated watershed meet land use requirements for post-closure mine sites that are set regionally and nationally (McCullough and van Etten 2011)?
2. *Objectives*—Will the pit lake meet functional targets that are achievable, desirable to stakeholders, and acceptable to regulators?
3. *Criteria*—Will the pit lake meet prescriptive criteria, such as site-specific water quality and toxicological thresholds?

There are several sources of information that can be used to define success measures, such as

- Corporate sustainability goals and targets (MMSD 2002)
- Commitments made by the mining company in environmental impact assessments (EIAs) and other applications, which include commitments made by previous property owners

- Numerical predictions that have been generated in EIAs and that have been used in ecological risk assessments
- Long-term pit lake water balance and water quality modeling (DITR 2016)
- Stakeholder expectations
- Regulatory requirements (Jones and McCullough 2011)
- Analogue lake studies (Van Etten et al. 2014)
- Observed water quality from existing pit lakes in similar geologic deposits (Johnson and Castendyk 2012)
- Leading, international mining-industry practice
- Prescriptive, site-specific objectives that are based on biological thresholds and relevant baseline conditions and ecological risk assessments (McCullough and Pearce 2014)

The importance of developing closure criteria for pit lakes early in the planning process cannot be overstated, because all mine closure design and mitigation should be directed toward ultimately meeting these criteria.

10.2.2 ANTICIPATING AND MEETING STAKEHOLDER AND REGULATOR EXPECTATIONS

Even though traditionally mine rehabilitation and closure have a poor record of finding stakeholder acceptance (Lamb et al. 2015), pit lake closures have been particularly poorly executed. As with the other components of mine operation and closure, all stakeholders should be identified early and consulted for their input on end of mine life quality and objectives, including objectives for pit lakes (Swanson 2011). Early engagement of stakeholders can lead to constructive input into the planning of pit lakes, reduced costs, fewer delays, and overall public/stakeholder/regulator acceptance.

Design for pit lakes is typically done by involving engineers and scientists, but not stakeholders (Swanson 2011). It is recommended to consult stakeholders on visions for pit lakes and potential beneficial end uses of pit lakes (McCullough and Lund 2006). Participation by communities in developing mine remediation targets leads to better decisions and, in some cases, to lower overall costs for mine remediation (NOAMI 2003). This is because the major stakeholders were involved from the beginning in decisions that could affect their enjoyment/use of the landscape (Figure 10.2). Information presented to communities on pit lake predictions can be complex, and thus information should be presented in an easy-to-understand format in order to engage the stakeholders in constructive discussions (Gerner and McCullough 2014, NOAMI 2003).

10.2.3 PREDICTING AND MANAGING WATER BALANCES

The time to refill pit lakes is site specific and must be determined on a case-by-case basis. In cases with high rates of evaporation or highly permeable aquifers, the pit lakes can refill in a few years. In arid regions, some pit lakes will never completely refill passively, and are termed terminal pit lakes (McCullough et al. 2013) because they act as an ongoing groundwater sink (Figure 10.3). Such lakes may be used as mitigation to prevent contaminated groundwater from migrating away from a mine site. In terminal lakes, evaporation is the only route through which water leaves a pit lake, so it can be expected (and predicted with mass balance models) that concentrations of solutes will increase over time (Castendyk and Eary 2009, Geller et al. 2013). The ultimate concentrations may be controlled by solubility, which can be predicted using geochemical software, for example, PHREEQC (Parkhurst and Appelo 1999).

Where net evaporation is low, it is expected that pit lakes will refill passively. In many cases, it is preferable to accelerate the filling process to reduce the closure management period. In arid regions, it may be necessary to rapidly fill lakes both to improve initial water quality and therefore long-term water quality (from AMD generation in particular) but also to achieve an equilibrium level lake.

FIGURE 10.2   "Because a final land use of farming has been decided, we suggest you all learn agriculture."

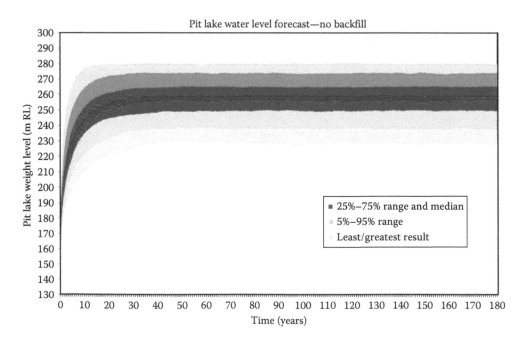

FIGURE 10.3   Water balances are a fundamental tool to understanding pit lake closure issues and should form the basis of all pit lake closure planning. *Note:* m RL, metres Relative Level.

This option should be evaluated as part of the closure planning process in consideration of regional surface hydrology and availability of water to be used for filling.

Connection of the pit lake to surrounding groundwater sources can play a large role in the water quality and hydrological cycle/budget of the pit lake; if a pit lake water surface is above the water table, water will flow out of the pit to the groundwater and thus provide a pathway to transport potential contaminants to a larger area (Castendyk and Eary 2009).

10.2.4 Understanding Long-Term Vertical Mixing Regimes

Pit lakes regularly display some form of water column stratification, even if only temporarily such as seasonal (Figure 10.4) or after significant climatic events such as storms. Among a host of other differences, compared to natural lakes, pit lakes are more prone to become meromictic (lower layers non-mixing) because they generally have smaller surface areas, steeper sides, greater depths, and higher salinities (Table 10.1) (Gammons and Duaine 2006). Vertical mixing in lakes is primarily driven by wind currents across the lake surface, and the smaller fetch of pit lakes provides less opportunity to translate wind energy into water currents that are necessary for lake turnover.

In pit lakes, as in natural lakes, the frequency and depth of vertical mixing will affect many other variables. These parameters must be defined in advance of developing geochemical predictions of water quality so that accurate volumes for epilimnion, hypolimnion, and monimolimnion layers can be accurately represented and the simulated parcels of water mixed at appropriate intervals. Vertical mixing transports oxygen to the lower portion of the lake, which in turn affects biological and chemical reactions. For example, oxidation state influences the mobilization of metals and cycling of nutrients. Of particular importance is the potential effect of oxidation state on sulfide minerals; under oxidizing conditions, sulfide minerals will react to form sulfuric acid and dissolved metals, whereas under reducing conditions, sulfide minerals will precipitate—a process that has been used to mitigate AMD in meromictic pit lakes (Pelletier et al. 2009). Given the influence of vertical

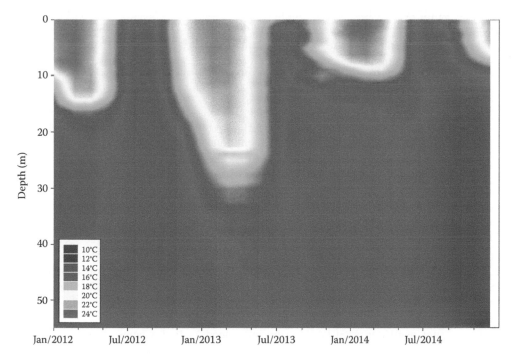

FIGURE 10.4 Changes to pit lake mixing may be a result of thermal or chemical changes, or both, in the lake water quality over time.

TABLE 10.1
Typical Key Differences between Natural Lakes and Pit Lakes

Aspect	Natural Lake	Pit Lake
Bank steepness	Shallow	Steep
Riparian vegetation	Well developed	Poorly developed or absent
Depth/area ratio	Low	High
Sediments	Well developed	Absent/poorly developed
Organic material	High	Low
Habitat availability	High	Low
Ecological connectivity	Medium to high	Low
Nutrient concentrations	Medium to high	Low
Metal concentrations	Typically low	Typically high
Salinity	Typically low	Often high
Water quality	Good	Moderate to poor
Bank stability	High	Low
Retention time	Medium to high	Very high
Hydroperiod	Seasonal to permanent	Permanent

mixing on these processes, the anticipated mixing behavior of a pit lake should be evaluated and understood as early as possible in the mine planning process.

There are a variety of guidelines that describe lake geometries that will affect lake mixing. The most common is the relative depth, defined as the maximum depth as a percentage of mean diameter. Natural lakes usually have relative depths of less than 2%, whereas pit lakes typically have relative depths of 10%–40% (Doyle and Runnels 1997). While measures such as relative depth provide useful descriptors of pit geometries, they are not predictive measures because they do not account for other important variables, such as water density and wind speed. The most reliable method for predicting lake mixing is through the use of numerical models (such as CE-QUAL-W2 or DYRESM) that mechanistically account for these variables.

10.2.5 IDENTIFYING CONTAMINANTS OF CONCERN

There are a wide range of contaminants of concern (COCs) in pit lakes. The most common COCs in hardrock pit lakes are low pH and elevated element concentrations caused by AMD. AMD is a phenomenon that occurs when sulfur-bearing waste rock, tailings, or other materials are weathered during mining and mine closure practices. Weathering of sulfide minerals can lead to release of acid and elevated concentrations of contaminants in runoff, groundwater, or pit lake water. These acidic waters often carry a high load of elements that are more soluble at low pH. AMD is commonly associated with coal and hard rock mines.

The COCs at a given mine are often, but not always, related to an obvious source such as the ore body or extraction chemicals. For example, the Berkeley Pit Lake in Montana, which is perhaps the most famous worst-case example of a pit lake, is a former copper mine pit that now contains levels of copper, zinc, and iron that exceed water quality guidelines by orders of magnitude (Gammons and Duaime 2006). Similar contamination has been observed at copper mines in California (Levy et al. 1997) and Sweden (Ramstedt et al. 2003).

Long-term water quality in a pit lake can be influenced by hydrochemical processes such as geoenvironmental characteristics, water balance, mineral solubility, and sediment biogeochemical processes (Geller et al. 2013). Constituents that most often exceed guidelines are copper, cadmium, lead, mercury, nickel and zinc, followed by arsenic, sulfate, and cyanide (Kuipers et al. 2006).

Blasting residues such as ammonia and nitrate are also often elevated in mine waters, and may persist into closure (Banks et al. 1997). In sub-Arctic Canadian mines, salinity and major ions are typical COCs because of saline groundwater that must be dewatered for mining. The saline groundwater may be disposed of in pit lakes, or saline groundwater may flow passively into pit lakes at closure when dewatering ceases. In oil sands pit lakes, the COCs are primarily organic constituents such as naphthenic acids, phenolics, and polycyclic aromatic hydrocarbons originating from process waters and tailings (CEMA 2012).

Less obvious COCs may be present as well. For example, at the proposed Gahcho Kué Diamond Mine (De Beers 2012), geochemical testing of pilot plant tailings identified phosphorus as a COC, which led to changes in the closure plan to mitigate runoff from mine wastes and to avoid eutrophication of closure waterbodies. Total suspended solids can be expected to be elevated during the early years of lake development, before vegetation becomes established in the littoral zone, but this should be a temporary phenomenon in a properly designed pit lake.

In summary, while there may be obvious COCs at a given mine, a full suite of metals, major ions, nutrients, and organics should be evaluated to determine site-specific COCs prior to mine development.

10.2.6 MITIGATING ACID AND METALLIFEROUS DRAINAGE (AMD)

Poor water quality degraded by AMD is the single biggest environmental risk and cause of beneficial end use loss for pit lakes (McCullough 2008) (Figure 10.5). Mine drainage may be acidic, neutral, or even alkaline as constituents such as metals and metalloids may be in elevated concentrations in all forms. Once begun, the process of AMD is very difficult to stop and can extend even for millennia (Leblanc et al. 2000, Wisotzky 2013). Hence, the emphasis on AMD management should always be first on preventing weathering of potentially acid generating (PAG) materials by exposure to water and oxygen (Castendyk and Webster-Brown 2007) as AMD is notoriously difficult to halt once started. This process begins by long-term geochemical characterization of all materials that may

FIGURE 10.5 Contamination of pit lake water quality from AMD may range from mild to extreme such as this. But it is a key determinant of final pit lake risk and opportunity.

contact pit lake water or water sources including above ground sources, such as waste rock dumps and tailings impoundments, and below ground sources, such as backfill and fractured geologies.

Disposal of PAG materials above the water table is usually best suited to arid climates where AMD production will be limited by water availability. However, a strategy often considered to reduce pit lake AMD issues is subaqueous disposal of PAG occurring in tailings, waste risk, and pit shell exposures (Dowling et al. 2004). However, subaqueous disposal of waste should not be thought of as a singular solution to PAG management. Rather, it is merely one consideration of a broader closure strategy that, when used appropriately and in certain circumstances, may reduce AMD production and long-term environmental and social liability.

Where AMD has not been prevented, a number of active and passive treatments are available, although all of these treatments should be considered requiring ongoing attention and maintenance (Gammons et al. 2009, Geller and Schultze 2013, Younger and Wolkersdorfer 2004). Active treatments may be simple limestone or lime putty additions to treat acidity (Schultze et al. 2013), although the ongoing cost, particularly in remote areas once mine infrastructure is closed, should not be underestimated. The economic liability to the remaining responsible jurisdiction is likely to exceed the economic benefit from mining with a few generations of treatment, which is why active treatment is only typically sought when there is a risk of off-site contamination exposure to social or environmental receptors.

Passive treatments may include initial or ongoing treatment with biologically active materials such as nutrients and organic matter (Frömmichen et al. 2004, Kumar et al. 2011, 2016, Wen et al. 2015).

In contrast to the typical isolate and management strategy for AMD-contaminated pit lakes, alternative treatments to long-term water quality may consider redirecting hydrological systems through one or more pit lakes. This use of riverine "flow-through" must be carefully considered and well understood first, but then may form a strategic catchment-scale diversion of inflows to attenuate and dilute pit lake waters. This strategy often has potential to both mitigate AMD contamination in pit lake water and often river water quality issues such as elevated nutrient concentrations and suspended sediments as well (McCullough 2015, McCullough and Schultze 2015).

10.2.7 SUBAQUEOUS DISPOSAL OF LIQUID AND SOLID MINE WASTE

The option to dispose of mine waste in pit lakes is often attractive to mining companies because it is more cost effective than other treatment or disposal technologies (Figure 10.6). Disposal of mine waste in pit lakes is an accepted practice in some industries and regions (Davé 2009, Dowling et al. 2004, Schultze et al. 2011). However, it is controversial and considered unproven until demonstrated at the field scale in the oil sands industry (OSTC 2012). If successful, several other companies in the region will likely apply water-capped tailings technology with a potential savings of billions of dollars for the industry as a whole compared to other disposal technologies. Deep pit disposal of fine tailings has also been approved for the diamond mining industry in Northern Canada (De Beers 2012).

FIGURE 10.6   Backfill of potentially acid forming (PAF) or other geochemically hazardous material should form part of a well-considered and informed closure strategy and not just "sweeping waste issues under a mat."

If subaqueous disposal of mine waste is contemplated, the following issues should be evaluated to reduce risks to closure water quality:

- *Effect on water mining regime and water balance*—Reduced pit lake surface area and depth can alter water balances leading to seepage or even decant. Mixing durations may be reduced or inhibited altogether, with contaminated monimolimnion waters being transported to the surface.
- *Tailings resuspension*—A hydrodynamic analysis should be completed to understand the potential for resuspension of fine particles and the formation of buoyant plumes.
- *Metal leaching and AMD*—Geochemical testing should be completed to predict the potential for acid generation and metal leaching and to understand which oxidation state would minimize these effects on water quality.
- *Sediment toxicity*—Standard bioassays should be conducted to predict the toxicity to benthic organisms.

10.2.8 HEALTH AND SAFETY ISSUES

The most significant acute health and safety risks for persons in and around pit lakes relate to falls and drowning. Pit lake high walls may often be unstable, particularly following rebounding groundwater pore pressures and decades of wave action. Unstable walls frequently result in slips that may endanger nearby structures and persons near the high wall (McCullough and Lund 2006). Where communities reside nearby, pit lakes may present risks for recreational swimmers where there is a risk of drowning with the steep lake margin typically of pit lake edges or by falls from high walls into water or submerged obstacles that have not been regraded (Ross and McCullough 2011).

Chronic health risks are not well understood, but there is a potential for health issues for recreational users in AMD-contaminated pit lake water, even in remote areas where pit lakes may be used as recreational opportunities. Low pH and elevated contaminant concentrations may lead to skin and eye damage and irritation, particularly due to regular exposure to vulnerable groups, such as children and the elderly (Hinwood et al. 2012) (Figure 10.7).

FIGURE 10.7 People will often swim in contaminated pit lakes, either unaware or in spite of health concerns regarding the water quality.

There are also human health risks where end uses include fisheries, either planned or unplanned. Aquatic ecosystem food chains have been found to accumulate contaminants such as selenium, mercury, and cadmium. These metals bioconcentrate in keystone predator sports fish and crustacea (McCullough et al. 2009a, Miller et al. 2013).

10.2.9 HISTORICAL RELIABILITY OF MODEL PREDICTIONS

The reliability and accuracy of mine water predictions were examined by Kuipers et al. (2006) in a comparison of water quality predictions made in environmental impact statements to operational water quality observed at hard rock mines. The mines that were examined included major mines across Western USA, but the issues they identified are applicable to mines worldwide. They found that in the majority of cases, water quality predictions did not perform well, and impacts were often underestimated (Figure 10.8). These poor predictive results were mainly due to the following:

- *Inadequate hydrologic characterization*—Inaccuracies arose from overestimating dilution potential, poor characterization of the hydrologic regime, and poor flood forecasting.
- *Inadequate geochemical characterization*—Inaccuracies arose from inadequate sampling of geologic materials, lack of proper geochemical testing of materials such as metal leaching and AMD potential, and improper application of test results to models.
- *Mitigation failure*—In many cases, mitigation was assumed to reduce concentrations, but the mitigation was either not effective or not implemented.

FIGURE 10.8 Model predictions should be used as a guide and not gospel to pit lake water quality expectations.

Although poor water quality prediction performance has been found at many mine pit lakes, present and future pit lake modeling efforts should be able to improve upon this record. Success in predicting water quality will be reliant on following leading modeling practices that were not adhered to in many of the case studies in Kuipers et al. (2006). Guidance for predicting pit lake water quality is provided in a companion document by Maest et al. (2005) as well as by Vandenberg et al. (2011, 2016).

In particular, a postaudit of water quality predictions is essential (Dunbar 2013) for identifying excursions from predictions early in the mine life and applying adaptive management strategies as soon as possible. Postaudits of modeling predictions should be available to stakeholders, reviewed by regulators, and ideally disseminated to the wider modeling community so that they can learn from the strengths and weaknesses of past experiences and continually improve their methods (Beddoes et al. 2016).

10.3 CONCLUSIONS

Pit lakes are an inevitability for many open-cut mining operations, regardless of proposals or external pressures for backfill. Yet pit lakes are complex and highly variable physico-chemico-biological systems, and their rehabilitation and closure success will only come with careful planning. Consequently, across commodity type and socio-environmental contexts, there is a wide range of outcomes observed worldwide in terms of chemical characteristics and suitability for aquatic habitat. While there are examples of very unsuccessful pit lakes, these still serve as lessons learned that can be followed to increase the likelihood of success in constructing future pit lakes (Castendyk 2011). The most important lessons learned are to develop a conceptual model of the pit lake and understand its processes as early as possible, engage stakeholders early in the process, and begin environmental monitoring at the exploration stage and conduct a postaudit of predictions to guide adaptive management (Castendyk 2011, Gammons et al. 2009).

The key issues described above should be considered in each of the planning, designing, commissioning, and abandonment stages of a pit lake (Vandenberg et al. 2015). The outcome of a decision made or an assessment completed during a previous stage of development may be found to be incorrect or no longer valid as environmental data or stakeholder or regulator requirements evolve (McCullough and Harkin 2015). Or the pit lake and its inflows may be altered by changing mine plans or mine closure plans in response to fluctuating commodity prices. Consequently, mining companies should anticipate an iterative process whereby assumptions and decisions are refined to reduce uncertainty related to the issues above. This may involve reconsidering options and revisiting strategies discounted earlier under different circumstances, such as understanding of the physicochemical context and of regulatory and other social constraints and expectations. This iterative process of pit lake closure planning refinement should form an explicit part of mine closure planning for the broader site (McCullough et al. 2009b).

Pit lake rehabilitation guidance manuals (e.g., CEMA 2012, McCullough 2011) and compilations of pit lake experiences and research (Castendyk and Eary 2009, Gammons et al. 2009, Geller et al. 2013) have been developed in the past 5 years, and these should be consulted throughout the planning, design, and construction process for additional details.

REFERENCES

Banks, D., P.L. Younger, R. Arnesen, E.R. Iversen, and S.B. Banks. 1997. Mine-water chemistry: The good, the bad and the ugly. *Environmental Geology* 32(3):157–174.

Beddoes, P., M. Herrell, J.A. Vandenberg, J. Richards, and R. Millar. 2016. Validation of Springer Pit lake water balance and water quality model, Mount Polley Mine, British Columbia, Canada. In *Proceedings of the IMWA 2016*, Leipzig, Germany, July 11–15, 2016.

Castendyk, D. 2011. Lessons learned from pit lake planning and development. In *Mine Pit Lake Closure and Management*, C.D. McCullough, ed., pp. 15–28. Perth, Western Australia, Australia: Australian Centre for Geomechanics.

Castendyk, D.N., and L.T. Eary. 2009. *Mine Pit Lakes: Characteristics, Predictive Modeling, and Sustainability.* Littleton, CO: Society for Mining, Metallurgy and Exploration (SME).

Castendyk, D.N., and J.G. Webster-Brown. 2007. Sensitivity analyses in pit lake prediction, Martha Mine, New Zealand 1: Relationship between turnover and input water density. *Chemical Geology* 244(1–2):42–55.

CEMA. 2012. *End Pit Lakes Guidance Document.* Calgary, Alberta, Canada: Cumulative Environmental Management Association.

Davé, N. 2009. Disposal of reactive mining waste in man-made and natural water bodies; Canadian Experience. In *Proceedings of the Marine and Lake Disposal of Mine Tailings and Waste Rock*, Egersund, Norway, September 7–10, 2009.

De Beers Canada, Inc. 2012. Gahcho Kué Project Environmental Impact Statement. Supplemental Information Submission. Calgary, Alberta, Canada: De Beers Canada, Inc.

DITR. 2016. *Leading Practice Sustainable Development Program for the Mining Industry—Risk Assessment and Management.* Canberra, Australian Capital Territory, Australia: Department of Industry, Tourism and Resources.

Dowling, J., S. Atkin, G. Beale, and G. Alexander. 2004. Development of the sleeper pit lake. *Mine Water and the Environment* 23:2–11.

Doyle, G.A., and D.D. Runnels. 1997. Physical limnology of existing mine pit lakes. *Mining Engineering* 49:76–80.

Dunbar, D.S. 2013. Modelling of pit lakes. In *Acidic Pit Lakes—Legacies of Surface Mining on Coal and Metal Ores*, W. Geller, M. Schultze, R.L.P. Kleinmann, and C. Wolkersdorfer, eds., pp. 186–224. Berlin, Germany: Springer.

Frömmichen, R., K. Wendt-Potthoff, K. Friese, and R. Fischer. 2004. Microcosm studies for neutralization of hypolimnic acid mine lake water (pH 2.6). *Environmental Science and Technology* 38:1877–1887.

Gammons, C.H., and T.E. Duaine. 2006. Long term changes in the limnology and geochemistry of the Berkeley Pit Lake, Butte, Montana. *Mine Water and the Environment* 25:76–85.

Gammons, C.H., L.N. Harris, J.M. Castro, P.A. Cott, and B.W. Hanna. 2009. Creating lakes from open pit mines: Processes and considerations, with emphasis on northern environments. Canadian Technical Report of Fisheries and Aquatic Sciences NWT Pit Lake Report 2836, Ottawa, Canada. 106p.

Geller, W., and M. Schultze. 2013. Remediation and management of acidified pit lakes and ouflowing waters. In *Acidic Pit Lakes—Legacies of Surface Mining on Coal and Metal Ores*, W. Geller, M. Schultze, R.L.P. Kleinmann, and C. Wolkersdorfer, eds., pp. 225–264. Berlin, Germany: Springer.

Geller, W., M. Schultze, R.L.P. Kleinmann, and C. Wolkersdorfer. 2013. *Acidic Pit Lakes—Legacies of Surface Mining on Coal and Metal Ores.* Berlin, Germany: Springer.

Gerner, M., and C.D. McCullough. 2014. Planning for the future: Development of beneficial end use from a quarry pit lake, Victoria, Australia. In *Proceedings of Life of Mine Conference*, Brisbane, Queensland, Australia, July 16–18, 2014.

Hinwood, A., J. Heyworth, H. Tanner, and C.D. McCullough. 2012. Recreational use of acidic pit lakes—Human health considerations for post closure planning. *Journal of Water Resource and Protection* 4:1061–1070.

Johnson, E., and Castendyk, D.N. 2012. The INAP Pit Lakes Database: A novel tool for the evaluation of predicted pit lake water quality. In *Proceedings of the Ninth International Conference on Acid Rock Drainage*, Ottawa, Ontario, Canada, May 20–26, 2012, Technical Paper 0037, W.A. Price, C. Hogan, and G. Tremblay eds., Mine Environment Neutral Drainage (MEND), pp. 1–12.

Jones, H., and C.D. McCullough. 2011. Regulator guidance and legislation relevant to pit lakes. In *Mine Pit Lakes: Closure and Management*, C.D. McCullough, ed., pp. 137–152. Perth, Western Australia, Australia: Australian Centre for Geomechanics.

Kuipers, J.R., A.S. Maest, K.A. MacHardy, and G. Lawson. 2006. *Comparison of Predicted and Actual Water Quality at Hardrock Mines: The Reliability of Predictions in Environmental Impact Statements.* Kuipers & Associates and Buka Environmental, Earthworks, Washtington, DC.

Kumar, N.R., C.D. McCullough, and M.A. Lund. 2011. Bacterial sulfate reduction based ecotechnology for remediation of acidic pit lakes. In *Mine Pit Lakes: Closure and Management*, C.D. McCullough, ed., pp. 121–134. Perth, Western Australia, Australia: Australian Centre for Geomechanics.

Kumar, R.N., C.D. McCullough, M.A. Lund, and S. Larrañãga. 2016. Assessment of factors limiting algal growth in acidic pit lakes—A case study from Western Australia, Australia. *Environmental Science and Pollution Research* 23:5915–5924.

Lamb, D., P.D. Erskine, and A. Fletcher. 2015. Widening gap between expectations and practice in Australian minesite rehabilitation. *Ecological Management and Restoration* 16(3):186–194.

Leblanc, M., J.A. Morales, J. Borrego, and F. Elbaz-Poulichet. 2000. 4,500-year-old mining pollution in Southwestern Spain: Long-term implications for modern mining pollution. *Economic Geology* 95:655–662.

Levy, D.B., K.H. Custis, W.H. Casey, and P.A. Rock. 1997. The aqueous geochemistry of the abandoned Spenceville Copper Pit, Nevada County, California. *Journal of Environmental Quality* 26:233–243.

Maest, A.S., J.R. Kuipers, C. Travers, and D.A. Atkins. 2005. *Predicting Water Quality at Hardrock Mines: Methods and Models, Uncertainties, and State-of-the-Art.* Kuipers & Associates and Buka Environmental, Earthworks, Washtington, DC.

McCullough, C.D. 2008. Approaches to remediation of acid mine drainage water in pit lakes. *International Journal of Mining, Reclamation and Environment* 22(2):105–119.

McCullough, C.D. 2011. *Mine Pit Lakes: Closure and Management.* Perth, Western Australia, Australia: Australian Centre for Geomechanics (ACG).

McCullough, C.D. 2015. Consequences and opportunities of river breach and decant from an acidic mine pit lake. *Ecological Engineering* 85:328–338.

McCullough, C.D., and C. Harkin. 2015. Engineered flow-through closure of an acid pit lake: A case study. In *Proceedings of the International Mine Closure 2015 Congress*, Vancouver, British Columbia, Canada.

McCullough, C.D., D. Hunt, and L.H. Evans. 2009a. Sustainable development of open pit mines: Creating beneficial end uses for pit lakes. In *Mine Pit Lakes: Characteristics, Predictive Modeling, and Sustainability*, D.N. Castendyk and L.E. Eary, eds., pp. 249–268. Littleton, CO: Society for Mining, Metallurgy, and Exploration (SME).

McCullough, C.D., N.R. Kumar, M.A. Lund, M. Newport, E. Ballot, and D. Short. 2012. Riverine breach and subsequent decant of an acidic pit lake: Evaluating the effects of riverine flow-through on lake stratification and chemistry. In *Proceedings of the International Mine Water Association (IMWA) Congress*, Bunbury, Western Australia, Australia, pp. 533–540.

McCullough, C.D., and M.A. Lund. 2006. Opportunities for sustainable mining pit lakes in Australia. *Mine Water and the Environment* 25(4):220–226.

McCullough, C.D., G. Marchand, and J. Unseld. 2013. Mine closure of pit lakes as terminal sinks: Best available practice when options are limited? *Mine Water and the Environment* 32(4):302–313.

McCullough, C.D., and J.I. Pearce. 2014. What do elevated background contaminant concentrations mean for AMD risk assessment and management in Western Australia? In *Proceedings of the Eighth Australian Workshop on Acid and Metalliferous Drainage*, Adelaide, South Australia, Australia, April 28–May 2, 2014.

McCullough, C.D., and M. Schultze. 2015. Riverine flow-through of mine pit lakes: Improving both mine pit lake and river water quality values? In *Proceedings of the Joint International Conference on Acid Rock Drainage ICARD/International Mine Water Association (IMWA) Congress*, Santiago, Chile.

McCullough, C.D., J. Steenbergen, C. te Beest, and M.A. Lund. 2009b. More than water quality: Environmental limitations to a fishery in acid pit lakes of Collie, south-west Australia. In *Proceedings of the International Mine Water Conference*, Pretoria, South Africa, October 19–23, 2009.

McCullough, C.D., and E.J.B. Van Etten. 2011. Ecological restoration of novel lake districts: New approaches for new landscapes. *Mine Water and the Environment* 30:312–319.

Miller, L.L., J.B. Rasmusssen, V.P. Palace, G. Sterling, and A. Hontela. 2013. Selenium bioaccumulation in stocked fish as an indicator of fishery potential in pit lakes on reclaimed coal mines in Alberta, Canada. *Environmental Management* 52(1):72–84.

MMSD (Mining, Minerals, and Sustainable Development). 2002. *Breaking New Ground.* Final Report of the Mining, Minerals, and Sustainable Development Project. London, U.K.: Earthscan Publications.

NOAMI. 2003. Lessons learned: On community involvement in the remediation of orphaned and abandoned mines; case studies and analysis. A report of the National Orphaned/Abandoned Mines Initiative (NOAMI), Ottawa, Ontario, Canada.

OSTC (Oil Sands Tailings Consortium). 2012. Technical guide for fluid fine tailings management. Technical guide prepared for OSTC and COSIA, August 2012. Canada. Canada's Oil Sands Innovation Alliance (COSIA), 131pp.

Parkhurst, D.L., and C.A.J. Appelo. 1999. User's guide to PHREEQC (Version 2)—A computer program for speciation, batch-reaction, one-dimensional transport, and inverse geochemical calculations. U.S. Geological Survey, Denver, CO.

Pelletier, C.A., M. Wen, and G.W. Poling. 2009. Flooding pit lakes with surface water. In *Mine Pit Lakes: Characteristics, Predictive Modeling, and Sustainability*, D.N. Castendyk and L.E. Eary, eds., pp. 187–202. Littleton, CO: Society for Mining, Metallurgy, and Exploration (SME).

Ramstedt, M., E. Carlsson, and L. Lovgren. 2003. Aqueous geochemistry in the Udden pit lake, northern Sweden. *Applied Geochemistry* 18:97–108.

Ross, T., and C.D. McCullough. 2011. Health and Safety working around pit lakes. In *Mine Pit Lakes: Closure and Management*, C.D. McCullough, ed., pp. 167–181. Perth, Western Australia, Australia: Australian Centre for Geomechanics.

Schultze, M., B. Boehrer, K. Friese, M. Koschorreck, S. Stasik, and K. Wendt-Potthoff. 2011. Disposal of waste materials at the bottom of pit lakes. In *Mine Closure 2011: Proceedings of the Sixth International Conference on Mine Closure*, Lake Louise, Alberta, Canada.

Schultze, M., M. Hemm, W. Geller, and F.-C. Benthaus. 2013. Remediation and management of acidified pit lakes and outflowing waters. In *Acidic Pit Lakes—Legacies of Surface Mining on Coal and Metal Ores*, W. Geller, M. Schultze, R.L.P. Kleinmann, and C. Wolkersdorfer, eds., pp. 225–264. Berlin, Germany: Springer.

Swanson, S. 2011. What type of lake do we want? Stakeholder engagement in planning for beneficial end uses of pit lakes. In *Mine Pit Lakes: Closure and Management*, C.D. McCullough, ed., pp. 29–42. Perth, Western Australia, Australia: Australian Centre for Geomechanics.

Van Etten, E.J.B., C.D. McCullough, and M.A. Lund. 2014. Setting goals and choosing appropriate reference sites for restoring mine pit lakes as aquatic ecosystems: Case study from south-west Australia. *Mining Technology* 123(1):9–19.

Vandenberg, J., N. Lauzon, S. Prakash, and K. Salzsauler. 2011. Use of water quality models for design and evaluation of pit lakes. In *Mine Pit Lakes: Closure and Management*, C.D. McCullough, ed., pp. 63–80. Perth, Western Australia, Australia: Australian Centre for Geomechanics.

Vandenberg, J., C. McCullough, and D. Castendyk. 2015. Key issues in mine closure planning related to pit lakes. In *Proceedings of the Joint International Conference on Acid Rock Drainage ICARD/International Mine Water Association (IMWA) Congress*, Santiago, Chile.

Vandenberg, J.A., K. Salzsauler, and S. Donald. 2016. Best practices checklist for modelling mine waters. In *Proceedings of the IMWA 2016*, Leipzig, Germany, July 11–15, 2016.

Wen, M.E., C.A. Pelletier, K. Norlund, G.W.R. Wolff, and D. Berthelot. 2015. Phytoremediation to improve pit lake water quality. In *Mine Closure 2015—Proceedings of the 10th International Conference on Mine Closure*, Vancouver, British Columbia, Canada.

Wisotzky, F. 2013. Avoidance and source treatment. In *Acidic Pit Lakes—Legacies of Surface Mining on Coal and Metal Ores*, W. Geller, M. Schultze, R.L.P. Kleinmann, and C. Wolkersdorfer, eds., pp. 258–264. Berlin, Germany: Springer.

Younger, P.L., and C. Wolkersdorfer. 2004. Mining impacts on the fresh water environment: Technical and managerial guidelines for catchment scale management. *Mine Water and the Environment* 23:S2–S80.

11 Carbon Sequestration Potential on Mined Lands

Sally Brown, Andrew Trlica, John Lavery, and Mark Teshima

CONTENTS

11.1 Introduction 189
    11.1.1 Mining-Related Disturbance 190
    11.1.2 Restoration Methods... 190
11.2 Rates of Carbon Sequestration 190
    11.2.1 Carbon Accrual Over Time or With Topsoil... 191
    11.2.2 Carbon Accrual With Residuals.. 191
11.3 Other Factors 194
    11.3.1 General Factors... 194
    11.3.2 End Use 194
    11.3.3 Amendment Factors.. 196
11.4 Conclusions... 198
References.. 199

11.1 INTRODUCTION

The terrestrial carbon cycle represents the third largest carbon pool, after the oceanic pool and fossil reserves. In fact all fossil reserves originated from photosynthesis, the process that drives the short-term carbon cycle. In the short-term cycle, a majority of the carbon that is fixed during photosynthesis is used as an energy source by animals and returned to the atmosphere as CO_2. The portion of this fixed carbon that remains is transformed into plant and animal biomass and soil organic matter. The richer or higher in organic matter the soil is, the more productive are the plants that grow on it. This high level of net primary productivity increases the rate of carbon fixation and returns a greater portion of the carbon to the soil. While the focus of carbon sequestration has often been on assuring that the specific carbon that has been sequestered remains in place, there is also the potential to store carbon by enriching soils. Increased soil organic matter will provide a higher level of carbon in the soil with a new or increasing steady state of fixation and decay.

Most of the conventionally managed agricultural soils have become depleted in organic matter (Amundson et al. 2015; Doran, 2002; Montgomery, 2007). In addition, much of the fixed carbon produced on these soils is removed each year with the harvest. As a result, their equilibrium cycles of carbon fixation and release are likely performing at sub-optimal levels. Increasing of carbon reserves on these soils has been suggested as both a means to combat climate change and as a way to provide increased resilience for the changes that are and will occur (Doran, 2002; Ussiri and Lal, 2005). Higher productivity soils will be more drought resistant and less dependent on mineral fertilizers. Mined soils represent an extreme example of soils that are depleted of organic matter. As a result, without restoration, these soils only provide for minimal productivity and plant growth. As the extreme on the spectrum of disturbed and/or depleted soils, these lands also represent a significant potential carbon sink. Restoring these lands to optimal productivity would both store significant quantities of carbon and provide additional land for a wide range of ecosystem services.

11.1.1 MINING-RELATED DISTURBANCE

Mined, disturbed soils fall under a relatively broad spectrum of types of materials and associated challenges for restoration. At one end of this spectrum, the material that has been mined is nontoxic and the topsoil that covered the material prior to extraction has been stored and can be used for restoration (Ussiri et al. 2006a,b). This case represents the least challenging scenario for restoring and potentially the smallest carbon sink. From here, other sites have effectively removed the surface soil horizon and left the mineral subsoil or noncontaminated tailings to function as topsoil (Pepper et al. 2012). In certain cases, the subsoil material that is exposed is high in reduced sulfur and iron and will oxidize when exposed to oxygen. These acidic wastes pose a much more significant challenge for restoration (Orndorff et al. 2008). At the other end of the spectrum, the residuals from mining contain phytotoxic concentrations of trace elements (Brown et al. 2014). Finally, in potentially the most challenging of cases, the surface residual material is both high in phytotoxic metals and potential acidity (Brown et al. 2005; Stuczynski et al. 2007). At each of these types of sites, there may also be challenges associated with the physical characteristics of the surface material and local climate and precipitation patterns. A detailed discussion of types of sites can be found in chapters of this text as well as in the published literature.

11.1.2 RESTORATION METHODS

Restoration of mined soils or residuals from mining can be accomplished using a range of restoration practices. Associated costs and time to reach optimal function will vary based on the type of site and restoration practice. The rate of carbon accrual will also vary. The Intergovernmental Panel on Climate Change (IPCC) estimates carbon sequestration potential across a wide range of sites and climates. A default average estimate rate of 0.25 MgC/ha/year accrual in degraded lands post reclamation is included (IPCC, 2000). Depending on the methods used for restoration, restoration of one site can also result in disturbance of additional sites (Brown et al. 2014). If nontoxic subsoil is left and allowed to develop into a functional soil naturally, the rate of carbon accrual will be very slow. One estimate of time required for soil formation suggested that approximately 1800 years is required for 15 cm of topsoil to develop (Montgomery, 2007). This development will be accelerated if stockpiled topsoil is used. However, if topsoil is harvested from an external site rather than stockpiled at the site in question, the restoration of the mined land will have an associated disturbance and carbon loss from a previously undisturbed site. For example, in many cases topsoil is excavated from one site for use elsewhere. The exploited area is typically exposed subsoil, itself requiring restoration. Another common practice is to restore a site using residual materials. These residuals typically include a wide range of organic wastes applied to the disturbed area alone, or in combination with mineral residuals. Using residue-derived amendments can be done to provide a source of organic matter to the existing surface material. This is a means to greatly accelerate the process of transforming these surface materials into functional soils. It can also be done as a means to create manufactured and functional soil horizons. For this approach, organic residuals are mixed with mineral materials so that the resulting amendment has similar characteristics to a naturally occurring soil. The newly created topsoil is typically placed over the disturbed material.

11.2 RATES OF CARBON SEQUESTRATION

The rate of carbon sequestration for different types of restoration practices will be a function of the speed at which the site attains equilibrium. If high rates of organic amendments have been added, there will be a period when a portion of the added organic matter mineralizes while newly fixed carbon is added to the soils. In other cases, increasing productivity over the course of multiple growing seasons will add carbon to the soil over time. Different examples and rates of carbon sequestration are discussed for different types of sites and approaches.

11.2.1 Carbon Accrual Over Time or With Topsoil

Though unweathered mine spoils can start to develop an A horizon enriched in organic matter and show other changes indicating soil formation in less than a decade (Haering et al. 1993; Thomas and Jansen, 1985), soil organic carbon (SOC) recovery after drastic disturbance from mining is typically very slow. In the absence of human intervention, many disturbed areas require on the order of hundreds to thousands of years to recover SOC to pre–disturbance levels (Insam and Domsch, 1988; Montgomery, 2007; Naeth et al. 1987). Soil organic carbon accumulates slowly in comparison to its total sink capacity in part because of smaller annual carbon inputs from diminished plant productivity. Efficient nutrient cycling may also be slow to recover because microbial biomass (a component of SOC) may take decades to rebound to pre–disturbance levels (Pepper et al. 2012). Diminished microbial communities (Pepper et al. 2012) and a general lack of soil invertebrates (Brown et al. 2005, 2014) may contribute to reduced decomposition of new plant carbon inputs, which in turn retards nutrient cycling and may explain why some reclaimed soils (transiently) show higher SOC content than adjacent undisturbed soil (Stahl et al. 2003).

Rates of SOC accumulation under conventional restoration practices (topsoil replacement and fertilizers) vary based on the quantity and quality of the topsoil used, the climate at the site, and the targeted end use of the site (Table 11.1). For example, some revegetated pipeline corridors on the Canadian prairie gained SOC at a rate of 1.31 Mg/ha/year (Naeth et al. 1987). Soil under a meadow established on a reclaimed coalmine (0–10 cm) doubled in SOC concentration over 20 years, from 14.2 to 28.7 MgC/ha at one site and 15.1 to 30.24 MgC/ha at another, an accumulation rate of 0.73 and 0.76 MgC/ha/year, respectively (Shukla et al. 2004). Schwenke et al. (2000) reported linear increases of 0.43 MgC/ha/year in the top 10 cm on reforested bauxite mine soils in Australia. Singh and Singh (2006) measured mine soils accreting SOC at a rate of 0.2–1.5 MgC/ha/year in the 0–20 cm depth under young tree plantations established on reclaimed mine soil in a dry tropical setting. Insam and Domsch (1988) reported that 50 years post reclamation, the top 15 cm of soils in lignite fields reclaimed to agriculture contained 0.8% SOC and had not yet reached their predicted equilibrium of 1% SOC (a long-term accretion rate of +0.011% C/year), while 50-year-old sites reclaimed to forest contained 1.7% SOC (a long-term accretion rate of +0.051% C/year). Soils reclaimed from a surface coalmine in Ohio and under pasture for 20 years showed a greater than 2.6-fold increase in SOC in the Ap horizon, from 0.86%–0.87% C in 1981 to 2.23%–2.38% C in 2001 (Underwood and Smeck, 2002).

11.2.2 Carbon Accrual With Residuals

The efficacy of amendments for accelerating the rate of mine soil restoration has been well documented (Brown and Chaney, 2016; Larney and Angers, 2012; Wijesekara et al. 2016). Use of amendments in lieu of or in combination with stockpiled topsoil can accelerate the rate of carbon storage. If the amendments contain high concentrations of carbon, there will be an immediate increase in surface carbon concentrations following amendment addition. This increase in carbon will typically be associated with an increase in net primary productivity, initiating the short-term cycle of carbon fixation and decay. A portion of the carbon added with the amendment will mineralize, while additional carbon will be added to the restored soil through plant growth. Over time, sites restored with amendments will also reach equilibrium carbon concentrations.

Albaladejo et al. (2008) found a onetime application of increasing rates of organic municipal waste produced increasingly higher SOC concentration, standing plant biomass, and annual net primary productivity on degraded semiarid lands, even 16 years after application, sequestering an additional 3.4–9.5 MgC/ha over conventionally reclaimed plots. Onetime application of paper de-inking sludge and N and P fertilizers improved soil properties and enhanced standing biomass for two consecutive growing seasons on a reclaimed sand pit (Fierro et al. 1999). Similar results were noted for paper mill sludge applied to reclaimed coal mine spoil (Li and Daniels, 1997). SOC was

TABLE 11.1

Selected SOC Accumulation Rates in Reclaimed Mine Soils, Conventional and Residuals-Amended Sites

Citation	Disturbance Type	Reclamation	SOC Accum. (Mg/ha/Year)	Site Age (Years)	Depth (cm)
Akala and Lal (2000, 2001)	Coal (Ohio)	Conv. w/ TS	0.7–2.3 (forest); 0.5–3.1 (pasture)	21–25	0–15
Stahl et al. (2003)	Coal (Wyoming)	Conv. w/ TS	0.17	19	0–15
Shukla et al. (2004)	Coal (Ohio)	Conv. w/ TS	0.73–0.76	20	0–10
Singh and Singh (2006)	Coal (Central India)	Conv. w/o TS	0.2–1.5	4–5	0–10
Schwenke et al. (2000)	Bauxite (Queensland, Australia)	Conv. w/ TS	0.43 (native trees); 0.75 (pasture)	10–15	0–10
Ussiri et al. (2006a)	Coal (Ohio)	Conv. w/ TS	0.6–2.4 (pasture to tree plantation)	10	0–50
Insam and Domsch (1988)	Lignite (W. Germany)	Conv. w/ and w/o TS	Ag. soils: 0.13 (<5 years) to 0.01 (<40 years) Forest soils: 0.25 (<5 years) to 0.3 (<40 years)	0–5; 5–40	0–15
Underwood and Smeck (2002)	Coal (Ohio)	Conv. w/ TS	1.5–1.6[a]	20	0–10
Shukla et al. (2005)	Acidic coal (Ohio)	FGD + 112 Mg/ha compost	Vs. initial spoil 1.67 (0–10 cm)/1.64 (10–20 cm) Vs. FGD only 0.64 (0–10 cm)/1.64 (10–20 cm)	10	0–10; 10–20
Albaladejo et al. (2008)	Desertification (Spain)	130 and 260 Mg/ha MSW	Vs. initial 1.33 (@130) 1.86 (@260) Vs. control 0.29 (@130) 0.81 (@260)	16	0–10
Tian et al. (2009)	Coal (Illinois)	BS (various rates)	2.98–4.71 (max vs. control) 1.73 ("net C sequestration") **Total C (Mg/ha)**	34	0–15

(Continued)

TABLE 11.1 (*Continued*)
Selected SOC Accumulation Rates in Reclaimed Mine Soils, Conventional and Residuals-Amended Sites

Citation	Disturbance Type	Reclamation	SOC Accum. (Mg/ha/Year)	Site Age (Years)	Depth (cm)
Brown et al. (2014)	Pb, Zn mine tailings (Missouri)	BS + lime	Control 12.7	12	0–15
			Biosolids low 80.3		
			Biosolids high 73		
			Compost 59		
Trlica and Brown (2013)	Cu mine (British Columbia)	BS	Control 4.6	8	0–15
			Biosolids 43		
	Gravel pit (British Columbia)		Control 11	9	0–15
			Biosolids 50 Mg/ha 35		
			Biosolids 102 Mg/ha 40		
			Biosolids + Pulp sludge 47		
	Coal mine (Washington)		Topsoil 40 Mg/ha	17	0–15
			Topsoil + Biosolids 54 Mg ha		
	Coal mine (Pennsylvania)		Topsoil 51	27	0–15
			Topsoil + Compost 71		

[a] Unless specified, bulk density values were estimated as 1.44 g/cm^3 bulk density.

FGD, Flue gas desulfurization by-product; TS, topsoil; BS, biosolids; MSW, municipal solid waste; Conv., conventional fertilizer.

elevated for 2 years after reclamation with compost and alfalfa at oil well sites in southern Alberta, gaining about 0.13 Mg SOC for every Mg of residual added (Zvomuya et al. 2007). A similar study found greater topsoil replacement depth, and amendment with compost or manure improved SOC in the top 15 cm (1.86%–2.13% C) over unamended control (1.62%–1.75% C) (Larney et al. 2005) and improved crop response (Larney et al. 2003).

There are only a few long-term studies on comparative SOC accumulation in disturbed soils reclaimed with organic residuals versus soils reclaimed conventionally. Benfeldt et al. (2001) reported increased organic matter in the top 10 cm versus conventionally reclaimed soils in plots applied with organic residuals after 5 years, but this difference had disappeared after 16 years since reclamation (all plots had equilibrated at about 10 Mg organic matter/ha, equivalent to 5.8 MgC/ha). Organic residuals applied to graded mine soil stored slightly less C than sites reclaimed with borrowed topsoil (Shukla et al. 2005).

In Fulton county, IL, a long-term study (Tian et al. 2009) of former strip mined lands reclaimed to agricultural soils repeatedly applied with high loading rates of biosolids showed that over 34 years SOC increased by an average of 1.73 MgC/ha/year compared with essentially static SOC levels observed in fields receiving only mineral fertilizer. Biosolids application rate was significantly correlated with SOC content, though fields that had not received biosolids for many years retained elevated SOC in comparison to fertilized soils. Peak SOC concentrations of 6.5%–7% SOC were found in fields recently receiving the highest loading rates of biosolids. Total gains in SOC across all fields receiving biosolids was 24.8–166 MgC/ha compared to 5.7 MgC/ha in fields receiving only synthetic fertilizer. Even after subtracting the carbon assuming to remain from applied biosolids, fields applied with biosolids contained 18.2–104 MgC/ha of extra SOC due to enhanced microbial humification efficiency and soil aggregation, equating to a "C sequestration efficiency" of 0.049–0.086 MgC sequestered as SOC per Mg of biosolids applied.

Trlica and Brown (2013) surveyed multiple long-term sites that had been restored with organic amendments, either solely or in combination with topsoil. Reference controls had had topsoil application or been fertilized. Across all sites, use of amendments resulted in increased carbon storage in comparison to conventionally reclaimed sites. This difference was much more pronounced among sites that had not received topsoil addition.

11.3   OTHER FACTORS

11.3.1   GENERAL FACTORS

Other research has shown that carbon storage is increased when C is added to soil in combination with appropriate quantities of N, P, and S (Kirkby et al. 2014; Wuest and Reardon, 2016). This is most easily done by using an amendment that has a well-balanced ratio of carbon to other nutrients. Examples include animal manures, municipal biosolids, and food/ yard-based composts (US EPA, 2011). Extent of protection of organic matter, either through aggregation or through association with inorganic oxides, may also impact the quantity of C that is stored (Stewart et al. 2012). Soil texture will also be a factor, with higher total C possible in finer-textured soils in comparison to coarser textured soils (Li and Evanylo, 2013). As with agricultural soils, management factors will also impact C storage. Sites that are conventionally tilled after restoration will have lower carbon storage than sites maintained as no till (Stewart et al. 2012).

11.3.2   END USE

In addition to the carbon dynamics within the developing soils, other factors can have an enormous impact on the final carbon balance for a restored site. These factors include the alternate end use or disposal for the materials used to restore the site and the final end use of the site. Materials can be identified and end use can be determined in order to maximize the carbon credits associated with site restoration.

SOC accumulation rate and final equilibrium SOC storage can also depend upon the plant species present (Akala and Lal, 2000, 2001; Berdense, 1990; Schwenke et al. 2000; Singh and Singh, 2006). Amichev et al. (2008) reported soil carbon storage after 20–55 years to a depth of 1.5 m on pre–Surface Mining Control and Reclamation Act (SMCRA) coalmines of 40 MgC/ha under pine forest and 47 MgC/ha under hardwoods, compared to 71 MgC/ha under undisturbed forest. Soils under different tree species in a dry tropical setting have also shown significant differences in SOC accumulation rate (Singh and Singh, 2006). White pine plantations were slightly more productive than native hardwood forests on mine spoil (Gorman et al. 2001), especially if grown on noncompacted mine spoil (Keltin et al. 1997).

Land management decisions post reclamation can impact SOC accumulation rates on reclaimed soils (Ussiri et al. 2006b) (Figure 11.1). SOC content, under reclaimed coal mine soils converted from pasture to black locust plantation, accrued at 2.4 MgC/ha/year—significantly higher than other tree types and pasture (Ussiri et al. 2006b). At a 10-year-old reclaimed coalmine site, Shukla and Lal (2005) found increased SOC in the top 10 cm in land under forage compared to land under corn or soybeans, and also higher SOC concentration in the mined sites overall (e.g., 3.33% C mined vs. 1.75% C unmined). Shrestha and Lal (2007) noted differences in carbon storage for mined lands with a similar history of disturbance, reclamation, and age. All sites were located in the same region and had been restored with a surface application of 30 cm of topsoil 28 years prior to sampling.

FIGURE 11.1 Restoration of a sand and gravel mine in British Columbia. Carbon sequestration is typically focused on soil considerations (a). Land use, in this case a poplar plantation (b), will also impact the quantity of carbon stored on the restored lands.

The authors found higher carbon storage in the site restored to pasture in comparison to hay and forest in the surface 0–5 cm depth. The rates of carbon storage were mirrored by associated increases in soil N. The authors looked at aggregate size as a factor for the observed differences, but did not consider additional inputs, such as animal manures on the pasture site as factors.

Sites maintained as open space, forest, or farm will have much greater carbon storage across the range of soil textures, tree species, and management types than soils restored and converted to residential development. Trlica and Brown (2013) calculated the impact of low-density subdivision construction on a hypothetical hectare of land restored with or without amendments to forestry in the Pacific Northwest over a 30-year time frame. Considering increases in above- and belowground biomass, restoring the site to forestry resulted in net sequestration of −293 and −475 Mg CO_2 for sites without and with amendment addition. Constructing four 293 m² homes and associated roadways and considering energy use in the homes, but not transportation related emissions, resulted in net emissions of 1269 Mg CO_2 over the same time frame. That figure includes use of amendments to establish a forest on about half of the site.

11.3.3 Amendment Factors

If the materials used for restoration are diverted from disposal sites or combustion facilities where they would otherwise emit significant quantities of fugitive gases, there will be carbon benefits associated with fugitive gas avoidance. Brown et al. (2014) used a mixture of biosolids and lime to restore large-scale plots on a metal-contaminated mine waste in Jasper County, MO. Over 50% of the biosolids currently generated in this area are incinerated. The authors calculated the total C benefits of diverting the material from incineration to use for restoration. For the biosolids required to restore 1 ha of land (336 Mg), credits totaled −156 Mg CO_2. In contrast, the current practice of combustion results in emissions of 485 Mg CO_2 for the same quantity of materials. (Figure 11.2).

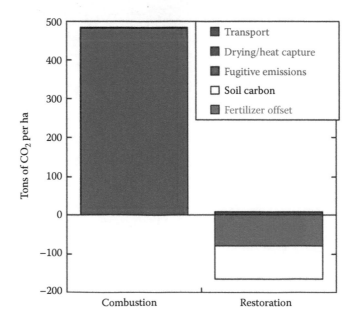

FIGURE 11.2 Emissions balance for biosolids disposal by combustion or use for land application to restore a mining impacted site. (Data from Brown, S. et al., *Sci. Total Environ.*, 485–486, 624, 2014.)

OPTIONS FOR ONE COAL MINE

The TransAlta coal mine in western Washington State started mining and burning coal in 1971. A total of 2900 ha has been disturbed with more than 2000 ha still to be restored. Portions of the disturbed areas were mined prior to the 1977 passage of the Federal Surface Mining Control and Reclamation Act when topsoil storage was not required. Much of the initial restoration at the site was done using topsoil harvested from other locations. Portions were also amended with municipal biosolids (Trlica and Brown, 2013). Current regulations require that restoration restore the depth of soil to a minimum of 45 cm and up to 90 cm. For 45 cm, that is a minimum of 1 Mg/m^3 or 10,000 Mg per ha of soil. It has been suggested that using by-products from coal combustion is one method to enhance carbon sequestration on degraded lands (Palumbo et al. 2004). Across the United States, more than half of the ash that is currently generated is disposed of. Beneficial use of bottom ash as an alternative to disposal is recommended by the USEPA (https://www.epa.gov/coalash/coal-ash-reuse). The coal burned at this facility is low in sulfur, and as a result does not generate by-products associated with treating emissions from high-sulfur coal. However, significant quantities of bottom ash, the mineral residual from coal combustion, are produced—approximately 1 kg of bottom ash for every 100 kg of coal. We tested different mixtures of the bottom ash in combination with biosolids ± sawdust for their ability to support grass and trees as a component of manufactured topsoil. Yield, plant nutrient, and metal concentrations and leachate characteristics were measured. A range of mixtures showed superior growth to topsoil + fertilizer. These results are in agreement with previously published studies (Palumbo et al. 2004; Ussiri and Lal, 2005). Despite Federal encouragement of the use of these by-products, it is not clear if this approach will meet with State regulatory approval. Whatever approach is used to restore the site, it is likely that soil carbon concentrations and aboveground biomass will increase as a result of the restoration. This will result in increased carbon sequestration on the 2000 ha that are currently devoid of plant growth (Figure 11.3).

Much of the work on soil carbon sequestration on disturbed lands has focused on the quantity of carbon stored on the lands themselves without a consideration for the carbon impacts of the restoration process. However, it is important to understand that different restoration techniques can have a significant impact on the carbon balance of the project as a whole. It is possible to gain some insight into this by using the bottom ash generated at the plant as an example. The power plant on-site has the capacity to generate 1340 MW of electricity. An approximate quantity of coal for that amount of electricity is 3.8 million tons (http://www.ucsusa.org/clean_energy/coalvswind/brief_coal.html#bf-toc-4). At a ratio of 1:100, this means that approximately 38,000 Mg of bottom ash are generated annually. This would be sufficient material to restore about 4 ha/year. Over the total operational years of the plant (power generation started in 1971), this material would have been sufficient to restore about 180 ha of the site. Rather than being used on-site, the material is currently landfilled. This means both that there are emissions associated with transport of the material to a landfill and transport of a substitute material to the site. A 300 km round trip haul distance is assumed for both landfilling the bottom ash and bringing appropriate alternative materials to the site. Estimates for rail and truck based emissions per year were calculated (see calculations below) (Bickford et al. 2014; Brown et al. 2010). Assuming a 300 km round trip haul, associated annual emissions for both landfill disposal and bringing alternative materials to the site range from 100 to 375 Mg CO_2/year considering both truck and rail transportation options.

FIGURE 11.3 An aerial view of the TransAlta coal mine in western Washington state and relative growth response of grass to different mixtures of biosolids and stockpiled topsoil, bottom ash, and silica sand.

Estimated emissions for truck and rail transport to and from the coal mine

	Landfill Ash	Alternative Material
	Mg CO_2/Year	
Truck	500	500
Rail	187.5	187.5

In addition to transport-associated emissions, use of a virgin material such as silica sand as a substitute for the bottom ash would result in additional land disturbance and requirements for restoration. This example illustrates that the methods and materials used for mine site restoration can have a significant impact on the total carbon storage potential of a site.

11.4 CONCLUSIONS

Lands disturbed by mining activities offer a clear potential for soil carbon sequestration. The rate of carbon accumulation will vary based on a number of factors, including local climate and precipitation. The approach used to restore the site will also impact the rate of recovery and associated rate of carbon accumulation. Topsoiling accelerates soil formation over fertilizer addition alone. However, in cases where sufficient topsoil has not been stockpiled, use of amendments has shown to accelerate soil development and increase carbon storage. Rates of carbon accumulation on these sites, particularly with the use of amendments, are high. However, it is critical to factor in the sources of amendments in these calculations. Amendments diverted from disposal may increase carbon offsets, whereas amendments that necessitate disturbance of other sites can eliminate or reduce benefits. Finally, end use of the site can be tailored to maximize benefits (reforestation) or may eliminate them (home construction).

REFERENCES

Akala, V.A., and R. Lal. 2000. Potential of mineland reclamation for soil carbon sequestration in Ohio. *Land Degrad. Develop.* 11:289–297.

Akala, V.A., and R. Lal. 2001. Soil organic carbon pools and sequestration rates in reclaimed minesoils in Ohio. *J. Environ. Qual.* 30:2098–2104.

Albaladejo, J., J. Lopez, C. Boix-Fayos, G.G. Barbera, and M. Martinez-Mena Akala. 2008. Long-term effect of a single application of organic refuse on carbon sequestration and soil physical properties. *J. Environ. Qual.* 37:2093–2099.

Amichev, B., J.A. Burger, and J.A. Rodrigue. 2008. Carbon sequestration by forests and soils on mined land in the Midwestern and Appalachian coalfields of the U.S. *Forest Ecol. Manage.* 256:1949–1959.

Amundson, R., A.A. Berhe, J.W. Hopmans et al. 2015. Soil and human security in the 21st century. *Science* 348:1261071-1–1261071-6.

Benfeldt, E.S., J.A. Burger, and W.L. Daniels. 2001. Quality of amended mine soil after sixteen years. *Soil Sci. Soc. Am. J.* 65:1736–1744.

Berdense, F. 1990. Organic matter accumulation and nitrogen mineralization during secondary succession in heathland ecosystems. *J. Environ. Qual.* 78:413–427.

Bickford, E., T. Holloway, A. Karambelas et al. 2014. Emissions and air quality impacts of truck-to-rail freight modal shifts in the Midwestern United States. *Environ. Sci. Technol.* 48:446–454.

Brown, S., A. Carpenter, and N. Beecher. 2010. Calculator tool for determining greenhouse gas emissions for biosolids processing and end use. *Environ. Sci. Technol.* 44:9505–9515.

Brown, S., M. Mahoney, and M. Sprenger. 2014. A comparison of the efficacy and ecosystem impact of residuals-based and topsoil-based amendments for restoring historic mine tailings in the tri-state mining district. *Sci. Total Environ.* 485–486:624–632.

Brown, S., M. Sprenger, A. Maxemchuk et al. 2005. Ecosystem function in alluvial tailings after biosolids and lime addition. *J. Environ. Qual.* 34:139–148.

Brown, S.L., and R.L. Chaney. 2016. Use of amendments to restore ecosystem function to metal mining-impacted sites: Tools to evaluate efficacy. *Curr. Pollut. Rep.* 2:91.

Doran, J. 2002. Soil health and global sustainability: Translating science into practice. *Agric. Ecosyst. Environ.* 88:119–127.

Fierro, A., D.A. Angers, and C.J. Beauchamp. 1999. Restoration of ecosystem function in an abandoned sand-pit: Plant and soil responses to paper de-inking sludge. *J. Appl. Ecol.* 36:244–253.

Gorman, J., J. Skousen, J. Sencindiver, and P. Ziemkiewicz. 2001. Forest productivity and minesoil development under a white pine plantation versus natural vegetation after 30 years. American Society for Surface Mining and Reclamation, Albuquerque, NM.

Haering, K.C., W.L. Daniels, and J.A. Roberts. 1993. Changes in mine soil properties resulting from overburden weathering. *J. Environ. Qual.* 22:194–200.

Insam, H., and K.H. Domsch. 1988. Relationship between soil organic carbon and microbial biomass on chronosequences of reclamation sites. *Microbial. Ecol.* 15:177–188.

IPCC (Intergovernmental Panel on Climate Change). 2000. *Land Use, Land-Use Change, and Forestry.* Cambridge, U.K.: Cambridge University Press, 373pp.

Keltin, D.L., C. Siegel, and J.A. Burger. 1997. Value of commercial forestry as a post-mining land use. In: *1997 National Meeting of the American Society for Surface Mining and Reclamation*, Austin, TX, May 10–15, 1997. J.E. Brandt (ed.). pp. 344–348.

Kirkby, C.A., A.E. Richardson, L.J. Wade et al. 2014. Nutrient availability limits carbon sequestration in arable soils. *Soil Biol. Biochem.* 68:402–409.

Larney, F.J., O.O. Akinremi, R.L. Lemke et al. 2003. Crop response to topsoil replacement depth and organic amendment on abandoned natural gas wellsites. *Can. J. Soil Sci.* 83:415–423.

Larney, F.J., O.O. Akinremi, R.L. Lemke et al. 2005. Soil responses to topsoil replacement depth and organic amendments in wellsite reclamation. *Can. J. Soil Sci.* 85:307–317.

Larney, F.J., and D.A. Angers. 2012. The role of organic amendments in soil reclamation: A review. *Can. J. Soil Sci.* 92:19–38.

Li, J., and G.K. Evanylo. 2013. The effects of long-term application of organic amendments on soil organic carbon accumulation. *Soil Sci. Soc. Am. J.* 77:964–973.

Li, R.S., and W.L. Daniels. 1997. Reclamation of coal refuse with a papermill sludge amendment. In: *1997 National Meeting of the American Society for Surface Mining and Reclamation*, Austin, TX, May 10–15, 1997. J.E. Brandt (ed.). pp. 277–290.

Montgomery, D.R. 2007. Soil erosion and agricultural sustainability. *Proc. Natl. Acad. Sci. USA* 104:13268–13272.

Naeth, M.A., W.B. McGill, and A.W. Bailey. 1987. Persistence of changes in selected soil chemical and physical properties after pipeline installation in solonetzic native rangeland. *Can. J. Soil Sci.* 67:747–763.

Orndorff, Z.W., W.L. Daniels, and D.S. Fanning. 2008. Reclamation of acid sulfate soils using lime-stabilized biosolids. *J. Environ. Qual.* 37:1447–1455.

Palumbo, A.V., J.F. McCarthy, J.E. Amonette et al. 2004. Prospects for enhancing carbon sequestration and reclamation of degraded lands with fossil-fuel combustion by-products. *Adv. Environ. Res.* 8:425–438.

Pepper, IL, H.G. Zerzghi, S.A. Bengson et al. 2012. Bacterial populations within copper mine tailings: Long-term effects of amendment with class A biosolids. *J. Appl. Microbiol.* 113:569–577.

Schwenke, G.D., L. Ayre, D.R. Mulligan et al. 2000. Soil stripping and replacement for the rehabilitation of bauxite-mined land at Weipa. II: Soil organic matter dynamics in mine soil chronosequences. *Aust. J. Soil Res.* 38:371–393.

Shrestha, R.K., and R. Lal. 2007. Soil carbon and nitrogen in 28-year-old land uses in reclaimed coal mine soils of Ohio. *J. Environ. Qual.* 36:1775–1783.

Shukla, M.K., and R. Lal. 2005. Soil organic carbon stock for reclaimed minesoils in northeastern Ohio. *Land Degrad. Develop.* 16:377–386.

Shukla, M.K., R. Lal, and M.H. Ebinger. 2005. Physical and chemical properties of a minespoil eight years after reclamation in northeastern Ohio. *Soil Sci. Soc. Am. J.* 69:1288–1297.

Shukla, M.K., R. Lal, J. Underwood, and M. Ebinger. 2004. Physical and hydrological characteristics of reclaimed minesoils in southeastern Ohio. *Soil Sci. Soc. Am. J.* 68:1352–1359.

Singh, A.N., and J.S. Singh. 2006. Experiments on ecological restoration of coal mine spoil using native trees in a dry tropical environment, India: A synthesis. *New Forests* 31:25–39.

Stahl, P.D., J.D. Anderson, L.J. Ingram et al. 2003. Accumulation of organic carbon in reclaimed coal mine soils of Wyoming. In: *National Meeting of the American Society of Mining and Reclamation and the 9th Billings Land Reclamation Symposium*, Billings, MT, June 3–6, 2003. pp. 1206–1215.

Stewart, C.E., R.F. Follett, J. Wallace et al. 2012. Impact of biosolids and tillage on soil organic matter fractions: Implications of carbon saturation for conservation management in the Virginia Coastal Plain. *Soil Sci. Soc. Am. J.* 76:1257–1267.

Stuczynski, T., G. Siebielec, W.L. Daniels et al. 2007. Biological aspects of metal waste reclamation with biosolids. *J. Environ. Qual.* 36:1154–1162.

Thomas, D., and I. Jansen. 1985. Soil development in coal mine spoils. *J. Soil Water Conserv.* 40:439–442.

Tian, G., T.C. Granato, A.E. Cox et al. 2009. Soil carbon sequestration resulting from long-term application of biosolids for land reclamation. *J. Environ. Qual.* 38:61–74.

Trlica, A., and S. Brown. 2013. Greenhouse gas emissions and the interrelation of urban and forest sectors in reclaiming one hectare of land in the Pacific Northwest. *Environ. Sci. Technol.* 47:7250–7259.

Underwood, J.F., and N.E. Smeck. 2002. Soil development in two Ohio minesoils under continuous grass cover for twenty-five years following reclamation. In: *National Meetings of the American Society of Mining and Reclamation*, Lexington, KY, June 9–13, 2002.

US EPA. 2011. Terrestrial carbon sequestration analysis of terrestrial carbon sequestration at three contaminated sites remediated and revitalized with soil amendments. EPA-542-R-10-003. https://clu-in.org/conf/tio/amendments_102711/. Accessed May 17, 2017.

Ussiri, D.A.N., and R. Lal. 2005. Carbon sequestration in reclaimed minesoils. *Crit. Rev. Plant Sci.* 24:151–165.

Ussiri, D.A.N., R. Lal, and P.A. Jacinthe. 2006a. Soil properties and carbon sequestration of afforested pastures in reclaimed minesoils of Ohio. *Soil Sci. Soc. Am. J.* 70:1797–1806.

Ussiri, D.A.N., R. Lal, and P.A. Jacinthe. 2006b. Post-reclamation land use effects on properties and carbon sequestration in minesoils of southeaster Ohio. *Soil Sci.* 171:261–271.

Wijesekara, H., N.S. Bolan, M. Vithanage et al. 2016. Utilization of biowaste for mine spoil rehabilitation. *Adv. Agron.* 138:97–173.

Wuest, S.B., and C.L. Reardon. 2016. Surface and root inputs produce different carbon/phosphorus ratios in soil. *Soil Sci. Soc. Am. J.* 80:463–471.

Zvomuya, F., F.J. Larney, P.R. DeMaere et al. 2007. Reclamation of abandoned natural gas wellsites with organic amendments: Effects on soil carbon, nitrogen and phosphorus. *Soil Sci. Soc. Am. J.* 71:1186–1193.

Section IV

Mine Site Revegetation Potential

12 Phytotechnologies for Mine Site Rehabilitation

Ramesh Thangavel, Rajasekar Karunanithi,
Hasintha Wijesekara, Yubo Yan,
Balaji Seshadri, and N.S. Bolan

CONTENTS

12.1 Introduction 203
12.2 Phytotechnologies 203
12.3 Processes Involved in Phytotechnologies...206
    12.3.1 Soil Cover...207
    12.3.2 Hydraulic Control 208
    12.3.3 Contaminant Removal 209
    12.3.4 Rhizosphere Modification 209
12.4 Summary and Conclusions...210
Acknowledgments...211
References..211

12.1 INTRODUCTION

Soils are a prime and very important natural resource, and soil fertility is a major concern for sustainable agriculture and economic development of any country. In recent decades, problems of contaminated land sites, water bodies, groundwater, and air worldwide have increased manyfold due to anthropogenic activities. Mining is one of the anthropogenic activities that cause pollution problems in, around, and outside of mining areas. It results in the mobilization of metals and organic and inorganic substances into the environment, which causes pollution of air, soils, sediments, vegetation, and surface and groundwater. It also increases the morbidity and mortality of plant and animal species and results in the loss of visual, aesthetic characteristics of landscapes (Bolan et al. 2003; Pavli et al. 2015).

Due to geochemical changes in mine site soils, the physical, chemical, and biological activities of the soils are changed. Soils of mine sites are often considered as drastically disturbed. They are nutritionally and microbiologically reduced habitats that have reduced crop productivity and food quality, and they need restoration for revegetation (Singh et al. 2004). The shortage of organic matter at mine sites results in poor aggregate stability and structure (Castillejo and Castello 2010). Low water-holding capacity, erosion of tailings by wind and water, crusting, and cracking are some of the adverse effects that result from mine site soil structure (Hossner and Hons 1992). Mine site soil also contains hazardous contaminants such as heavy metals, which may enter groundwater and food chains. Thus, poor soil structure, contaminant toxicity, and low microbial activity are major constraints for revegetation and restoration of mine sites.

12.2 PHYTOTECHNOLOGIES

Remediation of mine site soils is necessary to minimize their impact on ecosystems. This is a challenging job with respect to cost and technical complexity. Worldwide, different physical, chemical, and biological approaches have been used for the rehabilitation of mine sites to restore ecosystem structure and function, thereby improving the ecological integrity and sustainability of the system. It includes removal, isolation, incineration, solidification/stabilization, vitrification, thermal treatment, solvent extraction, and chemical oxidation. However, these methods have disadvantages like high cost, intensive labor, irreversible changes in soil properties, and disturbance of native soil microflora. Chemical methods also involve the movement of contaminated materials to treatment sites, thus adding risks of secondary contamination. Therefore, less environmentally disruptive and more cost-effective methods are needed for rehabilitation of mine sites soil.

Phytoremediation is a remediation strategy that is novel, cost-effective, efficient, environmentally eco-friendly, applicable *in situ*, and solar-driven (Vithanage et al. 2012). It can be used to improve the physical, chemical, and biological properties of mine sites. Phytotechnologies used for the remediation of contaminated sites are known as phytoremediation technologies, and they use higher plants to clean up and revegetate contaminated sites. These techniques differ in the processes by which plants can remove, immobilize, or degrade contaminants (Adriano et al. 2004; Robinson et al. 2009). Plants generally are used to manage the contaminants without affecting the topsoil, thus conserving its utility and fertility. In addition, plants may improve soil fertility with inputs of organic matter from aboveground and belowground biomass (Mench et al. 2009). Phytoremediation can be implemented *in situ* or *ex situ* to clean up a variety of organic and inorganic contaminants. The substances that may be subjected to phytoremediation include metals (Pb, Zn, Cd, Cu, Ni, Hg), metalloids (As, Sb), inorganic compounds (NO_3^-, NH_4^+, PO_4^{3-}), radioactive chemical elements (U, Cs, Sr), petroleum hydrocarbons (BTEX) [benzene, toluene, ethylbenzene and xylene], pesticides and herbicides (atrazine, bentazon, and chlorinated and nitroaromatic compounds), explosives (TNT, DNT) [trinitrotoluene, 2,4-dinitrotoluene], chlorinated solvents (TCE, PCE) [trichloroethylene, perchloroethylene], and industrial organic wastes (PCPs, PAHs) [polychlorinated pesticides, polycyclic aromatic compounds] (Ensley 2000).

Phytoremediation includes seven mechanisms, which are phytodegradation, phytostabilization, phytovolatilization, phytoextraction, phytofiltration, rhizodegradation, and phytodesalination, which plants use to avoid, partition, or remove toxic contaminants. Naturally evolved characteristics of plants can be used for cleanup purposes. Phytoremediation techniques include different modalities, depending on the chemical nature and properties of the contaminant and the plant characteristics. Thus, phytoremediation comprises seven different strategies, although more than one may be used by a plant simultaneously (Figure 12.1).

1. *Phytodegradation (Phytotransformation)*: In this technique, inside the plant cells, the organic contaminants are degraded, that is, metabolized or mineralized by some specific enzymes. For example, the nitroaromatic compounds, chlorinated solvents, and pesticides and anilines are degraded by nitroreductases, dehalogenases and laccases, respectively (Rylott and Bruce 2008). This technique is limited to the removal of only organic contaminants, because heavy metals are nonbiodegradable.

2. *Phytostabilization (Phytoimmobilization)*: It is the use of plants for the stabilization of organic or inorganic contaminants in soil to reduce their mobility and bioavailability in the environment, thus preventing their migration into groundwater or their entry into the food chain (Erakhrumen 2007). Heavy metals are immobilized through sorption by the roots, precipitation by the direct action of root exudates, complexation or valance reduction in the rhizosphere, and they are subsequently trapped in the soil matrix (Wuana and Okieimen 2011). By secreting redox enzymes, plants alter hazardous metals to a less toxic form and reduce damage to the environment. However, phytostabilization is not a permanent solution for heavy metal remediation, because it limits only the movement of the heavy metals in soil.

3. *Phytovolatilization*: It relies on the ability of some plants to absorb and volatilize certain metals/metalloids. Some elements, specifically Hg, Se, and As, are absorbed by roots, are converted into nontoxic forms, and then are released into the atmosphere. This technique can also be used for organic compounds (Ruiz and Daniell 2009; Ali et al. 2013). However,

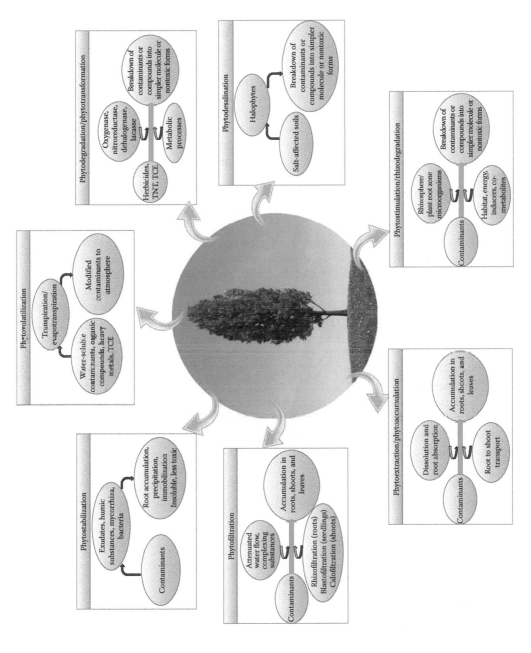

FIGURE 12.1 Schematic diagram illustrating the phytoremediation strategies.

the applicability of this technique is limited, because it does not remove the pollutants completely and transfers the pollutants from one place (soil) to another (atmosphere).

4. *Phytoextraction* (*Phytoaccumulation*, *Phytoabsorption*, or *Phytosequestration*): It involves the uptake of contaminants from soil or water by plant roots followed by their translocation and accumulation in aerial parts, that is, shoots (Rafati et al. 2011). It is mainly applied to metals such as Cd, Ni, Cu, Zn, and Pb, but it can also be used for other elements, like Se and As, and organic compounds (Pedron et al. 2009). Translocation of metals to the plant shoots is a vital biochemical process and is needed for an effective phytoextraction, because harvest of root biomass is generally not practicable (Tangahu et al. 2011).

5. *Phytofiltration*: Plants are used to remove, concentrate, and/or precipitate the contaminants, heavy metals, or, in particular, radioactive elements from an aqueous medium through their root system or submerged organs. Phytofiltration may be rhizofiltration (use of plant roots) or blastofiltration (use of seedlings), or caulofiltration (use of excised plant shoots) (Mukhopadhyay and Maiti 2010). The contaminants are adsorbed or absorbed and thus, their movement to underground water is reduced. Plants with high root biomass, or high absorption surface with more accumulation capacity and tolerance to contaminants, perform better (Pratas et al. 2012).

6. *Rhizodegradation* (*Phytostimulation*): Growing roots promote the proliferation of rhizosphere microorganisms that utilize exudates and metabolites of plants as a source of carbon and energy, and they create a nutrient-rich environment. By utilizing these carbon and energy sources, rhizospheric microorganisms degrade organic pollutants in the rhizosphere. Plants can stimulate microbial activity about 10–100 times more in the rhizosphere through the secretion of exudates containing carbohydrates, amino acids, and flavonoids (Yadav et al. 2010). In addition, plants may exude biodegrading enzymes themselves, capable of degrading organic contaminants in the soil. The application of phytostimulation is limited to organic contaminants (Frers 2009).

7. *Phytodesalination*: It is a recent and emerging biological approach that aims to rehabilitate sodic and saline-sodic soils by using Na-hyperaccumulating halophytes to enable these soils to support normal plant growth. Halophytic plants have been suggested to be naturally better adapted to cope with heavy metals compared to glycophytic plants (Manousaki and Kalogerakis 2011). For example, halophytes such as *Suaeda maritima* and *Sesuvium portulacastrum* removed 504 and 474 kg of sodium chloride, respectively, from 1 ha of saline soil in a period of 4 months and resulted in better crop production after a few repeated cultivations and harvests (Ravindran et al. 2007).

This chapter describes various processes involved in phytotechnologies and the role of these processes in improving physicochemical properties of mine sites.

12.3 PROCESSES INVOLVED IN PHYTOTECHNOLOGIES

Phytotechnology processes involved in rehabilitating mine sites, thereby reducing the mobility and bioavailability of contaminants, include the following:

- Mechanical stabilization of the site to minimize erosion by wind and water
- Enhancement of evapotranspiration, thereby reducing the leaching of contaminants
- Establishment of a vegetation barrier that reduces the likelihood of physical contact with the soil by animals and humans
- Uptake and sequestration of contaminants in the root system
- Alteration of soil factors that influence the speciation and immobilization of contaminants (pH, organic matter, redox levels)
- Root exudates that regulate the precipitation and immobilization of the contaminants

12.3.1 Soil Cover

Vegetative cover plays a vital role in stabilization by reducing the water flux through the soil profile and mechanically stabilizing the soil through root growth. This reduces the movement of soil particles and associated contaminants. However, the physical, chemical, and biological properties of mine soils that control vegetative growth determine the successful outcome of this technology. Soil erosion potential is increased if the soil has no or sparse vegetative cover of plants and/or plant residues. Plant and residue cover protects the soil from rain-splash and slows the movement of surface runoff, thereby increasing infiltration. Similarly, vegetation reduces the wind velocity, thereby mitigating the dispersion of soil and sediments. The vegetation-induced reduction in soil erosion likely reduces the movement of contaminants and subsequent off-site contamination. The erosion-reducing effectiveness of plant and/or residue covers depends on the type, extent, and quantity of cover. Vegetation and residue combinations that completely cover the soil, and which intercept all falling raindrops at and close to the surface, are most efficient in controlling soil erosion. Live ground covers provide the best protection against loss of soil, because they slow down runoff water after rain, allow water to infiltrate into the soil, and lessen evaporation losses. Tree roots are found to increase soil strength by 2–8 kPa depending on the tree species, while grass roots contribute 6–18 kPa, thereby decreasing the erosion potential of soils (Simon and Collison 2002). The selection of plants suitable for restoration is one of the key factors that accomplish revegetation of heavy metal-contaminated sites. Plant communities tolerant of the metals play a major role in restoration of heavy metal-contaminated sites (Banuelos and Ajwa 1999).

Often, contaminated site soils are not conducive for plant growth due to metal toxicity, lack of nutrients and microbial activity, and poor physical properties. For this reason, phytostabilization of contaminated sites may require soil amendments such as biosolids, lime, or green waste as a source of nutrients and also as a conditioner to improve soil properties and to stimulate plant growth. Further, these amendments have low concentrations of metal(loid)s. Some of these amendments also enhance the efficiency of phytostabilization by altering the solubility and bioavailability of contaminants. For example, biosolids are generally applied to cultivated land given that they contain a wide range of nutrients and carbon, thereby improving soil conditions for crop production. Liming of contaminated soil is a common practice to immobilize heavy metals and to ameliorate soils, thereby facilitating revegetation of contaminated soils. As demonstrated by Caille et al. (2004), addition of lime decreased As mobility and concentration due to the formation of Ca–As precipitates, but the effect was rather inconsistent. Rainfall and temperature also affect phytoremediation through their effects on plant growth, reactions of contaminants, and soil erosion. Because most contaminated sites may not have a ready connection to regular water supply for irrigation, rainfall plays a vital role in the establishment of vegetation. Rainfall and temperature control the leaching of contaminants and erosion of soil and sediments by affecting both the plant growth and soil surface characteristics such as cracking and crust formation. While cracking increases the leaching of contaminants, loose, dry, and bare soil is susceptible to wind erosion by dispersion.

Soil plays a significant role in controlling the immobilization and bioavailability of contaminants in the environment, thereby affecting the phytostabilization of contaminated sites. The primary soil factors influencing the immobilization and bioavailability of contaminants include soil pH, soil organic matter, cation and anion exchange capacities, texture (clay content), and soil type. Organic contaminants and most metals bind strongly to organic matter in soils, thereby reducing their bioavailability. Organic contaminants preferentially partition to the nonpolar domain of organic matter relative to the polar aqueous phase (Poerschmann and Kopinke 2001), while the organic acid functional groups typically present in organic matter have a high affinity to attract metal cations (Zaccone et al. 2009). Similarly, the promising effects observed when organic materials are added to mitigate inorganically contaminated soil may be due to fulvic and humic acid complexation, which improves the physical and chemical conditions of the soil (Kumpiene et al. 2008).

The major effects of vegetation on water and wind erosion include

*   Interception of the direct impact of raindrops and wind
*   Decreasing the velocity of runoff, and hence the cutting action of water and its capacity to entrain soil and sediment
*   Root-induced compaction and increases in soil strength, aggregation, and porosity
*   Enhancement of vegetation-induced biological activities and their influence on soil aggregation and porosity
*   Transpiration of water, leading to the subsequent drying out of the soil
*   Insulation of the soil against temperature variation that can result in cracking or "frost heave"

12.3.2   HYDRAULIC CONTROL

Hydraulic control is the term given to the use of plants to control the migration of subsurface water through the rapid uptake of large volumes of water by the plants through transpiration. Plants with desirable phenotypic and genotypic characteristics are selected for sustainable management of soil remediation. The density, morphology, and depth to which plant roots penetrate the soil are critical to potential application of this technology. The plants are effectively acting as natural hydraulic pumps, which, when a dense root network has been established near the water table in the soil, can transpire a large volume of water per day (e.g., 6 L of water/plant/m^2/day, equaling to 2190 mm per year) (Ashwath and Venkatraman 2010). This fact has been utilized to decrease the migration of contaminants from surface water into the groundwater and drinking water supplies. In addition, deep rooted plants with high transpiration rates, such as the hybrid poplar, can access soil depths of as much as 6 m (Unterbrunner et al. 2007). Fibrous roots offer a large surface area for contaminant absorption and plant–microbe interactions. They also facilitate the stabilization of soil and enhance the microbial volatilization of metals in the rhizosphere.

Plants are central to phytostabilization because plant characteristics regulate both the transformation of metals and binding of soil particles. Plants regulate the movement of contaminants through leaching and surface runoff by controlling the flow of water in soils (i.e., hydraulic control). For example, phytostabilization (i.e., phytocapping) of contaminated sites involves placing a layer of soil material and growing dense vegetation on top of the soil layer (Chen et al. 2007; Venkatraman and Ashwath 2007). The water holding capacity of the soil layer allows it to act as a "sponge" to reduce infiltration during rain events, particularly when plants are inactive. During the growing season, the evapotranspiration activity of the plants and soil surface acts as a "biopump" that reduces the moisture content of the soil layer during the events of rain and irrigation. Trapping and consuming of water in the root zone result in less volume of water acting as a vehicle to carry contaminants beyond the grasp of roots, thereby leading to their leaching into groundwater (Clothier and Green 1997). Plant-induced hydraulic control influences both diffusion and mass flow transport processes, thereby regulating the movement of contaminants in soils (Robinson et al. 2006).

Among the most extreme examples of metal-tolerant plants are the so-called hyperaccumulators, which can accumulate elements such as Zn, Mn, Ni, Co, Cd, or Se at high concentrations in their aboveground biomass (Yang et al. 2005). The major processes of metal hyperaccumulation by plants include (1) bioactivation of metals in the rhizosphere through root–microbe interaction; (2) enhanced uptake by metal transporters in the plasma membranes; (3) detoxification of metals by distributing them to the apoplast, such as binding of metals to cell walls and chelation of metals in the cytoplasm with various ligands (e.g., phytochelatins, metallothioneins, metal-binding proteins); and (4) sequestration of metals into the vacuole by tonoplast-located transporters (Yang et al. 2005). Plants suitable for restoration should be capable of developing an extensive root system and a large amount of biomass, while keeping the translocation of metals from roots to shoots as low as possible in soils with high metal concentrations (Rizzi et al. 2004). In addition, they must adapt to diverse site conditions, establish readily, and require little money and effort to maintain. Further, they must be able to survive and reproduce in contaminated soil (Flege 2000). Phytostabilization requires that the plants tolerate trace elements in the substrate. The species or varieties should also tolerate any nutrient imbalances in the substrate. Use of exotic species in this role is fraught with problems to ecosystems, as they may establish themselves as weeds. However, exotic species are less likely to suffer from native herbivores, thus increasing growth and reducing the amount of contaminant that enters into the food chain. Competition from weeds is often more problematic than soil contaminants in phytostabilization (Dickinson et al. 2009). Each contaminated site has a unique environment. Therefore, choosing the most suitable species requires a short planting trial to test several varieties in a small area of the site, particularly for non-soil media such as mine spoil or biosolids.

12.3.3 CONTAMINANT REMOVAL

Hyperaccumulating plants that are effective in the removal of metals can be used to enhance this technology (Wong 2003). However, chemical and biological redox reactions affect their bioavailability and mobility, namely, a number of reactions occur that include adsorption, complexation, precipitation, and reduction, which control their leaching, runoff losses, and bioavailability. Furthermore, these chemical interactions that contribute to metal retention by soil colloids include sorption and complexation with inorganic and organic ligands. For instance, at high soil pH and in the presence of SO_4^{2-}, CO_3^{2-}, OH^-, and HPO_4^{2-}, precipitation appears to be the predominant process when metal cation concentrations are high (Naidu et al. 1996). Nevertheless, plants perform a variety of functions intracellularly by facilitating other functions in order to uptake the metals from the soil. Shenker et al. (2001) observed the formation of phytosiderophores in roots that formed chelates with metals and initiated the solubilization of these metals in calcareous soils. Plants can release products such as organic exudates that decrease the pH, which increases the solubility of metals (Youssef and Chino 1989; Mench and Martin 1991). The role of mycorrhizal fungi is a prerequisite for soil rehabilitation, as they can increase nutrient uptake and reduce pollutants through phytoextraction (Entry et al. 1996; Meier et al. 2012). Volatilization through microbial conversion of metals into species of hydride and methylated forms is common. Among them, methylation is the most hazardous phenomenon, as it is capable of releasing poisonous methyl gas from As, Hg, and Se (Adriano et al. 2004). However, the biomethylation of Se is of interest, because methylated compounds, such as dimethyl selenide, are less toxic than dissolved Se oxyanions (Meyer et al. 2007). This may be due to the presence of organic matter in the soil that donates methyl species for biomethylation and abiotic methylation in soils and sediments. Furthermore, it has been observed that addition of phosphate leads to precipitation of metals (Zn, Pb) in the presence of other co-contaminants (He et al. 2005; Park et al. 2010). Regardless, contaminants affect phytostabilization by changing plant growth and associated microbial communities. Evidentially, due to the excess load of heavy metals, especially in mine tailings, phytostabilization and microbial association are adversely affected (Mendez and Maier 2008). In such cases, the establishment of vegetation requires soil amendments to reduce the bioavailability of phytotoxic metals.

12.3.4 RHIZOSPHERE MODIFICATION

Successful phytoremediation of organic pollutants by rhizodegradation has been confirmed for a wide range of organic and inorganic contaminants, such as hydrocarbons (Bramley-Alves et al. 2014), pesticides (Truu et al. 2015), chlorinated organics (Ghany et al. 2015), and heavy metals (Seshadri et al. 2015). Any phytoremediation process starts at the soil–plant interface, called the rhizosphere, and hence the process of heavy metal(loid) transformation mainly occurs in this region. With increasing demand for safe disposal of agricultural and industrial wastes, rhizosphere soil is not only considered as a source of nutrients for plant growth but also as a sink for the removal of contaminants from these waste materials. Rhizosphere-induced changes in soil biochemical properties regulate the transformation, mobility, and bioavailability of metals, thereby affecting the phytostabilization of contaminated sites. The major rhizosphere-induced biochemical properties that influence metal dynamics include direct effects of root-derived enzymes (Gramms and Rudeschko 1998) and indirect effects of enhanced aeration due to root burrowing and water consumption, enhanced microbial activity, and modified microbial composition due to C input from root exudates. Priming or triggering effects from metabolic precursors exuded by roots induce enzymatic activity and metabolic pathways that may attack the pollutant. In addition, unspecific effects of changes in pH, osmotic potential, redox potential, and partial pressures of O_2 and CO_2 may occur (Curl and Truelove 1986).

The major source of plant-based OH^-/H^+ fluxes affecting the pH in the rhizosphere is related to the differential uptake of cations and anions by plant roots (Tang and Rengel 2003). Acidification affects the solubility and speciation of metal ions in several ways, foremost of which include (1) modification of the surface charge in variable charge soils, (2) altering the speciation of metals, and (3) influencing the redox reactions of the metals (Adriano 2001). Adsorption of metals almost invariably decreases with increasing soil acidity (Yang et al. 2006). In the case of metalloids, such as As, the effect of soil acidity on adsorption is manifested through two interacting factors: the increasing negative surface potential on the plane of adsorption and the increasing amount of negatively charged As^{5+} species present in soil solution. While the first factor results in lower As^{5+} adsorption, the second factor is likely to increase adsorption.

In the rhizosphere, some prokaryotic (bacteria, archaea) and eukaryotic (algae, fungi) microorganisms excrete extracellular polymeric substances (EPS), such as polysaccharides, glycoprotein, lipopolysaccharide, and soluble peptides. These substances possess a substantial quantity of anion functional groups that can absorb metal ions. A number of microbes are involved in EPS production, namely, *Bacillus megaterium*, *Acinetobacter*, *Pseudomonas aeruginosa*, sulfate-reducing bacteria, and Cyanobacteria (Satpute et al. 2010). The cell wall of microbes also plays a major role in metal adsorption and redox reactions. Microbially mediated humification processes in the rhizosphere may have an important influence on the persistence and bioavailability of contaminants in surface soils.

12.4   SUMMARY AND CONCLUSIONS

Phytotechnologies are used for the rehabilitation of contaminated soils based on the combined action of plants and their associated microbial communities to degrade, remove, transform, or immobilize toxic compounds found in soils, sediments, groundwater, and surface water. These technologies have been used to treat many organically and inorganically contaminated soils and water. Phytostabilization is primarily aimed at containing the mobility of contaminants through their immobilization within the root zone of plants and "holding" soil and sediments, thereby preventing offsite contamination through their migration via wind and water erosion, leaching, and soil dispersion. Phytostabilization also results in the removal of contaminants through plant uptake and volatilization, and this technology can be enhanced by using soil amendments that are effective in the immobilization of metals. For example, this technology (i.e., phytocapping) has been found to be very effective in mitigating leachate and greenhouse gas generation in the management of landfill sites. The main advantage of this technology is that it reduces the mobility, and therefore the risk, of contaminants without necessarily removing them from their source location. Furthermore, this technology does not generate contaminated secondary waste that needs treatment, and it also provides ecosystem development to achieve biodiversity corridors. However, because the contaminants are left in place, the site requires regular monitoring in order to maintain the stabilizing conditions. If soil amendments are used to enhance immobilization, they may need to be periodically reapplied to maintain the effectiveness of the phytoimmobilization.

ACKNOWLEDGMENTS

The authors are very grateful to Abinandan Sudharsanam, PhD scholar, Global Centre for Environmental Remediation (GCER), Faculty of Science and Information Technology, University of Newcastle, Australia, for his help during the preparation of this chapter. The lead author is also very thankful to the Department of Biotechnology (DBT), Government of India, for providing financial support in the form of DBT Overseas Associateship for the year 2015–2016. He is indebted to the Indian Council of Agricultural Research (ICAR), New Delhi, India, and ICAR Research Complex for North-Eastern Hill (NEH) Region, Meghalaya, India, for allowing him to pursue postdoctoral research program at the Global Centre for Environmental Remediation (GCER), Faculty of Science and Information Technology, University of Newcastle, Australia.

REFERENCES

Adriano, D. C. 2001. *Trace Elements in Terrestrial Environments: Biogeochemistry, Bioavailability and Risks of Metals*, 2nd edn. Springer, New York.

Adriano, D. C., Wenzel, W. W., Vangronsveld, J., and Bolan, N. S. 2004. Role of assisted natural remediation in environmental cleanup. *Geoderma* 122:121–142.

Ali, H., Khan, E., and Sajad, M. A. 2013. Phytoremediation of heavy metals—Concepts and applications. *Chemosphere* 91:869–881.

Ashwath, N. and Venkatraman, K. 2010. Phytocapping: An alternative technique for landfill remediation. *Int. J. Environ. Waste Manag.* 6:51–70.

Bañuelos, G. S. and Ajwa, H. A. 1999. Trace elements in soils and plants: An overview. *J. Environ. Sci. Health Tox. Hazard. Subst. Environ. Eng.* 34:951–974.

Bolan, N. S., Adriano, D. C., Natesan, R., and Koo, B. J. 2003. Effects of organic amendments on the phytoavailability of chromate in mineral soil. *J. Environ. Qual.* 32:120–128.

Bramley-Alves, J., Wasley, J., King, C., Powell, S., and Robinson, S. A. 2014. Phytoremediation of hydrocarbon contaminants in subantarctic soils: An effective management option. *J. Environ. Manage.* 142:60–69.

Caille, N., Swanwick, S., Zhao, F. J., and McGrath, S. P. 2004. Arsenic hyperaccumulation by *Pteris vittata* from arsenic contaminated soils and the effect of liming and phosphate fertilisation. *Environ. Pollut.* 132:113–120.

Castillejo, J. M. and Castello, R. 2010. Influence of the application rate of an organic amendment (Municipal Solid Waste [MSW] Compost) on gypsum quarry rehabilitation in semiarid environments. *Arid Land Res. Manag.* 24:344–364.

Chen, L., Huang, Z., Gong, J., Fu, B., and Huang, Y. 2007. The effect of land cover/vegetation on soil water dynamic in the hilly area of the loess plateau, China. *Catena* 70:200–208.

Clothier, B. E. and Green, S. R. 1997. Roots: The big movers of water and chemical in soil. *Soil Sci.* 162:534–543.

Curl, E. A. and Truelove, B. 1986. *The Rhizosphere*. Springer Verlag, Berlin, Germany.

Dickinson, N., Baker, A., Doronila, A., Laidlaw, S., and Reeves, R. 2009. Phytoremediation of inorganics: Realism and synergies. *Int. J. Phytoremediat.* 11:97–114.

Ensley, B. D. 2000. Rationale for use of phytoremediation. In *Phytoremediation of Toxic Metals: Using Plants to Clean Up the Environment*, eds. I. Raskin and B. D. Ensley, pp. 3–11. John Wiley & Sons, Inc., New York.

Entry, J. A., Vance, N. A., Hamilton, M. A., Zabowsky, D., Watrud, L. S., and Adriano, D. C. 1996. Phytoremediation of soil contaminated with low concentrations of radionuclides. *Water Air Soil Pollut.* 88:167–176.

Erakhrumen, A. A. 2007. Phytoremediation: An environmentally sound technology for pollution prevention, control and remediation in developing countries. *Educ. Res. Rev.* 2:151–156.

Flege, A. 2000. Forest recultivation of coal-mined land: Problems and prospects. In *Reclaimed Land*, ed. M. J. Haigh, pp. 291–337. A.A. Balkema, Rotterdam, the Netherlands.

Frers, C. 2009. *El uso de plantas acuáticas en el tratamiento de aguas residuales*. El Planeta Azul, Carmen de Areco, Argentina.

Ghany, A., Al Abboud, M. A., Negm, M. E., and Shater, A. R. M. 2015. Rhizosphere microorganisms as inducers for phytoremediation—A review. *Int. J. Bioinform. Biomed. Eng.* 1(1):7–15.

Gramms, G. and Rudeschko, O. 1998. Activities of oxidoreductase enzymes in tissue extracts and sterile root exudates of three crop plants, and some properties of the peroxidase component. *New Phytol.* 138:401–409.

He, Z. L., Yang, X. E., and Stoffella, P. J. 2005. Trace elements in agroecosystems and impacts on the environment. *J. Trace Elem. Med. Biol.* 19:125–140.

Hossner, L. R. and Hons, F. M. 1992. Reclamation of mine tailings. *Adv. Soil Sci.* 17:311–350.

Kumpiene, J., Lagerkvist, A., and Maurice, C. 2008. Stabilization of As, Cr, Cu, Pb and Zn in soil using amendments—A review. *Waste Manage.* 28:215–225.

Manousaki, E. and Kalogerakis, N. 2011. Halophytes present new opportunities in phytoremediation of heavy metals and saline soils. *Ind. Eng. Chem. Res.* 50:656–660.

Meier, S., Borie, F., Bolan, N., and Cornejo, P. 2012. Remediation of metal polluted soils by arbuscular mycorrhizal fungi. *Crit. Rev. Environ. Sci. Technol.* 42:741–775.

Mench, M. and Martin, E. 1991. Mobilization of cadmium and other metals from 2 soils by root exudates of *Zea mays, Nicotiana tabacum* and *Nicotiana rustica. Plant Soil* 132:187–196.

Mench, M., Schwitzguebel, J. P., Schroeder, P., Bert, V., Gawronski, S., and Gupta, S. 2009. Assessment of successful experiments and limitations of phyto-technologies: Contaminant uptake, detoxification and sequestration, and consequences for food safety. *Environ. Sci. Pollut. Res.* 16:876–900.

Mendez, M. O. and Maier, R. M. 2008. Phytostabilization of mine tailings in arid and semiarid environments—An emerging remediation technology. *Environ. Health Perspect.* 116:278.

Meyer, J., Schmidt, A., Michalke, K., and Hensel, R. 2007. Volatilization of metals and metalloids by the microbial population of an alluvial soil. *Syst. Appl. Microbiol.* 30:229–238.

Mukhopadhyay, S. and Maiti, S. K. 2010. Phytoremediation of metal enriched mine waste: A review. *Global J. Environ. Res.* 4:135–150.

Naidu, R., Kookana, R. S., Sumner, M. E., Harter, R. D., and Tiller, K. G. 1996. Cadmium adsorption and transport in variable charge soils: A review. *J. Environ. Qual.* 26:602–617.

Park, J. H., Bolan, N. S., Mallavarapu, M., and Naidu, R. 2010. Comparative value of phosphate sources on the immobilization of lead, and leaching of lead and phosphorus in lead contaminated soils. *Sci. Total Environ.* 409(4):853–860.

Pavlik, T., Puzder, M., Benčo, G., and Mudarri, T. 2015. Analysis of suitable bi helios green as an effective business intelligence. *Int. J. Interdiscip. Theory Pract.* 6:72–77.

Pedron, F., Petruzzelli, G., Barbafieri, M., and Tassi, E. 2009. Strategies to use phytoextraction in very acidic soil contaminated by heavy metals. *Chemosphere* 75:808–814.

Poerschmann, J. and Kopinke, F. D. 2001. Sorption of very hydrophobic organic compounds (VHOCs) on dissolved humic organic matter (DOM). 2. Measurement of sorption and application of a Flory-Huggins concept to interpret the data. *Environ. Sci. Technol.* 35:1142–1148.

Pratas, J., Favas, P. J. C., Paulo, C., Rodrigues, N., and Prasad, M. N. V. 2012. Uranium accumulation by aquatic plants from uranium-contaminated water in Central Portugal. *Int. J. Phytoremediat.* 14:221–234.

Rafati, M., Khorasani, N., Moattar, F., Shirvany, A., Moraghebi, F., and Hosseinzadeh, S. 2011. Phytoremediation potential of *Populus alba* and *Morus alba* for cadmium, chromuim and nickel absorption from polluted soil. *Int. J. Environ. Res.* 5:961–970.

Ravindran, K. C., Venkatesan, K., Balakrishnan, V., Chellappan, K. P., and Balasubramanian, T. 2007. Restoration of saline land by halophytes for Indian soils. *Soil Biol. Biochem.* 39:2661–2664.

Rizzi, L., Petruzzelli, G., Poggio, G., and Guidi, G. V. 2004. Soil physical changes and plant availability of Zn and Pb in a treatability test of phytostabilization. *Chemosphere* 57:1039–1046.

Robinson, B., Schulin, R., Nowack, B., Roulier, S., Menon, M., Clothier, B., Green, S., and Mills, T. 2006. Phytoremediation for the management of metal flux in contaminated sites. *For. Snow Landsc. Res.* 80:221–234.

Robinson, B. H., Bañuelos, G., Conesa, H. M., Evangelou, M. W. H., and Schulin, R. 2009. The phytomanagement of trace elements in soil. *Crit. Rev. Plant Sci.* 28:240–266.

Ruiz, O. N. and Daniell, H. 2009. Genetic engineering to enhance mercury phytoremediation. *Curr. Opin. Biotech.* 20:213–219.

Rylott, E. L. and Bruce, N. C. 2008. Plants disarm soil: Engineering plants for the phytoremediation of explosives. *Trends Biotechnol.* 27(2):73–81.

Satpute, S. K., Banat, I. M., Dhakephalkar, P. K., Banpurkar, A. G., and Chopade, B. A. 2010. Biosurfactants, bioemulsifiers and exopolysaccharides from marine microorganisms. *Biotechnol. Adv.* 28:436–450.

Seshadri, B., Bolan, N. S., and Naidu, R. 2015. Rhizosphere-induced heavy metal(loid) transformation in relation to bioavailability and remediation. *J. Soil Sci. Plant Nutr.* 15(2):524–548.

Shenker, M., Fan, T. W. M., and Crowley, D. E. 2001. Phytosiderophores influence on cadmium mobilization and uptake by wheat and barley plants. *J. Environ. Qual.* 30:2091–2098.

Simon, A. and Collison, A. J. C. 2002. Quantifying the mechanical and hydrologic effects of riparian vegetation on streambank stability. *Earth Surf. Proc. Landf.* 27:527–546.

Singh, A. N., Raghubanshi, A. S., and Singh, J. S. 2004. Comparative performance and restoration potential of two *Albizia* species planted on mine spoil in a dry tropical region, India. *Ecol. Eng.* 22:123–140.

Tang, C. and Rengel, Z. 2003. Role of plant cation/anion uptake ratio in soil acidification. In *Handbook of Soil Acidity*, ed. Z. Rengel, pp. 57–81. Marcel Dekker, New York.

Tangahu, B. V., Abdullah, S. R. S., Basri, H., Idris, M., Anuar, N., and Mukhlisin, M. 2011. A review on heavymetals (As, Pb, and Hg) uptake by plants through phytoremediation. *Int. J. Chem. Eng.* 21:1–31.

Truu, J., Truua, M., Espenberga, M., Nõlvaka, H., and Juhanson, J. 2015. Phytoremediation and plant-assisted bioremediation in soil and treatment wetlands: A review. *Open Biotechnol. J.* 9:85–92.

Unterbrunner, R., Puschenreiter, M., Sommer, P., Wieshammer, G., Tlustos, P., Zupan, M., and Wenzel, W. W. 2007. Heavy metal accumulation in trees growing on contaminated sites in Central Europe. *Environ. Pollut.* 148:107–114.

Venkatraman, K. and Ashwath, N. 2007. Phytocapping: An alternative technique for reducing leachate and methane generation from municipal landfills. *Environmentalist* 27:155–164.

Vithanage, M., Dabrowska, B. B., Mukherjee, B., Sandhi, A., and Bhattacharya, P. 2012. Arsenic uptake by plants and possible phytoremediation applications: A brief overview. *Environ. Chem. Lett.* 10:217–224.

Wong, M. H. 2003. Ecological restoration of mine degraded soils, with emphasis on metal contaminated soils. *Chemosphere* 50:775–780.

Wuana, R. A. and Okieimen, F. E. 2011. Heavy metals in contaminated soils: A review of sources, chemistry, risks and best available strategies for remediation. *ISRN Ecol.* 2011:1–20.

Yadav, R., Arora, P., Kumar, S., and Chaudhury, A. 2010. Perspectives for genetic engineering of poplars for enhanced phytoremediation abilities. *Ecotoxicology* 19:1574–1588.

Yang, J. Y., Yang, X. E., He, Z. L., Li, T. Q., Shentu, J. L., and Stoffella, P. J. 2006. Effects of pH, organic acids, and inorganic ions on lead desorption from soils. *Environ. Pollut.* 143:9–15.

Yang, X., Feng, Y., He, Z., and Stoffella, P. J. 2005. Molecular mechanisms of heavy metal hyperaccumulation and phytoremediation. *J. Trace Elem. Med. Biol.* 18:339–353.

Youssef, R. A. and Chino, M. 1989. Root-induced changes in the rhizosphere of plants. 2. Distribution of heavy-metals across the rhizosphere in soil. *J. Plant Nutr. Soil Sci.* 35:609–621.

Zaccone, C., Soler-Rovira, P., Plaza, C., Cocozza, C., and Miano, T. M. 2009. Variability in As, Ca, Cr, K, Mn, Sr, and Ti concentrations among humic acids isolated from peat using NaOH, $Na_4P_2O_7$ and NaOH + $Na_4P_2O_7$ solutions. *J. Hazard. Mater.* 167:987–994.

13 Phytocapping of Mine Waste at Derelict Mine Sites in New South Wales

Dane Lamb, Peter Sanderson, Liang Wang,
Mohammed Kader, and Ravi Naidu

CONTENTS

13.1 Introduction 215
13.2 Issues at Derelict Mine Sites 216
13.3 Remediation and Management Options 217
13.4 Capping Approaches at Derelict Mines 222
    13.4.1 Webbs Consols, New South Wales 222
    13.4.2 Mole River, New South Wales 225
13.5 Phytocapping Technology 226
    13.5.1 Irrigation 227
    13.5.2 Cap-Waste Interactions 228
    13.5.3 Plant Species for Phytocapping 229
    13.5.4 Phytocapping Enhancement With Soil Amendments 230
13.6 Conclusions and Future Research 233
References 233

13.1 INTRODUCTION

Historically, mining of metalliferous ore bodies was a relatively dispersed activity, with numerous small mines occurring throughout many western countries including the United States, the United Kingdom, and Australia (Soucek et al. 2000, Grant et al. 2002, Mayes et al. 2009). Many metalliferous mine sites began operation in the late eighteenth and early nineteenth centuries and were abandoned in most instances before the environmental movement in Western countries. As such, there was very little recognition of the potential impacts caused by the dispersal of metal toxicants such as arsenic (As), cadmium (Cd), copper (Cu), lead (Pb), and zinc (Zn) into the surrounding environments from these sites. Many of these contaminants are cariogenic in humans (e.g., As), cause a range of human health–related impacts (Pb, Cd), and are toxic to ecological receptors in nearby streams and surrounding terrestrial environments (Cu, Zn, Mn, Ni). As a result of the lack of regard for potential impacts, much of the mining waste was discarded carelessly throughout mining sites, and in some cases, directly into nearby watercourses.

Such mine sites are considered to be "derelict." Derelict mine sites are defined broadly as sites with no clear attribution of responsibility for rehabilitation, resulting in government organizations taking on the costly burden (Grant et al. 2002). Bussière (2010) defined abandoned or derelict mines as any site having "… no solvable identifiable owners or operators for the facilities, or if the facilities have reverted to governmental ownership." Management and remediation of these sites are extremely challenging, as public funding is often highly limited.

One of the most cost-effective approaches to dealing with these abandoned mines in a reasonable time frame is to relocate and isolate the waste in landfills or cover them with clean fill. The nature of materials (e.g., highly acid, metal rich) means that, for many areas of a given mine, innovative remediation approaches, such as phytoremediation, are not cost effective, efficient, nor can they be achieved in a reasonable time frame. Wastes from abandoned metalliferous mines are often vast in volume, highly contaminated material (percentage concentrations), contain high levels of existing and latent acidity, and are located near sensitive environments.

Covering the waste on-site provides a simple and economic means of waste disposal. The benefits of capping contaminated material at old mine sites include removal of contaminants from water flow paths, minimization of wind and water erosion, and reduced risk of exposure to contaminants among human and ecological receptors. However, inadequate design and management may result in serious degradation of the environment and many of the benefits are reduced. Surface and groundwater contamination by leachate, which is enriched with a range of contaminants, occurs even in municipal landfill sites and can lead to the loss of a valuable resource (El-Fadel et al. 1997). Government agencies often cannot afford expensive engineered landfills for waste at all abandoned mines, thereby potentially resulting in environmental pollution due to failed containment of waste. The capping technologies applied to many derelict mine sites are often overly simplistic and are not adequately designed to avoid contaminant movement.

Landfill covers are used to reduce the quantity of water that percolates into and control the release of leachates and gases from closed landfills. They are designed to lower the risk of groundwater contamination from leachate generation and the environmental and health impacts of gaseous emissions into the environment. Traditionally, landfill covers have been designed to minimize water entry through the use of low permeability layers (e.g., compacted clay caps and geosynthetic liners). The construction of these covers is extremely costly and nonsustainable, as their performance can deteriorate with time (Albright et al. 2004). At derelict mine sites, the "capping" of waste may simply involve concentrating material in a defined zone and covering the waste with nearby clay or soil. Alternative cover systems, such as phytocaps, are increasingly being considered for use in a range of waste disposal sites, including municipal and mine sites. Phytocapping involves placing a layer of soil material and growing dense vegetation on top of a landfill. In this system, soil acts mainly as the "storage" to hold and store water, and releases it to the plants when they are in need.

Phytocapping encompasses phytoremediation technology that includes phytoextraction, rhizoextraction, phytovolatilization, and phytostabilization, and is gaining social acceptance as a green and environment-friendly technology (Lamb et al. 2013). Unlike municipal landfills, which are often located near urban zones, capping of materials at derelict mine sites are often located in isolated areas. The growing recognition of the role that plants and other biota can potentially play in sustainable management of waste disposal has resulted in an increased interest in phytocapping technology. To date, this has largely focused around municipal sites. The following section will firstly review the issues associated with derelict mine sites, detail two examples of current capping at sites in Australia, briefly review available remediation techniques for metal(loid)s, and finally review the literature on phytocaps and their potential applications at derelict mine sites.

13.2   ISSUES AT DERELICT MINE SITES

Historically, mining operations across Australia operated and closed with little regard for the fate of environmental pollutants. Between 500 and 20,000 derelict mine sites are estimated to exist in New South Wales alone (Grant et al. 2002, Archer and Caldwell 2004). The problem of derelict mines is international. There are more than 550,000 derelict mine sites in the United States alone (Soucek et al. 2000), >10,000 in Canada, 5,500 in Japan, and >10,000 abandoned metalliferous mines in the United Kingdom (Mayes et al. 2009). Serious human and environmental health impacts have been reported across all inhabited continents, in some cases with likely linkages between health problems and derelict mine sites (Zhuang et al. 2009). Many derelict mine sites around the world are metalliferous and/or pyritic, and as such have a range of inorganic contaminants, including high levels of acidity, cyanide, metalloids (arsenic, antimony, chromium), and metals (e.g., cadmium, copper, lead, zinc, iron, aluminium, and manganese).

Derelict metalliferous mine sites are typically characterized by (1) the presence of high to extremely high concentrations of metal(loid)s, (2) existing acid mine drainage (AMD) issues and subsequent high levels of acidity in soil and runoff, (3) latent acidity due to remaining pyritic and FeS mineral phases, and (4) high levels of soluble inorganic constituents arising from acidity, including sulphate and other toxicants due to the dissolution of soil/ore minerals such as Fe, Al, and Mn. Table 13.1 presents a selection of different derelict mine sites from around the world. The total concentrations vary between sites. However, the concentrations are often reported to be exceedingly high. Lim et al. (2008) reported As concentrations in the range of 3,500–143,000 mg kg^{-1}, Aslibekian and Moles (2003) and MacKenzie and Pulford (2002) report Pb concentrations in the range of 985–210,000 mg kg^{-1} in Ireland and Scotland. Similarly, in Australia, exceptionally high concentrations have been observed at many derelict mine sites (Ashley and Lottermoser 1999, Ashley et al. 2004).

Common issues associated with derelict metalliferous mines are issues associated with acid mine drainage (AMD). The extracted ore materials are commonly sulphidic in nature, containing metal sulphides, including pyrite. Upon extraction from the earth, pyritic material oxidizes upon initial exposure to oxygen, resulting in substantial generation of acidity (Nordstrom 2011). Oxidation of pyritic and metal-bearing sulphides is well known to be associated with low pH of tailings, seepages, and receiving streams in the range of 2–4, and in extreme cases, substantially higher H^+ activities. Indeed, pH values as low as −3.6 have been recorded at the Richmond Mountain mine (Iron Mountain Ming) in California, United States (Nordstrom et al. 2000). The high levels of acidity result in high solubility of most metals, including Al, Mn, and Fe. In some cases, such as the Cowarra mine (New South Wales, Australia), high acidity results in significant solubility of metals even at low total metal loadings (Table 13.1). Off-site movement of high acidity and metal loading represents a major risk factor to local communities. For example, the Dabaoshan mine in China has resulted in substantial contamination of the Hengshi River. Its use for irrigation in local gardens and drinking water has been associated with high rates of liver, oesophageal, and other cancers (Zhuang et al. 2009, 2013). In New South Wales, Australia, the problem of exposure to toxic and carcinogenic metal(loid)s from derelict sites also exists. However, in general the exposure pathway may or may not be as direct as experienced in other areas of the world.

Rehabilitation and "cleanup" of derelict mine sites is typically associated with extreme cost. Governments are invariably left with the cleanup cost and are faced with a huge expense with little available funds for rehabilitation. One of the primary concerns in derelict mine sites is public safety and human health impacts from contamination. Although the presence of contaminants at a given site is of concern, the urgency for rehabilitation and action depends primarily on exposure to human and ecological receptors of contaminants. It is thus important to recognize that although a hazard may exist, the associated risk to human health and the environment from contamination may be low. Unfortunately, the detailed information required to form informed decisions about risk at derelict mine sites is not available. Detailed information of contaminant loadings in soil and water are often not available, and certainly information on contamination of groundwater is uncommon.

13.3 REMEDIATION AND MANAGEMENT OPTIONS

Risk-based land management (RBLM) is based on a scientific approach that is focused around exposure of receptors to contaminants (Naidu et al. 2008a,b). Critical elements of risk-based management are source characterization, clear demarcation of potential receptors, and exposure pathways. Exposure of contaminants to a given receptor involves numerous processes, including

TABLE 13.1

Examples of Metal(loid) Contamination, pH, and Solubility at Derelict Mines Globally

Mine Name (Elements Extracted)	Country	Total Concentration (mg kg⁻¹)	Soluble Concentration (mg L⁻¹)	pH	Notes	References
Songcheon (Gold-silver)	South Korea	As 3,500–143,000 Cu 30–749 Pb 125–50,800 Mn n.a. Zn 580–7,540	As 0–0.8 Cu 0.007–0.017 Pb 0.0001–0.005 Mn n.a. Zn 0.002–5.4	n.a.		Lim et al. (2008)
Dabaoshan mine	China	As n.a. Cu 88.0–1,310 Pb 109–621 Mn n.a. Zn 170–1,663	n.a	4.8–6.9	Agricultural soils near mine	Zhuang et al. (2009)
Silvermines (Zinc, Lead, Barium)	Ireland	As n.a. Cu n.a Pb 985–4,840 Mn n.a. Zn 500–8,920	n.a.	4.8–5.7		Aslibekian and Moles (2003)
Tyndrum	United Kingdom, Scotland	Pb 13,000–210,000 Zn 5,500–34,000			Mine waste	MacKenzie and Pulford (2002)
	Pennsylvania, USA		As 0–0.06 Cu 0.0004–0.140 Pb 0.0–0.011 Mn 0.019–74 Zn 0.0006–10	n.a.		Cravotta III (2008)
Cowarra Gold	Australia, NSW	As 50–857 Cu 11–130 Pb 10–105 Mn 93–265 Zn 16–105	As 0–0.009 Cu 0.003–8.6 Pb 0–0.03 Mn 0.04–12.1 Zn 0.002–1.08	Soil pH 3.3–5.0		Unpublished
Webbs Consols (Silver, Lead, Zinc)	Australia, NSW	As 26–283,000 Cu 6–16,000 Pb 39–124,000 Mn 94–5,210 Zn 29–299,000	As 18.5 (mean) Cu 0.28 Pb 1.33 Mn 2.61 Zn 39.6 Mine waters	Stream pH 1.9–4.9		Ashley et al. (2004)

n.a., data not available.

transportation, uptake of contaminants, likelihood and frequency of contact, and bioavailability from environmental media. Risk assessment of contaminated sites involves characterization of potential impacts posed by contaminants to ecological and human health. Risk assessment of contaminated sites considers (1) groundwater and surface water resources, (2) human health, and (3) ecological health.

There are numerous examples of risk assessments conducted on derelict mine sites internationally. However, most risk assessments have focused specifically on detailed human health and ecological risk assessments (ERA), rather than the full process of risk-based management of sites. For example, Lim et al. (2008), Zhuang et al. (2009), and Aslibekian and Moles (2003) conducted human health risk assessments (HHRA) using bioaccessibility and plant accumulation data. Similarly, detailed ecological risk assessments have been reported for metalliferous abandoned mines (Jasso-Pineda et al. 2007, Antunes et al. 2008, Min et al. 2013). Many human health assessment studies exaggerate the actual risk, as the assumption of site multifunctionality is applied; this includes the use of derelict metalliferous mine sites for vegetable production. Although many studies report an unrealistic characterization of risk, the incorporation of bioavailability into assessments is indicative of the path forward. Similar to comprehensive rehabilitation, detailed risk assessments (ecological especially) are extremely costly and a tiered approach is needed to avoid unnecessary expenditure.

Risk-based land management is a suitable approach for management of derelict mine sites. The primary reason is that RBLM utilizes several stages of assessment within a single site or among sites, allowing prioritization of funds and assessment of remediation options over time. RBLM for derelict mine sites has been applied in the UK for prioritization of sites across the country. The methodology allows for identification of sites of high risk and selective use of public funds for remediation where possible or indeed necessary (Mayes et al. 2009). A similar approach has been applied in New South Wales for prioritization between sites. In both cases, within a given site further development and prioritization is needed. In particular, an approach is needed that underpins remedial options and decisions, especially in situations when high capital expenditure is being considered.

Following identification of high risk sites and areas, remediation within a site may be required. The available remediation technologies can be grouped into two categories: (1) *ex situ* techniques that require removal of the contaminated soil or groundwater for treatment either on-site or off-site, and (2) *in situ* techniques, that attempt to remediate without excavation of contaminated soils (Table 13.2). Generally *in situ* techniques are favored over *ex situ* techniques because of (1) reduced costs due to elimination or minimization of excavation, transportation to disposal sites, and sometimes treatment itself, (2) reduced health impacts on the public or the workers, and (3) the difficulty in remediating inaccessible sites, for example, those located at greater depths or under buildings. Although *in situ* techniques have been highly successful with organic contaminated sites, the success of *in situ* strategies with metal contaminants has been limited. Given that organic contamination often occurs as mixtures, a combination of more than one strategy is required to either successfully remediate or manage metal contaminated soils. Strategies used for the remediation and management of highly contaminated sites vary substantially depending on the nature of contamination, the threat contaminants pose to animal and human systems, and also the commercial value of the land. The full-scale *ex situ* or *in situ* remediation techniques (including the emerging new technologies) that have been employed to clean contaminated soils (Table 13.2) are noted.

In situ immobilization is now considered an attractive cost-effective technique for managing metal contaminated soils. The technique relies on the addition of an amendment to a soil to increase the proportion of metal sorbed in the soil solid phase through either precipitation or increased metal sorption (Oste et al. 2002). The aim of the immobilization process is not to remove the metal

TABLE 13.2

Cost-Effectiveness of Various Remediation Technologies for Metal(loid)s

Technology	Cost-Effectiveness	Scale of Operations	Range of Metals
Immobilization		Field scale-*in situ*	Broad
Electroremediation	$US50 to $US150 m^{-3}	Pilot scale	Broad
Phytoremediation	$US60,000 to $US100,000 ha^{-1}	Pilot scale	Broad
Solidification		Field	Broad
Excavation/washing		Largely ex situ on-site	Broad

contaminant from the soil, but to reduce its availability and activity. Since *in situ* immobilization was recognized to be an effective strategy for managing metal contaminated soils, many different amendments have been tested for the immobilization of heavy metals in soils. Such amendments include agricultural products such as lime (Geebelen et al. 2003), phosphate (Hettiarachchi and Pierzynski 2002, Melamed et al. 2003), and organic matter (Brown et al. 2003, 2005, Farfel et al. 2005), including various industrial products (Mench et al. 1994, Oste et al. 2002, Geebelen et al. 2003).

Phytoremediation technology is based on the capacity of the plants to remove, immobilize, or render the metal contaminants harmless. Numerous researchers (Baker et al. 1991, Brown et al. 1994, Raskin et al. 1994, Cunningham and Lee 1995) have suggested that higher plants, particularly the hyper-accumulator plants, have the potential to remove toxic metals from contaminated soils. While phytoremediation relies on hyper-accumulator plants to transport metals from contaminated soils to aboveground plant components, phytostabilization uses plants to limit the mobility and bioavailability of metals in soils. Such plants are capable of tolerating high metal concentrations in soil and are able to immobilize metals through precipitation, complexation, or the reduction/oxidation of the metal (e.g., Cr^{VI} to Cr^{III}). The rhizofiltration plants use roots to concentrate metals and these plants have been largely used to remediate metal containing wastes, often in constructed wetlands or reed beds (Ensley 1995). Table 13.3 lists the advantages and disadvantages of the various types of phytoremediation technologies. A substantial barrier to the applicability of phytoremediation at derelict mines is the common co-occurrence of extreme acidity and metal(loid) solubility (Nordstrom et al. 2000, Nordstrom 2011). Phytoremediation, therefore, when considered in terms of its different forms in Table 13.3, is applicable only to certain zones with a derelict mine site.

Electroremediation is an emerging technology that involves the application of a low density direct current between electrodes placed in the soil to mobilize contaminants in the form of a charged species (Will 1995). The technology has been found to be particularly useful to remove heavy metals

TABLE 13.3
Advantages and Disadvantages of Phytoremediation Technology

Type of Phytoremediation	Advantages	Disadvantages
Phytoextraction by trees	High biomass production	Potential for off-site migration and leaf transportation of metals to surface
		Metals are concentrated in plant biomass and must be disposed of eventually
Phytoextraction by crops	High biomass and increased growth rate	Potential threat to the food chain through ingestion by herbivores
Phytoextraction by grasses	High accumulation	Low biomass production and slow growth
		Metals are concentrated in plant biomass and must be disposed of eventually
Phytostabilization	No disposal of contaminated biomass required	Remaining liability issues, including maintenance for indefinite period of time (containment rather than removal)
Rhizofiltration	Readily absorbs metals	Application for treatment of water only
		Metals are concentrated in plant biomass and must be disposed of eventually

Source: Adapted from USEPA, *Recent Developments for In Situ Treatment of Metal Contaminated Soils*, Office for Solid Waste and Emergency Response, Washington, DC, 1997.

from contaminated fine grained soils with a low hydraulic permeability (Mulligan et al. 2001, Page and Page 2002). Acar and Malkoc (2004) demonstrated the critical role of soil pH and buffer capacity on Pb mobility in soils. At low pH, between 75% and 95% of Pb was removed from kaolin clay while very little Pb was removed when 13% Ca was present because of the high buffer capacity of Ca. Despite these studies and those quoted by Wise (1994), the application of this technique has been limited to only a few sites in Europe and South Korea. The USEPA (1997) summarized the performance of electrochemical remediation technology at five field sites in Europe. The lack of application of this technology in other areas may partly be related to the limited research on long-term contaminated soils and the refinement required for the specific properties of each contaminated site. Another major barrier to the employment of electrolysis-based cleanup methods is that (1) the large changes in pH at the cathode and anode and (2) the need for addition of a compound to promote mobility (Page and Page 2002). One potential benefit is the preexisting acidity, which promotes metal mobility and also avoids the costs of purchase and transportation of chemicals (Rojo et al. 2009, Hansen et al. 2013). There is a need to assess this technology further, especially using long-term contaminated soils under conditions similar to those encountered in the field. Presently the field application of electroremediation in isolated areas is restricted by demonstrated performance and costs of deployment.

The choice of remediation technology for the cleanup of contaminated land depends on a number of factors, including the extent of the contamination, concentrations of contaminant, health risks associated with the contaminants, the available remediation technologies, and socioeconomic factors. Within a given derelict mine zone, no single remediation technology is suitable. For instance, diffuse broad-acre contamination (e.g., from smelting works or alluvial deposition of contaminants) may be best dealt with by the immobilization processes where the bioavailability of the contaminants can be minimized. However, phytoremediation technology is now being considered as an alternative, given that it removes metals from the contaminated soil. The biomass production of metal accumulator plants are generally low and this may limit the capacity for the removal of metals to approximately $10–400$ kg ha^{-1} year^{-1} depending on the pollutant, plant species, climate, and other factors (USEPA 1997). Thus, for a depth of 30 cm, this capacity amounts to a reduction of $2.5–100$ mg kg^{-1} soil year^{-1} of soil contaminants. Therefore, the removal rate using phytoremediation may take decades for complete cleanup of contaminated soils (DOE 1994). Many derelict mine sites are not strongly regulated with regard to access, and the high plant shoot concentrations of metals may pose additional risks. The cost associated with erecting entry barriers add to additional practical difficulties associated with direct remediation with hyperaccumulators. As such, phytoremediation may not be practical for remediation of many areas of derelict mines, including highly contaminated areas, highly acidified zones, and also sites existing in arid areas (Arnold et al. 2015).

Alternative remedial options include excavation and disposal (expensive, and may pose risk to environment and human health), soil washing, or electroremediation. However, application of a soil washing process may depend on the range of metal contaminants and the cost can generally be greater than the phytoremediation technique and almost 50% of the pump-and-treat method (USEPA 1997). The factors that most significantly affect costs are the initial and target concentrations of contaminants, permeability of the soil, and the depth to which the contaminant has migrated. For a detailed estimate of the cost of soil-washing technique, readers are directed to Moore and Matsumoto (1993). The cost of applying soil-washing or electroremediation technologies varies with the nature of the contaminants and soil properties, although the cost of electricity is important for the latter process. Electroremediation could result in direct costs of \$US50 m^{-3} or more (USEPA 1997), although costs ranging from \$US90 to \$US150 are cited for projects carried out in Europe by Geokinetics International Inc. Due to cost restrictions for many derelict mining sites, isolation and storage is often the preferred option depending on the perceived risk at the site. However, the design and management of the integrity of the capping technologies in many cases may be improved considering emerging technologies.

13.4   CAPPING APPROACHES AT DERELICT MINES

Municipal landfill design is primarily focused on reducing water ingress into the cap to reduce gas and leachate generation (Lamb et al. 2013). Capping designs at derelict mines are more focused on reducing the dispersal via erosion and human exposure to contaminants associated with waste (Figure 13.1). Contaminated wastes are of concern due to direct exposure of people accessing the site, ecological impacts on-site, or in aquatic environments downstream. Similarly, dispersal of contaminations via erosion and hydrogeological transportation is a significant source which, left untreated, may or may not pose a risk to downstream users of the water. Illustrative examples of existing caps that have been erected to assist in restricting contaminant movement to streams, humans, and ecological receptors are provided below.

13.4.1   Webbs Consols, New South Wales

The Webbs Consols Mine is a derelict lead-zinc-silver mine, situated in the New England region of northern New South Wales. Small-scale mining at the site has been intermittent from the mid-1880s to the mid-1950s (Appleton 2003). The valley of the Webbs Consols Creek hosts four main shafts and two small processing areas. A fifth significant shaft (the Lucky Lucy) lies north of the valley, in a catchment which drains into Webbs Consols.

Associated waste lies in the vicinity of each shaft and processing area and consists of heterogeneous waste rock and tailings. The host rock includes granite, basalt, volcanics, and siltstone. The mineral ore contains high concentrations of lead, zinc, silver, copper, arsenic, and sulphides (Ashley et al. 2004).

With little environmental management or acid-buffering material, the mining waste has produced acid mine drainage and mobilized metals and metalloids to contaminate soil and water for approximately 15 ha surrounding the Webbs Consols Mine (Ashley et al. 2004). During an initial visit to the site in April 2016, a portable X-ray fluorescence spectrometer (XRF) was used to take preliminary contaminant measurements. The analysis revealed widespread and a high level of contamination throughout the site (see Figure 13.2). Concentrations were always highest around processing and

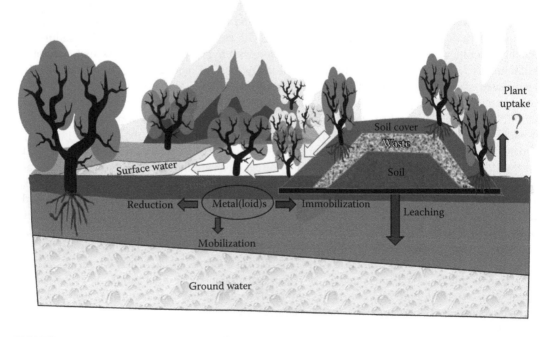

FIGURE 13.1   Schematic representation of processes controlling metal(loid) movement at derelict mine sites.

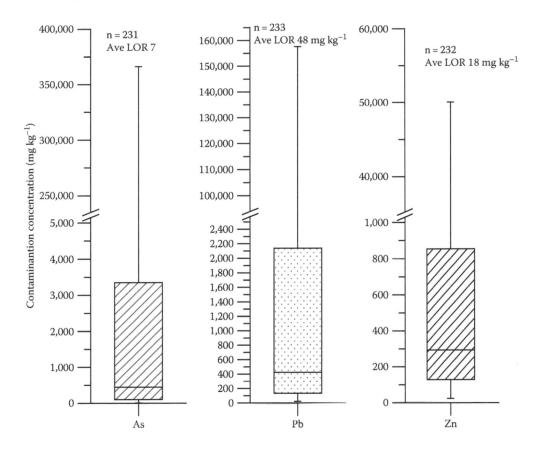

FIGURE 13.2 Arsenic, Pb, and Zn concentrations across the Webbs Consols derelict mine site, New South Wales, Australia. Values determined on-site using portable x-ray fluorescence instrumentation (Model: Olympus, Delta Premium). LOR, limit of reporting (error associated with XRF measurement). Analysis was performed on the soil surface.

extraction areas. In addition, the alluvial soils adjacent to the Webbs Consols Creek between the minor processing plant and main shaft showed very high concentrations (As, Pb, Zn > 2000 mg kg^{-1}). In vitro bioaccessibility measurements of selected samples (0.25 mm fraction) (Drexler and Brattin 2007) has shown high to very high bioaccessibilty of As and Pb on-site (As 5.5–1520 (mean 313) mg kg^{-1}; Pb 74–9990 (mean 1780) mg kg^{-1}).

The old tailings dam is located directly in the drainage line approximately 550 m from the Webbs Consols Creek; water from the mining area drains into Severn River. The material disposed in the old tailing dam was previously shown to be grossly contaminated (Ashley et al. 2004). Arsenic, Pb, and Zn concentrations in the old tailings were reported to be in the 4,140–283,200 (mean 124,000), 571–124,000 (mean 62,900), and 156–299,000 (mean 39,900) mg kg^{-1} range, respectively. Additionally, due to the waste being located directly in the drainage line of the sub catchment, it also represented a major source of contamination to downstream sediments (see Figure 13.3). Remediation options in this case were limited and the old tailing material was to be partially mitigated using environmental control measures. This consisted of removal of the tailings material upstream and placed within a capping. The work was completed in 2006 and the final cap is approximately 50 m × 40 m. The top and bottom of the cap included a nylon material with approximately 100 mm of lime. The top of the material had an additional 300 mm of locally sourced "clay." The top of the cover was seeded with a mix of grass species (see Image 13.1). No information is presently available on whether there is subsurface flow beneath the capping area.

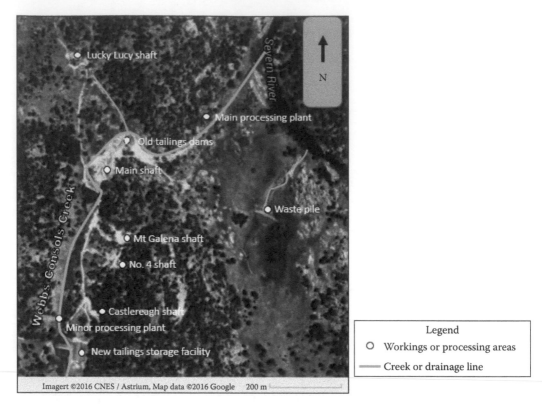

FIGURE 13.3   Map of Webbs Consols mine northern New South Wales, Australia.

IMAGE 13.1   A northeastern view of the Webbs Consols capping of mine tailings, New South Wales. Location of cap is upslope of the major area of contamination (see Image 13.2 for an aerial view of position within site).

13.4.2 Mole River, New South Wales

The Mole River Arsenic Mine is a derelict mine located in northern New South Wales, Australia on a 2413 ha grazing property. Intensive phases of mining and refining arsenopyrite ore occurred between 1924 and 1940. The mine produced both arsenic trioxide and arsenic pentoxide (arsenic is utilized in wood preservation, semiconductors, insecticides, alloys). During peak production, 200 tonnes of ore per week and 19,000 tonnes in total over the life of the mine were produced. In 1935, an explosion in a part of the processing plant released 5 tonnes of arsenic trioxide into the atmosphere, leading to widespread contamination of the surrounding pasture areas with arsenic. The mine was closed in 1940, but did not undergo any rehabilitation at the time.

More recently environmental assessments were undertaken to define the extent of As contamination at the site and adjacent areas. A study by Nguyen (2007) reported high concentrations of As in the waste rock heap and residual processed materials, which had resulted in contamination of soil and contamination of Sam's Creek from erosion runoff. The study concluded that the site was still very unstable and that potentially large amounts of water soluble arsenic species could be discharged into surrounding ecosystems by chemical and mechanical weathering.

Due to the contamination risk, the Mole River Arsenic Mine was prioritized for works by the Derelict Mines Program (DMP), which oversees mine sites where no company, organization, or individual can be found to be responsible for the rehabilitation of a site. Rehabilitation works were conducted on the site in 2008. The works program was designed to prevent ongoing contamination of Sam's Creek and Mole River and contain it on-site. Waste material was excavated from around the site (to a minimum depth of 500 mm). In total 1150 m³ of contaminated soil was stockpiled and mixed with 30 tonne of lime to stabilize the waste material. The stockpiled soil was shaped to a 4:1 horizontal:vertical batter angle. A drainage collection pond was constructed at the bottom of the site (Image 13.2). The stockpiled waste was capped with a total of 3000 m³ clay to a minimum of 300 mm spread over the contaminated stockpiles. The cap was fertilized and seeded with pasture (Image 13.2).

The rehabilitation plan resulted in improvements to surface drainage and the prevention of the movement of contaminants by encapsulating contaminated material on-site. All disturbed areas

IMAGE 13.2 Mole River capping of As-contaminated waste. Foreground shows drainage compound. Center-top of photo shows the completed capping works and drainage lines.

were revegetated to ensure surface stability. Six piezometers were installed to monitor water quality around the site and monitoring to date has shown a significant increase in surface stability and an improvement in water quality leaving the site.

13.5 PHYTOCAPPING TECHNOLOGY

Derelict mine sites differ from operational mines in that modern control measures were not implemented before or upon completion of mining. This includes a lack of modern storage infrastructure, lack of water storage or treatment, and as a result, highly active AMD (Bussière 2010). In terms of governance, state and national organizations are left with large numbers of sites, resulting in significant financial costs in the short to medium term. The covers noted in the Webbs Consols and Mole River sites provide some control over highly contaminated material. Covers of even 0.30 m provide a reduction in erosion of material, exposure to human and ecological receptors, and potentially a reduction in water and atmospheric exposure from wastes. The latter points are critical in the future generation of metal-rich AMD leaching into surface and groundwater. Unfortunately, there are limited monitoring data available on the ingress of oxygen and water into cover systems at derelict mine sites in New South Wales, and, in general, the design has not accounted for these properties.

Conventional covers for municipal solid waste (MSW) are well known to have deteriorated and failed over the medium to long term (Nixon et al. 2001, Martin and Stephens 2008, Lamb et al. 2013). Similarly, the long-term stability of covers at derelict mine sites are also of concern. Various modifications have therefore been proposed to the conventional design for waste cover systems. This has included bioreactor and oxygen-restrained cover designs, which contrast with the "dry-tomb" design and phytocapping technology (Blakey et al. 1997, Albright et al. 2003, Yang and Zhou 2008, Ashwath and Venkatraman 2010). Cover liners in the conventional MSW sites typically have a range of clay contents (Lamb et al. 2013). The clay content, bulk density, nutrient content, and other poor soil factors (pH, salinity) make productive growth of plants difficult due to low water availability, and poor soil structure and aeration. Therefore, it is preferable to design the cover system accounting for soil characteristics, plant species, climate, and other design features.

A definition of "phytocapping" is lacking. It can be considered a form of phytoremediation. Cunningham and Lee (1995) have defined phytoremediation "as the use of green plants to remove, contain, or render harmless environmental contaminants." In landfill soils where contamination has occurred, or is irrigated with landfill leachate, phytoremediation of soil contaminants may be necessary, or a part of the phytocapping technique. Phytocapping as a subdiscipline to phytoremediation may be defined as the use of higher plants to minimize percolation into waste of landfill sites and stabilize and mitigate off-site migration of soil, aqueous, or gaseous environmental contaminants (Lamb et al. 2013). Phytocapping in the literature, particularly in the United States, is known by different terms. Due to the alternative covers assessment project (ACAP), alternative covers have reached some acceptance in the United States. Other usages may include "alternative earthen covers," "agronomic covers" "alternative covers," "geologic covers," "natural covers," "evapotranspiration covers," or ET covers.

The primary aim of conventional landfill cover systems is to minimize moisture entering the waste (Vasudevan et al. 2003). Precipitation, evaporation, transpiration, and temperature are the essential elements to be considered while designing a capping system. Selection of capping earthen material and depth must also be considered (Berger and Melchoir 2009). In recent years, conventional capping systems for MSW sites in developed and many developing countries have been made of polyvinyl chloride plastics (PVC) (Levin and Hammond 1990), in addition to the more customary caps of compacted clay (Othman et al. 1994), GCLs (geosynthetic clay liners) (Benson 2000), PVC, and HDPE (high density polyethylene) (Simon and Müller 2004). Landfill capping systems maintain the hydrological balance by insertion of material with hydraulic conductivities of 10^{-7} to 10^{-9} cm s^{-1}, therefore, causing runoff rather than infiltration into the cover system (Bendz et al. 1997).

Phytocapping minimizes water entry through the cover by a variety of processes (Lamb et al. 2013). These include canopy evaporation in which part of the rain directly evaporates into the atmosphere without reaching the ground surface, water storage in soil matrix after rainfall or canopy throughfall reaches the ground, and evapotranspiration. The performance of phytocapping technology impinges on accurately accounting for the local climate, water storage capacity, and plant influences on hydrological cycle, particularly transpiration rate during critical wet periods (Molz and Browning 1977, Ettala 1988a, Anderson et al. 1993). Cover thickness in the literature ranges from 700 to 2500 mm (Lamb et al. 2013).

The approach to phytocapping of landfills depends primarily on the final end use that is desired. Factors which may influence the decision on end use may include legislative directives on the appropriateness of plant species on the landfill, the climate of the area, and the location relative to urban land use (Nixon et al. 2001). The majority of landfills in Australia have not been designed specifically for phytocapping technology. Phytocapping ideally requires a soil cover that is not compacted and possesses loam-like textures. Landfill designs—not explicitly including vegetative cover—tend to possess high clay content with high bulk density. The typical design of the landfill "cover" is the addition of a low-permeability clay lining that is compacted to reduce the hydraulic conductivity. Above the clay lining is placed an additional soil material, which may consist of a range of properties. Despite the extreme difficulty in obtaining soils with adequate nutrient content for plant growth, and the likelihood of high clay content of cover materials (Nixon et al. 2001), the texture of covers are often silty loam to clay loam. The characteristics of the cover materials limit the selection of options of post-closure activities unless specifically designed for phytocapping (Lamb et al. 2013).

13.5.1 IRRIGATION

Establishment and support of vigorous plant growth requires irrigation on phytocaps in most climates. Productive plant growth is an essential component of phytocapping systems. Plants must be of sufficient health when rain events occur to remove incoming precipitation. Plant selection is an essential element of phytocap designs and must be consistent with the climate of the site. However, irrigation is required at the beginning of a vegetation project. Irrigation is typically supplied through drip irrigation systems to support plant growth (Anderson et al. 1993, Albright et al. 2004, Scanlon et al. 2005, Venkatraman and Ashwath 2009, Ashwath and Venkatraman 2010). Municipal landfills have issues associated with irrigation because (1) landfill sites are located in areas away from developed areas, and (2) the use of public water for irrigation of landfill sites is a wasteful use of a critical resource. Derelict mines face additional issues, since in most cases they are located in highly remote areas and may or may not have local water supplies. As observed at the Webbs Consols and Mole River sites, water associated with the site is often contaminated with a range of metal(loid) contaminants, high acidity, and salinity (Ashley and Lottermoser 1999, Ashley et al. 2004).

At MSW landfills, landfill leachate generated onsite are possible sources of irrigation (Phillips et al. 2004, Phillips and Sheehan 2005). Despite the unfavorable chemical characteristics of landfill leachate, its reuse has been demonstrated to be beneficial, especially when compared to plots with no irrigation (Ettala 1987, 1988b). In a pot study using loam soil, Justin et al. (2010) found that landfill leachate increased plant growth in *Populus* sp. and *Salix* sp. by up to 155% compared to nutrient solution. MacDonald et al. (2008) reported increased plant productivity when plants were irrigated with landfill leachate over 3 years. The two grass species produced significantly more biomass under landfill leachate irrigation than the two tree species. By the second growth season, poplar and *Salix babylinica* were significantly impacted by the landfill leachate irrigation due to salt accumulation.

Ongoing irrigation with nondegradable contaminants such as metal(loid)s may result in additional soil contamination issues (Jones et al. 2006). The utilization of municipal landfill leachate has been shown to have some issues with metalloid contamination (Kjeldsen et al. 2002, Zupancic et al. 2009). Zupancic et al. (2009) reported irrigation waters with elevated Cr was used for capping irrigation (up to 2.75 mg L^{-1}). Nevertheless, irrigation was found not to result in elevated Cr in the soil capping material after 3 years. After 1 year, there appeared to be significant Cr concentrations in the soil (~70 mg kg^{-1}), which declined substantially during the remaining 2 years of the study. Most studies have not been conducted over a long time frame (e.g., >10 years) (Winant et al. 1981, Shrive et al. 1994). Sodic and saline waters may impact upon the health of phytocapping materials (Hernández et al. 1999, Phillips et al. 2004, Zupanc and Justin 2010). Phillips et al. (2004) found that sodic properties of landfill leachate in Queensland, Australia, resulted in decreased aggregate stability. In addition, deterioration in the hydraulic properties of the final capping material was shown to impair leachate infiltration, resulting in increased surface erosion in vegetated plots. Similar results have been reported by Zupancic et al. (2009), Hernández et al. (1999), and Chan (1982). Some studies have reported no detrimental effect to plants due to salt, sodicity, or contaminants (Smith et al. 1999, MacDonald et al. 2008, Loncnar et al. 2010). In the absence of a sodicity amelioration program, such as the addition of organic and gypsic amendments, severe structural issues will result in poor health of phytocaps (Cook et al. 1994). The issues associated with irrigation quality should be assessed relative to the need for irrigation. Zornberg et al. (2003) concluded that at a Superfund site in southern California, irrigation was not appropriate since it resulted in additional percolation into the waste based on model simulations with LEACHM.

13.5.2 Cap-Waste Interactions

Capping thickness of MSW landfills are defined by regulatory authorities for a given category of waste materials and conditions. In MSW designs, the capping thickness is relatively thin overlying a low permeability layer, ranging from 0.3 m to approximately 1 m, depending on the waste and design of the landfill. As shown in the Webbs Consols and Mole River examples at derelict mines, the overlying cap thickness is often thin (e.g., 0.3–0.5 m) and without the engineered excluding layers. At derelict mine sites, the capping material is limited by (1) the availability of adequate material of high clay content, (2) uncontaminated material, and (3) cost associated with extraction and or transportation. The cap depth is critical in the design not only in terms of percolation into the cap but also the selection of plant species and the likelihood of roots penetrating the cap (Lamb et al. 2013). The growth of shrub and tree species with large root systems on thin landfill caps may penetrate through the cap into the waste below. Root penetration may cause significant risks to the environment via preferential water flow paths to the waste. Vertical transmission of water in soil profiles due to root-generated preferential flow paths has been shown to occur in laboratory- and field-based studies (Camobreco et al. 1996, Gish et al. 1998, Devitt and Smith 2002, Green et al. 2006, Johnson and Lehmann 2006, Tsegaye et al. 2007). Bundt et al. (2000) has shown that old root paths increased transport of radionuclides and also the growth of microbes along old root channels. Tree roots can also influence soil structure (Glinski and Lipeic 1990), bulk density (Kalman et al. 1996), and pore size (Johnson et al. 2003), which subsequently impact flow vertically in capping profiles (Shouse et al. 1995). Therefore, plants with large adventitious and deep root systems, if able to penetrate to the waste, have the potential to seriously limit the effectiveness of the capping material.

For impermeable lining materials of landfills, available evidence has suggested an inability of plant roots to rupture clay or synthetic lining materials (Handel et al. 1997, Holl 2002). Holl (2002) investigated the ability of a shrub or heathland community roots to penetrate a geosynthetic liner in western United States. Despite the liner being at a depth of only 0.3 m, Holl (2002) reported that none of the 11 plant species studied penetrated the liner. Furthermore, Holl (2002) reported that the tap root declined vertically to the liner, wherefrom the root system spread laterally. Similarly, Handel et al. (1997) studied 22 tree and shrub species on the Staten Island Landfill, New York. No plant species was able to penetrate the clay cap, and few showed significant root growth above the clay cap due to anoxic and acidic conditions. As a result of plant growth on phytocaps in Rockhampton, Queensland, Australia, the storage capacity of the capping material was reduced by approximately 9%–14% (Venkatraman and Ashwath 2009).

Capping thickness and plant selection may also influence the chemistry in the capped waste due to moisture and oxygen ingress (Nicholson et al. 1989, Elberling et al. 1994, Aachib et al. 2004, Biglari et al. 2009). Nicholson et al. (1989) reported a cover designed to retard movement of atmospheric oxygen to the cap. Limited diffusion of oxygen was proposed to help mitigate metal solubility and acid production by limiting oxidation of pyritic waste. Furthermore, limited oxygen would promote the reduction of existing sulphate, resulting in reduction in metal solubility from metal sulphide precipitation (Biglari et al. 2009). The design reported by Nicholson et al. (1989) is dependent on the maintenance of moist cover material, and as noted by Barbour (1990), may be impractical in the long term. Considerable research has been conducted on oxygen diffusion rates through caps under various conditions and materials (Nicholson et al. 1989, Biglari et al. 2009, Bussière 2010). The role plant species play in increasing oxygen ingress into the waste via preferential pathways under dry conditions is an area requiring further work in derelict mines.

13.5.3 PLANT SPECIES FOR PHYTOCAPPING

Phytocapping is a new technology for managing waste disposal sites. Therefore, available data of suitable species is limited (Nixon et al. 2001, Venkatraman and Ashwath 2009). Selection of plant species depends heavily on the geographical location (i.e., country and climate), the approach to phytocapping, and the soil conditions on which phytocapping is to occur, and metal and acid tolerance among other factors. European and North American phytocapping works have been dominated by the genera *Salix* and *Populus* (Ettala 1988a,b, Ettala et al. 1988, Nixon et al. 2001). Ettala (1988b) studied the growth of *Salix aquatica*, *Betula pendula*, and *Populus rasumowskyana*. In a shallow cover soil (<0.3 m depth), it was found that *S. aquatica* produced the greatest biomass and was able to transpire the highest amounts of water. Hybrid poplars have continued to be common in evapotranspiration cover projects since their use is readily accepted by the U.S. regulatory authorities in phytoremediation projects (Licht et al. 2001, Licht and Isebrands 2005).

As a result of shallow cover soils, herbaceous species such as grasses, shrubs, and groundcovers have been routinely utilized in engineered and phytocaps in MSW landfills. Grasses have the benefit that they are able to form relatively dense covers close to the ground surface, and protect the ground from erosion (Lal 1976, Lal et al. 1979, Ngatunga et al. 1984). The dense covers are effective in reducing water splashing from rainfall (Lal 1998), water runoff, and wind velocity at the soil surface (Lal 1976, 1998, Lal et al. 1979, Ngatunga et al. 1984, Starr et al. 2000, Bielders et al. 2001). Furthermore, Albright et al. (2006) reported that successful phytocapping trials were dominated by systems consisting of grasses, forbs, and shrubs. The disadvantage of using herbaceous species is that their roots may not grow deep enough to remove moisture from soil depths >1 m, thus limiting their effectiveness as phytopumps (Preston and McBride 2004, Ruth et al. 2007). Similar results were reported on capped mining soils in Australia (Yunusa et al. 2010).

Selection of plant species is dependent on soil resources, climate, and the end use of the cover system. In MSW landfills, where relatively large surface areas may be utilized, Lamb et al. (2013) noted that additional benefits may be gained by selection of crops with additional uses. Apart from the notion of developing an aesthetic zone, high biomass crops could be produced for the purposes of bioenergy production. Indeed, in South Australia, *Arundo donax* growth was shown to be significantly enhanced on an MSW landfill with the assistance of biosolids (Lamb et al. 2012). While there is variability between derelicts with respect to size, the capping of contaminated materials tends to result in a small surface area for utilization. Therefore, plant selection will be mostly based on survival, transpiration rate, drought, acidity, and metal tolerance. In Australia, some studies have been conducted (Phillips et al. 2004, Phillips and Sheehan 2005, Venkatraman and Ashwath 2009, Ashwath and Venkatraman 2010). Ashwath and Venkatraman (2010) examined the performance of 21 plant species in Rockhampton (Queensland, Australia) at two different soil thickness depths (700 and 1400 mm). Of the 21 species, 19 showed good survival rates and grew adequately. *Populus* sp. and *Salix* sp. were the only species with poor survival rates. Using a generalized transpiration

TABLE 13.4
Plant Species Utilized in Phytocaps of Mining Waste in Australia, Including Average Rainfall Data, Where Known

Plant Species	Common Name	Average Annual Rainfall (mm)	Reference
Acacia rostellifera		742 (winter dom)	Gwenzi et al. (2012)
Melaleuca nesophila		742 (winter dom)	
Senna artemisioides	Silver cassia	400 (uniform)	Arnold et al. (2015)
Sclerolaena birchii	Galvanized burr	400 (uniform)	
Themeda triandra	Kangaroo grass		
Bothriochloa macra	Redgrass		
Chloris truncata	Short windmill grass		
Atriplex nummularia		800 mm	Yunusa et al. (2010)
Casuarina glauca			
Melaleuca linarifolia			
Syncarpia glomulifera			
Cynodon dactylon	Couch grass		
Axonopus affinis	Carpet grass		
Trifolium repens	White clover		
Pennisetum clandestinum	Kikuyu grass		

rate of 1.5 mm day^{-1}, HYDRUS 1D was used to estimate percolation rates into the waste of <17 and <24 mm year^{-1} for the 700 and 1400 mm deep caps, respectively. The conventional cover was 78 mm year^{-1}.

The aim of cover systems of contaminated mine waste at derelict mine sites is to limit further dispersal of contaminants in the long term. Therefore, the main function of a phytocap is to limit water percolation through the waste and maintain the integrity of the cover. An important consideration in the selection of plants is their ability to not only resist dry and wet periods but also high winds and storms (Lamb et al. 2013). Given that thin soil covers are likely to be used, tree species may be more prone to being uprooted in strong winds. Nevertheless, of the studies identified in Australian mine covers, a number of medium sized tree species have been utilized (Table 13.4). These include *Acacia* spp., *Melaleuca* spp., *Eucalyptus* spp., and *Casuarina* sp. The remaining plant species used were typically small shrubs, grasses, or groundcovers (Gwenzi et al. 2012, Arnold et al. 2015). Arnold et al. (2015) reported only two plant species (*Sclerolaena birchii, Senna artemisioides*) in the arid area of western New South Wales. It was concluded that in arid areas, and partially due to the low plot covering rate of these species (26%), evaporation made a greater contribution to moisture loss than transpiration. Yunusa et al. (2010) reported that types of vegetation (woodland species, groundcover, and shrubs) within one plot were most effective at reducing water ingress into waste materials.

13.5.4 PHYTOCAPPING ENHANCEMENT WITH SOIL AMENDMENTS

Capping material for derelict mines will generally be sourced locally using material available on-site or nearby, due to cost constraints in importing cap materials. Thus mineralogy and other physical and chemical properties can vary considerably from site to site, although little published material exists on the properties of caps to date. Many characteristics of metalliferous mine wastes are often unfavorable to successful vegetation establishment, most notably phytotoxic levels of residual heavy metals, low nutrient status, and poor physical structure of the substratum (Tordoff et al. 2000). Growth of vegetation above the cap depends critically on the proposed design, whether it is designed

(1) specifically to reduce water percolation into and through the waste, (2) to reduce oxygen diffusion into the waste, (3) to limit erosion processes of the waste, and (4) to support healthy plant growth. The composition of the cover material and application of other materials depend on the aims of the cover.

Amendments have been applied to ameliorate mine waste by neutralization of acidity, reduce metal solubility, and enhance properties of the material for establishment of a vegetative cover. Amendments are a relatively low-cost rehabilitation strategy. Lime is one of the most readily applied amendments to mine wastes because of its ability to neutralize acidic waste and reduce potential for metal leaching. Lime has been reported to restore the vegetative cover to a high metal mine waste due to increased pH and reduced phytoavailability of metals in some cases (Khan and Jones 2009). Alkaline amendments such as lime also inhibit pyrite oxidation (Doye and Duchesne 2005). Lime may be incorporated with the waste material, used as fertilizer in the cover layer, or used as a layer in a cap to neutralize percolation. Inorganic soil amendments are often associated with a particular soil issue, and are relatively expensive to apply over large areas. The large amount that is required to ameliorate the range of limiting soil factors in landfill covers makes it prohibitively expensive. Organic waste by-products are typically more cost effective and efficient at improving soil conditions for a wide range of issues, providing a plethora of benefits to mineral soils (Park et al. 2011b).

When applied to mine waste, organic materials improve the structure of the capping material, increase water holding capacity and nutrient retention, and provide plant nutrients in a slow-release form, facilitating vegetation establishment. A large diversity of organic wastes have been tested and utilized for this purpose including biosolids, MSW, pulp mill by-products and industrial wastes, and a range of composts (Tordoff et al. 2000, Jones and Haynes 2011, Wijesekara et al. 2016).

Large quantities of organic biowastes, including mixed MSW, biosolids, animal and poultry manure, and paper mill sludge are produced globally. The amount of biosolids produced in the United States and Australia have been estimated at 5645×10^3 tonnes year^{-1} and 407×10^3 tons year^{-1}, respectively (Thangarajan et al. 2013). The characteristics of the biosolids depend on the composition of the wastewater, the type of wastewater treatment used, and the type of subsequent treatment applied to the biosolids. Biosolids typically contain 40%–60% organic matter, 3%–5% nitrogen, 1%–4% phosphorus, other nutrients (K, Ca, Mg, S), and trace metals. The extent of recycling and reuse of such wastes vary considerably, depending on the regulations related to utilization of these wastes in local jurisdictions. Biosolids are alkaline stabilized to generate Class A biosolids for land application. More recently, biochar has been considered for utilization as a soil conditioner for mine wastes. Biochar is a carbon-rich product produced by biomass combustion (pyrolysis) under high temperatures and low oxygen conditions (Park et al. 2011a). Biochar may enhance plant growth by supplying and retaining nutrients and by improving the physical, chemical, and biological properties of the soil. Biochar applied to mine tailings has been shown to increase pH, nutrient- and water-retention capacity, and decrease bioavailability of metals (Fellet et al. 2011, Rodríguez-Vila et al. 2014).

Land application of municipal and industrial organic by-products generally increases the biological productivity of soils in agricultural and land reclamation settings (Ettala 1988a, Acea and Carballas 1999, Brown and Chaney 2000, Jones et al. 2010). Plant productivity has been shown to have improved by a large diversity of organic wastes, including biosolids, MSW, pulp mill by-products and industrial wastes (Kennedy et al. 1999), and a range of composts (Jones and Haynes 2011, Park et al. 2011b). Animal manures and municipal wastes have in the past been used as a source of essential nutrients and other benefits to soils. Waste water generation and intensification of livestock have generated increasingly large quantities of solid organic wastes with variable composition (Burton and Turner 2003). Substantial quantities of organic stockpiles are generated each year in most countries (Kennedy et al. 1999, Ghosh et al. 2010, Park et al. 2011b). The primary sources of manure are derived from chicken, swine, and cattle farms, although farm dairy effluent is also significant. The chemical, biological, and physical properties depend on factors such as animal type, animal attributes (age, size), water use, manure collection (floor type), season, bedding type, and storage and handling of manures (Miller et al. 1991, 2003). Composting of animal manure is often desirable to reduce the quantity of animal manure needing disposal and to stabilize the organic material. In addition, several industries generate significant quantities of organic waste, such as paper mills and olive mills (Edwards and Someshwar 2000).

Organic wastes are generally effective in increasing biological productivity of landfill cover soils. Recent work investigating the role of vegetation in landfill capping design has added organic materials to improve growth (Lamb et al. 2012, 2013, Seshadri et al. 2013). Compost materials from MSW or other sources are the most common by-products to date, and only a few cases have used biosolids or other (e.g., peanut hull) wastes. Application of organic residues is typically added at a ratio of 1:2 to 5:1 (organic by-product:mineral). However, in some cases (Börjesson et al. 2004, Einola et al. 2007), the top layer of the final cover has consisted entirely of organic amendments. Capping materials high in organic matter have been shown to enhance vegetative productivity within at least 20–30 cm depth (Ettala 1988a). Addition of organic by-products has been shown numerous times to enhance CH_4 oxidation and co-metabolism of trace contaminant gases in landfill caps. Soil erosion occurred due to a loss of physical properties of soil from landfill leachate irrigation, but amendment with organic waste was shown to reduce soil erosion and improve aggregate stability (Phillips et al. 2004).

AMD from derelict mines represents a substantial liability for governments and industry (Aubertin et al. 2006, Smirnova et al. 2009, Bussière 2010). AMD can continue long after mining operations are complete through the oxidation of sulphidic minerals in leftover tailings impoundments and waste rock piles. Source control measures are considered the preferable approach to control the formation of AMD, but not generally practiced at old abandoned mine sites (Johnson and Hallberg, 2005). These techniques aim to preclude the formation of AMD in the first instance. Various approaches that have been evaluated to prevent or minimize the generation of mine drainage waters include flooding/sealing of underground shafts, underwater tailings storage, storage in sealed waste heaps, blending of mineral wastes, solidification of tailings, application of anionic surfactants, and microencapsulation (Johnson and Hallberg 2005).

One of the proposed aims for safe disposal of AMD-producing waste at derelict mines is to limit oxygen ingress through the use of oxygen barriers. This can take the form of a water cover (Simms et al. 2000, Bussière et al. 2006) and covers containing oxygen consuming materials such as wood waste, straw mulch, or other organic residues (Tremblay 1994, Cabral et al. 2000). Limiting diffusion of oxygen assists to mitigate metal solubility and acid production by limiting oxidation of pyritic waste. Furthermore, limited oxygen would promote the reduction of existing sulphate, resulting in reduction in metal solubility from metal sulphide precipitation (Costa and Duarte 2005).

Covers based on high moisture retention and capillary barrier effects have been proposed to limit oxygen ingress, particularly in prevailing humid climates (Nicholson et al. 1989, Bussière et al. 2006, Bussière 2010). This design is dependent on the maintenance of moist cover material to retard movement of atmospheric oxygen through the cap (Nicholson et al. 1989, Bussière 2010). Measurement and prediction of the oxygen diffusion in different types of materials and degrees of saturation have been conducted in laboratory studies, which is an important parameter for the performance of these soil covers (Aachib et al. 2004). The hydraulic behavior of covers with capillary barrier effects (CCBE) is also important (Aubertin et al. 2006, Parent and Cabral 2006, Li et al. 2013). An 8-year monitoring of these covers has shown promise for their application (Bussière et al. 2006).

Vegetation succession and the role plant species play in CCBE performance in the long term may be important, increasing oxygen ingress into the waste via preferential pathways under dry conditions and reducing saturation of the top cap layer by water uptake (Smirnova et al. 2009). A biointrusion barrier using plant species with allelopathic effects may improve long term performance of CCBEs by excluding tree species which impact cover performance (Smirnova et al. 2009). This is an area requiring further work for application of covers to mitigate water and oxygen ingress interaction with waste at derelict mines.

13.6   CONCLUSIONS AND FUTURE RESEARCH

Derelict mines are widespread throughout developed and undeveloped countries. Derelict mines are frequently old and abandoned before environmental controls were required for closing of mine sites. Often, but not always, serious contamination and acidity issues arise as a result. Furthermore, due to the large numbers of sites state and national governmental bodies are required to manage, funding for cleanup cannot cover comprehensive cleanup approaches in the short to medium term. Metal(loid) remediation approaches such as *in situ* immobilization, phytoextraction, and electrore-mediation are not suitable for the full range of contamination issues at a given site. Contaminated material inevitably will need to be isolated in capping works. A limited budget means that comprehensive cover and isolation designs are often utilized. Phytocapping and other oxygen-limiting caps can provide additional benefits to designs currently implemented in the field. Further research is required to generate cost-effective cover systems combining phytocaps and oxygen retaining systems that are stable in the long term.

REFERENCES

Aachib, M., M. Mbonimpa, and M. Aubertin. 2004. Measurement and prediction of the oxygen diffusion coefficient in unsaturated media, with applications to soil covers. *Water, Air, and Soil Pollution* 156 (1): 163–193.

Acar, F., and E. Malkoc. 2004. The removal of chromium (VI) from aqueous solutions by *Fagus orientalis* L. *Bioresource Technology* 94 (1): 13–15.

Acea, M.J., and T. Carballas. 1999. Microbial fluctuations after soil heating and organic amendment. *Bioresource Technology* 67: 65–72.

Albright, W., C. Benson, G. Gee, T. Abichou, S. Tyler, and S. Rock. 2006. Field performance of three compacted clay landfill covers. *Vadose Zone Journal* 5 (4): 1157.

Albright, W., C. Benson, G. Gee, A.C. Roesler, T. Abichou, P. Apiwantragoon, B.F. Lyles, and S.A. Rock. 2004. Field water balance of landfill final covers. *Journal of Environmental Quality* 33 (6): 2317–2332.

Albright, W., C.H. Benson, G.W. Gee, T. Abichou, A.C. Roesler, and S.A. Rock. 2003. Examining the alternatives. *Civil Engineering* 73 (PNWD-SA-6048): 70–75.

Anderson, J.E., R.S. Nowak, T.D. Ratzlaff, and O.D. Markham. 1993. Managing soil moisture on waste burial sites in arid regions. *Journal of Environmental Quality* 22 (1):62–69.

Antunes, S.C., B.B. Castro, R. Pereira, and F. Gonçalves. 2008. Contribution for tier 1 of the ecological risk assessment of Cunha Baixa uranium mine (Central Portugal): II. Soil ecotoxicological screening. *Science of the Total Environment* 390 (2–3): 387–395.

Appleton, J. 2003. A re-assessment of the heritage significance of historical features at Webb's Consols Mine Emmaville, Northern NSW. Unpublished: New South Wales Department of Land and Water Conservation, Parramatta, NSW, Australia.

Archer, M.J.G., and R.A. Caldwell. 2004. Response of six australian plant species to heavy metal contamination at an abandoned mine site. *Water, Air, and Soil Pollution* 157 (1–4): 257–267.

Arnold, S., A. Schneider, D. Doley, and T. Baumgartl. 2015. The limited impact of vegetation on the water balance of mine waste cover systems in semi-arid Australia. *Ecohydrology* 8 (3): 355–367.

Ashley, P.M., and B.G. Lottermoser. 1999. Arsenic contamination at the Mole River mine, northern New South Wales. *Australian Journal of Earth Sciences* 46 (6): 861–874.

Ashley, P.M., B.G. Lottermoser, A.J. Collins, and C.D. Grant. 2004. Environmental geochemistry of the derelict Webbs Consols mine, New South Wales, Australia. *Environmental Geology* 46 (5): 591–604.

Ashwath, N., and K. Venkatraman. 2010. Phytocapping: An alternative technique for landfill remediation. *International Journal of Environment and Waste Management* 6 (1): 51–70.

Aslibekian, O., and R. Moles. 2003. Environmental risk assessment of metals contaminated soils at silvermines abandoned mine site, Co tipperary, ireland. *Environmental Geochemistry and Health* 25 (2): 247–266.

Aubertin, M., E. Cifuentes, V. Martin, S. Apithy, B. Bussière, J. Molson, R.P. Chapuis, and A. Maqsoud. 2006. An investigation of factors that influence the water diversion capacity of inclined covers with capillary barrier effects. *Geotechnical Special Publication* 147 (1): 613.

Baker, A.J.M., R.D. Reeves, and S.P. McGrath. 1991. In situ decontamination of heavy metal polluted soils using crops of metal-accumulating plants—A feasibility study. In *In Situ Bioreclamation*, Hinchey, R.E. and Offenbach, R.F. (Eds.), pp. 600–605. Boston, MA: Butterworth-Heinemann.

Barbour, S.L. 1990. Reduction of acid generation in mine tailings through the use of moisture-retaining cover layers as oxygen barriers: Discussion. *Canadian Geotechnical Journal* 27 (3): 398–401.

Bendz, D., V.P. Singh, and M. Akesson. 1997. Accumulation of water and generation of leachate in a young landfill. *Journal of Hydrology* 203 (1–4): 1–10.

Benson, C. 2000. Liners and covers for waste containment. In: *Proceedings of the Fourth Kansai International Geotechnical Forum, Creation of a New Geo-Environment*. Japanese Geotechnical Society, Kyoto, Japan, pp. 1–40.

Berger, K., and S. Melchoir. 2009. Landfill capping and water balance of cover systems. *12th Waste Management and Landfill Symposium*, Sardinia, Italy, October 5–9, 2009.

Bielders, C.L., S. Alvey, and N. Cronyn. 2001. Wind erosion: The perspective of grass roots communities in the Sahel. *Land Degradation & Development* 12 (1): 57–70.

Biglari, M., J.M. Scharer, R.V. Nicholson, and T.C. Charles. 2009. Evaluation of bacterial community structure and its influence on sulfide oxidation in a bio-leaching environment. *Geomicrobiology Journal* 26 (1): 44–54.

Blakey, N.C., K. Bradshaw, P. Reynolds, and K. Knox. 1997. Bio-reactor landfill—A field trial of accelerated waste stabilisation. *Sixth International Landfill Symposium*, Sardinia, Italy, October 13–17.

Börjesson, G., I. Sundh, and B. Svensson. 2004. Microbial oxidation of CH_4 at different temperatures in landfill cover soils. *FEMS Microbiology Ecology* 48 (3): 305–312.

Brown, S., B. Christensen, E. Lombi, M. McLaughlin, S. McGrath, J. Colpaert, and J. Vangronsveld. 2005. An inter-laboratory study to test the ability of amendments to reduce the availability of Cd, Pb, and Zn in situ. *Environmental Pollution* 138 (1): 34–45.

Brown, S.L., and R.L. Chaney. 2000. Combining by-products to achieve specific soil amendment objectives. In *Land Application of Agricultural, Industrial and Municipal By-Products*, edited by J.F. Power and W.A. Dick, pp. 343–360. Madison, WI: Soil Science Society of America Inc.

Brown, S.L., R.L. Chaney, J.S. Angle, and A.J.M. Baker. 1994. Phytoremediation potential of *Thlaspi caerulescens* and bladder campion for zinc-and cadmium-contaminated soil. *Journal of Environmental Quality* 23 (6): 1151–1157.

Brown, S.L., C.L. Henry, R. Chaney, H. Compton, and P.S. DeVolder. 2003. Using municipal biosolids in combination with other residuals to restore metal-contaminated mining areas. *Plant and Soil* 249 (1): 203–215.

Bundt, M., A. Albrecht, P. Froidevaux, P. Blaser, and H. Fluhler. 2000. Impact of preferential flow on radionuclide distribution in soil. *Environmental Science and Technology* 34 (18): 3895–3899.

Burton, C.H., and C. Turner. 2003. *Manure Management: Treatment Strategies for Sustainable Agriculture*, 2nd edn. Wrest Park, Bedford, U.K.: Silsoe Research Institute.

Bussière, B. 2010. Acid mine drainage from abandoned mine sites: Problematic and reclamation approaches. In Chen, Y., Zhan, L., Tang, X (eds)., *Advances in Environmental Geotechnics*, pp. 111–125. Springer Berlin, Heidelberg.

Bussière, B., A. Maqsoud, M. Aubertin, J. Martschuk, J. McMullen, and M. Julien. 2006. Performance of the oxygen limiting cover at the LTA site, Malartic, Quebec. *CIM Bulletin* 1 (6): 1–11.

Cabral, A., I. Racine, F. Burnotte, and G. Lefebvre. 2000. Diffusion of oxygen through a pulp and paper residue barrier. *Canadian Geotechnical Journal* 37 (1): 201–217.

Camobreco, V.J., B.K. Richards, T.S. Steenhuis, J.H. Peverly, and M.B. McBride. 1996. Movement of heavy metals through undisturbed and homogenized soil columns. *Soil Science* 161 (11): 740–750.

Chan, K.Y. 1982. Changes to a soil on irrigation with a sanitary landfill leachate. *Water, Air, and Soil Pollution* 17 (3): 295–304.

Cook, F.J., F.M. Kelliher, and S.D. McMahon. 1994. Changes in infiltration and drainage during wastewater irrigation of a highly permeable soil. *Journal of Environmental Quality* 23 (3): 476–482.

Costa, M.C., and J.C. Duarte. 2005. Bioremediation of acid mine drainage using acidic soil and organic wastes for promoting sulphate-reducing bacteria activity on a column reactor. *Water, Air, and Soil Pollution* 165 (1–4): 325–345.

Cravotta III, C.A. 2008. Dissolved metals and associated constituents in abandoned coal-mine discharges, Pennsylvania, USA. Part 1: Constituent quantities and correlations. *Applied Geochemistry* 23 (2): 166–202.

Cunningham, S.D., and C.R. Lee. 1995. Phytoremediation: Plant-based remediation of contaminated soils and sediments. In *Bioremediation*, edited by H.D. Slipper and R.F. Turco, pp. 145–156. Madison, WI: Soil Science Society of America Inc.

Devitt, D.A., and S.D. Smith. 2002. Root channel macropores enhance downward movement of water in a Mojave Desert ecosystem. *Journal of Arid Environments* 50 (1): 99–108.

DOE. 1994. *In Situ Remediation Integrated Program*, edited by Department of Environment. Washington, DC: U.S. Department of Energy, pp. 1–59.

Doye, I., and J. Duchesne. 2005. Column leaching test to evaluate the use of alkaline industrial wastes to neutralize acid mine tailings. *Journal of Environmental Engineering* 131 (8): 1221–1229.

Drexler, J.W., and W. Brattin. 2007. An in vitro procedure for estimation of lead relative bioavailability: With validation. *Human and Ecological Risk Assessment* 13 (2): 383–401.

Edwards, J.H., and A.V. Someshwar. 2000. Chemical, physical, and biological characteristics of agricultural and forest by-products for land application. In *Land Application of Agricultural, Industrial, and Municipal By-Products*, edited by J.F. Power and W.A. Dick, pp. 1–62. Madison, WI: Soil Science Society of America.

Einola, J.K.M., R. Kettunen, and J.A. Rintala. 2007. Responses of methane oxidation to temperature and water content in cover soil of a boreal landfill. *Soil Biology and Biochemistry* 39 (5): 1156–1164.

El-Fadel, M., A. Findikakis, and J.O. Leckie. 1997. Environmental impacts of solid waste landfilling. *Journal of Environmental Management* 50 (1): 1–25.

Elberling, B., R.V. Nicholson, and J.M. Scharer. 1994. A combined kinetic and diffusion model for pyrite oxidation in tailings: A change in controls with time. *Journal of Hydrology* 157 (1): 47–60.

Ensley, B.D. 1995. Will plants have a role in bioremediation? *Fourteenth Annual Symposium 1995 in Current Topics in Plant Biochemistry, Physiology and Molecular Biology*, 1–2.

Ettala, M. 1987. Influence of irrigation with leachate on biomass production and evapotranspiration on a sanitary landfill. *Aqua Fennica* 17 (1): 69–86.

Ettala, M. 1988a. Evapotranspiration from a *Salix aquatica* plantation at a sanitary landfill. *Aqua Fennica* 18 (1): 3–14.

Ettala, M.O. 1988b. Short-rotation tree plantations at sanitary landfills. *Waste Management and Research* 6 (3): 291–302.

Ettala, M.O., K.M. Yrjonen, and E.J. Rossi. 1988. Vegetation coverage at sanitary landfills in Finland. *Waste Management and Research* 6 (3): 281–289.

Farfel, M.R., A.O. Orlova, R.L. Chaney, P.S.J. Lees, C. Rohde, and P.J. Ashley. 2005. Biosolids compost amendment for reducing soil lead hazards: A pilot study of Orgro® amendment and grass seeding in urban yards. *Science of the Total Environment* 340 (1): 81–95.

Fellet, G., L. Marchiol, G. Delle Vedove, and A. Peressotti. 2011. Application of biochar on mine tailings: Effects and perspectives for land reclamation. *Chemosphere* 83 (9): 1262–1267.

Geebelen, W., D.C. Adriano, D. van der Lelie, M. Mench, R. Carleer, H. Clijsters, and J. Vangronsveld. 2003. Selected bioavailability assays to test the efficacy of amendment-induced immobilization of lead in soils. *Plant and Soil* 249 (1): 217–228.

Ghosh, S., P. Lockwood, N. Hulugalle, H. Daniel, P. Kristiansen, and K. Dodd. 2010. Changes in properties of sodic Australian Vertisols with application of organic waste products. *Soil Science Society of America Journal* 74 (1): 153–160.

Gish, T., D. Gimenez, and W. Rawls. 1998. Impact of roots on ground water quality. *Plant and Soil* 200 (1): 47–54.

Gliński, J., and J. Lipiec. 1990. *Soil Physical Conditions and Plant Roots*. Boca Raton, FL: CRC Press Inc.

Grant, C.D., C.J. Campbell, and N.R. Charnock. 2002. Selection of species suitable for derelict mine site rehabilitation in New South Wales, Australia. *Water, Air, and Soil Pollution* 139 (1–4): 215–235.

Green, J.C., I. Reid, I.R. Calder, and T.R. Nisbet. 2006. Four-year comparison of water contents beneath a grass ley and a deciduous oak wood overlying Triassic sandstone in lowland England. *Journal of Hydrology* 329 (1–2): 16–25.

Gwenzi, W., E. Veneklaas, T. Bleby, I. Yunusa, and C. Hinz. 2012. Transpiration and plant water relations of evergreen woody vegetation on a recently constructed artificial ecosystem under seasonally dry conditions in Western Australia. *Hydrological Processes* 26 (21): 3281–3292.

Handel, S.N., G.R. Robinson, W.F.J. Parsons, and J.H. Mattei. 1997. Restoration of woody plants to capped landfills: Root dynamics in an engineered soil. *Restoration Ecology* 5 (2): 178–186.

Hansen, H.K., V. Lamas, C. Gutierrez, P. Nuñez, A. Rojo, C. Cameselle, and L. Ottosen. 2013. Electroremediation of copper mine tailings. Comparing copper removal efficiencies for two tailings of different age. *Minerals Engineering* 41 (0): 1–8.

Hernández, A.J., M.J. Adarve, A. Gil, and J. Pastor. 1999. Soil salivation from landfill leachates: Effects on the macronutrient content and plant growth of four grassland species. *Chemosphere* 38 (7): 1693–1711.

Hettiarachchi, G., and G. Pierzynski. 2002. In situ stabilization of soil lead using phosphorus and manganese oxide. *Journal of Environmental Quality* 31 (2): 564–572.

Holl, K.D. 2002. Roots of chaparral shrubs fail to penetrate a geosynthetic landfill liner. *Ecological Restoration* 20 (2): 112.

Jasso-Pineda, Y., G. Espinosa-Reyes, D. González-Mille, I. Razo-Soto, L. Carrizales, A. Torres-Dosal, J. Mejía-Saavedra, M. Monroy, A. Ize, M. Yarto, and F. Díaz-Barriga. 2007. An integrated health risk assessment approach to the study of mining sites contaminated with arsenic and lead. *Integrated Environmental Assessment and Management* 3 (3): 344–350.

Johnson, A., I.M. Roy, G.P. Matthews, and D. Patel. 2003. An improved simulation of void structure, water retention and hydraulic conductivity in soil with the pore-core-three-dimensional network. *European Journal of Soil Science* 54: 477–489.

Johnson, D.B., and K.B. Hallberg. 2005. Acid mine drainage remediation options: A review. *Science of the Total Environment* 338 (1–2): 3–14.

Johnson, M.S., and J. Lehmann. 2006. Double-funneling of trees: Stemflow and root-induced preferential flow. *Ecoscience* 13 (3): 324–333.

Jones, B.E.H., and R.J. Haynes. 2011. Bauxite processing residue: A critical review of its formation, properties, storage, and revegetation. *Critical Reviews in Environmental Science and Technology* 41 (3): 271–315.

Jones, B.E.H., R.J. Haynes, and I.R. Phillips. 2010. Effect of amendment of bauxite processing sand with organic materials on its chemical, physical and microbial properties. *Journal of Environmental Management* 91 (11): 2281–2288.

Jones, D.L., K.L. Williamson, and A. Owen. 2006. Phytoremediation of landfill leachate. *Waste Management* 26 (8): 825–837.

Justin, M.Z., N. Pajk, V. Zupanc, and M. Zupančič. 2010. Phytoremediation of landfill leachate and compost wastewater by irrigation of *Populus* and *Salix*: Biomass and growth response. *Waste Management* 30 (6): 1032–1042.

Kalman, R., K. Sandor, M. van Genuchten, and J. Per Erik. 1996. Estimation of water retention characteristics from the bulk density and particle size distribution of Swedish soils. *Soil Science Society of America Journal* 161: 832–845.

Kennedy, M., Y. List, L.Y. Lu, A. Foo, A. Robertson, and R.H. Newman. 1999. Kiwifruit waste and novel products made from kiwifruit waste: Uses, composition and analysis. In *Analysis of Plant Waste Materials*, edited by H.F. Linskens and J.F. Jackson, pp. 121–147. Berlin, Germany: Springer-Verlag.

Khan, M.J., and D.L. Jones. 2009. Effect of composts, lime and diammonium phosphate on the phytoavailability of heavy metals in a copper mine tailing soil. *Pedosphere* 19 (5): 631–641.

Kjeldsen, P., M. Barlaz, A. Rooker, A. Baun, A. Ledin, and T. Christensen. 2002. Present and long-term composition of MSW landfill leachate: A review. *Critical Reviews in Environmental Science and Technology* 32 (4): 297–336.

Lal, R. 1976. Soil erosion on Alfisols in Western Nigeria:: III. Effects of rainfall characteristics. *Geoderma* 16 (5): 389–401.

Lal, R. 1998. Soil erosion impact on agronomic productivity and environment quality. *Critical Reviews in Plant Sciences* 17 (4): 319–464.

Lal, R., G.F. Wilson, and B.N. Okigbo. 1979. Changes in properties of an alfisol produced by various crop covers. *Soil Science Society of America Journal* 27: 377.

Lamb, D., S. Heading, N. Bolan, and R. Naidu. 2012. Use of biosolids for phytocapping of landfill soil. *Water, Air, and Soil Pollution* 223 (5): 2695–2705.

Lamb, D.T., K. Venkatraman, N.S. Bolan, N. Ashwath, G. Choppala, and R. Naidu. 2013. Phytocapping: An alternative technology for the sustainable management of landfill sites. *Critical Reviews in Environmental Science and Technology* 44: 561–637.

Levin, S.B., and M. Hammond. 1990. Examination of PVC in a top cap application. *Geosynthetic Testing for Waste Containment Application* STP1081:369–382.

Li, J.H., L. Du, R. Chen, and L.M. Zhang. 2013. Numerical investigation of the performance of covers with capillary barrier effects in South China. *Computers and Geotechnics* 48: 304–315.

Licht, L., E. Aitchison, W. Schnabel, M. English, and M. Kaempf. 2001. Landfill capping with woodland ecosystems. *Practice Periodical of Hazardous, Toxic, and Radioactive Waste Management* 5 (4): 175–184.

Licht, L.A., and J.G. Isebrands. 2005. Linking phytoremediated pollutant removal to biomass economic opportunities. *Biomass and Bioenergy* 28 (2): 203–218.

Lim, H., J. Lee, H. Chon, and M. Sager. 2008. Heavy metal contamination and health risk assessment in the vicinity of the abandoned Songcheon Au–Ag mine in Korea. *Journal of Geochemical Exploration* 96 (2–3): 223–230.

Loncnar, M., M. Zupančič, P. Bukovec, and M.Z. Justin. 2010. Fate of saline ions in a planted landfill site with leachate recirculation. *Waste Management* 30 (1): 110–118.

MacDonald, N.W., R.R. Rediske, B.T. Scull, and D. Wierzbicki. 2008. Landfill cover soil, soil solution, and vegetation responses to municipal landfill leachate applications. *Journal of Environmental Quality* 37 (5): 1974–1985.

MacKenzie, A.B., and I.D. Pulford. 2002. Investigation of contaminant metal dispersal from a disused mine site at Tyndrum, Scotland, using concentration gradients and stable Pb isotope ratios. *Applied Geochemistry* 17 (8): 1093–1103.

Martin, P., and W. Stephens. 2008. Willow water uptake and shoot extension growth in response to nutrient and moisture on a clay landfill cap soil. *Bioresource Technology* 99 (13): 5839–5850.

Mayes, W.M., D. Johnston, H.A.B. Potter, and A.P. Jarvis. 2009. A national strategy for identification, prioritisation and management of pollution from abandoned non-coal mine sites in England and Wales. I.: Methodology development and initial results. *Science of the Total Environment* 407 (21): 5435–5447.

Melamed, R., X. Cao, M. Chen, and L. Ma. 2003. Field assessment of lead immobilization in a contaminated soil after phosphate application. *Science of the Total Environment* 305 (1): 117–127.

Mench, M., J. Vangronsveld, V. Didier, and H. Clijsters. 1994. Evaluation of metal mobility, plant availability and immobilization by chemical agents in a limed-silty soil. *Environmental Pollution* 86 (3): 279–286.

Miller, J.J., B.W. Beasley, L.J. Yanke, F.J. Larney, T.A. McAllister, B.M. Olson, L.B. Selinger, D.S. Chanasyk, and P. Hasselback. 2003. Bedding and seasonal effects on chemical and bacterial properties of feedlot cattle manure. *Journal of Environmental Quality* 32: 1887–1894.

Miller, R.E., X. Lei, and D.E. Ullrey. 1991. Trace elements in Animal nutrition. In *Micronutrients in Agriculture*, edited by J.J. Mortvedt, pp. 593–662. Madison, WI: Soil Science Society of America Inc.

Min, X., X. Xie, L. Chai, Y. Liang, M. Li, and Y. Ke. 2013. Environmental availability and ecological risk assessment of heavy metals in zinc leaching residue. *Transactions of Nonferrous Metals Society of China* 23 (1): 208–218.

Molz, F.J., and V.D. Browning. 1977. Effect of vegetation on landfill stabilization. *Ground Water* 15 (6): 409–415.

Moore, R.E., and M.R. Matsumoto. 1993. *Investigations of the Use of In Situ Soil Flushing to Remediate a Pb Contaminated Site*. Lancaster, PA: Hazardous and Industrial Waste.

Mulligan, C.N., R.N. Yong, and B.F. Gibbs. 2001. Surfactant-enhanced remediation of contaminated soil: A review. *Engineering Geology* 60 (1): 371–380.

Naidu, R., S.J.T. Pollard, N.S. Bolan, G. Owens, and A.W. Pruszinski. 2008a. Bioavailability: The underlying basis for risk based land management. In *Developments in Soil Science*, edited by A.B. McBratney, A.E. Hartemink and N. Ravendra, pp. 53–72. Burlington, MA: Elsevier.

Naidu, R., K.T. Semple, M. Megharaj, A.L. Juhasz, N.S. Bolan, S. Gupta, B. Clothier, R. Schulin, and R. Chaney. 2008b. Bioavailability, definition, assessment and implications for risk assessment. In *Chemical Bioavailability in Terrestrial Environment*, edited by R. Naidu, pp. 39–52. Amsterdam, the Netherlands: Elsevier.

Ngatunga, E.L.N., R. Lal, and A.P. Uriyo. 1984. Effects of surface management on runoff and soil erosion from some plots at Mlingano, Tanzania. *Geoderma* 33 (1): 1–12.

Nguyen, A.D. 2007. Geochemical and physical factors controlling arsenic mobility at the Mole River arsenic mine Australia. MPhil, School of Physical Sciences, University of Queensland, Brisbane, Queensland, Australia.

Nicholson, R., R. Gillham, J. Cherry, and E. Reardon. 1989. Reduction of acid generation in mine tailings through the use of moisture-retaining cover layers as oxygen barriers. *Canadian Geotechnical Journal* 26 (1): 1–8.

Nixon, D.J., W. Stephens, S.F. Tyrrel, and E.D.R. Brierley. 2001. The potential for short rotation energy forestry on restored landfill caps. *Bioresource Technology* 77 (3): 237–245.

Nordstrom, D.K. 2011. Hydrogeochemical processes governing the origin, transport and fate of major and trace elements from mine wastes and mineralized rock to surface waters. *Applied Geochemistry* 26 (11): 1777–1791.

Nordstrom, D.K., C.N. Alpers, C.J. Ptacek, and D.W. Blowes. 2000. Negative pH and extremely acidic mine waters from iron mountain, California. *Environmental Science & Technology* 34 (2): 254–258.

Oste, L.A., T. Lexmond, and W. van Riemsdijk. 2002. Metal immobilization in soils using synthetic zeolites. *Journal of Environmental Quality* 31 (3): 813–821.

Othman, M.A., C.H. Benson, E.J. Chamberlain, and T.F. Zimmie. 1994. Laboratory testing to evaluate changes in hydraulic conductivity of compacted clays caused by freeze-thaw: State-of-the-art. *Hydraulic Conductivity and Waste Contaminant Transport in Soil* STP1142: 227–254.

Page, M., and C. Page. 2002. Electroremediation of contaminated soils. *Journal of Environmental Engineering* 128 (3): 208–219.

Parent, S.-É., and A. Cabral. 2006. Design of inclined covers with capillary barrier effect. *Geotechnical & Geological Engineering* 24 (3): 689–710.

Park, J., G. Choppala, N. Bolan, J.W. Chung, and T. Chuasavathi. 2011a. Biochar and black carbon reduce the bioavailability and phytotoxicity of heavy metals. *Plant and Soil* 348: 439–451.

Park, J., D. Lamb, P. Paneerselvam, G. Choppala, N. Bolan, and J. Chung. 2011b. Role of organic amendments on enhanced bioremediation of heavy metal(loid) contaminated soils. *Journal of Hazardous Materials* 185 (2–3): 549–574.

Phillips, I.R., M. Greenway, and S. Robertson. 2004. Use of phytocaps in remediation of closed landfills-correct selection of soil materials. *Land Contamination & Reclamation* 12 (4): 339–348.

Phillips, I.R., and K.J. Sheehan. 2005. Use of phytocaps in remediation of closed landfills-suitability of selected soils to remove organic nitrogen and carbon from leachate. *Land Contamination & Reclamation* 13 (4): 339–348.

Preston, G.M., and R.A. McBride. 2004. Assessing the use of poplar tree systems as a landfill evapotranspiration barrier with the SHAW model. *Waste Management and Research* 22: 291–305.

Raskin, I., P.B.A.N. Kumar, S. Dushenkov, and D. Salt. 1994. Bioconcentration of heavy metals by plants. *Current Opinion in Biotechnology* 5 (3): 285–290.

Rodríguez-Vila, A., E. Covelo, R. Forján, and V. Asensio. 2014. Phytoremediating a copper mine soil with *Brassica juncea* L., compost and biochar. *Environmental Science and Pollution Research* 21 (19): 11293–11304.

Rojo, A., H.K. Hansen, and P. Guerra. 2009. Electrodialytic remediation of copper mine tailing pulps. *Separation Science & Technology* 44 (10): 2234–2244.

Ruth, B., B. Lennartz, and P. Kahle. 2007. Water regime of mechanical biological pretreated waste materials under fast-growing trees. *Waste Management and Research* 25: 408–416.

Scanlon, B., R. Reedy, K. Keese, and St. Dwyer. 2005. Evaluation of evapotranspirative covers for waste containment in Arid and Semiarid Regions in the Southwestern USA. *Vadose Zone Journal* 4 (1): 55–71.

Seshadri, B., N. Bolan, G. Choppala, and R. Naidu. 2013. Differential effect of coal combustion products on the bioavailability of phosphorus between inorganic and organic nutrient sources. *Journal of Hazardous Materials* 261: 817–825.

Shouse, P.J., W.B. Russell, D.S. Burden, H.M. Selim, J.B. Sisson, and M. van Genuchten. 1995. Spatial variability of soil water retention functions in a silt loam soil. *Soil Science* 159: 1–12.

Shrive, S.C., R.A. McBride, and A.M. Gordon. 1994. Photosynthetic and growth responses of two broad-leaf tree species to irrigation with municipal landfill leachate. *Journal of Environmental Quality* 23 (3): 534–542.

Simms, P., E. Yanful, L. St-Arnaud, and B. Aubé. 2000. A laboratory evaluation of metal release and transport in flooded pre-oxidized mine tailings. *Applied Geochemistry* 15 (9): 1245–1263.

Simon, F., and W. Müller. 2004. Standard and alternative landfill capping design in Germany. *Environmental Science & Policy* 7 (4): 277–290.

Smirnova, E., B. Bussière, F. Tremblay, and J. Cyr. 2009. Bio-intrusion barrier made of plants with allelopathic effects to improve long term performance of covers with capillary barrier effects. *Proceedings of the CLRA Conference*, Quebec City, Quebec, Canada, April 23–26, 2009.

Smith, D.C., J. Sacks, and E. Senior. 1999. Irrigation of soil with synthetic landfill leachate—Speciation and distribution of selected pollutants. *Environmental Pollution* 106 (3): 429–441.

Soucek, D., D. Cherry, R. Currie, H. Latimer, and G. Trent. 2000. Laboratory to field validation in an integrative assessment of an acid mine drainage–impacted watershed. *Environmental Toxicology and Chemistry* 19 (4): 1036–1043.

Starr, G.C., R. Lal, R. Malone, D. Hothem, L. Owens, and J. Kimble. 2000. Modeling soil carbon transported by water erosion processes. Land Degradation and Development 11 (1): 83–91.

Thangarajan, R., N. Bolan, G. Tian, R. Naidu, and A. Kunhikrishnan. 2013. Role of organic amendment application on greenhouse gas emission from soil. *Science of the Total Environment* 465: 72–96.

Tordoff, G.M., A.J.M. Baker, and A.J. Willis. 2000. Current approaches to the revegetation and reclamation of metalliferous mine wastes. *Chemosphere* 41 (1): 219–228.

Tremblay, R.L. 1994. Controlling acid mine drainage using an organic cover: The case of the East Sullivan Mine, Abitibi, Quebec. *Proceedings of the International Land Reclamation and Mine Drainage Conference and Third International Conference on the Abatement of Acidic Drainage*, Pittsburgh, PA, April 24–29, 1994.

Tsegaye, T., A. Johnson, W. Mersie, S. Dennis, and K. Golson. 2007. Transport of atrazine through soil columns with or without switchgrass roots. *Journal of Food, Agriculture and Environment* 5 (2): 345–350.

USEPA. 1997. *Recent Developments for In Situ Treatment of Metal Contaminated Soils*. Washington, DC: Office for Solid Waste and Emergency Response.

Vasudevan, N.K., S. Vedachalam, and D. Sridhar. 2003. Study on the various methods of landfill remediation. *Workshop on Sustainable Landfill Management*, Chennai, India, December 3–5, 2003.

Venkatraman, K., and N. Ashwath. 2009. Phytocapping: Importance of tree selection and soil thickness. *Water, Air, & Soil Pollution: Focus* 9 (5): 421–430.

Wijesekara, H., N.S. Bolan, M. Vithanage, Y. Xu, S. Mandal, S.L. Brown, G.M. Hettiarachchi, G.M. Pierzynski, L. Huang, Y.S. Ok, M.B. Kirkham, C.P. Saint, and A. Surapaneni. 2016. Utilization of biowaste for mine spoil rehabilitation. *Advances in Agronomy* 138: 97–173.

Will, F.G. 1995. Removing toxic substances from soil using electrochemistry. *Chemistry and Industry* 10: 376–379.

Winant, W.M., H.A. Menser, and O.L. Bennett. 1981. Effects of sanitary landfill leachate on some soil chemical properties. *Journal of Environmental Quality* 10 (3): 318–322.

Wise, D. 1994. *Remediation of Hazardous Waste Contaminated Soils*. Vol. 8. CRC Press, Boca Raton, FL.

Yang, Z., and S. Zhou. 2008. The biological treatment of landfill leachate using a simultaneous aerobic and anaerobic (SAA) bio-reactor system. *Chemosphere* 72 (11): 1751–1756.

Yunusa, I., M. Zeppel, S. Fuentes, C. Macinnis-Ng, A. Palmer, and D. Eamus. 2010. An assessment of the water budget for contrasting vegetation covers associated with waste management. *Hydrological Processes* 24 (9): 1149–1158.

Zhuang, P., Z.A. Li, B. Zou, H.P. Xia, and G. Wang. 2013. Heavy metal contamination in soil and soybean near the Dabaoshan mine, South China. *Pedosphere* 23 (3): 298–304.

Zhuang, P., M.B. McBride, H.P. Xia, N.Y. Li, and Z.A. Lia. 2009. Health risk from heavy metals via consumption of food crops in the vicinity of Dabaoshan mine, South China. *Science of the Total Environment* 407 (5): 1551–1561.

Zornberg, J.G., LaFountain, L., and Caldwell, J.A. 2003. Analysis and design of evapotranspirative cover for hazardous waste landfill. *Journal of Geotechnical and Geoenvironmental Engineering*, 129 (5): 427–438.

Zupanc, V., and M.Z. Justin. 2010. Changes in soil characteristics during landfill leachate irrigation of *Populus deltoides*. *Waste Management* 30 (11): 2130–2136.

Zupancic, M., M. Justin, P. Bukovec, and V. Selih. 2009. Chromium in soil layers and plants on closed landfill site after landfill leachate application. *Waste Management* 29 (6): 1860–1869.

14 Rehabilitation of an Abandoned Mine Site with Biosolids

Abdulaziz Alghamdi, M.B. Kirkham, Deann R. Presley,
Ganga Hettiarachchi, and Leigh Murray

CONTENTS

14.1 Introduction... 241
14.2 Materials and Methods 243
14.3 Results and Discussion 248
Acknowledgments.. 255
References... 255

14.1 INTRODUCTION

Abandoned mine sites have left a legacy of contamination worldwide. The environmental problems associated with them are serious and global (Dybowska et al. 2006). The lead (Pb) and zinc (Zn) mines in the Tri-State Mining District of southeast Kansas, southwest Missouri, and northeast Oklahoma are such mines. The wastes from these mines have polluted groundwater, rivers, lakes, sediments, and soils (Abdel-Saheb et al. 1994; Carroll et al. 1998; Brown et al. 2004; Pierzynski and Gehl 2004; Schaider et al. 2007; Schwab et al. 2007; Juracek 2008; Pierzynski et al. 2010), as well as fish and mussels (Brumbaugh et al. 2005; Schmitt et al. 2005; Angelo et al. 2007). Methods to remediate the mine wastes are urgently needed (Johnson et al. 2016).

This District has a history of mining that goes back to the early 1800s, when Pb was mined by trappers and explorers for bullets (Pope 2005, p. 5; Baker 2008, pp. 12–13). Commercial mining in this region began about 1850 (Gibson 1972, p. 14) and rapidly expanded after the Civil War (Gibson 1972, p. 26). Mining operations were first limited, because of the lack of adequate transportation (Gibson 1972, p. 24) and heavy machinery. However, after the Civil War, around 1870, railroads extended lines into the Tri-State District. In the late 1800s, small, individually owned surface mines were bought out by larger mining companies (Gibson 1972, p. 68), which further improved opportunities for development. From 1850 to 1950, the District was the world's leading producer of Pb and Zn concentrates, accounting for 50% of the U.S. Zn production and 10% of its Pb production (Gibson 1972, p. 266). The mines in the District supplied the United States so it could fight four major wars (Civil War, World War I, World War II, and the Korean War). Industry in the United States and abroad used the Pb and Zn concentrates from the District to produce munitions, bearings, castings, pipe, galvanized metals, batteries, chains, nails, and numerous other products (Gibson 1972, pp. 266–267). The mines in the Tri-State Mining District lasted until 1970 (Pope 2005, p. 1; United States Environmental Protection Agency 2007). The District includes Galena, Kansas, where mines began to operate in 1876 (Pope 2005, p. 7). A century of mining operations in Galena has left Pb and Zn contamination throughout the city. The waste materials around the mines are highly polluted, not only with Pb and Zn, but also with cadmium (Cd), which co-occurs geologically with Zn.

Despite the high standard of living brought to Americans by the products from the Tri-State mines (Gibson 1972, p. 266), the miners and local population endured health problems from the beginning of the mining activities. The miners succumbed to silicosis, or miner's consumption, caused by the inhalation of the flint dust produced by drilling and blasting (Gibson 1972, p. 182). It predisposed the miners and townspeople to tuberculosis. In 1940, Cherokee County in Kansas, where Galena is located, recorded more cases of tuberculosis than any other county in Kansas, and in 1951, the county's death rate from tuberculosis was six times greater than for the rest of the state (Gibson 1972, p. 194). Residents of Galena have a higher incidence of kidney and heart diseases, skin cancer, and anemia compared to residents in control towns (Neuberger et al. 1990). These results suggest that environmental agents in Galena are associated with the causation of several chronic diseases among the residents.

Adding amendments to allow plant growth has been suggested as one way to remediate mine wastes. Amendments have been applied to the waste materials in the Tri-State Mining District, and Pierzynski et al. (1994) review the early literature about them. Pierzynski et al. (2002) added cattle manure as a soil amendment to the mine tailings at Galena to see if tall fescue (*Festuca arundinacea* Schreb.) would grow. After the first growing season, vegetative cover reached 71% but then steadily declined to 29% over the next two growing seasons. They attributed the poor growth to Zn toxicity.

Biosolids (sewage sludge) have often been used to remediate mine sites (Haering et al. 2000; Brown et al. 2005; Karathanasis et al. 2007; Stuczynski et al. 2007; Santibáñez et al. 2008; Sheoran et al. 2010; Madejón et al. 2012; Pepper et al. 2013; Mahar et al. 2015; Wijesekara et al. 2016). Biosolids are recommended for amelioration of degraded land, because they add nutrients—especially nitrogen and phosphorus—and organic matter to the soil for plant growth (Kirkham 1974; Lu et al. 2012). Recycling of biosolids to reclaim and revegetate areas disturbed by mining has long been promoted by the United States Environmental Protection Agency (1989).

However, little work has been done using biosolids to remediate the mine waste materials in the Tri-State Mining District. Brown et al. (2007) added amendments, including lime-stabilized biosolids and composted biosolids, on tailings from the Tar Creek National Priorities List Superfund Site in Oklahoma to see if they would restore vegetation and reduce availability of heavy metals that contaminate the tailings (Pb, Zn, and Cd). Plots were seeded with Bermuda grass (*Cynodon dactylon* Pers.). Bioaccessible Pb in the tailings was measured using a physiologically based extraction test. The biosolids did not reduce bioaccessible Pb in the tailings. When diammonium phosphate fertilizer was added with the biosolids, bioaccessible Pb was reduced. In general, 6 months after the amendments were added, growth was poor due, in large part, to the high electrical conductivity of the tailings (9.0 dS/m). But 18 months after they were added, all plots supported plant growth. Brown et al. (2007) reported plant Cd and Zn in the Bermuda grass, but not plant Pb.

Between 1998 and 2001, Brown et al. (2014) added biosolids plus lime to plots in Jasper County in southwestern Missouri, part of the Tri-State Mining District. They planted different grasses on the plots. They found that in 2012, 11–14 years after the amendments were added, dry matter of orchard grass (*Dactylis glomerata* L.), big bluestem (*Andropogon gerardii* Vitman), little bluestem (*Andropogon scoparius* Michx.), turkey foot (scientific name not given by Brown et al. 2014), Indian grass (*Sorghastrum nutans* L.), sideoats gamma (*Bouteloua curtipendula* Michx.), and fescue (*Festuca* sp.) were increased due to the biosolids with lime. In 2012, the average plant dry weight with biosolids (336 Mg/ha) and lime (48 Mg/ha) was 46 g/m^2, while minimal or no plant growth occurred at the control sites with non-amended mine waste materials. Brown et al. (2014) reported Pb, Zn, and Cd in the grass leaves, but did not report the concentrations of heavy metals in roots or in heads with grain.

In 2006, researchers at Kansas State University (Baker et al. 2011) added amendments (compost, lime, and bentonite) to mine waste materials at Galena, to see if they would change their microbial properties. They found that only high levels of compost increased microbial activity. Biosolids have never been applied to the mine waste materials at Galena, to see if they would reduce availability of the heavy metals to plants. Therefore, we sampled the waste materials in the plots established by Baker et al. (2011) and set up a greenhouse study with sudex, a sorghum-sudan grass hybrid, to determine the effect of biosolids on the growth of the sudex and transfer of heavy metals from roots to shoots and then to heads.

14.2 MATERIALS AND METHODS

The experiment was carried out between January and May 2015 in a greenhouse at Kansas State University in Manhattan, KS (39°12′N, 96°35′W, 325 m above sea level), using mine waste materials from abandoned mines located at Galena, KS (37°5′N; 94°38′W; 275 m above sea level). In 2006, Baker et al. (2011) established plots at two sites in Galena, called Site A and Site B, where mine waste materials had been collected and deposited on the surface for 100 years. The mine waste materials were a by-product in the initial processing of Zn- and Pb-containing ores. Both Site A and Site B were inside the city limits of Galena and were 2 km apart (Figure 14.1). Site B was near the center of the town and houses were around it, and Site A was on the outskirts of town and it had no buildings around it. Site A was established on May 8, 2006, and Site B on May 12, 2006. The sites were on level ground. Each experimental plot was 1 m × 2 m in size with three replications of seven different treatments, for a total of 21 plots at each site, or a total of 42 plots. Each plot had a galvanized steel border, 1 m × 2 m in size, to limit inter-plot contamination (Baker 2008, p. 138).

The seven treatments were (1) non-amended control plot, (2) a low compost treatment of 45 Mg/ha, (3) a high compost treatment of 269 Mg/ha, (4) low compost (45 Mg/ha) + lime as $Ca(OH)_2$ (11.2 Mg/ha), (5) high compost (269 Mg/ha) + lime as $Ca(OH)_2$ (11.2 Mg/ha), (6) low compost (45 Mg/ha) + lime as $Ca(OH)_2$ (11.2 Mg/ha) + bentonite applied at 50 g bentonite per kg compost, and (7) high compost (269 Mg/ha) + lime applied as $Ca(OH)_2$ (11.2 Mg/ha) + bentonite applied at 50 g bentonite per kg compost. The compost was composted beef (*Bos taurus*) manure, and the bentonite was a Wyoming bentonite obtained from Enviroplug Grout (Wyo-Ben, Inc., Billings, MT) (Baker 2008, p. 138). Treatments were applied and mixed to a depth of 30 cm (Baker 2008, p. 138). Switchgrass (*Panicum virgatum* L.) was seeded on the plots on May 26, 2006. Switchgrass did not grow on the plots in 2006 due to the high salinity of the compost and a lack of rainfall. In the fall of 2006, plots were seeded to annual ryegrass (*Lolium multiflorum* Lam.) as a winter cover crop, and the ryegrass was killed with glyphosate in the spring of 2007, when the plots were reseeded with switchgrass on April 19, 2007 (Baker 2008, pp. 138–139). The plots were sampled for biomass 535 and 841 days after Day 0, which Baker (2008, p. 139) designated as May 26, 2006. These days were November 12, 2007, and August 14, 2008.

No more amendments were added to the plots between May 2006, when the plots were established by Baker et al. (2011), and November 18–19, 2014, when we sampled the plots 8.5 years after the amendments had been added. At that time in 2014, we scooped up the top 13 cm (5 in.) of mine waste materials from each plot at Site A and Site B with a flat shovel and put them into 42 5-gallon (19 L) buckets with lids (Product Code 0 84305 3559 1, Home Depot, Atlanta, GA), one bucket for each plot. On November 19, 2014, we brought the buckets back to Manhattan, KS, where the mine waste materials were laid out on brown paper in the greenhouse to dry. On January 1, 2015, the mine waste materials were sieved using a sieve with 4 mm openings. On January 13, 84 plastic pots (each 22 cm in diameter and 22 cm in height) were filled with the mine waste materials. On January 13, liquid, aerobically digested biosolids from the Manhattan, KS, Wastewater Treatment Plant were obtained. On January 14, 1000 mL tap water was added to each pot and each pot drained. On January 21 and January 22, 500 mL of the liquid biosolids were applied each day to the surface of 42 pots (half of the pots). The 1000 mL of biosolids that were added made a layer about 1 cm thick. The percentage of dry solids of the biosolids was measured to be 2.35%, following the method described by the New York State Department of Environmental Conservation (c. 1965, p. 218). On January 26 and 27, 500 mL tap water was added each day to each pot that did not contain biosolids, and the pots drained each day due to the coarse nature of the mine waste materials, even though they had been sieved to 4 mm.

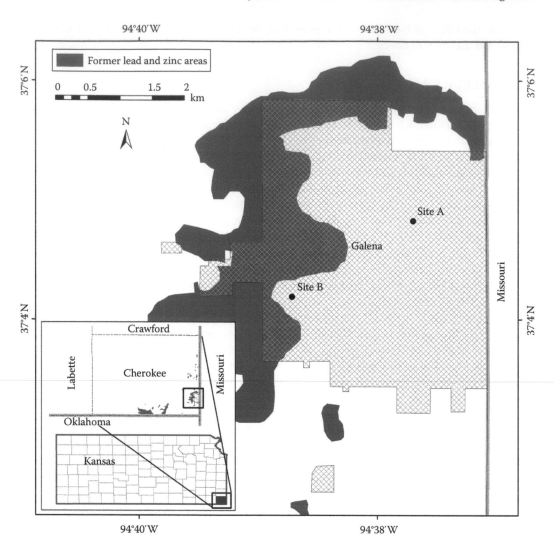

FIGURE 14.1 Map of Galena, Kansas, showing the location of Site A and Site B, where the mine waste materials were sampled. A map of Kansas is in the lower left-hand corner, and Cherokee County in southeastern Kansas is highlighted. Cherokee County is enlarged above the map for Kansas. Labette County is west of Cherokee County, and Crawford County is north of it. The state of Missouri is east of Cherokee County, and the state of Oklahoma is south of it. Galena is in the southeastern part of Cherokee County, and it is highlighted in the map of Cherokee County. In the map for Galena, the gray striations delineate the boundary of the town, and the areas in red show where the lead and zinc mines were located. (Courtesy of Ali Alghamdi, Department of Geography, Kansas State University.)

On January 28, 20 seeds of the forage crop sudex (*Sorghum bicolor* (L.) Moench × *S. sudanese* (Piper) Staph) (Chu and Kerr 1977; Summers et al. 2009) were planted in each of the 84 pots. We chose sudex because it is recommended for erosion control and to improve soil structure (Summers et al. 2009). On February 17, the plants in each pot were thinned to 10 plants per pot. All pots had 10 plants except for 9 pots. No plants germinated in two pots (both pots with mine waste material from the control plots, and one had biosolids and the other did not), and seven pots had between four and nine plants.

TABLE 14.1
Monthly Average of Day and Night Temperature (°C)
and Day and Night Humidity (%) at Four Locations
in the Greenhouse during the Experiment

Month (2015)	Temperature		Humidity	
	Day	Night	Day	Night
Northeast				
February	18.9	16.2	31.3	40.2
March	20.7	20.4	36.5	37.8
April	21.6	21.1	46.7	44.7
Southeast				
February	18.5	15.6	32.9	41.4
March	20.4	20.1	36.9	38.2
April	21.6	21.3	46.5	44.8
Northwest				
February	18.8	15.7	31.6	41.2
March	20.7	20.4	36.2	37.2
April	21.4	21.4	46.3	44.3
Southwest				
February	17.5	14.7	34.5	43.8
March	20.3	20.1	37.2	38.4
April	21.7	21.6	46.3	44.6

Between January 30 and May 15, the plants were kept well watered. The water content in the mine waste materials was monitored using inexpensive soil moisture meters (Faber et al. 1993). Due to the roughness of the mine waste materials, the rods on the meters broke, so two different types were used: Mini Moisture Tester (Luster Leaf Products, Inc., Woodstock, IL; Product Code 0 35307 01810 6) and HoldAll Moisture Meter (Panacea Products Corp., Columbus, OH; Product Code 0 70686 26002 9). Based on the measurements with the moisture meters, 500 mL tap water was usually added twice a week to each pot.

During the experiment, the temperature and relative humidity were measured hourly between February 9 and April 21 with a sensor for each in four boxes located in the southeast, southwest, northwest, and northeast corners of the greenhouse. The sensors were part of the Throckmorton Greenhouse Temperature Tracking System, an in-house built system. The sensors were hand-assembled by Arthur Selman, Network Specialist and Instructor at Kansas State University. Table 14.1 shows the average monthly day and night temperature and the average day and night relative humidity during the experiment at the four locations in the greenhouse. Pan evaporation rate between January 28 and May 15 averaged 0.30 cm/day. Natural daylight was used during the experiment.

Throughout the experiment, the height of the shoots was measured once a week by choosing at random 5 plants per pot. The height was measured from the surface of the mine waste materials to the tip of an extended leaf.

The plants were harvested on May 18–19 (110 and 111 days after planting) by cutting the culms just above the surface of the mine waste materials. Heads with grain were removed, if a plant produced a head. Leaves and culms were combined and labeled "shoots," and they were put into paper bags. Fresh weight of the shoots was measured. Dry weights of the shoots were determined by drying them to constant weight at 70°C. For the two pots that had no plants, their fresh and dry weights were recorded as zero. Roots were extracted by washing them in water to remove adhering mine waste materials. Because not all the roots were extracted, fresh and dry weights of the roots were not determined. The plant tissues were submitted to the Soil Testing Laboratory at Kansas State University for analysis.

The roots, shoots, and heads were digested using a nitric-perchloric acid digest (Kirkham 2000) and analyzed for P, K, Ca, Mg, Cd, Cu, Fe, Mn, Ni, Pb, and Zn using inductively coupled plasma-atomic emission spectroscopy (ICP-AES), also referred to as inductively coupled plasma-optical emission spectrometry (ICP-OES). Detection limits in mg/kg for the ICP-AES were Cd, 0.005; Cu, 0.003; Fe, 0.05; Mn, 0.003; Ni, 0.007; Pb, 0.003; and Zn, 0.003. These are the detection limits given by the manufacturer for optimal conditions. The detection limits for the elements in mg/kg of the plant samples were calculated as follows. Each plant sample had about 0.25 g, and the sample volume was 50 mL; 50 mL/0.25 g = 200. Each detection limit given by the manufacturer was multiplied by 200. This gave detection limits for the plant samples in mg/kg as follows: Cd, 1.0; Cu, 0.6; Fe, 10.0; Mn, 0.6; Ni, 1.4; Pb, 0.6; and Zn, 0.6. For the major elements, the practical quantitation limit of the ICP-AES is 1 mg/kg. Quality assurance/quality control was done by duplicating 10% of the samples, and the standard reference material came from the National Institute of Standards and Technology (NIST, SRM 1515, apple leaves).

For determination of total nitrogen in plant tissues, a salicylic-sulfuric acid digestion was used (Bremner and Mulvaney 1982, p. 621), and then the digest was analyzed for N by a colorimetric procedure using the Rapid Flow Analyzer (Model RFA-300) and RFA Methodology No. A303-S072 from Alpkem Corporation, Clackamas, OR.

The mine waste materials were analyzed for chemical constituents when they were brought back from Galena, KS, in November, 2014, and at the end of the greenhouse experiment. When the mine waste materials were sampled at the end of the experiment, the contents of each pot were dumped onto brown paper and then all the contents of each pot were mixed up. For the pots with biosolids, the crust was mixed up into the mine waste materials. The samples were placed in brown bags, which were submitted to the Soil Testing Laboratory for analysis.

The mine waste materials were analyzed for total concentrations of heavy metals using a nitric acid digest (Wahla and Kirkham 2008). Heavy metals in the mine waste materials were extracted using diethylenetriaminepentaacetic acid (DTPA) (Lindsay and Norvell 1978), and extractable concentrations were determined using the ICP-AES. Total P in the mine waste materials was determined in the same way as total N was determined for the plant samples (see earlier). Extractable P in the mine waste materials was determined using the Mehlich 3 test (Frank et al. 1998). The pH, electrical conductivity, and cation exchange capacity of the mine waste materials were determined using the methods described by Watson and Brown (1998), Whitney (1998), and Warncke and Brown (1998), respectively. Organic matter in the mine waste materials was determined using the loss of weight on ignition method described by Combs and Nathan (1998). Total nitrogen and total carbon in the mine waste materials, as well as total carbon in the heads, were determined using a LECO TruSpec CN Carbon/Nitrogen combustion analyzer, which reports total levels (inorganic and organic) of C and N on a weight percent basis, according to, for the mine waste materials, the TruSpec CN instrument method "Carbon and Nitrogen in Soil and Sediment," and, for plant tissue, the TruSpec CN instrument method "Carbon, Hydrogen, and Nitrogen in Flour and Plant Tissue," both published by LECO Corporation, St. Joseph, MI, in 2005. After determination of total C and N in the mine waste materials, total organic carbon was determined as follows. By pretreatment of a second LECO combustion sample with dilute phosphoric acid, carbon dioxide is released from calcium and magnesium carbonates in the mine waste materials, leaving only the total organic carbon present, which is then calculated.

The tap water used for watering the plants during the experiment was analyzed for pH, electrical conductivity, and elemental composition using the methods described earlier. The tap water had the following chemical characteristics: pH, 8.72; electrical conductivity, 0.40 dS/m; Ca, 23.51 mg/kg; K, 7.53 mg/kg; Mg, 14.54 mg/kg; Na, 45.02 mg/kg; and Cu, 0.01 mg/kg. Cadmium, Fe, Mn, Ni, Pb, and Zn were below detection levels (less than 0.01 mg/kg) in the tap water.

The biosolids were analyzed after they were collected on January 13, 2015. They were dried, and the sample was submitted to the Soil Testing Laboratory. It was analyzed for total concentrations of heavy metals, pH, organic matter, and total P using the same methods that were used for the mine waste materials. The biosolids sample had the following chemical characteristics: pH, 4.65; organic matter, 59.90%; total P, 3.40%; Cd, 9.1 mg/kg; Cu, 361.3 mg/kg; Fe, 3850.4 mg/kg; Mn, 90.1 mg/kg; Ni, 1.6 mg/kg; Pb, 13.1 mg/kg; and Zn, 577.0 mg/kg.

Another sample of the biosolids from the Manhattan, Kansas, Wastewater Treatment Plant was obtained on June 20, 2016, and a wet sample was submitted to the Soil Testing Laboratory for analysis of pH (Watson and Brown 1998), electrical conductivity (Whitney 1998), total suspended solids, total N, and total P. Total suspended solids were determined by filtering the biosolids through a 0.45 μm filter using a vacuum. The dry weight of the filter member was measured before and after filtration. The total suspended solids were calculated as mg/L (Csuros 1997). The total N and P in the liquid sample were determined by taking a 10 mL sample, which was then digested with a potassium persulfate reagent (Nelson 1987) in an autoclave and then analyzed according to Hosomi and Sudo (1986). The 2016 analyses showed that the wet biosolids had a pH of 6.10, an electrical conductivity of 1.75 dS/m, 18,200 mg/L total suspended solids, total N of 825.11 mg/kg, and total P of 685.52 mg/kg. An electrical conductivity less than 2 dS/m has negligible effects on crop growth (Bernstein, 1964). Suspended solids refer to small solid particles that remain in suspension in the liquid biosolids. The smaller the number, the better the biosolids are digested. Nitrogen and P, as documented in the literature, are given in the next paragraph. The sample obtained on June 20, 2016 was dried and extractable and the total concentrations of heavy metals were determined using the methods used on the mine waste materials. Extractable concentrations in mg/kg were Cd, 1.2; Cu, 96.5; Fe, 536.7; Mn, 79.7; Ni, 10.8; Pb, 7.4; and Zn, 406.6. Total concentrations in mg/kg were Cd, 1.1; Cu, 314.6; Fe, 6072.3; Mn, 165.1; Ni, 13.2; Pb, 14.0; and Zn, 442.5.

On a long-term basis (1995–2014), the Manhattan biosolids have 2.5% by weight of total solids; 1,053.9 mg/kg Kjeldahl N on a wet-weight basis; 44,143.4 mg/kg total Kjeldahl N on a dry-weight basis; and 25,153.9 mg/kg total P on a dry-weight basis. In 2015, the Manhattan biosolids were analyzed by the city four times (March, May, October, and November), and the average of the four analyses showed that the biosolids had 3.8% by weight of total solids; 1,814.1 mg/kg Kjeldahl N on a wet-weight basis; 48,575.0 mg/kg total Kjeldahl N on a dry-weight basis; and 30,800.0 mg/kg total P on a dry-weight basis (Dr. Abdu Durar, Water and Wastewater Division, Manhattan, KS, personal communication, June 22, 2016). The concentrations of N and P in the biosolids are similar to those reported in the literature (Dean and Smith 1973; Peterson et al. 1973; Vesilind 1975, p. 23). On a dry-weight basis, biosolids have between 1.8% and 6.4% N and 0.8% and 3.9% P.

The greenhouse had three benches oriented in the east-west direction. The door to the greenhouse, located on an interior hallway, was on the north wall of the greenhouse, which had a cooling pad. The south wall of the greenhouse had two fans and windows that faced outside. Near the ceiling of the greenhouse was a large plastic tube, about 80 cm in diameter, with holes through which air was pushed out to ventilate the greenhouse. The tube ran from the north wall to the south wall of the greenhouse. At the beginning of the experiment, the pots with the mine waste materials from the two sites and the three replications established by Baker et al. (2011) were placed on the three greenhouse benches, with one of the three replications from Site A and one of the three replications from Site B on each bench. See Alghamdi (2016, p. 61) for the layout. Pots with the seven original treatments of Baker et al. (2011) were placed randomly in a row. Pots with biosolids were lined up on the north side of each bench and the pots without biosolids were lined up on the south side of each bench. Therefore, on each half of a bench, there were two rows of seven pots.

On Mondays, Wednesdays, and Fridays of each week during the experiment, the pots were rotated. In each row of seven pots of a replication, either with or without biosolids, the pot at the eastern side was put in the location of the pot at the western side. Then each pot in a row was moved one pot toward the east. Consequently, each pot had a new position in the greenhouse three times a week.

The experimental design of Baker et al. (2011) was a complete block with treatment as the main factor at each site. They separated the sites (Site A and Site B) due to a significant site by treatment interaction ($p \leq 0.05$) for all measurements. We also analyzed the sites separately. The experiment was a randomized complete block with a split-plot design, in which site was the fixed blocking factor (whole plot treatment factor) for locations within sites (whole plot experimental units) and biosolids' method was the fixed split-plot treatment factor. There were a total of 84 pots (2 sites; 7 original treatments; 3 sample locations within a site; and 2 biosolids' methods, i.e., with and without biosolids). Because differences in measurements taken from plots from the seven original treatments established by Baker et al. (2011) generally were not significant at 0.05, the seven treatments were averaged together. Alghamdi (2016, Appendix B) shows the statistical analyses for the seven individual treatments. After averaging the 7 original treatments, the observations were reduced to 12 records (2 sites; 3 sample locations within a site; and 2 biosolids' methods). Statistical analyses were performed using PROC GLIMMIX of SAS Version 9.4 (Statistical Analysis System 2013). Least Square Means of biosolids' methods (with or without biosolids) were compared at the 0.05 level of significance within each site.

14.3   RESULTS AND DISCUSSION

As noted in the description of the statistical analysis (previous paragraph), differences in measurements taken from plots from the seven original treatments established by Baker et al. (2011) generally were not significant. This was true both for the measurements taken on the mine waste materials and on the plants. If differences did occur among treatments, they were not consistent. The measurements taken on plots with the amendments did not always have a lower or higher value than those from the control plots. Therefore, 8.5 years after the amendments were added, their effects were no longer evident based on our measurements. Baker et al. (2011) concluded that large amounts of organic matter were needed to support biomass in the mine waste materials over the 2-year period that they studied. Their results implied that their amendments would have to be added year after year to sustain growth on the mine waste materials. Gudichuttu (2014) suggested that high amounts of compost would be needed to maintain long-term sustainability of the plots established by Baker et al. (2011). In contrast to these results, differences among the treatments were observed nearly 8 years after Baker et al. (2011) added their amendments, as reported by Wijesekara et al. (2016, p. 154). The high compost treatment had higher microbial activity than the other treatments, although the activity had decreased substantially over the almost 8 years. Wijesekara et al. (2016, p. 155) said that, for maintenance of the effects of the amendments added by Baker et al. (2011), they would have to be added every 4–5 years.

The mine waste materials at both Site A and Site B were highly contaminated with Pb, Zn, and Cd (Table 14.2). They had at least 10 times more Pb, Zn, and Cd than non-contaminated soils. Total concentrations of these heavy metals in non-contaminated soils range from 2 to 200 mg/kg for Pb, 10 to 300 mg/kg for Zn, and 0.01 to 0.7 mg/kg for Cd (Kirkham 2008). In non-contaminated soils, extractable concentrations range from 0.05 to 46 mg/kg for Pb, 0.01 to 200 for Zn, and 0.01 to 0.5 mg/kg for Cd (Kirkham 2008). Total and extractable concentrations of Cu, Mn, and Ni in the mine waste materials were within normal concentration ranges. Total concentrations of Cu, Mn, and Ni in soils range from 2 to 200, 100 to 4000, and 5 to 5000 mg/kg, respectively (Kirkham 2008). Extractable concentrations of Cu, Mn, and Ni in soils range from 0.002 to 19.2, 0.001 to 4.8, and 0.01 to 403 mg/kg, respectively. The element Fe is abundant in soils, the amount ranging from 200 mg/kg to at least 10% (Sauchelli 1969, p. 40). Norrish (1975) gives an average concentration of Fe in soils, based on data from many different references, as 30,000 mg/kg. Therefore, the Fe in the mine waste materials was within normal ranges. The electrical conductivities of the mine waste materials at both Site A (0.21 dS/m) and Site B (0.32 dS/m) were low. At 0–2 dS/m (or 0–2 mmhos/cm), salinity effects are mostly negligible on crops (Bernstein 1964).

TABLE 14.2

Total and Extractable Concentrations (mg/kg) of Seven Heavy Metals in Mine Waste Materials Sampled on November 18–19, 2014, at Two Different Sites in Galena, Kansas

	Location	
Measurement	Site A	Site B
Cd, total	31.3a	40.6a
Cu, total	44.1a	40.9a
Fe, total	3618.9b	9057.0a
Mn, total	62.0b	212.4a
Ni, total	2.9b	3.6a
Pb, total	2643.0a	1126.6b
Zn, total	3480.3a	3070.5b
Cd, extractable	6.6a	1.5b
Cu, extractable	3.3b	8.1a
Fe, extractable	2.3b	33.5a
Mn, extractable	0.2b	2.4a
Ni, extractable	0.2a	0.1b
Pb, extractable	162.8a	67.2b
Zn, extractable	309.4a	143.2b
pH	7.16a	6.44b
EC, dS/m	0.21b	0.32a
CEC, meq/100 g	4.75a	5.71a
Organic matter, %	1.73a	1.56a
Total N, %	0.13a	0.11a
Total C, %	1.19a	1.03a
Total organic C, %	1.08a	0.98a
Total P, mg/kg	687a	717a
Extractable P, mg/kg	253a	175a

Site A was on the outskirts of town, and Site B was near the center of town. Also given are the pH, electrical conductivity (EC), cation exchange capacity (CEC), organic matter, total nitrogen, total carbon, total organic carbon, total phosphorus, and extractable phosphorus of the mine waste materials at the two sites. Within each row, values with the same letter do not differ significantly at 0.05. Each value is the average of 21 measurements. See text for description of statistical analyses.

Plants grown with biosolids grew taller than plants grown without biosolids (Figure 14.2). Plants grown on mine waste materials from Site B grew taller than plants grown on mine waste materials from Site A. At Site A and Site B, the shoots of the plants that grew with biosolids produced 5 and 8 times more fresh weight, and 7 and 13 times more dry weight, respectively, than shoots of plants without biosolids (Table 14.3). Roots in pots with biosolids grew to the bottom of the pots and penetrated the entire volume of the pots. Roots in pots without biosolids were short and were only on the surface of the pots. Only plants grown with biosolids produced heads with grain.

In addition to being stunted, the plants grown without biosolids were chlorotic and showed purple coloration. The symptoms were similar to photographs of toxicities caused by Pb and Cd (Dr. Douglas J. Jardine, Professor, Department of Plant Pathology, Kansas State University, personal communication, May 13, 2015). Hassett et al. (1976) found that radicle elongation of soil-grown corn

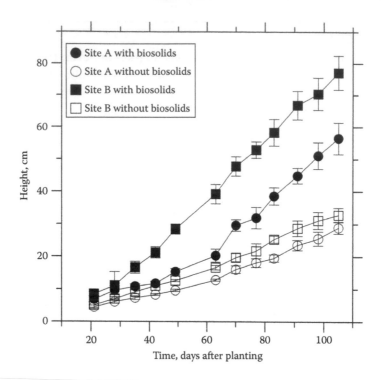

FIGURE 14.2   Height of sudex grown with and without biosolids in mine waste materials from two different sites in Galena, Kansas. Site A was on the outskirts of town, and Site B was near the center of town. The day of planting was January 28, 2015. Mean and standard deviations are shown for each data point (n = 105). If no standard deviation bars show, they fell within the data point.

TABLE 14.3

Fresh and Dry Weights (g/Pot) of Shoots and Heads of Sudex Grown with and without Biosolids in Mine Waste Materials from Two Different Sites in Galena, Kansas

	Site A		Site B	
	With Biosolids	**Without Biosolids**	**With Biosolids**	**Without Biosolids**
Fresh Weight				
Shoots				
	41.1a	8.4b	72.6A	9.2B
Heads				
	0.8	—[a]	2.0	—
Dry Weight				
Shoots				
	9.6a	1.3b	17.1A	1.3B
Heads				
	0.2	—	0.4	—

Site A was on the outskirts of town, and Site B was near the center of town. Within each row, values with the same lowercase letter do not differ significantly at 0.05 and values with the same capital letter do not differ significantly at 0.05. Each value is the average of 21 pots. See text for description of statistical analyses.

[a]   Plants grown without biosolids did not produce heads.

(*Zea mays* L.) seedlings was depressed by concentrations of 25 mg Cd/kg of soil or 250 mg Pb/kg of soil when the metals were added singly. But when Pb and Cd were added in combination, inhibition of radicle elongation occurred at lower concentrations. The effect of the metals when added in combination was greater than the sum of the effects when the metals were added singly. This shows that growth is more reduced when two toxic heavy metals are present as compared to one. The fact that Pb, Zn, and Cd were all elevated in the mine waste materials was a reason why they had an extremely deleterious effect on growth of the sudex without biosolids. As will be discussed later, the phosphorus in the biosolids appeared to immobilize the heavy metals, which allowed the plants with biosolids to grow tall and to maturity.

Roots at both Site A and Site B were highly contaminated with Pb, Zn, and Cd, but the contamination was less at Site B than at Site A (Table 14.4). Concentrations of these heavy metals far exceeded normal concentrations. The maximum concentrations of Pb, Zn, and Cd in plants

TABLE 14.4

Concentration (mg/kg) of Heavy Metals in Roots, Shoots, and Heads of Sudex Grown with and without Biosolids in Mine Waste Materials from Two Different Sites in Galena, Kansas

	Site A		Site B	
Heavy Metal	With Biosolids	Without Biosolids	With Biosolids	Without Biosolids
		Roots		
Cd	42.6b	56.4a	18.7B	42.0A
Cu	61.4a	41.3b	49.7A	60.6B
Fe	4842.7a	6465.9a	6465.9B	13,067.1A
Mn	72.2a	52.4a	116.3A	131.6A
Ni	8.0a	7.7a	6.9B	12.4A
Pb	1196.8b	1504.7a	585.0B	715.1A
Zn	7771.7a	6174.8b	3054.0A	5,045.6A
		Shoots		
Cd	12.1b	22.1a	6.6B	10.5A
Cu	5.5a	6.3a	8.9B	15.0A
Fe	443.1a	554.8a	186.9B	1,966.8A
Mn	50.6a	25.2b	57.3A	65.5A
Ni	1.6a	1.4a	1.1B	2.8A
Pb	59.7b	163.0a	22.2B	121.0A
Zn	2307.3a	1791.9a	711.5A	1,117.9A
		Heads		
Cd	2.2	—[a]	0.7	—
Cu	5.9	—	11.0	—
Fe	316.7	—	159.9	—
Mn	17.2	—	21.4	—
Ni	1.4	—	0.7	—
Pb	3.8	—	1.8	—
Zn	116.4	—	57.0	—

Site A was on the outskirts of town, and Site B was near the center of town. Within each row, values with the same lowercase letter do not differ significantly at 0.05 and values with the same capital letter do not differ at 0.05. Each value is the average of 21 measurements. See text for description of statistical analyses.

[a] Plants grown without biosolids did not produce heads.

grown under non-contaminated conditions are 5.0, 150, and 0.20 mg/kg, respectively (Liphadzi and Kirkham 2006). Except for Mn in roots with and without biosolids at Site A, Cu, Fe, Mn, and Ni were elevated above normal levels in the roots. Normal maximum concentrations of Cu, Fe, Mn, and Ni in plants are 15, 300, 100, and 1.0 mg/kg, respectively (Liphadzi and Kirkham 2006).

Concentrations of heavy metals in the shoots were less than in the roots (Table 14.4). Concentrations of Cu and Mn in all shoots were within normal ranges. Concentrations of Cd, Fe, Ni, Pb, and Zn in the shoots were higher than those normally found in plants.

Except for Ni at Site A and Cd at both Sites A and B, concentrations of heavy metals in the heads of plants were within normal concentration ranges (Table 14.4). And Ni was only slightly elevated above normal levels in the heads at Site A (1.4 versus 1.0 mg/kg for the normal maximum concentration). Even though the roots and shoots were highly contaminated with Pb and Zn, the concentrations of these heavy metals were within normal concentration ranges in the heads.

Other studies have shown that limited movement of heavy metals through plants offers a method to reduce their toxicity. Kirkham (1975) found that roots of corn grown in plots that had been treated for 35 years with biosolids contained high concentration of heavy metals, but only Cd and Cu were elevated in the leaves, and the grain had normal concentrations of heavy metals.

Cadmium was the heavy metal that was elevated above normal levels in the heads, especially at Site A. Site A and Site B had 2.2 and 0.7 mg/kg Cd in the heads, respectively, and the normal limit for Cd in plants is 0.2 mg/kg (Liphadzi and Kirkham 2006). Unlike Pb, which is highly immobile in plants (Liphadzi and Kirkham 2006), Cd is known to be mobile in plants and move in the transpiration stream (Jaakkola and Yläranta 1976). Both the sudex leaves and heads could not be used for forage, because of the elevated Cd. Other studies have shown that Cd is the heavy metal of most concern in plants that are eaten (Kirkham 1974; Liphadzi and Kirkham 2006; Clemens and Ma 2016).

Concentrations of N, P, K, and Mg in the roots, shoots, and heads were within normal concentration ranges (Table 14.5). Normal concentration ranges for N and K are 0.5%–5%, and for P and Mg they are 0.1%–1% (Liphadzi and Kirkham 2006). Normal concentration ranges for Ca in plants range from 0.5% to 5%. Calcium was low in the roots grown with and without biosolids at both Sites A and B.

In general, at the end of the experiment, total concentrations of heavy metals in the mine waste materials both with and without biosolids (Table 14.6) were similar to those at the beginning of the experiment (Table 14.2). Also, at the end of the experiment, extractable concentrations of heavy metals in the mine waste materials both with and without biosolids were similar to those at the beginning of the experiment. The Fe and Mn in biosolids affect the availability of heavy metals like Cd (Hettiarachchi et al. 2003b, 2006). However, in this experiment, there were no differences in total concentrations of Fe and Mn in the mine waste materials with and without biosolids except at Site B, where the mine waste materials with biosolids had a lower concentration of Mn than those without biosolids (Table 14.6).

At both Sites A and B with and without biosolids, the electrical conductivities of the mine waste materials at the end of the experiment (Table 14.7) were greater than at the beginning of the experiment (Table 14.2). At both Site A and Site B, the waste materials with biosolids had a higher electrical conductivity than those without biosolids (Table 14.7). However, the electrical conductivity of the mine waste materials with biosolids was still low (c. 1 dS/m) and below the threshold when electrical conductivity begins to decrease growth of salt-sensitive crops (2 dS/m) (Bernstein 1964).

Both with and without biosolids, soil organic matter at the end of the experiment (Table 14.7) was greater than that at the beginning of the experiment (Table 14.2). The increase may be due to the roots that were present in the soil at the end of the experiment. Total N was not changed during the experiment (compare Tables 14.2 and 14.7). At Site A, total N was slightly increased by the addition of biosolids, but it was not increased at Site B (Table 14.7). Usually, biosolids add N to soil (Kirkham 1974), but this was evident only at Site A. Therefore, the increase in growth of the plants grown with biosolids was probably not due to differences in N between the pots with biosolids and the pots without biosolids.

TABLE 14.5

Concentration (%) of N, P, K, Mg, and Ca in Roots, Shoots, and Heads of Sudex Grown with and without Biosolids in Mine Waste Materials from Two Different Sites in Galena, Kansas

	Site A		Site B	
Heavy Metal	With Biosolids	Without Biosolids	With Biosolids	Without Biosolids
		Roots		
N	1.15a	0.82b	0.78A	0.59B
P	0.30a	0.14b	0.17A	0.14B
K	0.78a	0.88a	0.57B	0.84A
Ca	0.50a	0.41b	0.27B	0.36A
Mg	0.29a	0.27a	0.19B	0.26A
		Shoots		
N	1.57a	1.05b	1.31A	0.93B
P	0.326a	0.21b	0.332A	0.26A
K	1.19a	1.31a	1.31B	1.75A
Ca	0.94b	1.42a	0.59B	0.86A
Mg	0.32a	0.36a	0.40A	0.31B
		Heads		
N	1.45	—[a]	1.40	—
P	0.252	—	0.249	—
K	0.780	—	0.768	—
Ca	0.160	—	0.133	—
Mg	0.217	—	0.256	—
C	44.21	—	45.27	—

Site A was on the outskirts of town, and Site B was near the center of town. Total carbon was determined only in the heads. Within each row, values with the same lowercase letter do not differ significantly at 0.05 and values with the same capital letter do not differ at 0.05. Each value is the average of 21 measurements. See text for description of statistical analyses.

[a] Plants grown without biosolids did not produce heads.

At the end of the experiment, total C and total organic C at both Site A and Site B were increased due to the presence of biosolids (Table 14.7). Digested biosolids are outstanding in their ability to increase the organic content of soils (Kirkham 1974), and this was evident in our experiment. Except for total organic C at Site B, total C and total organic C were higher at the end of the experiment than the beginning of the experiment (Tables 14.2 and 14.7). Karna (2014) found that organic C immobilized Pb and Zn. The C also increases aggregation, which makes the mine waste materials a better medium for plant growth. For both Sites A and B without biosolids, total P and total extractable P in the mine waste materials were slightly higher at the end of the experiment than at the beginning of the experiment (Tables 14.2 and 14.7). At both Sites A and B, total P and extractable P were increased after the addition of biosolids (Table 14.7).

Thus the biosolids increased P, needed for plant growth. The P added by the biosolids also may have reduced the availability of heavy metals. It has been known for a long time that P is effective in reducing heavy metal availability in soils (Chaney 1973; Kirkham 1977). Many studies have shown that P can stabilize heavy metals like Pb in soil (Hettiarachchi et al. 2001, 2003a; Hettiarachchi and Pierzynski 2002; Baker et al. 2014). A patented method for immobilization of metal availability in contaminated soils depends upon the addition of P (Pierzynski and Hettiarachchi 2002). The method is particularly useful for reducing the bioavailability of Pb and

TABLE 14.6

Total and Extractable Concentrations (mg/kg) of Heavy Metals in Mine Waste Materials with and without Biosolids

| Heavy Metal | Site A | | Site B | |
	With Biosolids	Without Biosolids	With Biosolids	Without Biosolids
		Total Concentration		
Cd	34.9a	33.4a	36.9A	34.3A
Cu	58.1a	48.0b	47.8A	46.3A
Fe	3931.2a	3764.2a	8428.5A	8095.5A
Mn	82.9a	78.8a	197.9B	220.9A
Ni	3.8a	3.5a	2.9A	3.0A
Pb	3406.5a	2499.5a	1288.7A	1228.9A
Zn	4041.0a	3981.3a	4558.1A	5365.3A
		Extractable Concentration		
Cd	6.6a	6.9a	1.1A	1.3A
Cu	3.4a	2.7a	8.6A	8.4A
Fe	3.5a	2.9a	37.4A	31.3B
Mn	0.5a	0.2a	3.9A	2.5B
Ni	0.4a	0.2b	0.1A	0.1A
Pb	113.1b	164.9a	60.0A	60.8A
Zn	319.4a	311.0a	118.6A	117.2A

The mine waste materials came from two different sites in Galena, Kansas. Site A was on the outskirts of town, and Site B was near the center of town. Within each row, values with the same lowercase letter do not differ significantly at 0.05 and values with the same capital letter do not differ significantly at 0.05. Each value is the average of 21 measurements. See text for description of statistical analyses.

TABLE 14.7

The pH, Electrical Conductivity (EC), Cation Exchange Capacity (CEC), Organic Matter, Total Nitrogen, Total Carbon, Total Organic Carbon, Total Phosphorus, and Extractable Phosphorus in Mine Waste Materials with and without Biosolids

| Measurement | Site A | | Site B | |
	With Biosolids	Without Biosolids	With Biosolids	Without Biosolids
pH	6.7b	7.5a	6.3B	6.7A
EC, dS/m	1.07a	0.89b	1.25A	0.97B
CEC, meq/100 g	6.9a	6.0a	7.6A	7.7A
Organic matter, %	2.73a	2.26b	2.24A	2.00B
Total N, %	0.15a	0.13b	0.11A	0.10A
Total C, %	1.71a	1.36b	1.11A	1.01B
Total organic C, %	1.44a	1.24b	1.08A	0.96B
Total P, ppm	1106a	777b	887A	705B
Extractable P, ppm	551a	349b	311A	208B

The mine waste materials came from two different sites in Galena, Kansas. Site A was on the outskirts of town, and Site B was near the center of town. Within each row, values with the same lowercase letter do not differ significantly at 0.05 and values with the same capital letter do not differ significantly at 0.05. Each value is the average of 21 measurements. See text for description of statistical analyses.

Group IIB metals (e.g., Zn and Cd) and Group VIII metals (e.g., Fe and Ni). When P is added to waste materials, it decreases the bioavailability of the metal contaminant by forming essentially irreversibly adsorbed metals and by causing the metal to react with the P source to form insoluble metal phosphates (e.g., Pb phosphate minerals or pyromorphites), thus rendering the metal contaminant non-bioavailable. Our results agree with those of Brown et al. (2007), who found that diammonium phosphate fertilizer reduced bioaccessible Pb in tailings on a site in Oklahoma in the Tri-State Mining District.

In conclusion, the increased growth of the plants grown with biosolids appeared to be due to the total C, total organic C, and P that they added to the mine waste materials. The total C and total organic C apparently immobilized the Pb and Zn. The P not only was an essential nutrient, but it also may have bound the heavy metals and made them less available for uptake. The results suggest that biosolids, which are readily available from any town and continually produced, should be added to mine waste materials to revegetate the degraded land.

ACKNOWLEDGMENTS

King Saud University in Riyadh, Saudi Arabia, funded the research assistantship of the senior author. We thank the following people: Dr. Yuxin (Jack) He and Ms. Cathryn Davis for helping to get the mine waste materials; Mr. Mosaed A. Majrashi for help throughout the experiment; Ms. Terri L. Branden, Facilities Maintenance Supervisor, Plant Sciences Greenhouse Complex, for help in the greenhouse during the experiment; Mr. Arthur Selman for constructing the temperature and humidity sensors used in the greenhouse; Mr. Jay Yaege, Laboratory Manager, Manhattan, KS, Wastewater Treatment Plant, for supplying the biosolids; Dr. Abdu Durar, Environmental Compliance Manager, City of Manhattan, Public Works Department, Water and Wastewater Division, Manhattan, KS, for information on the N and P in the biosolids; Ms. Zhining Ou, statistical consultant in the Statistics Laboratory, Kansas State University, for help with the statistical analyses; and Ms. Kathleen M. Lowe and Mr. Jacob (Jake) A. Thomas in the Soil Testing Laboratory at Kansas State University for doing the analyses, which were paid for by two awards to M.B. Kirkham (Dr. Ron and Rae Iman Outstanding Faculty Award and the Higuchi-University of Kansas Endowment Research Achievement Award) and by Hatch Grant No. 371047 to M.B. Kirkham.

REFERENCES

Abdel-Saheb, I., A.P. Schwab, M.K. Banks, and B.A. Hetrick. 1994. Chemical characterization of heavy-metal contaminated soil in southeast Kansas. *Water Air Soil Pollut.* 78: 73–82.

Alghamdi, A. 2016. Rehabilitation of waste materials near lead and zinc mining sites in Galena, Kansas. PhD dissertation, Kansas State University, Manhattan, KS, xxvii + 244pp.

Angelo, R.T., M.S. Cringan, D.L. Chamberlain, A.J. Stahl, S.G. Haslouer, and C.A. Goodrich. 2007. Residual effects of lead and zinc mining on freshwater mussels in the Spring River Basin (Kansas, Missouri, and Oklahoma, USA). *Sci. Total Environ.* 384: 467–496.

Baker, L.R. 2008. *In situ* remediation of Pb/Zn contaminated materials: Field- and molecular-scale investigations. PhD dissertation, Kansas State University, Manhattan, KS, xxxvi + 357pp.

Baker, L.R., G.M. Pierzynski, G.M. Hettiarachchi, K.G. Scheckel, and M. Newville. 2014. Micro-x-ray fluorescence, micro-x-ray absorption spectroscopy, and micro-x-ray diffraction investigation of lead speciation after the addition of different phosphorus amendments to a smelter-contaminated soil. *J. Environ. Qual.* 43: 488–497.

Baker, L.R., P.M. White, and G.M. Pierzynski. 2011. Changes in microbial properties after manure, lime, and bentonite application to a heavy metal-contaminated mine waste. *Appl. Soil Ecol.* 48: 1–10.

Bernstein, L. 1964. Salt tolerance of plants. Agriculture Information Bulletin No. 283. U.S. Department of Agriculture, Washington, DC, 23pp.

Bremner, J.M., and C.S. Mulvaney. 1982. Nitrogen-total. In Page, A.L., Miller, R.H., and Keeney, D.R., Eds., *Methods of Soil Analysis. Part 2. Chemical and Microbiological Properties*, 2nd edn. American Society of Agronomy and Soil Science Society of America, Madison, WI, pp. 595–624.

Brown, S., R. Chaney, J. Hallfrisch, J.A. Ryan, and W.R. Berti. 2004. In situ soil treatments to reduce the phyto- and bioavailability of lead, zinc, and cadmium. *J. Environ. Qual.* 33: 522–531.

Brown, S., M. Mahoney, and M. Sprenger. 2014. A comparison of the efficacy and ecosystem impact of residual-based and topsoil-based amendments for restoring historic mine tailings in the Tri-State mining district. *Sci. Total Environ.* 485–486: 624–632.

Brown, S., M. Sprenger, A. Maxemchuk, and H. Compton. 2005. Ecosystem function in alluvial tailings after biosolids and lime addition. *J. Environ. Qual.* 34: 139–148.

Brown, S.L., H. Compton, and N.T. Basta. 2007. Field test of in situ soil amendments at the Tar Creek National Priorities List Superfund Site. *J. Environ. Qual.* 36: 1627–1634.

Brumbaugh, W.G., C.J. Schmitt, and T.W. May. 2005. Concentrations of cadmium, lead, and zinc in fish from mining-influenced waters of northeastern Oklahoma: Sampling of blood, carcass, and liver for aquatic biomonitoring. *Arch. Environ. Contam. Toxicol.* 49: 76–88.

Carroll, S.A., P.A. O'Day, and M. Piechowski. 1998. Rock-water interactions controlling zinc, cadmium, and lead concentration in surface waters and sediments, U.S. Tri-State Mining District. 2. Geochemical interpretation. *Environ. Sci. Technol.* 32: 956–965.

Chaney, R.L. 1973. Crop and food chain effects of toxic elements in sludges and effluents. In *Proceedings of the Joint Conference on Recycling Municipal Sludges and Effluents on Land.* National Association of State Universities and Land-Grant Colleges, Washington, DC, pp. 129–141.

Chu, A.C.P., and J.P. Kerr. 1977. Leaf water potential and leaf extension in a sudex crop. *N. Z. J. Agric. Res.* 20: 467–470.

Clemens, S., and J.F. Ma. 2016. Toxic heavy metal and metalloid accumulation in crop plants and foods. *Annu. Rev. Plant Biol.* 67: 489–512.

Combs, S.M., and M.V. Nathan. 1998. Soil organic matter. In Brown, J.R., Ed., *Recommended Chemical Soil Test Procedures for the North Central Region.* SB 1001. Missouri Agricultural Experiment Station, Columbia, MO, pp. 53–58.

Csuros, M. 1997. *Environmental Sampling and Analysis Lab Manual.* Lewis Publishers, CRC Press, Boca Raton, FL, 400pp.

Dean, R.B., and J.E. Smith, Jr. 1973. The properties of sludges. In *Proceedings of the Joint Conference on Recycling Municipal Sludges and Effluents on Land.* National Association of State Universities and Land-Grant Colleges, Washington, DC, pp. 39–47.

Dybowska, A., M. Farago, E. Valsami-Jones, and I. Thornton. 2006. Remediation strategies for historical mining and smelting sites. *Sci. Prog.* 89: 71–138.

Faber, B., J. Downer, and L. Yates. 1993. Portable soil moisture meters. *HortTechnology* 3: 195–197.

Frank, K., D. Beegle, and J. Denning. 1998. Phosphorus. In Brown, J.R., Ed., *Recommended Chemical Soil Test Procedures for the North Central Region.* SB 1001. Missouri Agricultural Experiment Station, Columbia, MO, pp. 21–26.

Gibson, A.M. 1972. *Wilderness Bonanza—The Tri-State District of Missouri, Kansas, and Oklahoma.* University of Oklahoma Press, Norman, OK, 362pp.

Gudichuttu, V. 2014. Phytostabilization of multi-metal contaminated mine waste materials: Long-term monitoring of influence of soil amendments on soil properties, plants, and biota and the avoidance response of earthwormse. MS thesis, Kansas State University, Manhattan, KS, xiv + 173pp.

Haering, K.C., W.L. Daniels, and S.E. Feagley. 2000. Reclaiming mined lands with biosolids, manures, and papermill sludges. In Barnhisel, R.I., Darmody, R.G., and Daniels, W.L., Eds., *Reclamation of Drastically Disturbed Lands.* American Society of Agronomy, Crop Science Society of America, and Soil Science Society of America, Madison, WI, pp. 615–644.

Hassett, J.J., J.E. Miller, and D.E. Koeppe. 1976. Interaction of lead and cadmium on maize root growth and uptake of lead and cadmium by roots. *Environ. Pollut.* 11: 297–302.

Hettiarachchi, G.M., and G.M. Pierzynski. 2002. In situ stabilization of soil lead using phosphorus and manganese oxide: Influence of plant growth. *J. Environ. Qual.* 31: 564–572.

Hettiarachchi, G.M., G.M. Pierzynski, F.W. Oehme, O. Sonmez, and J.A. Ryan. 2003a. Treatment of contaminated soil with phosphorus and manganese oxide reduces lead absorption by Sprague-Dawley rats. *J. Environ. Qual.* 32: 1335–1345.

Hettiarachchi, G.M., G.M. Pierzynski, and M.D. Ransom. 2001. In situ stabilization of soil lead using phosphorus. *J. Environ. Qual.* 30: 1214–1221.

Hettiarachchi, G.M., J.A. Ryan, R.L. Chaney, and C.M. La Fleur. 2003b. Sorption and desorption of cadmium by different fractions of biosolids-amended soils. *J. Environ. Qual.* 32: 1684–1693.

Hettiarachchi, G.M., K.G. Scheckel, J.A. Ryan, S.R. Sutton, and M. Newville. 2006. μ-XANES and μ-XRF investigations of metal binding mechanisms in biosolids. *J. Environ. Qual.* 35: 342–351.

Hosomi, M., and R. Sudo. 1986. Simultaneous determination of total nitrogen and total phosphorus in freshwater samples using persulfate digestion. *Int. J. Environ. Stud.* 27: 267–275.

Jaakkola, A., and T. Yläranta. 1976. The role of the quality of soil organic matter in cadmium accumulation in plants. *J. Sci. Agric. Soc. Finland* 48: 415–425.

Johnson, A.W., M. Gutiérrez, D. Gouzie, and L.R. McAliley. 2016. State of remediation and metal toxicity in the Tri-State Mining District, USA. *Chemosphere* 144: 1132–1141.

Juracek, K.E. 2008. Sediment storage and severity of contamination in a shallow reservoir affected by historical lead and zinc mining. *Environ. Geol.* 54: 1447–1463.

Karathanasis, A., C. Johnson, and C. Matocha. 2007. Subsurface transport of heavy metals mediated by biosolid colloids in waste-amended soils. In Frimmel, F.H., von der Kammer, F., and Flemming, H.-C., Eds., *Colloidal Transport in Porous Media*. Springer Verlag, Berlin, Germany, pp. 175–201.

Karna, R.R. 2014. Mechanistic understanding of biogeochemical transformations of trace elements in contaminated minewaste materials under reduced conditions. PhD dissertation, Kansas State University, Manhattan, KS, xxi + 198pp.

Kirkham, M.B. 1974. Disposal of sludge on land: Effect on soils, plants, and ground water. *Compost Sci.* 15(2): 6–10.

Kirkham, M.B. 1975. Trace elements in corn grown on long-term sludge disposal site. *Environ. Sci. Technol.* 9: 765–768.

Kirkham, M.B. 1977. Trace elements in sludge on land: Effect on plants, soils, and ground water. In Loehr, R.C., Ed., *Land as a Waste Management Alternative*. Ann Arbor Science, Ann Arbor, MI, pp. 209–247.

Kirkham, M.B. 2000. EDTA-facilitated phytoremediation of soil with heavy metals from sewage sludge. *Int. J. Phytoremediation* 2: 159–172.

Kirkham, M.B. 2008. Trace elements. In Chesworth, W., Ed., *Encyclopedia of Soil Science*. Encyclopedia of Earth Sciences Series, Springer, Dordrecht, the Netherlands, pp. 786–790.

Lindsay, W.L., and W.A. Norvell. 1978. Development of a DTPA soil test for zinc, iron, manganese, and copper. *Soil Sci. Soc. Am. J.* 42: 421–428.

Liphadzi, M.S., and M.B. Kirkham. 2006. Physiological effects of heavy metals on plant growth and function. In Huang, B., Ed., *Plant-Environment Interactions*. 3rd edn. CRC Press, Taylor & Francis Group, Boca Raton, FL, pp. 243–269.

Lu, Q., Z.L. He, and P.J. Stoffella. 2012. Land application of biosolids in the USA: A review. *Appl. Environ. Soil Sci.* 2012(2): 1–11. doi: 10.1155/2012/201462.

Madejón, E., A.I. Doronila, P. Madejón, A.J.M. Baker, and I.E. Woodrow. 2012. Biosolids, mycorrhizal fungi and eucalypts for phytostabilization of arsenical sulphidic mine tailings. *Agroforest. Syst.* 84: 389–399.

Mahar, A., P. Wang, R. Li, and Z. Zhang. 2015. Immobilization of lead and cadmium in contaminated soil using amendments: A review. *Pedosphere* 25: 555–568.

Nelson, N.S. 1987. An acid-persulfate digestion procedure for determination of phosphorus in sediments. *Commun. Soil Sci. Plant Anal.* 18: 359–369.

Neuberger, J.S., Mulhall, M., Pomatto, M.C., Sheverbush, J., and Hassanein, R.S. 1990. Health problems in Galena, Kansas: A heavy metal mining superfund site. *Sci. Total Environ.* 94: 261–272.

New York State Department of Environmental Conservation, c. 1965. *Manual of Instruction for Sewage Treatment Plant Operators*. New York State Department of Environmental Conservation, Albany, NY, 247pp.

Norrish, K. 1975. Geochemistry and mineralogy of trace elements. In Nicholas, D.J.D., and Egan, A.R., Eds., *Trace Elements in Soil-Plant-Animal Systems*. Academic Press, New York, pp. 55–81.

Pepper, I.L., H.G. Zerzghi, S.A. Bengson, and E.P. Glenn. 2013. Revegetation of copper mine tailings through land application of biosolids: Long-term monitoring. *Arid Land Res. Manage.* 27: 245–256. doi: 10.1080/15324982.2012.719578.

Peterson, J.R., C. Lue-Hing, and D.R. Zenz. 1973. Chemical and biological quality of municipal sludge. In Sopper, W.E., and Kardos, L.T., Eds., *Recycling Treated Municipal Wastewater and Sludge through Forest and Croplands*. The Pennsylvania State University Press, University Park, PA, pp. 26–37.

Pierzynski, G.M., L.R. Baker, G.M. Hettiarachchi, K.G. Scheckel, V. Gudichuttu, and R. Pannu. 2010. The Tri-State Mining Region USA: Twenty years of trace element research. In *19th World Congress of Soil Science, Soil Solutions for a Changing World*, August 1–6, 2010, International Union of Soil Science, Brisbane, Queensland, Australia, pp. 254–256. (Published on DVD).

Pierzynski, G.M., and K.A. Gehl. 2004. An alternative method for remediating lead-contaminated soils in residential areas: A decision case study. *J. Nat. Resour. Life Sci. Educ.* 33: 63–69.

Pierzynski, G.M., and G.M. Hettiarachchi. 2002. Method of in-situ immobilization and reduction of metal bioavailability in contaminated soils, sediments, and wastes. United States Patent No. 6,383,128. U.S. Patent and Trademark Office, Alexandria, VA.

Pierzynski, G.M., M. Lambert, B.A.D. Hettrick, D.W. Sweeney, and L.E. Erickson 2002. Phytostabilization of metal mine tailings using tall fescue. Practice periodical hazardous toxic radioact. *Waste Manage* 6: 212–217.

Pierzynski, G.M., J.L. Schnoor, M.K. Banks, J.C. Tracy, L.A. Licht, and L.E. Erickson. 1994. Vegetative remediation at Superfund Sites. In Hester, R.E. and Harrison, R.M., Eds., *Mining and Its Environmental Impact*. Issues in Environmental Science and Technology 1, Royal Society of Chemistry, Cambridge, U.K., pp. 49–69.

Pope, L.M. 2005. Assessment of contaminated streambed sediment in the Kansas part of the historic tri-state lead and zinc mining district, Cherokee County, 2004. USGS Scientific Investigations Report 2005-5251. U.S. Geological Survey, Reston, VA, 61pp.

Santibáñez, C., C. Verdugo, and R. Ginocchio. 2008. Phytostabilization of copper mine tailings with biosolids: Implications for metal uptake and productivity of *Lolium perenne*. *Sci. Total Environ.* 395: 1–10.

Sauchelli, V. 1969. *Trace Elements in Agriculture*. Van Nostrand Reinhold, New York, 248pp.

Schaider, L.A., D.B. Senn, D.J. Brabander, K.D. McCarthy, and J.P. Shine. 2007. Characterization of zinc, lead, and cadmium in mine waste: Implications for transport, exposure, and bioavailability. *Environ. Sci. Technol.* 41: 4164–4171.

Schmitt, C.J., J.J. Whyte, W.G. Brumbaugh, and D.E. Tillitt. 2005. Biochemical effects of lead, zinc, and cadmium from mining on fish in the Tri-States District of northeastern Oklahoma, USA. *Environ. Toxicol. Chem.* 24: 1483–1495.

Schwab, P., D. Zhu, and M.K. Banks. 2007. Heavy metal leaching from mine tailings as affected by organic amendments. *Bioresour. Technol.* 98: 2935–2941.

Sheoran, V., A.S. Sheoran, and P. Poonia. 2010. Soil reclamation of abandoned mine land by revegetation: A review. *Int. J. Soil Sediment Water*, 3(2), Article 13 http://scholarworks.umass.edu/intljssw/vol3/iss2/13. Accessed May 17, 2017.

Statistical Analysis System. 2013. SAS Version 9.4. SAS Institute, Cary, NC.

Stuczynski, T., G. Siebielec, W.L. Daniels, G. McCarty, and R.L. Chaney. 2007. Biological aspects of metal waste reclamation with biosolids. *J. Environ. Qual.* 36: 1154–1162.

Summers, C.G., J.P. Mitchell, T.S. Prather, and J.J. Stapleton. 2009. Sudex cover crops can kill and stunt subsequent tomato, lettuce, and broccoli transplants through allelopathy. *California Agric.* 63(1): 35–40.

United States Environmental Protection Agency. 1989. EPA's policy promoting the beneficial use of sewage sludge and the new proposed technical sludge regulations. Office of Water, Office of Municipal Pollution Control, WH-595. United States Environmental Protection Agency, Washington, DC, 24pp.

United States Environmental Protection Agency. 2007. Tri-state mining district—Chat mining waste. United States Environmental Protection Agency, EPA530-F-07-016B. National Environmental Publications Information System, Online Database, US EPA, Washington, DC, http://nepis.epa.gov/Exe/ZyPDF.cgi/P1003J3X.PDF?Dockey=P1003J3X.PDF. Accessed May 17, 2017.

Vesilind, P.A. 1975. *Treatment and Disposal of Wastewater Sludges*. Ann Arbor Science, Ann Arbor, MI, 236pp.

Wahla, I.H., and M.B. Kirkham. 2008. Heavy metal displacement in salt-water-irrigated soil during phytoremediation. *Environ. Pollut.* 155: 271–283.

Warncke, D., and J.R. Brown. 1998. Potassium and other basic cations. In Brown, J.R., Ed., *Recommended Chemical Soil Test Procedures for the North Central Region*. SB 1001. Missouri Agricultural Experiment Station, Columbia, MO, pp. 31–33.

Watson, M.E., and J.R. Brown. 1998. pH and lime requirement. In Brown, J.R., Ed., *Recommended Chemical Soil Test Procedures for the North Central Region*. SB 1001. Missouri Agricultural Experiment Station, Columbia, MO, pp. 13–16.

Whitney, D.A. 1998. Soil salinity. In Brown, J.R., Ed., *Recommended Chemical Soil Test Procedures for the North Central Region*. SB 1001. Missouri Agricultural Experiment Station, Columbia, MO, pp. 59–60.

Wijesekara, H., N.S. Bolan, M. Vithanage, Y. Xu, S. Mandal, S.L. Brown, G.M. Hettiarachchi, G.M. Pierzynski, L. Huang, Y.S. Ok, M.B. Kirkham, C.P. Saint, and A. Surapaneni. 2016. Utilization of biowaste for mine spoil rehabilitation. *Adv. Agron.* 138: 97–173.

15 Dynamics of Heavy Metal(loid)s in Mine Soils

Anitha Kunhikrishnan, N.S. Bolan, Saikat Chowdhury,
Jin Hee Park, Hyuck Soo Kim, Girish Choppala,
Bhupinder Pal Singh, and Won Il Kim

CONTENTS

15.1 Introduction 259
15.2 Mine Soil as a Source of Heavy Metal(loid)s 265
15.3 Mine Soil as a Sink of Heavy Metal(loid)s 266
15.4 Transformation of Heavy Metal(loid)s in Mine Soils 267
    15.4.1 Physicochemical Processes 267
    15.4.2 Biological Processes 268
15.5 Acid Mine Drainage and (Im)mobilization of Heavy Metal(loid)s 272
15.6 Case Studies 274
    15.6.1 Brazil 274
    15.6.2 Republic of Korea 276
15.7 Summary and Conclusions 278
References 279

15.1 INTRODUCTION

Mine sites can be a potential threat to public health due to the risk of polluting nearby groundwater and soils. During the early mining period, mining companies had less strict remediation codes than those in place now, and waste material was customarily disposed in heaps (tailings) in the direct vicinity of the mine (Johnson et al. 2016; Pascaud et al. 2015). Once the ore was exhausted, companies either closed down or moved out, many of them leaving their mining waste behind. These abandoned wastes are considered among the worst environmental problems and a serious hazard to ecosystems and human health (Anawar 2015; Fields 2003; Hudson-Edwards et al. 2011). Tailing deposits generated from mining activities pose a potential risk for the soil and aquatic environments through the release of potentially toxic metal(loid)s occurring in a variety of minerals present in the tailings (Anawar 2015; Hudson-Edwards et al. 2011). Heavy metal(loid)s include both biologically essential (e.g., cobalt [Co], copper [Cu], chromium [Cr], manganese [Mn], and zinc [Zn]) and non-essential (e.g., arsenic [As], cadmium [Cd], lead [Pb], and mercury [Hg]) elements. The nonessential elements are highly toxic; however, at excessive concentrations, both groups are toxic to plants, animals, and/or humans (Adriano 2001; Alloway 1990).

High concentrations of heavy metal(loid)s can be found in and around abandoned and active mines due to the discharge and dispersion of mine waste materials into nearby soils, food crops, and stream sediments (Gutiérrez et al. 2016a; Jung 2001; McKenzie and Pulford 2002; Qi et al. 2016). This will eventually lead to a loss of biodiversity and be a potential health risk to residents in the vicinity of the mining area (Galán et al. 2003; Hudson-Edwards et al. 2011; Lee et al. 2001; Wong et al. 2002). Lead and Zn mines, for instance, are a common occurrence worldwide, and while approximately 240 mines are active, the vast majority of them have been abandoned for decades (Gutiérrez et al. 2016a). Kierczak et al. (2013) found that soils in the areas around historic smelters, which were active between the fourteenth and sixteenth centuries, are still highly polluted with metal(loids)s (up to 4000 mg kg^{-1} Cu, 1500 mg kg^{-1} Zn, 300 mg kg^{-1} As, and 200 mg kg^{-1} Pb), particularly due to the centuries-long dissolution of smelter wastes into the soils. In addition, the contamination of soil by heavy metal(loid)s could also be caused by the emission of coal dust produced from coal mine exploitation (Bhuiyan et al. 2010; Dang et al. 2002). The contents of Cu, Zn, Cd, and Pb, and their availability in the soil of a coalmine in Tongchuan, Shanxi Province, were significantly enhanced, among which Cd was the most severe pollutant, as demonstrated by Guo et al. (2012). Recently, Qi et al. (2016) observed the release of Pb and Zn at concentrations of 3,518.4 ± 896.1 and 10,413 ± 2,973.2 mg kg^{-1} dry weight, respectively, from a Pb–Zn mining area with an exploitation history of 60 years.

Mining-impacted soils are classified as poorly structured soils and lack vegetative cover due to the high toxicity associated with heavy metal(loid)s (Nawab et al. 2016). Once introduced through mining activities, metal(loid)s persist in the environment; however, speciation of metal(loid)s can influence metal(loid) distribution, transport, and bioavailability within the environment. Factors that can influence speciation include pH, redox conditions, availability of inorganic and organic ligands, soil organic matter, and competition from other ions (Anawar 2015; Bolan et al. 2014). Metal(loid) sorption is strongly pH-dependent and is influenced by metal(loid)-complex formation and ionic strength. For redox-sensitive elements, such as Cr, As, and selenium (Se), a change in mobility will occur under different redox conditions. Microbial/biotransformation also plays a key role in the behavior and fate of toxic trace elements, especially As, Cr, Hg, and Se in soils (Bolan et al. 2013; Kunhikrishnan et al. 2016). Biotransformation processes alter the speciation and oxidation/reduction state of these metal(loid)s in soils, thereby controlling their solubility and subsequent mobility (Gadd 2010).

Physicochemical and mineralogical characteristics of mine tailings such as total concentrations of chemical elements, pH, ratio of acid-producing to acid-neutralizing minerals, and primary and secondary mineral phases are also very important factors that control the actual release of potentially toxic metal(loid)s from the tailings into the environment. At many mining sites, there are abundant iron (Fe) and aluminum (Al) oxide precipitates that can act as effective sorbents for a variety of metal(loid)s. Tailings containing high concentrations of toxic metal(loid)s are bound primarily to sulfide minerals (Anawar 2015; Lottermoser 2007). Sulfides within the tailings are susceptible to oxidative weathering due to their long-term interaction with the atmosphere. Oxidation of sulfides may result in the formation of mine drainages with low pH, known as acid mine drainage (AMD), and high concentrations of sulfates and heavy metal(loid)s (Blowes et al. 2003; Rice and Herman 2012; Zobrist et al. 2009). This is mainly due to the absence of minerals in the tailings that are able to neutralize the acidity generated by sulfide oxidation. The major minerals within tailings decreasing the acidity are carbonates such as calcite, dolomite, ankerite, and siderite (Blowes et al. 1998; Bortnikova et al. 2012). Table 15.1 shows total metal(loid) concentrations and major minerals present in mine sites from various countries. Acid neutralization in the tailings occurs through the dissolution of these carbonate minerals, which consume hydrogen ions and generate neutral conditions within the tailings. Sulfide oxidation and/or neutralization processes, therefore, have a great influence on the mobility of heavy metal(loid)s in the tailings and their retention to the primary and secondary mineral phases within tailings (Drahota et al. 2016; Heikkinen and Räisänen 2008). Under acid pH conditions, many toxic heavy metals are generally more soluble and mobile than in near-neutral pH conditions (Conesa et al. 2008; Da Pelo et al. 2009), although metal(loid)s such as antimony (Sb) and As may exhibit high mobility within near-neutral tailings (Ashley et al. 2003; Hiller et al. 2012).

According to the report of the Korea Mine Reclamation Corporation, approximately 2000 metal(loid) mines are located across the country, and only 3% of these are currently operating. They reported that 70% of soil samples collected from abandoned mine sites were found to be contaminated with high concentrations of metal(loid)s such as Cu, Zn, As, Cd, and Pb (MIRECO 2013). Studies in Korea reported that leachate originating from abandoned mine sites was frequently found to be enriched with contaminants above the permissible level; however, in a soil–water system, not

TABLE 15.1

Selected Studies Showing Total Metal(loid) Concentrations and Major Minerals Present in Mine Sites from Various Countries

Mine and Country	Range of Total Metal(loid) Concentrations (mg kg⁻¹)	Major Minerals Present	Method of Determination	Reference
Cu/Pb/Zn mine, Australia (sediment)	Cd: 0.13–8.7 Cu: 10–628 Pb: 23–1,796 Zn: 697–6,818	Pyrite, sphalerite, galena, chalcopyrite, arsenopyrite, and gold	ICP-MS[*]	Wadige et al. (2016)
Pb/Zn mine, Thailand	Mine sites: As: 76[a] Cd: 1.3 Cu: 46 Pb: 853 Zn: 78 Near (<1 km) from mine sites: As: 41 Cd: 0.5 Cu: 35 Pb: 397 Zn: 161 Far (>1 km) from mine sites: As: 24 Cd: <0.1 Cu: 34 Pb: 114 Zn: 61	Sphalerite, galena	ICP-OES[*]	Intamo et al. (2016)
A[b] mine, Korea	As: 152,867 ± 440 Cd: 56 ± 1 Cu: 208 ± 5 Pb: 4,107 ± 39 Zn: 2,170 ± 64	Quartz, clinochlore, nacrite, sulfide/ sulfate minerals	ICP-MS,[*] ICP-OES, XRD[*]	Kim and Hyun (2015)
B[b] mine, Korea	As: 15,312 ± 222 Cd: 22 ± 3 Cu: 2,126 ± 40 Pb: 20,036 ± 432 Zn: 3,326 ± 225	Quartz, segnitite, kaliophilite, sulfide/ sulfate minerals	ICP-MS, ICP-OES, XRD	Kim and Hyun (2015)
Cu/W mine, Korea	As: 8.95–56.7 Cd: 0.63–3.03 Cu: 46.7–243 Pb: 30.1–93.1 Zn: 89.7–482	Chalcopyrite, wolframite	AAS, HG-AAS[*]	Kwon et al. (2017)
Au/Ag/(Cu)/ (Pb) mine, Korea	As: 6.66–71.0 Cd: 0.01–12.8 Cu: 14.7–147 Pb: 27.4–885 Zn: 104–1,600	Quartz	AAS, HG-AAS	Kwon et al. (2017)
Au/Ag mine, Korea	As: 22.2–704 Cd: 0.416–4.21 Cu: 16.9–32.9 Pb: 54.6–465 Zn: 95.6–670	—[b]	AAS, HG-AAS	Kwon et al. (2017)

(Continued)

TABLE 15.1 (*Continued*)

Selected Studies Showing Total Metal(loid) Concentrations and Major Minerals Present in Mine Sites from Various Countries

Mine and Country	Range of Total Metal(loid) Concentrations (mg kg⁻¹)	Major Minerals Present	Method of Determination	Reference
Pb/Zn mine, Korea	As: 2.85–9.71 Cd: 0.001–1.55 Cu: 21.7–44.6 Pb: 26.3–105 Zn: 131–557	—[b]	AAS, HG-AAS	Kwon et al. (2017)
Pb/Zn mine, USA (sediment)	Cd: ND[c]–281 Pb: 9–8,200 Zn: 21–51,600	Sphalerite, galena, chalcopyrite	ICP-AES[a]	Gutiérrez et al. (2016b)

Pb/Zn mine, France	Spoils and tailings:
	  As: 103–4,247
	  Cd: 386–1,605
	  Pb: 3,686–88,472
	  Zn: 56,999–125,456
	Soil:
	  As: 75
	  Cd: 8.3
	  Pb: 1,787
	  Zn: 2,879
		Galena	ICP-MS	Saunier et al. (2013)

Pb mine, Portugal	As: 2.8–208 Cr: 61–196 Cu: 21–193 Ni: 7.7–87 Pb: 24–9,330 Zn: 30–517	Sphalerite, galena	AAS, HG-AAS	Pratas et al. (2013)
Sn/As₂O₃ mine, Portugal	Cd: 0.03–7.35 Cu: 29.8–534 Cr: 5.55–237 Ni: 6.00–109 Pb: 18–328 Zn: 74.8–576	Cassiterite	ICP-MS	Favas et al. (2011)

Pb/(Ag)/Zn mine, Spain	Tailings:
	  As: 216–2,660
	  Cd: 2.6–40.7
	  Cu: 44–740
	  Pb: 1,700–21,600
	  Sb: 32.7–118
	  Zn: 2,410–13,100
	Soils:
	  As: 58–680
	  Cd: 0.7–32.8
	  Cu: 27–259
	  Pb: 235–6,010
	  Sb: 10.3–174
	  Zn: 331–5,250
		Pyrite, sphalerite, galena	ICP-MS	Oyarzun et al. (2011)

(Continued)

TABLE 15.1 (*Continued*)

Selected Studies Showing Total Metal(loid) Concentrations and Major Minerals Present in Mine Sites from Various Countries

Mine and Country	Range of Total Metal(loid) Concentrations (mg kg⁻¹)	Major Minerals Present	Method of Determination	Reference
Pb/Zn mine, Spain	Tailings: Cd: 2.88–54.47 Cu: 44.10–716.58 Pb: 1,243.24–93,900.87 Zn: 470.59–20,911.78 Arable land: Cd: ND^c–4.89 Cu: 9.56–48.26 Pb: 143.56–970.08 Zn: 54.46–451.65 Pasture land: Cd: ND^c–28.61 Cu: 10.13–331.68 Pb: 53.90–11,873.50 Zn: 63.69–3,639.71	Sphalerite, galena	ICP-AES	Rodríguez et al. (2009)
Cu/S mine, Italy	Cd: ND^c–5.14 Cr: 11–113 Cu: 411–4,098 Pb: 196–20,977 Zn: 394–2,513	Cupriferous pyrite, pyrite, chalcopyrite	AAS	Wahsha et al. (2012)
Pb/Zn mine, Greece	Tailings: As: 85.5–369.8 Cd: 15.4–174.1 Cu: 151.5–1,201 Pb: 3,162–12,567 Zn: 2,504–22,292 Soils: As: 4.7–14.1 Cd: 0.9–5.8 Cu: 12.8–42.7 Pb: 28.5–219 Zn: 47.4–474 Sediments: As: 7.5–24.6 Cd: 0.7–34.7 Cu: 14.9–130.7 Pb: 39.7–302 Zn: 101.2–3,986	Pyrite, sphalerite, galena, wurtzite, jordanite	GF-AAS, HG-AAS	Nikolaidis et al. (2010)
Pb/Zn mine, India	As: 9–367 Cd: 2–36 Cr: 20–76 Cu: 9–112 Ni: 24–90 Pb: 85–17,773 Zn: 410–6,521	Sphalerite, galena, pyrite, silver (native), chalcopyrite, arsenopyrite, pyrrhotite, magnetite	ICP-AES	Anju and Banerjee (2012)

(Continued)

TABLE 15.1 (*Continued*)

Selected Studies Showing Total Metal(loid) Concentrations and Major Minerals Present in Mine Sites from Various Countries

Mine and Country	Range of Total Metal(loid) Concentrations (mg kg^{-1})	Major Minerals Present	Method of Determination	Reference
C[b] mine, China	Cd: 1.3–7.6 Cu: 88–1,313 Pb: 105–621 Zn: 170–1,663	Pyrite, pyrrhotite, chalcopyrite	AAS	Zhuang et al. (2009)
Sb mine, China	As: 13–267 Cd: 0.7–96.9 Cr: 81–316 Cu: 23.2–261.1 Hg: 0.16–33.7 Ni: 29.1–85.9 Pb: 27–423 Sb: 10–2,159 Zn: 68–4,218	Stibnite	ICP-MS, ICP-AES, AFS[*]	Wang et al. (2010b)
Pb/Zn mine, China	As: 11.0–75.8 Cd: 0.28–93.88 Cr: 70.6–481.9 Cu: 30.7–111.3 Hg: 0.057–6.733 Ni: 19.1–145.7 Pb: 37.6–4,610 Zn: 103.6–16,400	Sphalerite, galena, pyrite, chalcopyrite	GF-AAS, HG-AAS, AAS, 1,5-diphenyl-carbohydrazide spectrophotometric method for Cr	Huang et al. (2013)
Cu mine, Iran	Cu: 30–1,330 Ni: 6–38 Pb: 20–770 Zn: 80–1,500	Chalcocite, chalcopyrite, covellite, bornite, molybdenite	AAS	Ghaderian and Ravandi (2012)

ICP-MS, Inductively Coupled Plasma-Mass Spectrometry; ICP-OES, Inductively Coupled Plasma-Optical Emission Spectrometry; ICP-AES, Inductively Coupled Plasma-Atomic Emission Spectrometry; AAS, Atomic Absorption Spectrometry; XRD, X-Ray Diffraction; GF-AAS, Graphite Furnace Atomic Absorption Spectroscopy; HG-AAS, Hydride Generation Atomic Absorption Spectroscopy; AFS, Atomic Fluorescence Spectroscopy.

[a] Median values of As and heavy metal concentrations.

[b] Not available.

[c] Not detected.

all elements are equally susceptible to leaching upon interacting with water (Hyun et al. 2012; Nam et al. 2010). Readily leachable or recalcitrant elements vary depending on the type/strength of the bonds through which these elements are associated with soil components. From the work conducted on several mine tailing samples, Kim et al. (2002) reported that only <0.4% of the total As is water-soluble and most of the recalcitrant fraction is present as arsenopyrite (FeAsS). Therefore, contamination of the adjacent environment with metal(loid)s present in tailing impoundments is dependent on many geochemical processes, such as sorption–desorption, oxidation–reduction, and mineral precipitation–dissolution reactions, which control their actual mobilization.

The forms of metal(loid)s, their transformation processes, and geochemical environments are very important to evaluate the potential metal(loid) mobility at mining sites. Understanding the geochemical processes that influence metal(loid) mobility, distribution, and bioavailability can aid in forecasting the potential ecological effects of metal(loid)s in mining environments. This chapter examines the dynamics of heavy metal(loid) in mine soils and how these soils can act as a source and as a sink of metal(loid)s. The major physicochemical and biological processes in relation to metal(loid) dynamics in mine soils are explained. Acid mine drainage and the mobilization/immobilization processes related to metal(loid) dynamics are described, and finally a few recent case studies in mine sites are detailed in this chapter.

15.2   MINE SOIL AS A SOURCE OF HEAVY METAL(LOID)s

Heavy metal(loid)s reach the soil environment through both geogenic and anthropogenic processes. Most of the metal(loid)s occur naturally in soil parent materials, mainly in forms that are not readily bioavailable for plant uptake. Heavy metal(loid)s added through anthropogenic activities, unlike geogenic inputs, typically have a high bioavailability (Bolan et al. 2014; Lamb et al. 2009; Naidu and Bolan 2008). Anthropogenic activities such as mining and smelting of metal(loid) ores have increased the occurrence of heavy metal(loid) contamination of soil and water sources. Opencast mining activities have a serious environmental impact on soils and water streams and have generated millions of tons of sulfide-rich tailings (Bhattacharya et al. 2012; Olías and Nieto 2015). Acidic drainage resulting from the oxidation of sulfides from metalliferous mine spoils leads to the leaching of large quantities of metal(loid)s including Fe^{2+}, manganese (Mn^{2+}), Pb^{2+}, Cu^{2+}, and Zn^{2+} (Muhammad et al. 2011; Vega et al. 2006). The Fe sulfides such as pyrite, marcasite, and pyrrhotite are the most common sources of AMD production, because they are ubiquitous in metal(loid) sulfide ores and not the target of ore beneficiation processes. Thus, metal(loid) contamination and AMD are major environmental concerns where waste materials containing metal(loid)-rich sulfides from mining activity have been stored or abandoned (Concas et al. 2006; Simate and Ndlovu 2014).

Pirrone et al. (2010) estimated that ore mining and processing are responsible for 13% of global mercury (Hg) emissions. Zhang et al. (2011) estimated that until the year 2007, cumulative emissions from mining and smelting activities in China were about 1.62 mega tonnes (Mt) Pb and 3.32 Mt Zn, with the contribution of the smelting processes accounting for 19% and 27%, respectively. It is widely known that large-scale nonferrous metal(loid) smelters are also important local-to-regional sources of pollution. For instance, Li et al. (2011b) reported that the largest Pb/Zn smelter in China, located in Zhuzhou (Hunan), emitted 77.82 tonnes of Cd into the atmosphere in the period 1991–2000, accounting for 95% of total emissions from the city.

In South Korea, large amounts of mine wastes including tailings have been left without proper environmental treatment, thus becoming an important point source of toxic elements such as As, Cd, Cu, Pb, and Zn in the environment. These materials are dispersed downslope by surface erosion, wind action, and effluent draining from the mine wastes contaminating the low-lying arable lands. Twenty-one percent of arable soil near mining and industrial areas was found to be contaminated by heavy metal(loid)s (NIAST 1997). The Daduk gold (Au), silver (Ag), lead (Pb), and zinc (Zn) mine, which is located in the middle part of Korea, is a major source of heavy metal(loid) contamination of arable soils (Lee et al. 2001). This mine was one of the largest Au–Ag–Pb–Zn mines in Korea. The mine ceased production in 1984 and large amounts of mine wastes have been left without proper environmental treatment. The mine tailings contained high concentrations of metal(loid)s, and they were dispersed downslope by erosion and effluent draining into low-lying land mainly used for paddy cultivation. Elevated levels of Cd were found in soils sampled in paddy fields and the forest area.

Contaminants associated with particulates emitted from mining operations are usually concentrated in the fine fraction (<2 mm), and those from smelting even concentrated in the ultrafine particle fraction (<0.5 mm), which may travel greater distances into the environment (Csavina et al. 2011, 2012, 2014; Ettler 2016; Ettler et al. 2005; Sorooshian et al. 2012; Uzu et al. 2011). This was demonstrated by Hou et al. (2006), who studied Pb concentrations and isotope compositions in soils affected by the Horne Cu smelter (Rouyn-Noranda, Canada), and noticed that the emission signature was to be found as far as 116 km downwind. However, highly volatile contaminants such as Hg can be deposited in soils to a lesser extent. For instance, Wu et al. (2014) calculated that a large-scale Pb/Zn smelter in Zhuzhou (China) emitted 105 tonnes of Hg during the period 1960–2011, with only 14% of this amount deposited locally in the soil, and the remainder emitted into the global pool.

15.3   MINE SOIL AS A SINK OF HEAVY METAL(LOID)s

In a soil–water system, when sources of a given contaminant are controlled, it will eventually accumulate in long-term sinks through time and the concentrations will constantly decrease unless the capacity of the sink is exceeded or external changes occur. A significant fraction of the metal(loid)s released by sulfide oxidation within waste rock dumps or mine tailings is retained within the deposits and underlying soils as secondary mineral precipitates. Iron is usually the most reactive chemical species, and as pH increases in acid sulfate systems, dissolved Fe can precipitate as secondary oxyhydroxysulfate minerals such as jarosite and schwertmannite in the pH range 2–5, or oxyhydroxide minerals such as goethite and ferrihydrite at high pH values (Bigham and Nordstrom 2000; Burton et al. 2006). Schwertmannite and jarosite are the two main secondary Fe(III) minerals formed from the oxidation of iron sulfides and incomplete hydrolysis of Fe(III), and are known to contain a substantial amount of retained acidity (Fitzpatrick et al. 2009; Sullivan and Bush 2004). While goethite and jarosite are well crystallized products, ferrihydrite is a poorly-ordered precipitate.

Figueiredo and da Silva (2011) investigated waste materials from two abandoned old mines in the Iberian Pyrite Belt in southern Portugal and reported that jarosite was efficient in immobilizing metal(loid)s particularly Pb. Many of these secondary Fe(III) minerals and their transformed products (i.e., goethite) are ochreous in color and therefore are indicative of previous and future acidity discharge in acid sulfate soils (Bigham et al. 1996). In a mineralogical study of Cu cliff tailings, McGregor et al. (1998) confirmed the presence of goethite, jarosite, gypsum, native sulfur, and a vermiculite-type clay mineral. Goethite, jarosite, and native sulfur form alteration rims and pseudo-morphs of the sulfide minerals. Interstitial cements, composed of goethite, jarosite, and gypsum, locally bind the tailings particles, forming hardpan layers. Microprobe analyses of goethite indicated that it contained up to 0.6 wt.% nickel (Ni), suggesting that the goethite is a repository for Ni. Other sinks included jarosite and a vermiculite-type clay mineral, which locally contained up to 1.6 wt.% Ni.

Valente et al. (2015) in a recent study demonstrated that jarosite and aggregates of As–Fe-rich spherical nanoparticles were the main sinks of As identified in the waste dumps. These aggregates were amorphous ferric arsenates, but, in conditions of appropriate Fe:As ratio, they may be a precursor of a crystalline arsenate (As[V]), namely scorodite. They suggested that dilution associated with rainfall may promote instability of jarosite. However, they found that jarosite was the dominant secondary mineral in the three sites, exhibiting well-defined euhedral habits. Lin and Herbert (1997) found weathering of sulfides in the leached zone had resulted in the migration of most heavy metal(loid)s to the accumulation zone or underlying soils, where they were retained in more stable phases such as secondary ferric minerals, including goethite and jarosite. They further noticed some metal(loid)s being temporarily retained in hydrated ferrous sulfates (e.g., melanterite, rozenite).

Under acidic conditions, Mn^{2+} has been found to remain in largely dissolved forms, or oxidize and precipitate as hydrous Mn oxides (Davies and Morgan 1989). If redox potential decreases, hydrous Mn oxides may also undergo reduction back to Mn^{2+}. At low pH (<5), Al^{3+} is also predominantly in dissolved form although hydroxysulfate minerals such as basaluminite have been noted in AMD (Bigham and Nordstrom 2000). Dissolved Al^{3+} and its inorganic complexes can hydrolyze and precipitate as hydrous Al(III) oxides, which are not redox sensitive like Fe and Mn oxides. Near-neutral AMD can form from rock that contains little pyrite or can originate as acidic AMD that has been neutralized by reaction with calcite ($CaCO_3$) and other minerals containing calcium, magnesium, potassium, and sodium. For example, dissolution of $CaCO_3$ neutralizes acid and can increase the pH and alkalinity of AMD. As the pH of initially acidic AMD increases to near-neutral values, concentrations of dissolved Fe^{III}, Al, and other metal(loid)s can decline as Fe^{III} and Al hydroxides precipitate (Cravotta and Kirby 2004).

15.4 TRANSFORMATION OF HEAVY METAL(LOID)s IN MINE SOILS

Metal(loid)s are retained in soil by sorption, precipitation, and complexation reactions, and removed through plant uptake, leaching, and volatilization. Most metal(loid)s are not subject to volatilization losses; however, As, Hg, and Se tend to form gaseous compounds (Bolan et al. 2013, 2014; Kunhikrishnan et al. 2016). The fate of metal(loid)s in soil is dependent on both soil properties and environmental factors.

15.4.1 Physicochemical Processes

Retention of charged metal(loid) solute species by charged surfaces of soil components is grouped into specific and nonspecific retention (Bolan et al. 1999; Sparks 2003). Nonspecific adsorption is a process in which the charge on the ions balances the charge in the soil particles through electrostatic attraction, whereas specific adsorption involves chemical bond formation between the ions in the solution and those in the soil surface (Li et al. 2006; Sposito 1984; Zenteno et al. 2013).

Both soil properties and soil solution composition determine the dynamic equilibrium between metal(loid)s in solution and solid phase. The concentration of metal(loid)s in soil solution is affected by the nature of both organic and inorganic ligand ions and soil pH through their influence on metal(loid) sorption processes (Bolan et al. 2003; Harter and Naidu 1995). The major reasons for the effect of inorganic anions on the sorption of metal(loid) cations is that, firstly, inorganic anions form ion pair complexes with metal(loid)s, thereby reducing metal(loid) sorption. Secondly, the specific sorption of ligand anions increases the negative charge on soil particles, hence increasing the sorption of metal(loid)s (Lackovic et al. 2003; Naidu et al. 1994).

The effect of pH values >6 in lowering free metal(loid) ion activities in soils has been attributed to the increase in pH-dependent surface charge on oxides of Fe, Al, and Mn, chelation by organic matter, or precipitation of metal(loid) hydroxides (Mouta et al. 2008; Stahl and James 1991). The activity of heavy metal(loid)s in solution in naturally acidic soils is found to decrease with increasing pH, which is attributed to increasing cation exchange capacity (CEC) (Naidu et al. 1997; Shuman 1986; Violante et al. 2010). Covelo et al. (2007) noticed that the sorption of Pb and Cu in acid soils correlated with organic matter content, while the retention of Cd, Cr, Ni, and Zn depended on clay minerals, especially kaolinite, gibbsite, and vermiculite. Complexation reactions between metal(loid)s and the inorganic and organic ligand ions also contribute to metal(loid) retention by colloid particles (Bolan et al. 2011; Harter and Naidu 1995). The extent of metal(loid)-organic complex formation, however, varies with a number of factors including temperature and metal(loid) concentration. All these interactions are controlled by solution pH and ionic strength, nature of metal(loid) species, dominant cation, and inorganic and organic ligands present in the soil solution.

Precipitation appears to be the predominant process in high pH soils and in the presence of anions, such as sulfates, carbonates, hydroxides, and phosphates, and when the concentration of the metal(loid) ion is high (Bolan et al. 2014). Coprecipitation of metal(loid)s especially in the presence of Fe oxyhydroxides has also been reported, and often such interactions lead to significant changes in the surface chemical properties of the substrate. Lu et al. (2011) confirmed that coprecipitation of Pb^{2+} with ferric oxyhydroxides occurred at ~pH 4 and is more efficient than adsorption in removing Pb^{2+} from aqueous solutions at similar sorbate/sorbent ratios and pH. Arsenate sorption onto ferrihydrite and Ni^{2+} and Cr^{3+} sorption onto hydrous Fe oxides showed that coprecipitation was a more efficient process than sorption for metal(loid) removal from aqueous solutions.

Secondary precipitates remove metal(loid)s from mine-impacted soils/waters through adsorption and/or coprecipitation reactions. The extent to which dissolved contaminants will sorb to these precipitates as outer sphere or inner sphere complexes will vary as a function of the contaminant species, the secondary precipitate, particle size and surface area, pH, and the presence of other competing ions. Examples of secondary precipitates that form in mine-impacted sites include oxyhydroxides (e.g., FeOOH[s]; [s] stands for "solid"), hydroxysulfates (e.g., $Fe_8O_8[OH]_6[SO_4][s]$), sulfates (e.g., $PbSO_4[s]$), and sulfides (e.g., $ZnS[s]$). Gutiérrez-Ruiz et al. (2012) noticed insoluble Pb arsenates in highly polluted soils near a Mexican Pb smelter. Due to high sulfate concentrations in mine-impacted soils/waters, metal(loid)s that form strong bonds (and relatively insoluble precipitates) with sulfates would be expected to precipitate and be relatively immobile. Anglesite ($PbSO_4$) and barite ($BaSO_4$) are examples of insoluble sulfate minerals (Smith 2007).

Hiller et al. (2013) investigated the geochemical and mineralogical characteristics of tailings deposited in voluminous impoundments and identified the processes controlling the mobility of selected toxic metals (Cu, Hg) and metalloids (As, Sb). The total solid-phase concentrations of metal(loid)s decreased in the order of Cu > Sb > Hg > As and reflected the proportions of sulfides present in the tailings. Secondary Fe phases were found and they retained relatively high amounts of metal(loid)s (up to 57.6 wt.% Cu, 1.60 wt.% Hg, 23.8 wt.% As, and 2.37 wt.% Sb). The batch leaching tests and lysimeter results revealed less mobility of these elements in the tailings, which they attributed to retention by Fe oxyhydroxides precipitation and metal(loid)-bearing sulfides.

The small size and exposure of tailings to rain stimulate the weathering rate of the ore minerals and associated sulfides (e.g., pyrite) into secondary minerals and ions in solution. The dissolution process depends on the nature of the mineral, pH, and liquid/solid ratio, among other factors. Palumbo-Roe et al. (2009) applied this approach to two nearby Pb-Zn mines in Wales, and found that dissolution of Pb was controlled by carbonates in one mine and by surface complexation to oxyhydroxides in the second mine, with a 10-fold increase of dissolved Pb in the former one, although both mines had similar bulk Pb concentration.

15.4.2 BIOLOGICAL PROCESSES

Oxidation–reduction reactions may be chemically or biologically driven. Metal(loid)s, including As, Cr, Hg, and Se, are most commonly subjected to microbial oxidation/reduction reactions, thereby influencing their speciation and mobility (Table 15.2). For instance, metals generally are less soluble in their higher oxidation state, whereas the solubility and mobility of metalloids depend on both the oxidation state and the ionic form (Ross 1994). Chromium dissolves as it is oxidized to chromate (Cr[VI]) and precipitates upon reduction to chromite (Cr[III]), and this is vital because Cr(VI) is more toxic than Cr(III). Similarly, As in soils and sediments can be oxidized to As(V) by arsenite-oxidizing bacteria, and since As(V) is strongly retained by inorganic soil components, microbial oxidation results in As immobilization (Bachate et al. 2012; Battaglia-Brunet et al. 2002). Under well-drained conditions, As is present as As(V), whereas under reduced conditions, As(III) dominates in soils, but elemental arsenic (As[0]) and arsine (H_2As) can also be present. Iron and Mn may be soluble under reducing conditions; consequently, metal(loid)s sorbed onto Fe oxides and Mn oxides can be released under reducing conditions.

The redox reactions are grouped into two categories, assimilatory and dissimilatory (Madigan and Brock 1991). In assimilatory reactions, the metal(loid) substrate is involved in the metabolic functioning of the organism by acting as terminal electron acceptor. In contrast, for dissimilatory reactions the metal(loid) substrate has no known role in the metabolic functioning of the species responsible for the reaction, and indirectly initiates redox reactions. Microbial oxidation-reduction in some cases can be many times faster than abiotic reactions. The oxidation of As(III) to As(V) by a *Thermus* species, for example, was found to be approximately 100 times faster than abiotic rates (Gihring and Banfield 2001). Studies demonstrated that oxidation of arsenopyrite by acidophilic Fe- and sulfur-oxidizing bacteria is considerably more effective than abiotic oxidation. For example, the oxidative dissolution of arsenopyrite was shown to be much more rapid and extensive when it occurred in the presence of *Leptospirillum ferrooxidans* than under abiotic conditions (Corkhill et al. 2006).

Dhal et al. (2010) using a *Bacillus* sp. bacterium from Cr(III) mine soils in Boula-Nuasahi mine in Orissa, India, demonstrated that the strain reduced >90% of 100 mg L^{-1} Cr(VI) in 144 h at pH 7. Mishra et al. (2010) have also reported the reduction of Cr(VI) in solution using bacterial

TABLE 15.2

Microorganisms Responsible for Metal(loid) Mobilization and Immobilization in Mine Soils/Tailings

Metal(loid)	Microorganisms	Mechanisms of Metal(loid) Mobilization/Immobilization	References
As	*Desulfuromonas palmitatis*	The bacterium mobilized As by Fe oxide reduction	Vaxevanidou et al. (2015)
As	Bacteria: *Shewanella* sp. OM1, *Pseudomonas* sp. OM2, *Aeromonas* sp.OM4, and *Serratia* sp.OM17	Bacteria mobilized As by microbial reductive dissolution of As(V) under reducing conditions	Drewniak et al. (2014)
As	*Sinorhizobium*-related isolate	Immobilization by microbial oxidation of As(III) to As(V)	Hamamura et al. (2013)
As	*Arabidopsis thaliana* *Escherichia coli*	Immobilization of As through bacterial accumulation	Dhankher et al. (2002); Song et al. (2003); Tsai et al. (2009)
As	Bacteria: *Clostridium collagenovorans*, *Desulfovibrio vulgaris* and *Desulfovibrio gigas*	Some bacteria mobilized As by methylation pathway involving reduction of As(V)	Bentley and Chasteen (2002); Paez-Espino et al. (2009)
As	Mn-oxidizing bacteria	Immobilization by microbial oxidation of As(III) to As(V)	Tebo et al. (2005); Miyata et al. (2007)
As	Strain WAO, an autotrophic inorganic-sulfur and As(III)-oxidizer	The microbe increased As mobilization from the solid phase by the oxidation of aqueous As (III) to As (V)	Rhine et al. (2008)
As	Bacteria: *Acidithiobacillus ferrooxidans*, *Acidithiobacillus thiooxidans*, and *Leptospirillum ferrooxidans*	Microbes mobilized As by oxidation of associated Fe or sulfur	Hackl and Jones (1997); Drewniak and Sklodowska (2013)
As	Sulfate-reducing bacteria	Immobilization of As by precipitation	Tsai et al. (2009)
Cd	Ectomycorrhizal fungi	Immobilization of Cd by fungal fixation	Krupa and Kozdroj (2004)
Cd	Vesicular arbuscular mycorrhizae	Immobilization of Cd by fungal stabilization	Tullio et al. (2003)
Cd	Bacteria: *Burkholderia cepacia*	Bacteria mobilized Cd by dissolution in the presence of organic acids produced by the microbe (decreasing pH)	Li and Wong (2010)
Co	Fungi: *Aspergillus* sp. and *Penicillium* spp.,	Fungi mobilized Co and bioleaching through organic acids produced by the fungi	Brandl (2001); Mulligan and Galvez-Cloutier (2003); Santhiya and Ting (2005)
Co	Fe(III)-reducing bacteria: *Shewanella alga*; *Geobacter sulfurreducens*	Co immobilization by microbes and microbial reduction of Co(III) to Co(II) in the presence of Mn(IV) oxide	Lloyd (2003); Gorby and Bolton (1998)
Cr	Sulfate-reducing bacterial biofilms: *Desulfotomaculum reducens*	Immobilization of Cr(VI) to Cr(III) by the bacteria and microbial sulfate reduction followed by precipitation	Smith and Gadd (2000)
Cr	Bacterium: *Desulfovibrio desulfuricans*	Immobilization of Cr(VI) by microbial lactate reduction followed by precipitation	Tucker et al. (1998); Lloyd (2003)

(Continued)

TABLE 15.2 (*Continued*)

Microorganisms Responsible for Metal(loid) Mobilization and Immobilization in Mine Soils/Tailings

Metal(loid)	Microorganisms	Mechanisms of Metal(loid) Mobilization/Immobilization	References
Cr	*Escherichia coli*, pseudomonads, *S. oneidensis*, and *Aeromonas* species	Immobilization of Cr by microbial reduction of Cr(VI) to Cr(III)	Lloyd (2003); Wang (2000)
Cu	Fungi: *Aspergillus* sp. and *Penicillium* sp.	Fungi mobilized Co and bioleaching through organic acids produced by the fungi	Brandl (2001); Mulligan and Galvez-Cloutier (2003); Santhiya and Ting (2005)
Cu	Lichens (Fungi)	Immobilization through Cu accumulation by fungi	Purvis and Pawlik-Skowronska (2008)
Cu	Archaea: *Sulfolobus*, *Acidianus*, *Metallosphaera*, and *Sulfurisphaera* spp., Bacteria: *Acidithiobacillus thiooxidans*, *Leptospirillum ferrooxidans*, *Sulfolobus* spp., and *Acidianus brierleyi*	The archaea mobilized Cu by microbial solubilization (sulfur oxidation) of insoluble copper sulfide	Rawlings (2002); Rawlings et al. (2003); Jerez (2009)
Fe	Bacteria: *Geothrix fermentans*; *Vibrio* sp.; *Geovibrio ferrireducens*; *Deferribacter thermophilus*; *Ferribacter limneticum*; *Wolinella succinogenes* and Archaea: *Archaeoglobus fulgidus*, *Pyrococcus furiosus*, and *Pyrodictium abyssi*	Microbes mobilized Fe by microbial reduction of Fe(III) to Fe(II) during the oxidation of hydrogen and short-chain fatty acids	Lloyd (2003); Gadd (2010)
Fe	Bacteria: *Gallionella* sp., *Leptothrix* sp.	Immobilization of Fe(III) oxides and hydroxides by bacteria, and precipitation and deposition around the bacterial cells caused Fe immobilization	Ehrlich and Newman (2009)
Hg	Bacteria and fungi (*Thiobacillus ferrooxidans*; *Geobacter metallireducens*)	Microbes mobilized Hg^{2+} by methylation to more toxic CH_3Hg^+. CH_3Hg^+ can be enzymatically reduced to volatile metallic Hg^0	Barkay and Wagner-Dobler (2005)
Mn	Bacteria: *Shewanella oneidensis*; *Geobacter metallireducens*	Some bacteria mobilized Mn by microbial reduction of Mn(IV) to Mn(II) during oxidation of organic compounds such as acetate	Lloyd (2003); Gadd (2010)
Mn	Fungi including *Acremonium* spp.	Immobilization of Mn(II) by microbial oxidation	Miyata et al. (2004, 2006, 2007); Saratovsky et al. (2009)
Mo	Bacterium: *Desulfovibrio desulfuricans*	Immobilization of Mo(VI) by microbial lactate reduction followed by precipitation	Tucker et al. (1998)

(*Continued*)

TABLE 15.2 (*Continued*)

Microorganisms Responsible for Metal(loid) Mobilization and Immobilization in Mine Soils/Tailings

Metal(loid)	Microorganisms	Mechanisms of Metal(loid) Mobilization/Immobilization	References
Pb	*Azotobacter chroococcum, Bacillus megaterium*, and *Bacillus mucilaginosus*	Immobilization of Pb by the bacterial adsorption to cell wall	Wu et al. (2006)
Pb	Ectomycorrhizal fungi	Immobilization of Pb by fungal fixation	Krupa and Kozdroj (2004)
Pb	Rhizobia	Immobilization of Pb by the bacteria and fixation via adsorption and precipitation on cell wall	Huang et al. (2002)
Sb	*Pseudomonas*- and *Stenotrophomonas*-related isolates	Microbial oxidation of Sb(III) to Sb(V) caused immobilization	Hamamura et al. (2013)
Sb	*Agrobacterium tumefaciens* and eukaryotic acidothermophilic Cyanidiales alga isolate	Microbial oxidation of Sb(III) to Sb(V) caused immobilization	Lehr et al. (2007)
Se	Fungi and bacteria	Some fungi and bacteria mobilized Se and Se methylation involved in volatilization of Se	Thompson-Eagle and Frankenberger (1992)
Se	*Desulfomicrobium norvegicum*	Immobilization of Se by bacteria through precipitation	Hockin and Gadd (2003)
Se	Bacterium: *Desulfovibrio desulfuricans*	Immobilization of Se(VI) by microbial lactate reduction followed by precipitation	Tucker et al. (1998)
Se	*Wolinella succinogenes, Desulfovibrio desulfuricans, Pseudomonas stutzeri, Escherichia cloacae*, and *Escherichia coli*	Immobilization of Se by enzymatic reduction of Se(VI) (nitrate reductases)	Lloyd (2003); Tomei et al. (1992, 1995)
U	Bacterium: *Desulfovibrio desulfuricans*	Immobilization of U(VI) by microbial lactate reduction followed by precipitation	Tucker et al. (1998)
U	*Bacillus sphaericus* JG-A12	Immobilization of U(VI) by bacterial bioaccumulation	Merroun et al. (2005)
U	Sulfate-reducing bacteria: Desulfobacterales, Desulfovibrionales, Syntrophobacteraceae, and Clostridiales	Immobilization of U by sulfate reduction	Sitte et al. (2010)
U	Gram-negative betaproteobacterium: *Acidovorax facilis*	Immobilization of U(VI) by biosorption and bioaccumulation	Gerber et al. (2016)
V	*Micrococcus lactilyticus, Desulfovibrio desulfuricans*, and *Clostridium pasteurianum*	Microbes immobilized V by microbial reduction of V(V) to V(III) during oxidation of organic compounds such as sugars and amino acids	Lloyd (2003)
Zn	Bacteria: *Burkholderia cepacia*	The bacteria mobilized Zn by pH decrease	Li and Wong (2010)
Zn	Mycorrhizal fungi	Immobilization of Zn by fungal stabilization	Christie et al. (2004)

strains isolated from 12 to 66 mg kg^{-1} Cr(VI) containing mine soil. Oxidation of Cr(III) to Cr(VI) is primarily mediated abiotically through oxidizing agents such as Mn(IV), and to a lesser extent by Fe(III), whereas reduction of Cr(VI) to Cr(III) is mediated through both abiotic and biotic processes (Choppala et al. 2012; Dhal et al. 2013). Chromate can be reduced to Cr(III) in environments where a ready source of electrons (Fe[II]) is available and microbial Cr(VI) reduction occurs in the presence of organic matter as an electron donor (Choppala et al. 2012; Hsu et al. 2009). In living systems, Se tends to be reduced rather than oxidized, and reduction occurs under both aerobic and anaerobic conditions. Dissimilatory Se(IV) reduction to Se(0) by chemical reductants such as sulfide or hydroxylamine, or biochemically by glutathione reductase, is the major biological transformation for remediation of Se oxyanions in anoxic sediments (Zhang and Frankenberger 2003). Therefore, the precipitation of Se(0), which has been associated with bacterial dissimilatory Se(VI) reduction, has an important environmental significance (Oremland et al. 1989, 2004).

Methylation is a biological mechanism for the removal of toxic metal(loid)s by converting them to methyl derivatives that are subsequently removed by volatilization (Frankenberger and Losi 1995). Methylated derivatives of As, Hg, and Se can originate from chemical and biological mechanisms and this frequently results in altered solubility, volatility, toxicity, and mobility. Although methylation of metal(loid)s occurs through both abiotic and biological processes, biological methylation (biomethylation) is considered to be the dominant process in soils and aquatic environments. Biomethylation may result in metal(loid) detoxification, since methylated derivatives may be excreted readily from cells, and are often volatile and may be less toxic, for example, organoarsenicals (Bolan et al. 2013, 2014).

Microorganisms in soils and sediments act as biologically active methylators (Frankenberger and Arshad 2001; Loseto et al. 2004). Organic matter provides the methyl-donor source for both biomethylation and abiotic methylation in soils and sediments. Lambertsson and Nilsson (2006) suggested that organic matter and alternative electron acceptors influenced methylation of Hg in the sediments. Some organisms also detoxify inorganic As by methylation. This produces monomethyl arsonate (MMA[V]), methylarsonite (MMA[III]), dimethylarsinate (DMA[V]), dimethylarsenite (DMA[III]), and trimethylarsine oxide (TMAO) (Stolz et al. 2010). Selenium biomethylation represents a potential mechanism for removing Se from contaminated environments as methylated compounds, such as dimethyl selenide (DMSe), since they are less toxic than dissolved Se oxyanions. Dimethyl selenide can be demethylated in anoxic sediments as well as anaerobically by an obligate methylotroph similar to *Methanococcoides methylutens* in pure culture. While an anaerobic demethylation reaction may form a toxic and reactive hydrogen selenide from less toxic DMSe, aerobic demethylation of DMSe yields selenite (Se[VI]), thereby retaining Se in the system (Frankenberger and Arshad 2001).

15.5   ACID MINE DRAINAGE AND (IM)MOBILIZATION OF HEAVY METAL(LOID)s

The geochemical forms of metal(loid)s in Fe-Mn oxides, organic fractions, and the residual phase of mining soils affect their mobility and bioavailability in AMDs (Anawar 2015; Garcia et al. 2005; Wang et al. 2010a). The strong binding of metal(loid)s, mainly with hydrous and crystalline oxide minerals of Fe, Mn, and Al, and the binding of Al-silicate through adsorption/precipitation processes are responsible for the low solubility of metal(loid)s including As and Sb (Anawar 2015; Bolan et al. 2014; Casiot et al. 2007; Casado et al. 2007; Filella et al. 2002). However, dissolution and colloidal transport can mobilize them from the mine wastes. Mining activities produce huge quantities of tailings. Sulfide tailings, an important source of metal(loid) contamination, are usually disposed of in open-air impoundments and thus are exposed to microbial oxidation. Microbial activities greatly enhance sulfide oxidation and result in the release of heavy metal(loid)s and the precipitation of Fe (oxy)hydroxides and sulfates (Coupland and Johnson, 2008; Johnson and Hallberg, 2009).

These secondary minerals in turn influence the mobility of dissolved metal(loid)s and play important roles in the natural attenuation of heavy metal(loid)s (Table 15.2).

The sulfide oxidation and precipitation of Al–Fe secondary minerals associated with AMD from an abandoned Cu mine waste pile at Touro, Spain, have been studied by Civeira et al. (2016), and they observed that the concentrations of amorphous Fe and Al phases influenced the chemical fractionation of hazardous elements more than the pH of AMD. Ferric Fe is ubiquitous in AMD-contaminated environments. Its precipitation via various pathways is a significant process in the attenuation of metal(loid)s on concentrations in mine drainage via coprecipitation reactions. Arenas-Lago et al. (2014) evaluated the distribution and the interactions among Pb, Zn, and Cd from a depleted mine at Rubiais (Lugo, Spain), and the soil geochemical phases by means of sequential chemical extraction, x-ray diffraction (XRD), field emission scanning electron microscopy/energy-dispersive x-ray spectroscopy (FESEM/EDS), and time-of-flight secondary ion mass spectrometry (TOF–SIMS). The concentration of Pb, Zn, and Cd were in the range of 850–6,761, 1,754–32,287, and 1.8–43.7 mg kg^{-1}, respectively, and the highest proportion was in the residual fraction. While the Mn oxides highly influenced the retention of Cd, Pb retention was mainly influenced by Fe oxides. Zinc was uniformly distributed among the residual fraction and the Fe and Mn oxides. TOF–SIMS and SEM/EDS techniques showed that Pb and Zn were present as sulfides and associated with Fe and Mn oxides. However, the authors warned against the sulfide oxidation processes of oxides and sulfides or alteration of the oxides causing leaching, which would result in the contamination of the protected ecosystem.

Natural attenuation, however, rarely is the single solution to AMD since a rainstorm will (re) mobilize the precipitates (Kothe and Büchel 2014). Drahota et al. (2016) examined As mobility in two 50-year-old sulfide-rich mining waste dumps (Jedová jáma and Dlouhá Ves) in the Czech Republic. Originally, all of the As in the mining wastes was present as arsenopyrite, but with time it has been replaced by secondary As phases. Most of As precipitated as amorphous ferric arsenate at Jedová jáma. They also noticed the accumulation of As in the scorodite and Fe (hydr)oxide (up to 3.2 wt.% As$_2$O$_5$) that is particularly represented by hematite. Mining wastes at Dlouhá Ves contained only trace amounts of scorodite, and As was primarily bound to Pb-jarosite and Fe (hydr) oxides (especially goethite) with up to 1.6 and 1.8 wt.% As$_2$O$_5$, respectively. The pore water collected after rainfall events indicated high concentrations of As (~4600 µg L^{-1}) at Jedová jáma, which they attributed to the dissolution of amorphous ferric arsenate and simultaneous precipitation of Fe (hydr)oxides under mildly acidic conditions (pH ~ 4.4). However, aqueous As at Dlouhá Ves was negligible (up to 1.5 µg L^{-1}). This was attributed to the efficient adsorption of As on the Fe (hydr) oxides and hydroxosulfates under acidic pH of ~2.8. The authors recommend that the mine wastes should be kept under acidic conditions or with high aqueous Fe(III) concentrations to prevent the release of As from dissolution of ferric arsenates. In another study, two sulfidic mine tailings within the Zambian Copperbelt in the north of Zambia were studied by Sracek et al. (2010). The principal secondary minerals at both sites were gypsum, poorly crystalline Fe(III) phases, and hematite. They noted that Cu and Co incorporated in hematite was immobilized within the mine tailings with no threat of AMD formation even over the long term.

Several As-bearing minerals found in mine wastes have been shown to undergo microbial-assisted reductive dissolution (Table 15.2). The dissimilatory As(V) reducers *Shewanella* sp. O23S and *Aeromonas* sp. O23A play a significant role in releasing As(V) from As-bearing minerals in a gold mine as observed by Drewniak et al. (2014). The experimental anaerobic dissolution of Pb-As-jarosite (PbFe$_3$(SO$_4$,AsO$_4$)$_2$(OH)$_6$) by *Shewanella putrefaciens* at circumneutral pH resulted in immediate Fe(III) reduction followed by As(V) reduction after 72 h (Smeaton et al. 2012). España et al. (2008) reported very high biomass of microbiological populations, which consisted of acidophilic and Fe-oxidizing archaea, and minor numbers of acidophilic bacteria in a pyrite pile of the San Telmo mine, Spain. Microbial leaching could extract metal(loid)s from mine wastes by employing Fe/sulfur-oxidizing microorganisms such as *Acidithiobacillus ferrooxidans*, *Acidithiobacillus thiooxidans*, *Leptospirillum ferrooxidans*, *Sulfolobus* sp., and thermophilic bacteria including *Sulfobacillus thermosulfidooxidans*, *Acidithiobacillus caldus*, *Sulfobacillus acidophilus*, and *Leptospirillum ferriphilum* (Fu et al. 2008; Johnson 2008; Li et al. 2011a; Watling 2006; Zhou et al. 2009).

15.6 CASE STUDIES

15.6.1 BRAZIL

In this study, Perlatti et al. (2014) evaluated the mineralogical and geochemical composition of waste rocks and determined total Cu concentration and speciation in soils from three different areas of an abandoned Cu mine in Viçosa do Ceará (northeastern Brazil). Their objectives were to assess geochemical transfer mechanisms of Cu from waste rock to soil and its influence on the physico-chemical properties and the bioavailability of Cu, establishing the bases to predict the real eco-toxicological risks, and to estimate the expected results for mine site restoration. The authors noted several villages and agricultural fields near the mine. The area consists of three geological formations, from the base to the top: São Joaquim, Mambira, and Ubari. The São Joaquim formation is mainly characterized by quartzite. The Mambira formation is again subdivided into three parts and the highest concentrations of Cu sulfide ores (chalcopyrite [$CuFeS_2$], chalcocite [Cu_2S], bornite [Cu_5FeS_4], and covellite [CuS], with arsenopyrite [$FeAsS$], galena [PbS], and sphalerite [ZnS] occurring in smaller proportions) occur in the upper formation (Pedra Verde). The Ubari formation comprises conglomerates of mudstone, sandstone, phyllite, quartzite, and quartz dispersed in a feld-spathic matrix, which also contains Cu in oxidized ores (malachite and cuprite) (Korpershoeck et al. 1979). The red Fe oxide zone between the Pedra Verde phyllites and Ubari conglomerates contains, on average, 0.1% of Cu with hematite (Fe_2O_3) as one of the main constituents.

The mine was exploited in the 1980s for the extraction and processing of Cu from sulfide (chalcopyrite and chalcocite) and carbonate (malachite) minerals. After the mine was closed in 1987, the processed minerals were dumped in piles on terraces constructed on the slope of the plateau, and due to its lack of maintenance, some of these piles have been eroded, spreading waste rock over distances of up to 1.5 km along the slope. Soils were sampled in three different mine compartments:

1. Ore processing area (Pr), site with terraces and leaching tanks; vegetation is sparse, with few shrub and tree species.
2. Waste rock area (Wr), site containing large amounts of mine waste, and the vegetation is sparse with few shrub species.
3. Border area (Bd), site characterized by the transition between waste rock area and adjacent vegetation. The rock waste is more widely dispersed and less abundant, and tree and shrub species are commonly present.

X-ray fluorescence (XRF) analysis of the waste rock samples demonstrated that Cu was the only element present at high concentrations and all other trace elements were present at usual concentrations for sedimentary and metamorphic rocks (Thornton 1995). The total Cu concentration in the waste rock ranged between 63.3 and 350,331 mg kg^{-1}, which indicated that this waste material represents a serious risk of environmental contamination associated with the Cu mobilization potential. The mineralogical and geochemical characteristics of the waste rock determined by XRD and SEM-EDS indicated that the main forms of Cu present are carbonates, oxides, and sulfides.

The pH ranged between 4.2 and 7.2, with higher values in the waste rock (Bd) area (mean pH = 6.6), which indicated that AMD was not an active process despite the presence of various Cu sulfides. The total organic carbon (TOC) contents were significantly higher in Bd soils than in the Pr and Wr soils due to higher vegetation cover. Similar to Bd, the total concentrations of most trace elements in Pr and Wr soils remained at normal levels, except for Cu. Copper was found at very high levels (46.4–11,904 mg kg^{-1}) in the following order: Pr > Wr > Bd, which indicated contamination of soils in the three study areas. The mean total Cu concentration in soils from Pr (11,180 mg kg^{-1})

exceeded the reference value for soil quality used by the current environmental legislation in Brazil (60.0 mg kg^{-1} for Cu; CONAMA 2009) by more than 185 times. In Wr (4683 mg kg^{-1}) and Bd (1086 mg kg^{-1}) soils, the mean values exceeded the reference value by 78 and 18 times, respectively. The waste rock from Pr was not subjected to leaching with sulfuric acid and thus contained higher concentrations of Cu than the waste rock from Wr and Bd. Although a large amount of waste rock was present at the surface of Wr soils, those materials had already been processed by leaching and, hence, contained lower levels of Cu. In contrast, although the soils from the border areas (Bd) contained higher levels of Cu than the reference values, the low total Cu was attributed to the smaller amount of waste rock at the soil surface.

The amount of Cu removed from the soils by sequential extraction was higher than the total Cu values obtained by XRF; however, they noticed a positive and significant correlation (r = 0.972, p < 0.001) between the amounts of Cu obtained by both methods. The results of the sequential extraction showed two contrasting geochemical scenarios (Figure 15.1). In the Pr and Wr soils, most of the Cu was associated with the carbonate fraction which they attributed to the presence of malachite and pseudomalachite. A significant proportion of Cu was also associated with the amorphous oxides in the Pr and Wr soils, with values ranging from 27.5% to 32.3% of the total Cu. The authors noted that the high proportion of Cu associated with amorphous oxides acts as a sink for soluble Cu, given the strong affinity between amorphous Fe oxides and Cu (Cerqueira et al. 2011).

In Bd soils, a clearly different pattern of Cu distribution was observed, particularly in the highest part of the slope (points B1, B2, and B3) (Figure 15.1). The Cu distribution patterns resembled those observed in Pr and Wr soils and were mainly related to higher levels of Cu associated with the organic fraction and higher levels of exchangeable Cu, and the latter associated with the lower pH and total sulfur. In the lower part of the slope (B4, B5, and B6), the distribution of Cu was more heterogeneous, mainly due to low proportions of Cu associated with carbonates and higher proportions of Cu associated with the organic fraction. The authors pointed out that the greater presence of plants in those sites may accelerate the dissolution of carbonates via the action of microorganisms and organic root exudates from the rhizosphere (Uroz et al. 2009), thus mobilizing Cu in the soil. Further, they noted that the higher levels of TOC and the strong affinity with organic matter would have enhanced the proportion of Cu associated with organic fraction, as shown by the positive and significant correlation between Cu associated with organic matter and TOC (r = 0.813; p < 0.01).

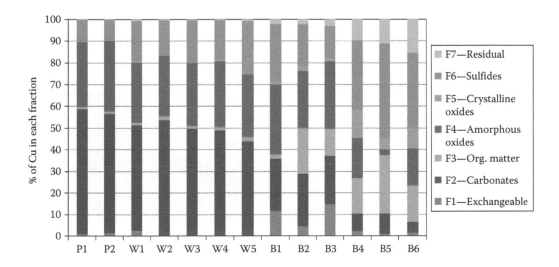

FIGURE 15.1 Percentage of copper in different soil fractions from Pedra Verde, Brazil (P, ore processing area; W, waste rock area; B, border area). (From Perlatti, F. et al., *Sci. Total Environ.*, 500–501, 91, 2014. With permission.)

The authors suggested that the use of plants or organic amendments in mine sites with high concentration of Cu carbonate-containing wastes should be viewed with caution, since it may enhance Cu mobilization due to an increase in the rate of carbonates dissolution.

15.6.2 REPUBLIC OF KOREA

Ilkwang mine is one of the abandoned metal(loid) mines in Busan, South Korea, in which Au, Ag, and Cu were mined. The pH of Ilkwang mine tailings was 2.65 and displayed Cd, Cu, and Pb concentrations of 4, 400, and 1826 mg kg^{-1}, respectively (Kim et al. 2001). Because of the acidic nature of mine tailings, strong AMD was discharged from the mine adit (Figure 15.2). Although there was a mine water treatment facility, it was not under operation (Figure 15.2). Therefore, heavy metal(loid)s were also released with average daily concentrations of 53.4 mg L^{-1} Fe, 6.25 mg L^{-1} Zn, 2.13 mg L^{-1} Mn, 0.14 mg L^{-1} As, and 0.04 mg L^{-1} Cd (Han et al. 2015). Han et al. (2015) evaluated the toxicity of mine water and sediments using the Microtox bioluminescent assay. The toxicity of stream water nearby the Ilkwang mine showed good correlation with pH and dissolved metal(loid) concentrations of water. However, the toxicity of the stream sediment did not show a close correlation with toxic elements such as As.

The geo-accumulation index showed that the sediment was highly contaminated with As, but not contaminated or moderately contaminated with Cu, Pb, and Zn. The X-ray near edge structure (XANES) spectra of the sediment showed that As in the sediment mainly presented as As(V) because of oxidation of dissolved As after discharging from the mine adit. Iron minerals contributed to the adsorption of As(V) in the sediments. The mine sediment comprised ferrihydrite, schwertmannite, and goethite. Ferrihydrite and schwertmannite were initially formed when Fe^{2+} was oxidized to Fe^{3+} in the acidic mine stream and when coprecipitation of As(V) with the Fe minerals occurred. Arsenic K-edge XANES spectra of the Ilkwang mine sediment precipitate showed that As was composed of 76.3% coprecipitated As in schwertmannite, 15.2% As adsorbed onto goethite, and 8.2% coprecipitated As in ferrihydrite (Park et al. 2016). The results suggested the change of precipitated ferrihydrite and schwertmannite to goethite and the adsorption of As on the surface of goethite (Figure 15.3). The processes resulted in the removal of As from mine water, which was regarded as a natural attenuation of As (Park et al. 2016).

Imgi is another abandoned metal(loid) mine located in Busan, South Korea. Mining activity in the Imgi mine started in 1980 and ended in 1992 and waste materials were left along the natural

(a)

(b)

FIGURE 15.2   (a) Mine adit and (b) abandoned mine water treatment facility of Ilkwang mine.

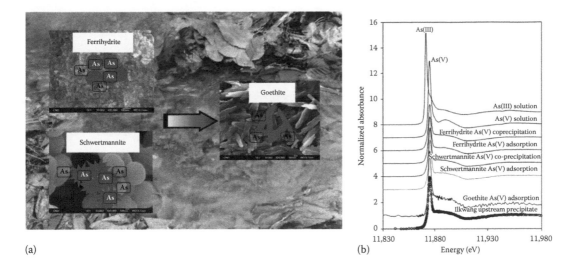

(a)   (b)

FIGURE 15.3 (a) Mechanism of As coprecipitation and adsorption with Fe minerals in mine sediments; (b) arsenic K-edge XANES spectra of reference compounds and Ilkwang mine precipitate and linear combination fitting result (open circles) of Ilkwang mine precipitate fitted using the reference compounds. (From Park, J.H. et al., *Water Res.*, 106, 295, 2016. With permission.)

(a)   (b)

FIGURE 15.4 (a) Waste dumps and (b) mine stream of Imgi mine.

slope (Figure 15.4). The Imgi mine stream was contaminated with Al, Fe, Mn, Pb, Cu, Zn, and Cd, and left with an acidic pH because of meteoric water interaction with tailing dumps (Jeong 2015). Because of the instability of tailing dumps, an attempt to prevent dispersion of mine tailings to the surrounding environment was made. Prior to field application, a column experiment to immobilize heavy metal(loid)s in mine tailings using *Canavalia ensiformis* extract was conducted. Crude extract of *C. ensiformis* induced the precipitation of $CaCO_3$ by the hydrolysis of urea. Arsenic, Mn, Zn, Pb, Cr, and Cu concentrations in leachates from the mine tailings treated with *C. ensiformis* crude extract were reduced by 31.7%, 65.7%, 52.3%, 53.8%, 55.2%, and 49.0%, respectively, which is attributed to $CaCO_3$ formation within mine tailings (Nam et al. 2016). A product (BioNeutro-GEM or BNG) produced from *C. ensiformis* extract was applied in a field pilot-scale experiment. The treatment of BNG enhanced the establishment of plants in the tailings (Ahn 2014) (Figure 15.5).

(a)   (b)

FIGURE 15.5   Tailing dumps (a) before and (b) after the treatment with BioNeutro-GEM. (Courtesy of Korea Institute of Geoscience and Mineral Resources, Development of Control Technology of Geoenvironmental Hazards Propagation Induced by Mining Activity, Project report, 2014.)

15.7   SUMMARY AND CONCLUSIONS

Mining is better regulated nowadays, due to tougher legislation and to technical advances used for environmental control. However, the lack of such control in the past has led to numerous abandoned mines around the world without any proper recovery measures (Anawar 2015; Bradshaw, 1997; Ersoy et al. 2008; Otero et al. 2012). Mining activities have contaminated the surrounding forests, downstream areas, and farmlands through AMD, mine tailings, and waste rocks (Anawar 2015) Waste rock containing metal(loid)s, when disposed of in open pits, react with water and oxygen and may be oxidized or dissolved, depending on the geochemical composition of the material, thereby transferring toxic metal(loid)s to the environment and causing a serious risk to environmental and public health (Lottermoser 2007). The presence of high concentrations of sulfides generates AMD, thus increasing the risk of contamination via large decreases in soil and water pH, which increases metal(loid) mobility and bioavailability (Kimball et al. 2009).

Ecological risk assessments are becoming increasingly important in evaluating the effects of historical mining activities as well as in predicting the potential effects of present and future mining. It is important to identify environmental controlling factors, such as pH, organic matter, and redox conditions, and their effect on metal(loid) concentration and exposure to biota. Although metal(loid)s persist in the environment once introduced through mining activities, their speciation can influence the mobility and bioavailability of metal(loid)s within the environment. Knowledge of how metal(loid) speciation responds to changing environmental conditions can be used to predict their mobility. Understanding the geochemical processes that influence metal(loid) mobility, distribution, and bioavailability can aid in forecasting the potential ecological effects of metal(loid)s in mining environments. The combination of traditional approaches, based on contaminant concentrations and extractabilities in soils, should be combined with detailed mineralogical investigations as well as applications of advanced synchrotron-based techniques, which are valuable for determining the release of toxic elements into the soil and their subsequent binding to individual soil components. Also, the formation and composition of a mineral deposit are essential for predicting its environmental impact. The geochemical and mineralogical composition of the ores will ultimately determine the chemical transformations of heavy metal(loid)s in the mine wastes (Jamieson 2011).

Secondary metal(loid) minerals play a major role in the solubility of metal(loid)s together with adsorption, coprecipitation, and related reactions. Roper et al. (2012) reviewed Sb and mentions that many of the known secondary Sb minerals remain to be studied chemically, and solubility and related thermochemical data are available for only a few of the more commonly occurring minerals.

They stress about the dearth of information on the kinetics of crystallization and transformation of secondary Sb minerals. Different mining wastes are enriched with variable amounts of sulfide and carbonate minerals that produce a variable degree of contamination controlled by a natural attenuation process.

Biotransformation processes can influence the solubility and subsequent mobility of metal(loid)s, especially As, Cr, Hg, and Se in soils. These processes can be readily managed and enhanced for efficient removal of contaminants, and thus a greater understanding of biotransformation processes will help to monitor the environmental fate of toxic elements and will help to develop suitable remediation technologies. The oxidative dissolution of these minerals occurs not only by abiotic reactions but also by microbial-catalyzed oxidation and bioleaching, which, therefore, can mobilize the toxic concentration of acidity, sulfate, and metal(loid)s in soil and aquatic environments. Shedding further light on the microbe–mineral interactions in tailings will also help us mitigate the environmental impacts of mining activities. The specific role of bacteria in the dissolution of metal(loid)-bearing minerals including As and formation of secondary metal(loid)-bearing minerals is still not well understood. Research into the use of microorganisms, for example, sulfate-reducing bacteria, will require identification and isolation of bacterial strains suited for each contaminated site. Additional work involving the use of novel surface mineral characterization techniques, such as low-vacuum scanning electron microscopy, atomic force microscopy, and synchrotron-based techniques such as scanning X-ray microscopy, is required to advance knowledge in this field.

The control mechanisms of metal(loid) concentrations in mine sites and tailings can be complex and are highly specific to the metal(loid) and the site. At the start of the AMD formation process, heavy metal(loid) concentrations are essentially controlled by the amount of metal(loid)-bearing minerals as well as their solubilities and dissolution rates. However, the threat of AMD to the environment cannot be solved by a single solution and will require the integrated implementation of a range of measures and processes. In order to optimize an appropriate remediation technique, it is important to understand the mining geological texture, mineralization types, geochemistry of mine waste rocks/tailings, secondary minerals, presence of toxic metal(loid)s, soil characteristics, and the local climate. Manipulating both the physicochemical and biological transformation processes of heavy metal(loid)s by controlling the factors affecting them will enable the sustainable management of metal(loid) contamination to mitigate their environmental and health impacts.

REFERENCES

Adriano, D. C. 2001. *Trace Elements in the Terrestrial Environments: Biogeochemistry, Bioavailability, and Risks of Heavy Metals.* Springer-Verlag, New York.

Ahn, J. S. 2014. Development of control technology of geoenvironmental hazards propagation induced by mining activity. Project report (No: 1711021742). Korea Institute of Geoscience and Mineral Resources (KIGAM), Daejeon, Republic of Korea.

Alloway, B. J. 1990. Soil processes and the behaviour of metals. In *Heavy Metals in Soils*, ed. B. J. Alloway, Blackie & Son Ltd., Glasgow, Scotland, pp. 7–28.

Anawar, H. M. 2015. Sustainable rehabilitation of mining waste and acid mine drainage using geochemistry, mine type, mineralogy, texture, ore extraction and climate knowledge. *Journal of Environmental Management* 158:111–121.

Anju, M. and D. K. Banerjee. 2012. Multivariate statistical analysis of heavy metals in soils of a Pb–Zn mining area, India. *Environmental Monitoring and Assessment* 184:4191–4206.

Arenas-Lago, D., M. Lago-Vila, A. Rodrigues-Seijo, M. L. Andrade, and F. A. Vega. 2014. Risk of metal mobility in soils from a Pb/Zn depleted mine (Lugo, Spain). *Environmental Earth Sciences* 72:2541–2556.

Ashley, P. M., D. Craw, B. P. Graham, and D. A. Chappell. 2003. Environmental mobility of antimony around mesothermal stibnite deposits, New South Wales, Australia and southern New Zealand. *Journal of Geochemical Exploration* 77:1–14.

Bachate, S. P., R. M. Khapare, and K. M. Kodam. 2012. Oxidation of arsenite by two β-proteobacteria isolated from soil. *Applied Microbiology and Biotechnology* 93:2135–2145.

Barkay, T. and I. Wagner-Dobler. 2005. Microbial transformations of mercury: Potentials, challenges, and achievements in controlling mercury toxicity in the environment. *Advances in Applied Microbiology* 57:1–52.

Battaglia-Brunet, F., M. C. Dictor, F. Garrido, C. Crouzet, D. Morin, K. Dekeyser, M. Clarens, and P. Baranger. 2002. An arsenic (III)-oxidizing bacterial population: Selection, characterization, and performance in reactors. *Journal of Applied Microbiology* 93:656–667.

Bentley, R. and T. G. Chasteen. 2002. Microbial methylation of metalloids: Arsenic, antimony, and bismuth. *Microbiology and Molecular Biology Reviews* 66:250–271.

Bhattacharya, S., K. Gupta, S. Debnath, U. C. Ghosh, D. Chattopadhyay, and A. Mukhopadhyay. 2012. Arsenic bioaccumulation in rice and edible plants and subsequent transmission through food chain in Bengal basin: A review of the perspectives for environmental health. *Toxicological & Environmental Chemistry* 94:429–441.

Bhuiyan, M. A., L. Parvez, M. A. Islam, S. B. Dampare, and S. Suzuki. 2010. Heavy metal pollution of coal mine-affected agricultural soils in the northern part of Bangladesh. *Journal of Hazardous Materials* 173:384–392.

Bigham, J. M. and D. K. Nordstrom. 2000. Iron and aluminum hydroxysulfates from acid sulfate waters. *Reviews in Mineralogy and Geochemistry* 40:351–403.

Bigham, J. M., U. Schwertmann, S. J. Traina, R. L. Winland, and M. Wolf. 1996. Schwertmannite and the chemical modelling of iron in acid sulfate waters. *Geochimica et Cosmochimica Acta* 60:2111–2121.

Blowes, D. W., J. L. Jambor, C. J. Hanton-Fong, L. Lortie, and W. D. Gould. 1998. Geochemical, mineralogical and microbiological characterization of a sulphide-bearing carbonate-rich gold-mine tailings impoundment, Joutel, Québec. *Applied Geochemistry* 13:687–705.

Blowes, D. W., C. J. Ptacek, J. L. Jambor, and C. G. Weisener. 2003. The geochemistry of acid mine drainage. In *Environmental Geochemistry*, ed. B. S. Lollar, vol. 9. Elsevier Ltd., Amsterdam, the Netherlands, pp. 149–204.

Bolan, N. S., D. C. Adriano, A. Kunhikrishnan, T. James, R. McDowell, and N. Senesi. 2011. Dissolved organic matter: Biogeochemistry, dynamics, and environmental significance in soils. *Advances in Agronomy* 110:1–75.

Bolan, N. S., D. C. Adriano, P. Mani, A. Duraisamy, and S. Arulmozhiselvan. 2003. Immobilization and phytoavailability of cadmium in variable charge soils: I. Effect of phosphate addition. *Plant and Soil* 250:83–94.

Bolan, N. S., G. Choppala, A. Kunhikrishnan, J. H. Park, and R. Naidu, 2013. Microbial transformation of trace elements in soils in relation to bioavailability and remediation. *Reviews in Environmental Contamination and Toxicology* 225:1–56.

Bolan, N. S., A. Kunhikrishnan, R. Thangarajan, J. Kumpiene, J. Park, T. Makino, M. B. Kirkham, and K. Scheckel. 2014. Remediation of heavy metal (loid)s contaminated soils—To mobilize or to immobilize?. *Journal of Hazardous Materials* 266:141–166.

Bolan, N. S., R. Naidu, J. K. Syers, and R. W. Tillman. 1999. Surface charge and solute interactions in soils. *Advances in Agronomy* 67:87–140.

Bortnikova, S., E. Bessonova, and O. Gaskova. 2012. Geochemistry of arsenic and metals in stored tailings of a Co-Ni arsenide-ore, Khovu-Aksy area, Russia. *Applied Geochemistry* 27:2238–2250.

Bradshaw, A. D. 1997. Restoration of mined lands—Using natural processes. *Ecological Engineering* 8:255–269.

Brandl, H. 2001. Heterotrophic leaching. In *Fungi in Bioremediation*, ed. G. M. Gadd. Cambridge University Press, Cambridge, U.K., pp. 383–423.

Burton, E. D., R. T. Bush, and L. A. Sullivan. 2006. Sedimentary iron geochemistry in acidic waterways associated with coastal lowland acid sulfate soils. *Geochimica et Cosmochimica Acta* 70:5455–5468.

Casado, M., H. M. Anawar, A. Garcia-Sanchez, and I. Santa-Regina. 2007. Antimony and arsenic uptake by plants in an abandoned mining area. *Communications in Soil Science and Plant Analysis* 38:1255–1275.

Casiot, C., M. Ujevic, M. Munoz, J. L. Seidel, and F. Elbaz-Poulichet. 2007. Antimony and arsenic mobility in a creek draining an antimony mine abandoned 85 years ago (upper Orb basin, France). *Applied Geochemistry* 22:788–798.

Cerqueira, B., E. F. Covelo, L. Andrade, and F. A. Vega. 2011. The influence of soil properties on the individual and competitive sorption and desorption of Cu and Cd. *Geoderma* 162:20–26.

Choppala, G. K., N. S. Bolan, M. Megharaj, Z. Chen, and R. Naidu. 2012. The influence of biochar and black carbon on reduction and bioavailability of chromate in soils. *Journal of Environmental Quality* 41:1175–1184.

Christie, P., X. L. Li, and B. D. Chen. 2004. Arbuscular mycorrhiza can depress translocation of zinc to shoots of host plants in soils moderately polluted with zinc. *Plant and Soil* 261:209–217.

Civeira, M., M. L. Marcos, J. C. Oliveira, J. C. Hower, D. M. Agudelo-Castañeda, S. R. Taffarel, C. G. Ramos, R. M. Kautzmann, and L. F. Silva. 2016. Modification, adsorption, and geochemistry processes on altered minerals and amorphous phases on the nanometer scale: Examples from copper mining refuse, Touro, Spain Matheus. *Environmental Science and Pollution Research* 23:6535–6545.

CONAMA, Conselho Nacional do Meio Ambiente/Brasil. 2009. *Resolução N°420—Dispõe sobre critérios e valores orientadores de qualidade do solo*. Ministério do Meio Ambiente, Brasília/DF.

Concas, A., C. Ardau, A. Cristini, P. Zuddas, and G. Cao. 2006. Mobility of heavy metals from tailings to stream waters in a mining activity contaminated site. *Chemosphere* 63:244–253.

Conesa, H. M., B. H. Robinson, R. Schulin, and B. Nowack. 2008. Metal extractability in acidic and neutral mine tailings from the Cartagena-La Unión mining district (SE Spain). *Applied Geochemistry* 23:1232–1240.

Corkhill, C. L., P. L. Wincott, J. R. Lloyd, and D. J. Vaughan. 2006. The oxidative dissolution of arsenopyrite (FeAsS) and enargite (Cu_3AsS_4) by *Leptospirillum ferrooxidans*. *Geochimica et Cosmochimica Acta* 70:3593–3612.

Coupland, K. and D. B. Johnson. 2008. Evidence that the potential for dissimilatory ferric iron reduction is widespread among acidophilic heterotrophic bacteria. *FEMS Microbiology Letters* 279:30–35.

Covelo, E. F., F. A. Vega, and M. L. Andrade. 2007. Simultaneous sorption and desorption of Cd, Cr, Cu, Ni, Pb, and Zn in acid soils. I. Selectivity sequences. *Journal of Hazardous Materials* 147:852–861.

Cravotta, C. A. and C. S. Kirby. 2004. Acidity and alkalinity in mine drainage: Practical considerations. *Proceedings of the American Society of Mining and Reclamation, National Meeting of the American Society of Mining and Reclamation and the 25th West Virginia Surface Mine Drainage Task Force*, April 18–24, 2004. ASMR, Lexington, KY.

Csavina, J., J. Field, M. P. Taylor, S. Gao, A. Landázuli, E. A. Betterton, and A. E. Sáez. 2012. A review on the importance of metals and metalloids in atmospheric dust and aerosol from mining operations. *Science of the Total Environment* 433:58–73.

Csavina, J., A. Landázuli, A. Wonaschütz, K. Rine, P. Rheinheimer, B. Barbaris, W. Conant, A. E. Sáez, and E. A. Betterton. 2011. Metal and metalloid contaminants in atmospheric aerosols from mining operations. *Water, Air, & Soil Pollution* 221:145–157.

Csavina, J., M. P. Taylor, O. Félix, K. P. Rine, A. E. Sáez, and E. A. Betterton. 2014. Size resolved dust and aerosol contaminants associated with copper and lead smelting emissions: Implications for emission management and human health. *Science of the Total Environment* 493:750–756.

Dang, Z., C. Liu, and M. J. Haigh. 2002. Mobility of heavy metals associated with the natural weathering of coal mine spoils. *Environmental Pollution* 118:419–426.

Da Pelo, S., E. Musu, R. Cidu, F. Frau, and P. Lattanzi. 2009. Release of toxic elements from rocks and mine wastes at the Furtei gold mine (Sardinia, Italy). *Journal of Geochemical Exploration* 100:142–152.

Davies, S. H. R. and J. J. Morgan. 1989. Manganese(II) oxidation kinetics on metal oxide surfaces. *Journal of Colloid and Interface Science* 129:63–77.

Dhal, B., H. N. Thatoi, N. N. Das, and B. D. Pandey. 2010. Reduction of hexavalent chromium by *Bacillus* sp. isolated from chromite mine soils and characterization of reduced product. *Journal of Chemical Technology and Biotechnology* 85:1471–1479.

Dhal, B., H. N. Thatoi, N. N. Das, and B. D. Pandey. 2013. Chemical and microbial remediation of hexavalent chromium from contaminated soil and mining/metallurgical solid waste: A review. *Journal of Hazardous Materials* 250–251:272–291.

Dhankher, O. P., Y. J. Li, B. P. Rosen, J. Shi, D. Salt, J. F. Senecoff, N. A. Sashti, and R. B. Meagher. 2002. Engineering tolerance and hyperaccumulation of arsenic in plants by combining arsenate reductase and gamma-glutamylcysteine synthetase expression. *Nature Biotechnology* 20:1140–1145.

Drahota, P., M. Knappová, H. Kindlová, A. Culka, J. Majzlan, M. Mihaljevič et al. 2016. Mobility and attenuation of arsenic in sulfide-rich mining wastes from the Czech Republic. *Science of the Total Environment* 557–558:192–203.

Drewniak, L., L. Rajpert, A. Mantur, and A. Sklodowska. 2014. Dissolution of arsenic minerals mediated by dissimilatory arsenate reducing bacteria: Estimation of the physiological potential for arsenic mobilization. *BioMed Research International* Article ID 841892, 12pp.

Drewniak, L. and A. Sklodowska. 2013. Arsenic-transforming microbes and their role in biomining processes. *Environmental Science and Pollution Research International* 20(11):7728–7739.

Ehrlich, H. L. and D. K. Newman. 2009. *Geomicrobiology*, 5th edn. CRC Press/Taylor & Francis, Boca Raton, FL.

Ersoy, A., T. Y. Yunsel, and U. Atici. 2008. Geostatistical conditional simulation for the assessment of contaminated land by abandoned heavy metal mining. *Environmental Toxicology* 23:96–109.

Espãna, J. S., E. G. Toril, E. L. Pamo, R. Amils, M. D. Ercilla, E. S. Pastor, and P. S. MartínÚriz. 2008. Biogeochemistry of a hyperacidic and ultraconcentrated pyrite leachate in San Telmo mine (Iberian Pyrite Belt, Spain). *Water, Air, and Soil Pollution* 194:243–257.

Ettler, V. 2016. Soil contamination near non-ferrous metal smelters: A review. *Applied Geochemistry* 64:56–74.

Ettler, V., Z. Johan, A. Barronnet, F. Jankovský, C. Gilles, M. Mihaljevič, O. Šebek, L. Strnad, and P. Bezdička. 2005. Mineralogy of air-pollution-control residues from a secondary lead smelter: Environmental implications. *Environmental Science and Technology* 39:9309–9316.

Favas, P. J. C., J. Pratas, M. E. P. Gomes, and V. Cala. 2011. Selective chemical extraction of heavy metals in tailings and soils contaminated by mining activity: Environmental implications. *Journal of Geochemical Exploration* 111:160–171.

Fields, S. 2003. The Earth's open wounds: Abandoned and orphaned mines. *Environmental Health Perspectives* 111:A154–A161.

Figueiredo, M. O. and T. P. da Silva. 2011. The positive environmental contribution of jarosite by retaining lead in acid mine drainage areas. *International Journal of Environmental Research and Public Health* 8:1575–1582.

Filella, M., N. Belzile, and Y. W. Chen. 2002. Antimony in the environment: A review focused on natural waters. I. Occurrence. *Earth-Science Reviews* 57:125–176.

Fitzpatrick, R. W., P. Shand, and R. H. Merry. 2009. Acid sulfate soils. In *Natural History of the Riverland and Murrayland*, ed. J. T. Jennings. Royal Society of South Australia (Inc.), Adelaide, South Australia, Australia, pp. 65–111.

Frankenberger, Jr., W. T. and M. Arshad. 2001. Bioremediation of selenium-contaminated sediments and water. *Biofactors* 14: 241–254.

Frankenberger, W. T. and M. E. Losi. 1995. Application of bioremediation in the cleanup of heavy elements and metalloids. In *Bioremediation: Science and Applications*, eds. H. D. Skipper and R. F. Turco. Soil Science Special Publication No. 43, Soil Science Society of America Inc., Madison, WI, pp. 173–210.

Fu, B., H. Zhou, R. Zhang, and G. Qiu. 2008. Bioleaching of chalcopyrite by pure and mixed cultures of *Acidithiobacillus* spp. and *Leptospirillum ferriphilum*. *International Biodeterioration & Biodegradation* 62:109–115.

Gadd, G. M. 2010. Metals, minerals and microbes: Geomicrobiology and bioremediation. *Microbiology* 156:609–643.

Galán, E., J. L. Gómez-Ariza, I. Gozález, J. C. Fernández-Caliani, E. Morales, and I. Giráldez. 2003. Heavy metal partitioning in river sediments severely polluted by acid mine drainage in the Iberian Pyrite Belt. *Applied Geochemistry* 18:409–421.

Garcia, G., A. L. Zanuzzi, and A. Faz. 2005. Evaluation of heavy metal availability prior to an in situ soil phytoremediation program. *Biodegradation* 16:187–194.

Gerber, U., I. Zirnstein, E. Krawczyk-Barsch, H. Lunsdorf, T. Arnold, and M. L. Merroun. 2016. Combined use of flow cytometry and microscopy to study the interactions between the gram-negative betaproteobacterium *Acidovorax facilis* and uranium(VI). *Journal of Hazardous Materials* 317:127–134.

Ghaderian S. M. and A. A. G. Ravandi. 2012. Accumulation of copper and other heavy metals by plants growing on Sarcheshmeh copper mining area, Iran. *Journal of Geochemical Exploration* 123:25–32.

Gihring, T. M. and J. F. Banfield. 2001. Arsenite oxidation and arsenate respiration by a new Thermus isolate. *FEMS Microbiology Letters* 204:335–340.

Gorby, Y. A. F. C. and H. Bolton. 1998. Microbial reduction of cobalt[III]EDTA$^-$ in the presence and absence of manganese(IV) oxide. *Environmental Science and Technology* 32:244–250.

Guo, X. L., J. Guo, Z. X. Chen, H. H. Gao, Q. J. Qin, W. Sun, and W. J. Zhang. 2012. Effects of heavy metals pollution on soil microbial communities metabolism and soil enzyme activities in coal mining area of Tongchuan, Shaanxi Province of Northwest China. *Chinese Journal of Applied Ecology* 23:798–806.

Gutiérrez, M., K. Mickus, and L. M. Camacho. 2016a. Abandoned Pb—Zn mining wastes and their mobility as proxy to toxicity: A review. *Science of the Total Environment* 565:392–400.

Gutiérrez, M., S. S. Wu, J. R. Rodriguez, A. D. Jones, and B. E. Lockwood. 2016b. Assessing the state of contamination in a historic mining town using sediment chemistry. *Archives of Environmental Contamination and Toxicology* 70:747–756.

Gutiérrez-Ruiz, M. E., A. E. Ceniceros-Gómez, M. Villalobos, F. Romero, and P. Santiago. 2012. Natural arsenic attenuation via metal arsenate precipitation in soils contaminated with metallurgical wastes: II. Cumulative evidence and identification of minor processes. *Applied Geochemistry* 27:2204–2214.

Hackl, R. P. and L. Jones.1997. Bacterial sulfur oxidation pathways and their effect on the cyanidation characteristics of biooxidised refractory gold concentrates. In *Proceedings of the 16th International Biohydrometallurgy Symposium*, Australian Mineral Foundation, Glenside, South Australia, Australia.

Hamamura, N., K. Fukushima, and T. Itai. 2013. Identification of antimony- and arsenic-oxidizing bacteria associated with antimony mine tailing. *Microbes and Environment* 28:257–263.

Han, Y. S., S. J. Youm, C. Oh, Y. C. Cho, and J. S. Ahn. 2015. Geochemical and eco-toxicological characteristics of stream water and its sediments affected by acid mine drainage. *Catena* 148:52–59.

Harter, R. D. and R. Naidu. 1995. Role of metal-organic complexation in metal sorption by soils. *Advances in Agronomy* 55:219–263.

Heikkinen, P. M. and M. L. Räisänen. 2008. Mineralogical and geochemical alteration of Hitura sulphide mine tailings with emphasis on nickel mobility and retention. *Journal of Geochemical Exploration* 97:1–20.

Hiller, E., B. Lalinská, M. Chovan, Ľ. Jurkovič, T. Klimko, M. Jankulár, R. Hovorič, P. Šottník, R. Fľaková, Z. Ženišová, and I. Ondrejková. 2012. Arsenic and antimony contamination of waters, stream sediments and soils in the vicinity of abandoned antimony mines in the Western Carpathians, Slovakia. *Applied Geochemistry* 27:598–614.

Hiller, E., P. Marián, R. Tóth, B. Lalinská-Voleková, Ľ. Jurkovič, G. Kučerová, A. Radková, P. Šottník, and J. Vozár. 2013. Geochemical and mineralogical characterization of a neutral, low-sulfide/high-carbonate tailings impoundment, Markušovce, eastern Slovakia. *Environmental Science and Pollution Research* 20:7627–7642.

Hockin, S. L. and G. M. Gadd. 2003. Linked redox-precipitation of sulfur and selenium under anaerobic conditions by sulfate-reducing bacterial biofilms. *Applied and Environmental Microbiology* 69:7063–7072.

Hou, X., M. Parent, M. M. Savard, C. Tassé, C. Bégin, and J. Marion. 2006. Lead concentrations and isotope ratios in the exchangeable fraction: Tracing soil contamination near a copper smelter. *Geochemistry: Exploration, Environment, Analysis* 6:229–236.

Hsu, N. H., S. L. Wang, Y. C. Lin, G. D. Sheng, and J. F. Lee. 2009. Reduction of Cr (VI) by crop-residue-derived black carbon. *Environmental Science and Technology* 43:8801–8806.

Huang, L. M., C. B. Deng, N. Huang, and X. J. Huang. 2013. Multivariate statistical approach to identify heavy metal sources in agricultural soil around an abandoned Pb–Zn mine in Guangxi Zhuang Autonomous Region, China. *Environmental Earth Sciences* 68:1331–1348.

Huang, Q. Y., W. L. Chen, and X. J. Guo. 2002. Sequential fractionation of Cu, Zn and Cd in soils in the absence and presence of rhizobia. In *Proceedings of the 17th WCSS*, August 14–21, Bangkok, Thailand, p. 1453.

Hudson-Edwards, K. A., H. E. Jamieson, and B. G. Lottermoser. 2011. Minewastes: Past, present, future. *Elements* 7:375–380.

Hyun, S., J. Kim, D. Y. Kim, and D. H. Moon. 2012. Effect of seepage conditions on chemical attenuation of arsenic by soils across an abandoned mine site. *Chemosphere* 87:602–607.

Intamo, P., A. Suddhiprakarn, I. Kheoruenromne, S. Tawornpruek, and R. J. Gilkes. 2016. Metals and arsenic concentrations of Ultisols adjacent to mine sites on limestone in Western Thailand. *Geoderma Regional* 7:300–310.

Jamieson, H. 2011. Geochemistry and mineralogy of solid mine waste: Essential knowledge for predicting environmental impact. *Elements* 7:381–386.

Jeong, S. W. 2015. Geotechnical and rheological characteristics of waste rock deposits influencing potential debris flow occurrence at the abandoned Imgi Mine, Korea. *Environmental Earth Sciences* 73:8299–8310.

Jerez, C. A. 2009. Metal extraction and biomining. In *Encyclopedia of Microbiology*, 3rd edn., ed. M. Schaechter. Elsevier, Amsterdam, the Netherlands, pp. 407–420.

Johnson, A. W., M. Gutiérrez, D. Gouzie, L. R. McAlily. 2016. State of remediation and metal toxicity in the tri-state mining district, USA. *Chemosphere* 144:1132–1141.

Johnson, D. B. 2008. Biodiversity and interactions of acidophiles: Key to understanding and optimizing microbial processing of ores and concentrates. *Transactions of Nonferrous Metals Society of China* 18:1367–1373.

Johnson, D. B. and K. B. Hallberg. 2009. Carbon, iron and sulfur metabolism in acidophilic microorganisms. *Advances in Microbial Physiology* 54:202–256.

Jung, M. C. 2001. Heavy metal contamination of soils and waters in and around the Imcheon Au-Ag mine, Korea. *Applied Geochemistry* 16:1369–1375.

Kierczak, J., A. Potysz, A. Pietranik, R. Tyszka, M. Modelska, C. Néel, V. Ettler, and M. Mihaljevič. 2013. Environmental impact of the historical Cu smelting in the Rudawy Janowickie Mountains (south-western Poland). *Journal of Geochemical Exploration* 124:183–194.

Kim, J. and S. Hyun. 2015. Nonequilibrium leaching behavior of metallic elements (Cu, Zn, As, Cd, and Pb) from soils collected from long-term abandoned mine sites. *Chemosphere* 134:150–158.

Kim, K. K., K. W. Kim, J. Y. Kim, I. S. Kim, Y. W. Cheong, and J. S. Min. 2001. Characteristics of tailings from the closed metal mines as potential contamination source in South Korea. *Environmental Geology* 41:358–364.

Kim, M. J., K. H. Ahn, and Y. Jung. 2002. Distribution of inorganic arsenic species in mine tailings of abandoned mines from Korea. *Chemosphere* 49:307–312.

Kimball, B. E., R. Mathur, A. C. Dohnalkova, A. J. Wall, R. L. Runkel, and S. L. Brantley. 2009. Copper isotope fractionation in acid mine drainage. *Geochimica et Cosmochimica Acta* 73:1247–1263.

Korpershoeck, H. R., J. A. C. Mendonça, and J. R. F. Torquato. 1979. A geologia da região de Pedra Verde, Ceará. Simpósio de Geologia do Nordeste, 9, Natal, Sociedade Brasileira de Geologia. pp. 349–363 (in Portuguese).

Kothe, E. and G. Büchel. 2014. UMBRELLA: Using MicroBes for the REgulation of heavy metaL mobiLity at ecosystem and landscape scAle. *Environmental Science and Pollution Research* 21:6761–6764.

Krupa, P. and J. Kozdroj. 2004. Accumulation of heavy metals by ectomycorrhizal fungi colonizing birch trees growing in an industrial desert soil. *World Journal of Microbiology and Biotechnology* 20: 427–430.

Kunhikrishnan, A., G. Choppala, B. Seshadri, J. H. Park, K. Mbene, Y. Yan, and N. Bolan. 2016. Biotransformation of heavy metal(loid)s in relation to the remediation of contaminated soils. In *Handbook of Metal-Microbe Interactions and Bioremediation*, eds. S. Das and H. R. Dash. Taylor & Francis, London, U.K., pp. 67–87.

Kwon, J. C., Z. D. Nejad, and M. C. Jung. 2017. Arsenic and heavy metals in paddy soil and polished rice contaminated by mining activities in Korea. *Catena* 148:92–100.

Lackovic, K., M. J. Angove, J. D. Wells, and B. B. Johnson. 2003. Modelling the adsorption of Cd(II) onto Muloorina illite and related clay minerals. *Journal of Colloid and Interface Science* 257:31–40.

Lamb, D. T., H. Ming, M. Megharaj, and R. Naidu. 2009. Heavy metal (Cu, Zn, Cd and Pb) partitioning and bioaccessibility in uncontaminated and long-term contaminated soils. *Journal of Hazardous Materials* 171:1150–1158.

Lambertsson, L. and M. Nilsson. 2006. Organic material: The primary control on mercury methylation and ambient methyl mercury concentrations in estuarine sediments. *Environmental Science and Technology*, 40:1822–1829.

Lee, C. G., H. T. Chon, and M. C. Jung. 2001. Heavy metal contamination in the vicinity of the Daduk Au-Ag-Pb-Zn mine in Korea. *Applied Geochemistry* 16:1377–1386.

Lehr, C. R., D. R. Kashyap, and T. R. McDermott. 2007. New insights into microbial oxidation of antimony and arsenic. *Applied and Environmental Microbiology* 73:2386–2389.

Li, H., J. Wang, Y. Teng, and Z. Wang. 2006. Study on the mechanism of transport of heavy metals in soil in western suburb of Beijing. *Chinese Journal of Geochemistry* 25:173–177.

Li, Q., R. Yu, Z. Sun, Y. Liu, M. Chen, H. Yin, Y. Zhang, Y. Liang, L. Xu, L. Sun, G. Qiu, and X. Liu. 2011a. Column bioleaching of uranium embedded in granite porphyry by a mesophilic acidophilic consortium. *Bioresource Technology* 102:4697–4702.

Li, W. C. and M. H. Wong. 2010. Effects of bacteria on metal bioavailability, speciation, and mobility in different metal mine soils: A column study. *Journal of Soils and Sediments* 10:313–325.

Li, Z., X. Feng, G. Li, X. Bi, G. Sun, J. Zhu, H. Qin, and J. Wang. 2011b. Mercury and other metal and metalloid soil contamination near a Pb/Zn smelter in east Hunan province, China. *Applied Geochemistry* 26:160–166.

Lin, Z. and R. B. Herbert, Jr. 1997. Heavy metal retention in secondary precipitates from a mine rock dump and underlying soil, Dalarna, Sweden. *Environmental Geology* 33:1–12.

Lloyd, J. R. 2003. Microbial reduction of metals and radionuclides. *FEMS Microbiology Reviews* 27:411–425.

Loseto, L. L., S. D. Siciliano, and D. R. Lean. 2004. Methylmercury production in High Arctic wetlands. *Environmental Toxicology and Chemistry* 23:17–23.

Lottermoser, B. G. 2007. *Mine Wastes—Characterization, Treatment, Environmental Impacts*, 2nd edn. Springer, Berlin, Germany.

Lu, P., N. T. Nuhfer, S. Kelly, Q. Li, H. Konishi, E. Elswick, and C. Zhu. 2011. Lead coprecipitation with iron oxyhydroxide nano-particles. *Geochimica et Cosmochimica Acta* 75:4547–4561.

Madigan, M. I. and T. D. Brock. 1991. *Biology of Microorganisms*, 6th edn. Prentice-Hall, Englewood Cliffs, NJ, p. 874.

McGregor, R. G., D. W. Blowes, J. L. Jambor, and W. D. Robertson. 1998. Mobilization and attenuation of heavy metals within a nickel mine tailings impoundment near Sudbury, Ontario, Canada. *Environmental Geology* 36:305–319.

McKenzie, A. B. and I. D. Pulford. 2002. Investigation of contaminant metal dispersal from a disused mine site at Tyndrum, Scotland, using concentration gradient and Pb isotope ratios. *Applied Geochemistry* 17:1093–1103.

Merroun, M. L., J. Raff, A. Rossberg, C. Hennig, T. Reich, and S. Selenska-Pobell. 2005. Complexation of uranium by cells and S-layer sheets of *Bacillus sphaericus* JG-A12. *Applied and Environmental Microbiology* 71:5532–5543.

MIRECO, Mine Reclamation Corporation. 2013. *Yearbook of MIRECO Statistics*. Mine Reclamation Corporation, Seoul, Korea (in Korean).

Mishra, V., D. P. Samantaray, S. K. Dash, B. B. Mishra, and R. K. Swain. 2010. Study on hexavalent chromium reduction by chromium resistant bacterial isolates of Sukinda mining area. *Our Nature* 8:63–71.

Miyata, N., Y. Tani, K. Iwahori, and M. Soma. 2004. Enzymatic formation of manganese oxides by an Acremonium-like hyphomycete fungus, strain KR21-2. *FEMS Microbiology Ecology* 47:101–109.

Miyata, N., Y. Tani, K. Maruo, H. Tsuno, M. Sakata, and K. Iwahori. 2006. Manganese(IV) oxide production by *Acremonium* sp. strain KR21-2 and extracellular Mn(II) oxidase activity. *Applied and Environmental Microbiology* 72:6467–6473.

Miyata, N., Y. Tani, M. Sakata, and K. Iwahori. 2007. Microbial manganese oxide formation and interaction with toxic metal ions. *Journal of Bioscience and Bioengineering* 104:1–8.

Mouta, E. R., M. R. Soares, and J. C. Casagrande. 2008. Copper adsorption as a function of solution parameters of variable charge soils. *Journal of the Brazilian Chemical Society* 19:996–1009.

Muhammad, S., M. T. Shah, and S. Khan. 2011. Heavy metal concentrations in soil and wild plants growing around Pb–Zn sulfide terrain in the Kohistan region, northern Pakistan. *Microchemical Journal* 99:67–75.

Mulligan, C. N. and R. Galvez-Cloutier. 2003. Bioremediation of metal contamination. *Environmental Monitoring and Assessment* 84:45–60.

Naidu, R. and N. S. Bolan. 2008. Contaminant chemistry in soils: Key concepts and bioavailability. In *Chemical Bioavailability in Terrestrial Environment*, ed. R. Naidu. Elsevier, Amsterdam, the Netherlands, pp. 9–38.

Naidu, R., N. S. Bolan, R. S. Kookana, and K. G. Tiller. 1994. Ionic-strength and pH effects on the sorption of cadmium and the surface charge of soils. *European Journal of Soil Science* 45:419–429.

Naidu, R., R. S. Kookana, M. E. Sumner, R. D. Harter, and K. G. Tiller. 1997. Cadmium sorption and transport in variable charge soils: A review. *Journal of Environmental Quality* 26:602–617.

Nam, I. H., S. B. Roh, M. J. Park, C. M. Chon, J. G. Kim, S. W. Jeong, H. Song, and M. H. Yoon. 2016. Immobilization of heavy metal contaminated mine wastes using *Canavalia ensiformis* extract. *Catena* 136:53–58.

Nam, S. M., M. Kim, S. Hyun, and S. H. Lee. 2010. Chemical attenuation of arsenic by soils across two abandoned mine sites in Korea. *Chemosphere* 81:1124–1130.

Nawab, J., S. Khan, M. Aamir, I. Shamshad, Z. Qamar, I. Din, and Q. Huang. 2016. Organic amendments impact the availability of heavy metal(loid)s in mine-impacted soil and their phytoremediation by *Penisitum americanum* and *Sorghum bicolor*. *Environmental Science and Pollution Research* 23:2381–2390.

NIAST, National Institute of Agricultural Science and Technology, Korea. 1997. Survey of heavy metals contamination degree of arable soil located in mining area. Annual report, Department of Agricultural Environment, National Institute of Agricultural Science and Technology, Republic of Korea, pp. 237–243 (in Korean).

Nikolaidis, C., I. Zafiriadis, V. Mathioudakis, and T. Constantinidis. 2010. Heavy metal pollution associated with an abandoned lead–zinc mine in the Kirki Region, NE Greece. *Bulletin of Environmental Contamination and Toxicology* 85:307–312.

Olías, M. and J. M. Nieto. 2015. Background conditions and mining pollution throughout history in the Río Tinto (SW Spain). *Environments* 2:295–316.

Oremland, R. S., M. J. Herbel, J. S. Blum, S. Langley, T. J. Beveridge, P. M. Ajayan, T. Sutto, A. V. Ellis, and S. Curran. 2004. Structural and spectral features of selenium nanospheres produced by Se-respiring bacteria. *Applied Environmental Microbiology* 70:52–60.

Oremland, R. S., J. T. Hollibaugh, A. S. Maest, T. S. Presser, L. G. Miller, and C. W. Culbertson. 1989. Selenate reduction to elemental selenium by anaerobic bacteria in sediments and culture: Biogeochemical significance of a novel, sulfate-independent respiration. *Applied and Environmental Microbiology* 55:2333–2343.

Otero, X. L., E. Álvarez, M. J. Fernandez-Sanjurjo, and F. Macias. 2012. Micronutrients and toxic trace metals in the bulk and rhizospheric soil of the spontaneous vegetation at an abandoned copper mine in Galicia (NW Spain). *Journal of Geochemical Exploration* 112:84–92.

Oyarzun, R., J. Lillo, J. A. López-García, J. M. Esbrí, P. Cubas, W. Llanos, and P. Higueras. 2011. The Mazarrón Pb–(Ag)–Zn mining district (SE Spain) as a source of heavy metal contamination in a semiarid realm: Geochemical data from mine wastes, soils, and stream sediments. *Journal of Geochemical Exploration* 109:113–124.

Paez-Espino, D., J. Tamames, V. de Lorenzo, and D. Canovas. 2009. Microbial responses to environmental arsenic. *Biometals* 22:117–130.

Palumbo-Roe, B., B. Klinck, V. Banks, and S. Quigley. 2009. Prediction of the long-term performance of abandoned lead zinc mine tailings in a Welsh catchment. *Journal of Geochemical Exploration* 100:169–181.

Park, J. H., Y. S. Han, and J. S. Ahn. 2016. Comparison of arsenic co-precipitation and adsorption by iron minerals and the mechanism of arsenic natural attenuation in a mine stream. *Water Research* 106:295–303.

Pascaud, G., S. Boussen, M. Soubrand, E. Joussein, P. Fondaneche, and B. H. Abdeljaouad. 2015. Particulate transport and risk assessment of Cd, Pb and Zn in a Wadi contaminated by runoff from mining wastes in a carbonated semiarid context. *Journal of Geochemical Exploration* 152:27–36.

Perlatti, F., X. L. Otero, F. Macias, and T. O. Ferreira. 2014. Geochemical speciation and dynamic of copper in tropical semi-arid soils exposed to metal-bearing mine wastes. *Science of the Total Environment* 500–501:91–102.

Pirrone, N., S. Cinnirella, X. Feng, R. B. Finkelman, H. R. Friedli, J. Leaner, R. Mason, A. B. Mukherjee, G. B. Stracher, D. G. Streets, and K. Telmer. 2010. Global mercury emissions to the atmosphere from anthropogenic and natural sources. *Atmospheric Chemistry and Physics* 10:5951–5964.

Pratas J., P. J. C. Favas, R. D'Souza, M. Varun, and M. S. Paul. 2013. Phytoremedial assessment of flora tolerant to heavy metals in the contaminated soils of an abandoned Pb mine in Central Portugal. *Chemosphere* 90:2216–2225.

Purvis, O. W. and B. Pawlik-Skowronska. 2008. Lichens and metals. In *Stress in Yeasts and Filamentous Fungi*, eds. S. Avery, M. Stratford, and P. V. West. Elsevier, Amsterdam, the Netherlands, pp. 175–200.

Qi, J., H. Zhang, X. Li, J. Lu, and G. Zhang. 2016. Concentrations, spatial distribution, and risk assessment of soil heavy metals in a Zn-Pb mine district in southern China. *Environmental Monitoring and Assessment* 188:413. doi:10.1007/s10661-016 5406-0.

Rawlings, D. E. 2002. Heavy metal mining using microbes. *Annual Reviews Microbiology* 56:65–91.

Rawlings, D. E., D. Dew, and C. du Plessis. 2003. Biomineralization of metal-containing ores and concentrates. *Trends in Biotechnology* 21:38–44.

Rhine, E. D., K. M. Onesios, M. E. Serfes, J. R. Reinfelder, and L. Y. Young. 2008. Arsenic transformation and mobilization from minerals by the arsenite oxidizing strain WAO. *Environmental Science and Technology* 42:1423–1429.

Rice, K. C. and J. S. Herman. 2012. Acidification of Earth: An assessment across mechanisms and scales. *Applied Geochemistry* 27:1–14.

Rodríguez, L., E. Ruiz, J. Alonso-Azcárate, and J. Rincón. 2009. Heavy metal distribution and chemical speciation in tailings and soils around a Pb–Zn mine in Spain. *Journal of Environmental Management* 90:1106–1116.

Roper, A., P. A. Williams, and M. Filella. 2012. Secondary antimony minerals: Phases that control the dispersion of antimony in the supergene zone. *Chemie der Erde—Geochemistry* 72:9–14.

Ross, S. M. 1994. Retention transformation and mobility of toxic metals in soils. In *Toxic Metals in Soil–Plant Systems*, ed. S. M. Ross. Wiley, New York, pp. 63–152.

Santhiya, D. and Y. P. Ting. 2005. Bioleaching of spent refinery processing catalyst using *Aspergillus niger* with high-yield oxalic acid. *Journal of Biotechnology* 116:171–184.

Saratovsky, I., S. J. Gurr, and M. A. Hayward. 2009. The structure of manganese oxide formed by the fungus *Acremonium* sp. strain KR21-2. *Geochimica et Cosmochimica Acta* 73:3291–3300.

Saunier, J. B., G. Losfeld, R. Freydier, and C. Grison. 2013. Trace elements biomonitoring in a historical mining district (*les Malines*, France). *Chemosphere* 93:2016–2023.

Shuman, L. M. 1986. Effect of ionic strength and anions on zinc adsorption by two soils. *Soil Science Society of America Journal* 50:1438–1442.

Simate, G. S. and S. Ndlovu. 2014. Acid mine drainage: Challenges and opportunities. *Journal of Environmental Chemical Engineering* 2:1785–1803.

Sitte, J., D. M. Akob, C. Kaufmann, K. Finster, D. Banerjee, E. M. Burkhardt, J. E. Kostka, A. C. Scheinost, G. Buchel, and K. Kusel. 2010. Microbial links between sulfate reduction and metal retention in uranium- and heavy metal-contaminated soil. *Applied and Environmental Microbiology* 76:3143–3152.

Smeaton, C. M., G. E. Walshe, A. M. L. Smith, K. A. Hudson-Edwards, W. E. Dubbin, K. Wright, A. M. Beale, B. J. Fryer, and C. G. Weisener. 2012. Simultaneous release of Fe and As during the reductive dissolution of Pb-As jarosite by *Shewanella putrefaciens* CN32. *Environmental Science and Technology* 46:12823–12831.

Smith, K. S. 2007. Strategies to predict metal mobility in surficial mining environments. In *Understanding and Responding to Hazardous Substances at Mine Sites in the Western United States*, ed. J. V. DeGraff, Reviews in Engineering Geology, Geological Society of America, vol. XVII, pp. 25–45.

Smith, W. L. and G. M. Gadd. 2000. Reduction and precipitation of chromate by mixed culture sulphate-reducing bacterial biofilms. *Journal of Applied Microbiology* 88:983–991.

Song, W. Y., E. J. Sohn, E. Martinoia, Y. J. Lee, Y. Y. Yang, M. Jasinski, C. Forestier, I. Hwang, and Y. Lee. 2003. Engineering tolerance and accumulation of lead and cadmium in transgenic plants. *Nature Biotechnology* 21:914–919.

Sorooshian, A., J. Csavina, T. Shingler, S. Dey, F. J. Brechtel, A. E. Sáez, and E. A. Betterton. 2012. Hygroscopic and chemical properties of aerosols collected near a copper smelter: Implications for public and environmental health. *Environmental Science and Technology* 46:9473–9480.

Sparks, D. L. 2003. *Environmental Soil Chemistry*, 2nd edn. Academic Press, San Diego, CA.

Sposito, G. 1984. *The Surface Chemistry of Soils*. Oxford University Press, New York.

Sracek, O., M. Mihaljevič, B. Křibek, V. Majer, and F. Veselovský. 2010. Geochemistry and mineralogy of Cu and Co in mine tailings at the Copperbelt, Zambia. *Journal of African Earth Sciences* 57:14–30.

Stahl, R. S. and B. R. James. 1991. Zinc sorption by B horizon soils as a function of pH. *Soil Science Society of America Journal* 55:1592–1597.

Stolz, J. F., P. Basu, and R. S. Oremland. 2010. Microbial arsenic metabolism: New twists on an old poison. *Microbe* 5:53–59.

Sullivan, L. A. and R. T. Bush. 2004. Iron precipitate accumulations associated with waterways in drained coastal acid sulfate landscapes of eastern Australia. *Marine and Freshwater Research* 55:727–736.

Tebo, B. M., H. A. Johnson, J. K. McCarthy, and A. S. Templeton. 2005. Geomicrobiology of manganese(II) oxidation. *Trends in Microbiology* 13:421–438.

Thompson-Eagle, E. T. and W. T. Frankenberger. 1992. Bioremediation of soils contaminated with selenium. In *Advances in Soil Science*, ed. R. Lal, and B. A. Stewart. Springer, New York, pp. 261–309.

Thornton, I. 1995. Metals in the global environment: Facts and misconceptions. International Council on Metals and the Environment, Ottawa, Canada.

Tomei, F. A., L. L. Barton, C. L. Lemanski, and T. G. Zocco. 1992. Reduction of selenate and selenite to elemental selenium by *Wolinella succinogenes*. *Canadian Journal of Microbiology* 38:1328–1333.

Tomei, F. A., L. L. Barton, C. L. Lemanski, T. G. Zocco, N. H. Fink, and L. O. Sillerud. 1995. Transformation of selenate and selenite to elemental selenium by *Desulfovibrio desulfuricans*. *Journal of Industrial Microbiology* 14:329–336.

Tsai, S. L., S. Singh, and W. Chen. 2009. Arsenic metabolism by microbes in nature and the impact on arsenic remediation. *Current Opinion in Biotechnology* 20:659–667.

Tucker, M. D., L. L. Barton, and B. M. Thomson. 1998. Reduction of Cr, Mo, Se and U by *Desulfovibrio desulfuricans* immobilised in polyacrylamide gels. *Journal of Industrial Microbiology and Biotechnology* 20:13–19.

Tullio, M., F. Pierandrei, A. Salerno, and E. Rea. 2003. Tolerance to cadmium of vesicular arbuscular mycorrhizae spores isolated from a cadmium-polluted and unpolluted soil. *Biology and Fertility of Soils* 37:211–214.

Uroz, S., C. Calvaruso, M. P. Turpault, and P. Frey-Klerr. 2009. Mineral weathering by bacteria: Ecology, actors and mechanisms. *Trends in Microbiology* 17:378–387.

Uzu, G., S. Sobanska, G. Sarret, J. J. Sauvain, P. Pradère, and C. Dumat. 2011. Characterization of lead-recycling facility emissions at various workplaces. Major insights for sanitary risk assessment. *Journal of Hazardous Materials* 186:1018–1027.

Valente, T., P. Gomes, M. A. Sequeira Braga, A. Dionísio, J. Pamplona, and J. A. Grande. 2015. Iron and arsenic-rich nanoprecipitates associated with clay minerals in sulfide-rich waste dumps. *Catena* 131:1–13.

Vaxevanidou, K., C. Christou, G. F. Kremmydas, D. G. Georgakopoulos, and N. Papassiopi. 2015. Role of indigenous arsenate and iron(III) respiring microorganisms in controlling the mobilization of arsenic in a contaminated soil sample. *Bulletin of Environmental Contamination and Toxicology* 94:282–288.

Vega, F. A., E. F. Covelo, and M. L. Andrade. 2006. Competitive sorption and desorption of heavy metals in mine soils: Influence of mine soil characteristics. *Journal of Colloid and Interface Science* 298:582–592.

Violante, A., V. Cozzolino, L. Perelomov, A. G. Caporale, and M. Pigna. 2010. Mobility and bioavailability of heavy metals and metalloids in soil environments. *Journal of Soil Science and Plant Nutrition* 10:268–292.

Wadige, C. P. M., A. M. Taylor, F. Krikowa, and W. A. Maher. 2016. Sediment metal concentration survey along the mine-affected Molonglo River, NSW, Australia. *Archives of Environmental Contamination and Toxicology* 70: 572–582.

Wahsha, M., C. Bini, E. Argese, F. Minello, S. Fontana, and H. Wahsheh. 2012. Heavy metals accumulation in willows growing on Spolic Technosols from the abandoned Imperina Valley mine in Italy. *Journal of Geochemical Exploration* 123:19–24.

Wang, J., C. B. Zhang, S. S. Ke, and B. Y. Qian. 2010a. Different spontaneous plant communities in Sanmen Pb/Zn mine tailing and their effects on mine tailing physico-chemical properties. *Environmental Earth Sciences* 62:779–786.

Wang, X., M. He, J. Xie, J. Xi, and X. Lu. 2010b. Heavy metal pollution of the world largest antimony mine-affected agricultural soils in Hunan province (China). *Journal of Soils and Sediments* 10:827–837.

Wang, Y. T. 2000. Microbial reduction of chromate. In *Environmental Microbe-Metal Interactions*, ed. D. R. Lovley. ASM Press, Washington, DC, pp. 225–235.

Watling, H. R. 2006. The bioleaching of sulphide minerals with emphasis on copper sulphides: A review. *Hydrometallurgy* 84:81–108.

Wong, S. C., X. D. Li, G. Zhang, S. H. Qi, and Y. S. Min. 2002. Heavy metals in agricultural soils of Pearl river delta, South China. *Environmental Pollution* 119:33–34.

Wu, Q., S. Wang, L. Wang, F. Liu, L. Che-Jen, L. Zhang, and F. Wang. 2014. Spatial distribution and accumulation of Hg in soil surrounding a Zn/Pb smelter. *Science of the Total Environment* 496:668–677.

Wu, S. C., Y. M. Luo, K. C. Cheung, and M. H. Wong. 2006. Influence of bacteria on Pb and Zn speciation, mobility and bioavailability in soil: A laboratory study. *Environmental Pollution* 144:765–773.

Zenteno, M. C., R. C. A. de Freitas, R. B. A. Fernandes, M. P. F. Fontes, and C. P. Jordão. 2013. Sorption of cadmium in some soil amendments for in situ recovery of contaminated soils. *Water, Air, and Soil Pollution* 224:1418.

Zhang, X., L. Yang, Y. Li, H. Li, W. Wang, and Q. Ge. 2011. Estimation of lead and zinc emissions from mineral exploitation based on characteristics of lead/zinc deposits in China. *Transactions of Nonferrous Metals Society of China* 21:2513–2519.

Zhang, Y. and W. T. Frankenberger. 2003. Factors affecting removal of selenate in agricultural drainage water utilizing rice straw. *Science of the Total Environment* 305:207–216.

Zhou, H. B., W. M. Zeng, Z. F. Yang, Y. J. Xie, and G. Z. Qiu. 2009. Bioleaching of chalcopyrite concentrate by a moderately thermophilic culture in a stirred tank reactor. *Bioresource Technology* 100:515–520.

Zhuang, P., M. B. McBride, H. Xia, N. Li, and Z. Li. 2009. Health risk from heavy metals via consumption of food crops in the vicinity of Dabaoshan mine, South China. *Science of the Total Environment* 407:1551–1561.

Zobrist, J., M. Sima, D. Dogaru, M. Senila, H. Yang, C. Popescu, C. Roman, A. Bela, L. Frei, B. Dold, and D. Balteanu. 2009. Environmental and socioeconomic assessment of impacts by mining activities—A case study in the Certej River catchment, Western Carpathians, Romania. *Environmental Science and Pollution Research* 16:S14–S26.

Section V

Case Studies of Successful Mine Site Rehabilitation

16 Mine Site Reclamation in Canada
Overview and Case Studies

Jin-Hyeob Kwak, Abimbola Ojekanmi, Min Duan, Scott X. Chang, and M. Anne Naeth

CONTENTS

16.1   Introduction 291
16.2   Regulations of Mining Operations in Canada 293
16.3   Environmental Impacts of Mining 294
16.4   Mining, Reclamation, and Sustainable Resource Development 295
16.5   Oil Sands Reclamation Case Studies 296
      16.5.1   Gateway Hill Reclamation 297
      16.5.2   Wapisiw Tailings Reclamation 299
16.6   Coal Mine Reclamation Case Studies 300
      16.6.1   Whitewood Coal Mine Reclamation 301
      16.6.2   Fording River Coal Mine Reclamation 301
16.7   Metal and Diamond Mine Reclamation Case Studies 302
      16.7.1   Diavid Diamond Mine Reclamation 302
      16.7.2   Cluff Lake Uranium Mine Reclamation 303
      16.7.3   Giant Gold Mine Reclamation 304
16.8   Lessons Learned and Future Challenges 304
References 305

16.1   INTRODUCTION

Canada has the second largest land area in the world and has a vast store of natural resources that can be mined. New and current mining activities are an important part of the Canadian economy. In 2014, mining associated activities contributed 57 billion Canadian dollars to the Canadian gross domestic product and natural resources accounted for 18.2% of Canadian goods exported (Mining Association of Canada 2015). The industry contributed approximately 71 billion Canadian dollars to the national economy in taxes, royalty payments, and employment over the last decade (2003–2012). The operations (and the decommissioning when mines reach the end of their lifespan) of these mines are regulated by the government to ensure long-term profitability and environmental sustainability through various federal, provincial, and territorial acts or regulations, including the Canadian Environmental Protection Act (Government of Canada 1999).

Mining operations for coal, oil sands, metals, uranium, diamonds and other gemstones, and a variety of other natural resources such as graphite, potash, bentonite, and silica are common across the provinces and territories of Canada (Figure 16.1). Alberta is home to one of the world's largest deposits of oil sands or bitumen resources, similar in size to major oil and gas resources such as those found in Venezuela and Saudi Arabia. Saskatchewan is well known for its potash and has the

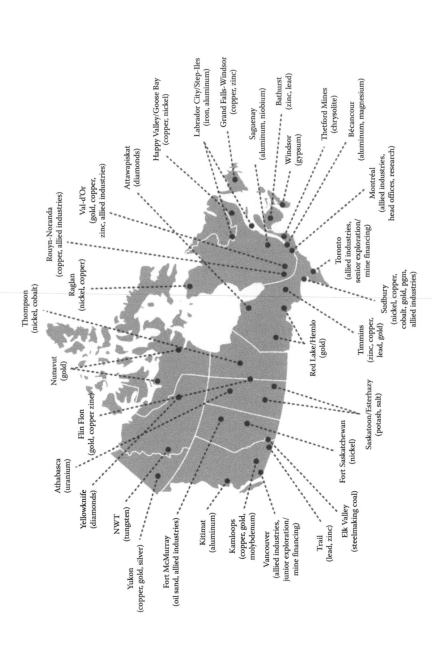

FIGURE 16.1 Representative distribution of mines in Canada. (Courtesy of Mining Association of Canada, Facts and figures of the Canadian mining industry 2015, The Mining Association of Canada, Ottawa, Ontario, Canada, 2015.)

largest high-grade uranium deposits in the world. The Northwest Territories, Nunavut, and Yukon have a number of high-quality diamond mines.

While exploration and development of natural resources provide considerable economic opportunities, they also pose significant environmental challenges. The operation of mining activities will cause disturbance; the size of the disturbance or environmental impact caused by any mining operation in Canada depends on the resource involved, whether the mine is surface or underground, and the need for ancillary facilities to operate, such as pipelines, specialized ports, rail lines, and roads. For example, in British Columbia, polycyclic aromatic hydrocarbons from an aluminum smelter increased concentrations of polycyclic aromatic hydrocarbons in Dungeness crabs (*Cancer magister*) (Eickhoff et al. 2003); lead and zinc emitted from a smelter increased metal concentrations in the surrounding environment (Goodarzi et al. 2002); and mercury and antimony concentrations in soils and nonvascular plants were elevated near decommissioned mercury mines, the Pinchi Lake and Bralorne Takla mines (Plouffe et al. 2004). Gold mine tailings produced in northern New Brunswick discharged mercury into the surrounding environment, with total dissolved mercury in groundwater and in stream sediments increasing up to 4000 times higher than pre-mining levels (Al et al. 2006). Mining activities have disturbed a large area of the landscape across Canada. For example, open-pit mining of oil sands resources in northern Alberta has disturbed approximately 895 km^2 of the boreal region as of 2013 (Alberta Energy 2015) and oil sands mining has caused massive loss of forests, peatlands, and stored carbon (Rooney et al. 2012).

Disturbed land must be reclaimed according to federal and provincial acts or regulations. Land reclamation, the process of converting disturbed land to its former or other productive uses (Powter 2002), involves mitigation activities to restore the quality of soil, water, air, and the overall ecosystem that is lost due to disturbance. The general components of land reclamation activities in Canada are often grouped into soil reclamation or reconstruction, contaminant remediation, and revegetation. The term rehabilitation is less commonly used in Canada, but implies that the land is returned to a form and productivity in conformity with a prior land use plan (Powter 2002). In this chapter, three broad classes of mine operations in Canada—oil sands, metals, and coal—are discussed. Emphasis is on specifics of each mine type, with case studies of successful mine reclamation practices to address relevant environmental impacts.

16.2   REGULATIONS OF MINING OPERATIONS IN CANADA

Canada's regulatory system for the mining industry is highly complex, comprising federal and provincial departments, with specific mandates for overall sustainability of mining and extraction operations. Both regulatory systems define specific requirements and criteria for successful implementation of the mine development plan through mine design, operation, and eventual site closure and reclamation. The federal regulatory departments provide Canada-wide or interprovincial mandates complemented with more stringent or localized requirements in the provincial system that is responsible for natural resource extraction.

Federal regulatory departments mainly address environmental impacts of mining with the Canadian Environmental Assessment Act, the Metal Mining Effluent Regulations (MMER) under the Fisheries Act, the International River Improvements Act, the Migratory Birds Convention Act, the Species at Risk Act, and the Canadian Environmental Protection Act. Other federal regulatory requirements associated with mining include the Canadian Environmental Protection Act, Environmental Code of Practice for Metal Mines, Guidelines for the Assessment of Alternatives for Mine Waste Disposal, and Streamlining the Approvals Process for Metal Mines with Tailings Impoundment Areas. The Mining Association of Canada, the national organization of the Canadian mining industry, is actively involved in all aspects of mining to address economic, environmental, and other aspects of the industry. The Canadian Environmental Assessment Agency focuses on developing environmental assessment standards and ensures compliance of the mining industry with the Canadian Environmental Assessment Act. The mandate of the National Research Council is to encourage and operationalize research, innovations, and application of new technology in mining operations and site closure. These organizations and other relevant stakeholders collectively define the Canada-wide framework for successful mine development, operation, and closure.

Provincial and territorial regulatory systems include various government departments with mandates for environmental management, research and development, policy development, local stakeholder representations and consultation, creation of public awareness, mine financial risk and financial security assessment, and approval of operation permits (Alberta Environment 2003; Powter et al. 2012). Emphasis of provincial and territorial regulatory systems is on local mines to ensure compliance with provincial and territorial regulations. For example, in the Province of Alberta, the Environmental Protection and Enhancement Act (EPEA) provides comprehensive compliance criteria required to operate and reclaim mine sites, including those in the oil sands region (Province of Alberta 2014). This act governs all aspects of Alberta mine operations related to water, air, soil, vegetation, wildlife, anthropogenic effects, and engineering. The EPEA is complemented by other technical requirements to guide soil salvage and conservation (Soil Quality Working Group 1987), manage remediation (Alberta Government 2016b; CCME 2006), define vegetation-based criteria for assessing reclamation success in boreal forests (Alberta Environment 2003), assess land capability (Leskiw 1998), and the final reclamation certification process (CEMA 2012). In the Northwest Territories, the mine regulatory department is Indian and Northern Affairs Canada with mandates to regulate mine site reclamation (INAC 2002). To ensure remediation of contaminated sites, Aboriginal Affairs and Northern Development Canada developed the contaminated site management policy for land areas within the first nation treaty in the Northwest Territories (AANDC 2011).

16.3 ENVIRONMENTAL IMPACTS OF MINING

Mining has significant impact on all components of an ecosystem. Surface mining is accompanied by intense deforestation, the displacement of wildlife and the disturbance of their habitat, draining of wetlands, disturbances to the soil and landscape, and introduction of mine wastes into the environment. For example, surface oil sands and diamond mining have disturbed approximately 895 km^2 (Alberta Energy 2015) and more than 150 km^2 (Dominion Diamond Corporation 2016; Rio Tinto 2016; Wikipedia Mine 2016), respectively, of land by 2013, with complete vegetation and soil removal. Underground mining requires extensive blasting, drilling, and subsurface infrastructure to reach the ores of interest. Aboveground disturbances are intense, and underground activities can change local and regional hydrogeology and impact groundwater quality and quantity. Deep geological drilling increases potential for oxidation of metals with release of acidic water from metal and coal mines. Acidic rock drainage increases the potential for material loss, structural instability, potential for subsidence, and water pollution.

Mine waste management and containment facilities are major sources of organic and inorganic contaminants with the potential to negatively impact the environment. For example, the biotic environment can be negatively impacted by polycyclic aromatic hydrocarbons (Eickhoff et al. 2003), lead, zinc (Goodarzi et al. 2002), mercury (Plouffe et al. 2004), and volatile organics such as benzene, toluene, ethylbenzene, and xylenes from tailings pond (Alberta Government 2016a). Mining effluent and waste increased arsenic (Koch et al. 2000; Zheng et al. 2003), lead, and zinc (Pugh et al. 2002) concentrations in plants. Mining waste caused metal contamination in the surrounding water and soil (Zheng et al. 2003).

Other indirect environmental impacts of mining include excessive water consumption, air quality degradation, and loss of biodiversity. For example, oil sands mining requires a large amount of water: 3 and 0.4–0.6 tonnes of water are required for each barrel of bitumen production in surface and in situ mining, respectively (Alberta Energy 2015; Oil Sands Magazine 2016). In 2013, oil sands operations in northern Alberta emitted the equivalent of 62 mega tonnes of carbon dioxide, which accounts for the largest part of total greenhouse gas emission in Alberta (38.2%) and

Canada (8.5%) (Alberta Energy 2015). Sustainable and safe operation of mines without significant impact to the environment has been a major and contentious issue of interest among all stakeholders of the mining industry.

16.4 MINING, RECLAMATION, AND SUSTAINABLE RESOURCE DEVELOPMENT

Sustainable resource development goes beyond addressing environmental impacts of mining operations through reclamation after the mining ceases. It requires that the mining operations themselves be conducted with a minimal impact on the environment. Canadian mining procedures and reclamation approaches attain this sustainable state through implementing various impact mitigation practices at all stages of mining operations, and balancing the utilization of resources for economic gain with the protection of the environment.

In Canada, topsoil salvage and appropriate capping or management of tailings and unsuitable subsoils are requirements that are a part of the mining operations. For example, Alberta has the Soil Quality Criteria Relative to Disturbance and Reclamation (Soil Quality Working Group 1987) guideline to address the question of what is a suitable topsoil or subsoil under a variety of disturbance and reclamation scenarios.

Progressive reclamation from mining design and operations to final mine closure and reclamation is emphasized upon in Canada. Progressive reclamation is specifically focused on minimizing the footprint for mines at any given time, as land is reclaimed as soon as possible. This involves regulated activities such as conservation of soil and water resources, monitoring of surface water and groundwater quality, surface soil stockpiling, landscape reconstruction, revegetation, safe management of waste streams, introduction and management of wildlife, and environmental monitoring and compliance audits. Progressive reclamation is made possible through removing, segregating, and storing soil during mine construction and mining. If ecological restoration is a goal, progressive reclamation may include collecting seeds from local, native grasses, shrubs, and trees and properly storing them for future use in land reclamation.

The impact mitigation plans or operations discussed above are needed to address issues related to corporate social license and the public interests/support. An example in the Province of Alberta is the need to address concerns of first nations or aboriginal communities regarding oil sands development by ensuring that the first nation's traditional knowledge is incorporated into mine reclamation plans or by ensuring that these communities are represented in stakeholder input opportunities. A clear and well-developed mine reclamation plan using inclusive methods is necessary to address such public concerns (CEMA 2012).

Mine closure planning and reclamation regulations are relatively new to the mining industry in Canada, paralleling an increase in technical knowledge to manage mine operations and in lessons learned from past mine reclamation activities. Prior to the 1970s and the introduction of reclamation regulations, many mines were abandoned without appropriate reclamation (Mining Fact 2016). The Canadian Council of Ministers of the Environment sets national science-based guidelines for a safe threshold of substances in the atmosphere, soil, and water (CCME 2006). Provincial regulatory requirements set pollution prevention procedures and policies on management and control of toxic substances and hazardous waste to reduce threats to ecosystem function and biological diversity (Government of Canada 1999).

Reclamation seeks to restore disturbed lands to pre-disturbance land uses or other productive land uses after mining. In Alberta, the term equivalent land capability is being used in terms of land reclamation requirements. This term means that the ability of the land to support various land uses after reclamation of the disturbed landscape is similar to the ability that existed prior to an activity being conducted on the land, but that the land uses before and after the disturbance will not necessarily be identical.

The general sequence of sustainable mining activities is as follows: surveying undisturbed eco-systems including soils, vegetation, and landscape before mining; salvaging and stockpiling rec-lamation materials on sites; landscape and soil reconstruction, revegetation with appropriate plant species; and maintenance and monitoring of the reconstructed soils and vegetation in a landscape context. A critical step is to reconstruct a functional soil system to support vegetation, hydrologic processes, microorganism habitats, and nutrient supply. Soil reconstruction requires a suitable sub-strate and a suitable cover soil. In forest ecosystems, for example, peat and/or forest floor materials with underlying mineral soils are salvaged before mining, stockpiled and applied on substrates as the cover soil during land reclamation. Overburden, the material below the surface soil that meets the soil quality criteria, and tailings sand from extraction plants are common substrates used in landscape design for oil sands reclamation, and waste rock or tailings waste is a common substrate in diamond and metal mine reclamation. In grasslands or other non-forest ecosystems, the organic component of the cover soil may be composed of amendments such as manure, compost, or other organic waste materials.

Once a soil (20–100 cm deep) is reconstructed, native plant species will be established through manual or mechanical planting of seeds or seedlings. Management of undesirable weeds or invasive species may be required. Plant species selection is an important part of land reclamation, being sig-nificantly more complicated for natural area land uses than agricultural or forestry operations. In the former, plant propagule availability is a limiting factor and such limitations are often mitigated through collection of seeds and other propagules prior to disturbance, with such plant propagules being used later in land reclamation operations.

Monitoring and ongoing assessment are required to determine reclamation success as part of the reclamation permit and reclamation certification process. Aboveground biomass and species com-position are typically monitored and assessed at 3- to 5-year intervals. Monitoring programs also can include measurement of land form and stability, soil depth, and other soil physical and chemi-cal properties. Parameters to be measured and measurement protocols, sampling procedures, and statistical analyses are often standardized. The reclamation certification process is used to confirm that mining disturbed land has been reclaimed to equivalent land capability and that it can sustain a functional ecosystem (Alberta Energy Regulator 2016).

In most Canadian jurisdictions, operators are required to present a conceptual mine closure plan prior to receiving approval to mine. This plan includes a time line, the expected spatial extent of mining impact, possible progression of ecosystem disturbance, soil salvage and stockpiling plans, landscape reconstruction and groundwater management plans, projected ecosystem designs for spe-cific reclamation protocols, and extent and projected end land use for areas expected to be disturbed (CEMA 2012). The closure plans are long-range plans that are usually revised in the Province of Alberta every 4 years (Alberta Energy Regulator 2016). On the basis of the long-range plans approved for particular mine sites, short-range plans are developed annually and approved by rel-evant regulatory bodies, with emphasis on site specifics not captured in the conceptual mine clo-sure plan. This includes current year soil salvage plans, revegetation plans, groundwater monitoring plans and reporting, landscape reconstruction plans, soil stockpiling plans, cover soil reconstruction plans, and annual reclamation reports on the amount of land reclaimed relative to the amount of land planned to be reclaimed (Alberta Energy Regulator 2016).

16.5   OIL SANDS RECLAMATION CASE STUDIES

Oil sands deposits in northern Alberta are primarily distributed in the watersheds of the Athabasca River, Peace River, and Cold Lake. The Athabasca oil sands region (AOSR) is the largest single bitumen deposit in the world and the region with the most active surface mine in northern Alberta (Mossop et al. 1982). Alberta's oil sands occupy approximately 140,200 km^2 of land area and only 20% of deposits can be recovered by surface mining. Surface mining disturbed more than 895 km^2 of boreal forests by 2013 (Alberta Energy 2015) and will disturb a further 4800 km^2 of lands.

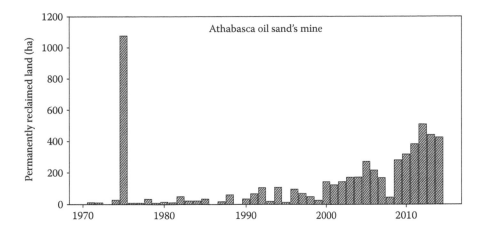

FIGURE 16.2   Annual size of land permanently reclaimed in the Alberta Oil Sands Region disturbed due to surface oil sands mining. (Courtesy of Oil Sands Information Portal, Age of permanent reclamation in the mineable oil sands region: 2014, 2015, http://osip.alberta.ca/library/Dataset/Details/28, accessed on June 1, 2016.)

In situ operations may result in a potential loss of 6500 km^2 of boreal forest (Alberta Government 2016a). In 2014, daily production of crude bitumen was 2.3 million barrels and is expected to reach 4 million barrels per day in 2024 (Alberta Energy 2015; Alberta Government 2016b).

The oil sands deposits are located in the mixedwood boreal forest region, where local geology includes surficial Pleistocene materials underlying various soil materials mostly of Luvisolic and Brunisolic soil types (Soil Classification Working Group 1998). Site hydrology is characterized by surficial and deep groundwater formations that are sometimes connected to the Athabasca River through subsurface sand channels or slow infiltration through clay till zones (RAMP 2016). Dominant tree species are trembling aspen (*Populus tremuloides*) and jack pine (*Pinus banksiana*).

The land permanently reclaimed in the AOSR increased from 20 ha per year in 1971 to 400 ha per year in 2014 (Figure 16.2). The 104 ha Syncrude Gateway Hill project was the first oil sands site to be issued a reclamation certificate (Syncrude Canada Ltd. 2016a,b). Suncor Energy Ltd. successfully reclaimed a tailings site, the Wapisiw Lookout project (Suncor Energy 2016). Most successfully reclaimed AOSR sites have not been certified and are permanently reclaimed subleases within the overall mine site (Figure 16.2).

16.5.1   Gateway Hill Reclamation

Syncrude Canada Ltd.'s Gateway Hill is a terrestrial land reclamation project located in the Alberta Oil Sands Region (Figure 16.3a through c) (Syncrude Canada Ltd. 2016a,b). The 104 ha of disturbed land was an overburden dump that was reconstructed to an upland ecosystem. The land had no recorded history of site contamination, negating the need for soil remediation. Selection of cover soil type and thickness of cover soil application and choice of plant species for reclamation of upslope positions were based on requirements for a natural boreal forest dominated by trembling aspen on fine-textured substrates. The downslope positions of the reconstructed landscape have a wetland system that existed prior to disturbance.

Peat mineral soil mix and LFH (litter-fibric-humic) mineral soil mix of 30–50 cm in thickness were applied as cover soils on overburden substrates (materials below soil parent material) after landscape reconstruction. For landscape stability, the highest slope angle was restricted to 30%, incorporating a properly designed drainage system. Only soils of suitable soil quality for revegetation were selectively salvaged. Soils with pH greater than 8.5 or less than 5 and electrical

FIGURE 16.3   (a) Distribution of permanently reclaimed (certified and non-certified) land of Syncrude Canada Ltd. and Suncor Energy Ltd. in the Alberta Oil Sands Region in 2014 (Courtesy of Testa, B.M., *Reclaiming Alberta's oil sands mines*, *Earth Magazine*, Published by American Geoscience Institute, 2010, http://www.earthmagazine.org/tags/restoration, accessed on June 11, 2016), (b) a typical mining disturbance west of the Gateway Hill overburden (left) and the same site after reclamation (right), (c) Gateway Hill after reclamation (b and c: Courtesy of Syncrude Canada Ltd., Database of reclamation pictures, 2016c, https://www.flickr.com/photos/syncrudecanada, accessed on November 24, 2016), (d) fluid tailings containment for Wapisiw remediation and reclamation, (e) fluid tailings containment after remediation and during landscape and cover soil construction, and (f) Wapisiw lookout after revegetation (d–f: Courtesy of Suncor Energy, Wapisiw lookout reclamation, 2016, http://www.suncor.com/sustainability/environment/land/wapisiw-lookout-reclamation, accessed on June 1, 2016).

conductivity (EC) greater than 4.0 ds m^{-1} were excluded; this ensures that cover soils with appropriate soil chemistry, including nutrient supply capacity to support a healthy rhizospheric microbial community, were used for land reclamation. Revegetation included planting of trembling aspen and jack pine at a stand density that is optimum for tree growth. By directly placing cover soil after salvage instead of stockpiling, a plant propagule bank was provided to supply native species as seeds and vegetative propagules.

Management included annual audits of soil conditions to ensure that the site met soil quality requirements (Soil Quality Working Group 1987). Vegetation assessments were conducted annually to ensure healthy tree growth in accordance with Alberta revegetation standards (Alberta Environment 2003). After a healthy soil-vegetation system was established on the site, the site was opened to wildlife access. The site was monitored every 5 years from 1978 to 2003, until the reclamation certification application filed by the company was approved. The site is still subjected to further monitoring to support final mine site closure and has hosted much research to better understand longer term requirements to support reclamation operations (Syncrude Canada Ltd. 2016a,b).

The reclamation certification process involved review of comprehensive compliance criteria by local stakeholders and relevant provincial regulatory bodies (CEMA 2012). Criteria for successful reclamation include demonstration of landscape stability, free of soil or water contamination, adequate site drainage, proper thickness and suitability of soil cover to support vegetation, healthy vegetation growth and performance, meeting local community expectations, existence of a healthy wildlife community, and provisions to ensure the site is not exposed to further impact from adjacent mining operations. The site was certified as a functional ecosystem in 2003 and was the first case of certified land reclamation in the AOSR that is recognized as having its function and ecosystem services fully restored.

16.5.2   Wapisiw Tailings Reclamation

The Wapisiw reclamation project of Suncor Energy Ltd. is a first tailings pond site in the oil sands to be reclaimed (Suncor Energy 2016). It is located in the Alberta Oil Sands Region and served as a tailings storage area between 1967 and 1997 (Figure 16.3d). The plateau surface was 220 ha in size with a circumference of 3 km; the 91 m high dyke increased total area of the site to 358 ha. The pond was decommissioned in 2006 and reclaimed by 2010 to a developing mixedwood forest with multiple streams and a small marsh wetland.

The tailings were pumped and dredged out of the pond and relocated to another pond for treatment. Remediation of the contaminated fluids was conducted with a pump-and-treat system of fluid tailings and other colloidal settling enhancement technology to separate the mixture of waste bitumen and water from floating (fine particles) or settled (coarse sandy particles) extracts (Figure 16.3d). The fine and coarse textured extracts were the remains of bitumen impregnated tar sands after removal of the oil or organic components in the extraction plants. They eventually formed part of the base materials in the site landscape design, reconstruction, and contouring (Suncor Energy 2016). The tailings water in the pond was replaced with 30 million tonnes of clean sand and then the clean sand was covered with 1.2 million m^3 of topsoil that had been removed from the site in the 1960s; the thickness of the topsoil applied was 50 cm. Drainage systems, hummocks, and swales were developed to manage water runoff from the reclaimed site. Cover soil was of peat mineral soil mix and LFH mineral soil mix (Figure 16.3e).

Revegetation included the planting of a cover crop of native grasses, oats (*Avena sativa*) and barley (*Hordeum vulgare*). The native grasses used were Canada wild rye (*Elymus canadensis*), fringed brome (*Bromus ciliatus*), June grass (*Koeleria macrantha*), Rocky Mountain fescue (*Festuca saximontana*), slender wheat grass (*Agropyron trachycaulum*), and tufted hair grass (*Deschampsia cespitosa*). Seeds were gathered from local trees and shrubs, then grown to seedlings in a greenhouse before transplanting 620,000 seedlings on the site at an appropriate planting density (approximately 2,820 trees ha^{-1}) (Figure 16.3f). Trees were balsam poplar

(*Populus balsamifera*), black spruce (*Picea mariana*), jack pine, trembling aspen, white birch (*Betula papyrifera*), and white spruce (*Picea glauca*). The shrubs used included beaked hazelnut (*Corylus cornuta*), bearberry (*Arctostaphylos uva-ursi*), blueberry (*Vaccinium myrtilloides*), bog cranberry (*Vaccinium oxycoccos*), buffaloberry (*Shepherdia canadensis*), chokecherry (*Prunus virginiana*), green alder (*Alnus viridis*), low bush cranberry (*Vaccinium vitis-idaea*), pin cherry (*Prunus pensylvanica*), prickly rose (*Rosa acicularis*), raspberry (*Rubus idaeus*), red osier dogwood (*Cornus stolonifera*), and willow (*Salix* spp.). Planted woody seedlings were fertilized and monitored (Suncor Energy 2016).

The wetland was designed to retain water throughout the year. It was designed for swales to conduct precipitation to the pond area. Weeping tiles at the base of the swales and the wetlands collect water coming from below, which may contain some process water. Wetland areas were seeded with 12 common wetland species. Species were awned sedge (*Carex utriculata*), beaked sedge (*Carex rostrata*), bladderwort (*Utricularia* species), dwarf birch (*Betula nana*), hornwort (*Ceratophyllum demersum*), marsh cinquefoil (*Potentilla palustris*), rat root (*Acorus calamus*), seaside arrow grass (*Triglochin maritima*), bog cranberry, water sedge (*Carex aquatilis*), and yellow pond lily (*Nuphar lutea*).

Access points were constructed to allow both wildlife and human access, providing alternate end land use options including recreation. Snags were included for birds and rock piles for small mammals and insects. Coarse woody debris piles were used in drainage swales to decrease erosion. The Wapisiw land reclamation operation is in the active monitoring stage with potentials for full reclamation certification in the future (Suncor Energy 2016).

16.6   COAL MINE RECLAMATION CASE STUDIES

Canada has 6.6 billion tonnes of proven recoverable coal reserves, which are mainly located in the western provinces of British Columbia, Alberta, and Saskatchewan (Coal Association of Canada 2016a). There are 24 permitted coal mines throughout Canada, with 19 currently in operation; 10 are located in British Columbia, 9 in Alberta, 3 in Saskatchewan, and 2 in Nova Scotia (Coal Association of Canada 2016a). Coal can be extracted through underground and surface mining, with the majority of coal in Canada produced by surface mining due to safety and work conditions. Although coal mining has played an important role in the Canadian economy in the past several decades, Alberta announced that the use of coal as a source of energy is planned to be completely phased out by 2030 to reduce greenhouse gas emissions and improve air quality (Mertz 2016).

Surface mining destroys the landscape and plant and wildlife habitats, generating significant environmental impacts on the soil, water, flora, fauna, and microorganisms. Surface mining has led to a dramatic alteration of fungal community structure and a decrease in soil microbial biomass because of disruptions to the upper soil horizons (Jurgensen 1979; Visser 1984; Durall and Parkinson 1991). Drainage systems, such as lakes, rivers, and streams, often intersect coal mined lands, and thus coal companies have responsibilities to maintain water quality and quantity in and around their mine sites, minimize the release of pollutants into local waterways, and ensure that the water used in the mining process is cleaned before being returned to the natural system.

Great efforts have been made by coal companies to complete land reclamation throughout the life of mines. There are many examples of successful reclamation of coal mines in Canada, such as the reclamation of the Capital Power Corporation's Genesee Mine, TransAlta Corporation's Whitewood Mine, Teck Resources Limited's Fording River Mine, and Westmoreland Coal Company's Estevan Mine. More than 75% of the land disturbed by coal mining in Alberta had been reclaimed by 2014 (Coal Association of Canada 2016b). The Alberta government has issued 19 coal mine reclamation certificates covering an area of approximately 2100 ha (Alberta Chamber of Resources 2011). The reclaimed land supports a variety of land uses, including agriculture, woodlands, wildlife habitat, recreation, and wetlands.

16.6.1 Whitewood Coal Mine Reclamation

The Whitewood mine, owned by TransAlta Corporation, was located approximately 70 km west of Edmonton, Alberta. It began operations in 1962 and was decommissioned in 2010. As one of Canada's largest coal mine reclamation programs, mined lands of 1900 ha have been reclaimed and issued reclamation certificates (TransAlta Corporation 2016).

Reclamation involved filling open pits with salvaged overburden material, contouring the landscape, and revegetating the land. Depending on agricultural capability classification, 0.35–1.0 m of subsoil was applied on the overburden, and then 20 cm of topsoil was applied. Graminoid species were seeded, usually with a cereal cover crop such as oats. Species mixes commonly comprised creeping red fescue (*Festuca rubra*), crested wheat grass (*Agropyron cristatum*), and white clover (*Trifolium repens*). Tree seedlings including trembling aspen, balsam poplar, jack pine, and white spruce were planted. More than 280,000 trees had been planted on reclaimed land since mine closure. Three end pit lakes and numerous wetland and wooded areas were arranged amid large tracts of agricultural lands.

Throughout the reclamation process, agricultural production, surface water and groundwater quality, rate of erosion, and development of vegetation and weeds were monitored to ensure the sustainability of the reclaimed land. A variety of end land uses including agriculture, wetland, wildlife habitat, and recreation are present within the mine footprint. Most reclaimed land was returned to private ownership. Some of the reclaimed land was leased to local farmers for agricultural use, while some was donated to fish and game clubs and other organizations to conserve reclaimed wetlands and natural areas for wildlife habitat (TransAlta Corporation 2016).

16.6.2 Fording River Coal Mine Reclamation

Fording River Mine, owned by Teck Resources Limited, was located in the Elk Valley of British Columbia. It began production in 1971 and provides approximately 8.5 million tonnes of metallurgical coal every year for export to the Asia-Pacific region. More than 4000 ha of disturbed land has been reclaimed to date.

One successful reclamation practice was to reclaim the open pit to fish habitat. One of the coal mine sites had to temporarily divert the water in a popular fishing stream called Henrietta Creek into a culvert for extraction of the coal and in the process a new lake was created downstream. After the coal mine was completed, Henrietta Creek was restored to a reconstructed channel to provide the fish habitat. The waterway supports a larger fish population than it did before the creek was moved (Erickson 1995).

Many reclamation researches have been conducted on the mine, and the results have been successfully integrated into reclamation practices (Straker et al. 2004, 2005). For example, mining at the Fording River Operations disturbed the habitat for elk and bighorn sheep, which resided year round in the Fording River Valley before mining. Reclamation to high-elevation wildlife habitat required the planting of conifers to provide cover, preferred browse shrub species, and native grass species for forage. Alpine larch (*Larix lyallii*), subalpine fir (*Abies lasiocarpa*), Engelmann spruce (*Picea engelmannii*), and lodgepole pine (*Pinus contorta*) were planted at 2.1 m spacing on sites with 2050–2250 m elevation. Each seedling was planted with a 10 g (20:10:5 nitrogen:phosphorus:potassium) fertilizer tablet. Monitoring was conducted twice annually for the first 2 years, once a year thereafter.

To test preferred elk browse, eleven shrub species were planted in two adjacent rows with 25 shrubs per row. Species included buffaloberry, chokecherry, prickly rose, red osier dogwood, Douglas maple (*Acer glabrum*), saskatoon (*Amelanchier alnifolia*), snowberry (*Symphoricarpos albus*), trembling aspen, birch-leaved spirea (*Spiraea betulifolia*), black cottonwood (*Populus trichocarpa*), willow (*Salix* spp.), and wolf willow (*Elaeagnus commutata*). Each shrub was planted with a 10 g (20:10:5 nitrogen:phosphorus:potassium) fertilizer tablet. Survival and growth of the planted species were monitored twice annually from 1995 to 2003. Native grass species trials were conducted using native seed developed by the Alberta Environmental Centre, including alpine bluegrass (*Poa alpina*), slender wheatgrass, awned slender wheatgrass (*Agropyron trachycaulum* var. *secundum*), and broadglumed wheatgrass (*Agropyron trachycaulum* var. *violaceus*). Seed was applied at approximately 2000 seeds ha^{-1} with fertilizer (11:55:0 nitrogen:phosphorus:potassium) applied at 400 kg ha^{-1}.

Results indicated that fall-planted spruce was the best choice for reforestation on the high-elevation coal spoil mine site. Shrub species had species-specific planting seasons in terms of yield performance; for example, spring planting for wolf willow, red osier dogwood, snowberry, prickly rose, and Douglas maple and fall planting for Engelmann spruce, lodgepole pine, and chokecherry yielded the best results. Most of the native grass species established successfully and were recommended for use for land reclamation on high-elevation sites.

Research has been conducted to evaluate the feasibility of using mine waste materials as growth media, relative to reclamation with a capping material of glacial till overburden. In the late 1970s, the growth of agronomic grasses and legumes was compared on waste rock and rock capped with glacial till. Coal waste was suitable for direct reclamation for vegetation cover establishment, legume content, and biomass production, providing more options for soil reconstruction. However, the glacial till required high phosphorus application rates (approximately double) relative to waste rock and was more susceptible to erosion (Straker et al. 2005).

16.7   METAL AND DIAMOND MINE RECLAMATION CASE STUDIES

Canadian metal mines are commonly classified by primary commodity type, including base metals (copper, zinc, lead, and nickel), precious metals (gold, silver, and platinum), uranium, iron, and other metals (Government of Canada 2016). Base metal, precious metal, and iron mines are mainly found in British Columbia, Ontario, Quebec, and the Northwest Territories (Natural Resources Canada 2016). These metals are extracted by open-pit and underground mining depending on the location of the ore. Northern Saskatchewan has one of the largest high-grade uranium deposits in the world and uranium mining started in 1952 (Government of Saskatchewan 2012). Open-pit uranium mining has disturbed approximately 500 ha of mixedwood boreal forests.

Canada was the fourth by volume (third by value) diamond producing country in the world in 2014 (Kimberley Process Rough Diamond Statistics 2016). There are four active diamond mines in Canada: three in the Northwest Territories and one in Ontario. Open-pit and underground diamond mining disturbed more than 150 km^2 of land by 2015 (Dominion Diamond Corporation 2016; Rio Tinto 2016; Wikipedia 2016).

16.7.1   Diavid Diamond Mine Reclamation

The Diavik Diamond Mine, Canada's largest diamond mine, is located in East Islands of Lac de Gras in the Northwest Territories. It was constructed in 2000 and began production in 2002. It disturbed approximately 1170 ha of the landscape by 2010. To access diamonds under the lake, two dikes were built for open-pit mining, transitioning to underground mining in 2012. Mining disturbances introduced major challenges through the removal of shallow soils, causing soil compaction, alterations to nutrient and hydrologic cycling, and changes in soil temperature (Drozdowski et al. 2012). Gravel pads, built to protect permafrost, are difficult to revegetate due to the limited seedbank, slow and low seed germination, low nutrient availability, low water holding capacity, and soil compaction. They commonly block natural drainage, are disconnected from the hydrologic system, and alter snow drift, resulting in water stress (Naeth and Wilkinson 2004). During the mine development, the top 7 cm of the soil and lake sediment materials were salvaged before mining and stored on-site.

Development of a suitable substrate or soil is most important in reclamation of the Diavik Diamond Mine, which is located in a cold region where soil development is very slow and the availability of suitable soil material from local sources is very limited. Amendments and substrate materials can be used to improve soil properties for revegetation (Reid and Naeth 2005a,b), and materials available locally must be used due to great transportation expense. Suitable soil amendments are structure improving (e.g., peat and lake sediment) and nutrient providing (e.g., fertilizer and gypsum) (Reid and Naeth 2005a,b). The fine-textured materials have high erosion risk (Naeth and Wilkinson 2011) and coarse-textured materials can reduce water and nutrient retention and cation exchange capacity (Drozdowski et al. 2012). Organic matter is essential for the development of self-sustaining plant communities as it increases soil water and nutrient availabilities, water and nutrient holding capacities, and cation exchange capacity; improves soil texture, pH, and electrical conductivity; and provides sources of energy for microorganisms (Naeth and Wilkinson 2010; Drozdowski et al. 2012).

At the Diavik Diamond Mine, revegetation experiments assessed combinations of substrate materials, including processed kimberlite, glacial till, mix of till and kimberlite (50:50 and 25:75), and gravel, and amendments including topsoil, sewage, fertilizer, and sludge from a water treatment facility (Drozdowski et al. 2012). After 2 years, plant density and health was the highest on 25:75 of till-kimberlite mix with sludge; unamended kimberlite had little plant cover. After 5 years, sewage treatments had significantly the greatest plant cover, followed by the till and fertilizer and unamended 50:50 till-kimberlite mix treatments (Naeth and Wilkinson 2010). Species richness was the highest in gravel fertilizer, gravel topsoil, till topsoil, and 50:50 till-kimberlite mix with topsoil. Plant growth and species richness were the lowest in the kimberlite and sludge treatments. As natural recovery in the arctic is very slow, adding target plant species to restoration sites is recommended (Naeth and Wilkinson 2014).

Management and contingency plans include activities to address factors such as aerated lagoon and biopile efficiency, presence of undesirable species, and overutilization of plants by wildlife. Reclamation monitoring included measurements of soil pH, compaction, available macronutrients, water content, pathogens, heavy metals, and erosion. Vegetation monitoring included seeding establishment, species composition, plant health, plant vigor, and wildlife utilization.

16.7.2 Cluff Lake Uranium Mine Reclamation

The Cluff Lake uranium mine in northern Saskatchewan's Athabasca basin operated from 1980 to 2002 by AREVA Resources Canada Inc. Decommissioning began in 2004 and most work was done by 2006, with environmental monitoring each year. The waste rock pile was constructed between 1982 and 1989 and was approximately 30 m high and 26.4 ha in size. A minimum depth of 20 cm of waste rock pile was compacted using a Caterpillar roller and 100 cm of non-compacted silty-sand till was applied over the compacted waste rock surface (AREVA Resources Canada, Inc. 2009). The tailings management area, which contained approximately 2.67 Mm³ of tailings, was covered by a 100 cm thick layer of non-compacted silty-sand till. A commercially available grass and legume mix was drill seeded after cover soil establishment to minimize erosion of grass seeds and to maximize availability of fertilizer. Deep rooted species were not considered for the seed mixtures to prevent the roots from reaching the tailings. Cover soil establishment and revegetation were conducted to reduce percolation of meteoric waters into the waste rock pile, attenuate radiation emanating from stored waste to acceptable levels, and provide a growth medium for development of a sustainable vegetation cover. Decommissioning of the waste rock pile and tailings management area was completed in 2006. Net percolation rates and vegetation growth were monitored from 2007 and cover systems were stable supporting growth of productive native plant species and radiation emanating level from stored waste materials was acceptable (Barber et al. 2015).

16.7.3 Giant Gold Mine Reclamation

Through the Departments of Aboriginal Affairs and Northern Development Canada, the federal government decided to reclaim the Giant Mine, one of the orphan mines in Yellowknife, Northwest Territories. Eight open-pit mines were planned to be developed into end pit lakes and 95 ha of tailings pond will be reclaimed by 2100 (INAC and GNT 2010). A target land use after reclamation is a recreational territorial park and the first reclaimed area will be open to the public in 2050.

Arsenic trioxide dust will be frozen in place through a ground freezing technique, called the frozen block method to prevent release of soluble arsenic into groundwater around the mine. A new water treatment plant will be constructed to eliminate off-site migration of contaminants in groundwater and treated water will be discharged into Great Slave Lake. Tailings water will be treated using bioreactors for 15 years and solid tailings will be encapsulated when tailings water treatment is completed. Tailings and sludge containment areas will be recontoured and covered with rock and soil to promote drainage and potential revegetation.

Contaminated soils and waste rock will be excavated and backfilled into the frozen zone and mineral substrates collected from preexisting borrow pits on site will be applied for land reclamation. The surface of each tailings area will be regraded and ditches and spillways will be constructed to limit erosion and redirect runoff. Tailings containment areas will be covered by a layer of quarried rock followed by an upper layer of fine-textured soil. Revegetation will include willow cuttings, seeding, planting tree plugs, and vegetation islands, using materials collected from the lease. To stabilize the soil and reduce erosion on the banks, a creek and adjacent floodplain will be planted with native and non-native plants.

A comprehensive monitoring program will be implemented that includes provision for monitoring groundwater and surface water, air quality, environmental effects, and ground temperatures within and around the frozen arsenic trioxide. The program will include inspections of pit walls, tailings covers, ditches and spillways, and other physical works (INAC and GNT 2010).

16.8 LESSONS LEARNED AND FUTURE CHALLENGES

Mine reclamation will continue to develop as the science and engineering related to this subject area advance. Reclamation is strongly affected by local conditions and thus experience from successful examples will help refine the techniques required in specific areas. Developing better techniques is necessary to accomplish full ecological restoration of mine sites within the shortest time period possible.

Reconstructing functional soils for land reclamation is a significant challenge (Fung and Macyk 2000; Naeth et al. 2013; Naeth and Wilkinson 2014). Organic matter is the main source of nutrients, but it is in limited supply in many mining areas, and reclaimed soils can have undesirable pH levels, high salinity, low nutrient availability, and low water holding capacity (Fung and Macyk 2000; Howat 2000). This becomes more challenging when native soils available for soil salvage have one or two of these negative chemical and physical properties, in which case operators do not have better options to meet the soil quality guidelines required to support revegetation. Therefore, research into pre- and post-reclamation soil quality should be encouraged. Topsoil must be salvaged before mining as it is usually the most fertile soil that can be used for land reclamation.

Establishment of trees and other vegetation is often a challenge in the reclamation of the land disturbed by mining. Harsh climate conditions, long winter seasons, and low precipitation in northern Canada limit revegetation. The current focus has been on restoring tree species in upland ecosystems due to the ease of constructing such an ecosystem with available mechanical equipment. Developing an understory species community is also necessary to increase plant species diversity. However, seeds are not commercially available for the majority of native understory species (Alberta Native Plant Council 2010), and usually little information can be obtained on required conditions for successful seed germination of native plant species (Harrington et al. 1999). Therefore, developing techniques for effectively growing native plant species should be considered a critical subject of reclamation research. One possibility that exists is to use direct soil placement techniques, in which case soils containing viable propagules are used for propagating such understory species. More effort needs to be put into the research of lowland or wetland systems and related plant species. The topsoil has the propagules for natural vegetation to be established after land reclamation (Mackenzie and Naeth 2010). Therefore, proper salvage and storage techniques are required to improve the success of establishing native vegetation cover.

Mine reclamation is time consuming and expensive and needs to be implemented using progressive options in which mine sites are continually reclaimed during the mine operation phase. Many mining sites are located in remote areas, especially diamond mines in the Northwest Territories. Proper preparation of mine closure and land reclamation plans before mining commences is essential to minimize cost and time for reclamation and revegetation. There is a need to keep research cost low and techniques for land reclamation need to be developed timely for remote sites to lower the risk of mine abandonment.

Criteria for mine reclamation should be determined involving stakeholder and interested parties such as regulators, researchers, local communities, and industries. The criteria for mine reclamation should be clearly defined at the initial phase of mining planning and possibly incorporated into best management practices (BMP).

Technical mine reclamation practices are built on the existing body of scientific knowledge. Transformation of this technical knowledge into operational practices always demands site-specific experience to understand the specific ecological characteristics of the mining site. This calls for continual update of mine rehabilitation procedures based on the feedback from mine operators and the stakeholders, and continued research.

REFERENCES

Aboriginal Affairs and Northern Development Canada (AANDC). 2011. DRAFT. Guidelines for the closure and reclamation of advanced mineral exploration and mine sites in the Northwest Territories. Land and Water Boards of the Mackenzie Valley and AANDC. Yellowknife, Northwest Territories, Canada, 92pp.

Al, T. A., A. C. Maprani, K. T. MacQuarrie et al. 2006. Effects of acid-sulfate weathering and cyanide-containing gold tailings on the transport and fate of mercury and other metals in Gossan Creek: Murray Brook mine, New Brunswick, Canada. *Applied Geochemistry* 21, 1969–1985.

Alberta Chamber of Resources. 2011. Caring for the land. Alberta Chamber of Resources, Edmonton, Alberta, Canada, pp. 1–20.

Alberta Energy. 2015. Facts and statistics. http://www.energy.alberta.ca/oilsands/791.asp (accessed July 4, 2016).

Alberta Energy Regulator. 2016. Reclamation process and criteria. https://www.aer.ca/abandonment-and-reclamation/reclamation-process-and-criteria (accessed June 10, 2016).

Alberta Environment. 2003. Revegetation using native plant materials: Guidelines for industrial development sites. R&R/03-03. Edmonton, Alberta, Canada.

Alberta Government. 2016a. Alberta tier 1 soil and groundwater remediation guidelines. Land Policy Branch, Policy and Planning Division, Alberta Environment and Parks, Edmonton, Alberta, Canada, 197pp.

Alberta Government. 2016b. Alberta's oil sands: Reclamation. http://www.oilsands.alberta.ca/reclamation.html (accessed May 1, 2016).

Alberta Native Plant Council. 2010. Native plant source list. http://www.anpc.ab.ca/content/index.php (accessed May 20, 2016).

AREVA Resources Canada, Inc. 2009. Cliff Lake project-detailed decommissioning plan. http://kiggavik.ca/wp-content/uploads/2013/04/Cluff-Lake-Detailed-Decommissioning-Plan-V2-Feb2009.pdf (accessed May 15, 2016).

Barber, L. A., B. K. Ayres, and B. Schmid. 2015. Performance evaluation of reclamation soil cover systems at Cluff Lake mine in northern Saskatchewan. In *Mine Closure 2015*, eds. A. B. Fourie, M. Tibbett, L. Sawatsky, and D. van Zyl, pp. 1–11. *Mine Closure 2015*, Mine Closure, Vancouver, British Columbia, Canada.

Canadian Council of Ministers of the Environment (CCME). 2006. Canadian environmental quality guidelines. Canadian Council of Ministers of the Environment. Environment Canada, Hull, Quebec, Canada, 1999, updated July 2006.

Coal Association of Canada. 2016a. Coal basics. http://www.coal.ca/coal-basics/ (accessed April 10, 2016).

Coal Association of Canada. 2016b. Alberta coal industry fact sheet. http://www.coal.ca/wp-content/uploads/2015/04/Alberta-Coal-Fact-Sheet_March-2015.pdf (accessed July 1, 2016).

Cumulative Environment Management Association (CEMA). 2012. Report—Criteria and indicators framework for oil sands mine reclamation certification. Fort McMurray, Alberta, Canada. Completed by Charlette Pell Poscente, Environmental Corp. CEMA Contract 2010-0028.

Dominion Diamond Corporation. 2016. Overview and strategy. http://www.ddcorp.ca/operations/overview-strategy (accessed May 10, 2016).

Drozdowski, B. L., M. A. Naeth, and S. R. Wilkinson. 2012. Evaluation of substrate and amendment materials for soil reclamation at a diamond mine in the Northwest Territories, Canada. *Canadian Journal of Soil Science* 92:77–88.

Durall, D. M. and D. Parkinson. 1991. Initial fungal community development on decomposing timothy (*Phleum pratense*) litter from a reclaimed coal-mine spoil in Alberta, Canada. *Mycological Research* 95:14–18.

Eickhoff, C. V., S. X. He, F. A. Gobas, F. A. P. C. Gobas, and F. C. P. Law. 2003. Determination of polycyclic aromatic hydrocarbons in dungeness crabs (*Cancer magister*) near an aluminum smelter in Kitimat Arm, British Columbia, Canada. *Environmental Toxicology and Chemistry* 22(1):50–58.

Erickson, D. L. 1995. Policies for the planning and reclamation of coal-mined landscapes: An international comparison. *Journal of Environmental Planning and Management* 38:453–468.

Fung, M. Y. P. and T. M. Macyk. 2000. Reclamation of oil sands mining areas. In *Reclamation of Drastically Disturbed Lands*, eds. R. I. Barnhisel, R. G. Darmody, and W. L. Daniels, pp. 755–774. ASA, CSSA and SSSA, Madison, WI.

Goodarzi, F., H. Sanei, M. Labonte, M. Labonte, and W. F. Duncan. 2002. Sources of lead and zinc associated with metal smelting activities in the Trail area, British Columbia, Canada. *Journal of Environmental Monitoring* 4:400–407.

Government of Canada. 1999. Canadian environmental protection act. http://laws-lois.justice.gc.ca/PDF/C-15.31.pdf (accessed May 15, 2016).

Government of Canada. 2016. Environmental code of practice for metal mines. https://www.ec.gc.ca/lcpe-cepa/default.asp?lang=En&n=CBE3CD59-1&offset=3&toc=show#s1_2 (accessed April 10, 2016).

Government of Saskatchewan. 2012. Sector fact sheets. http://www.economy.gov.sk.ca/mineralsindustry (accessed April 5, 2016).

Harrington, C. A., J. M. McGrath, and J. M. Mraft. 1999. Propagating native species: Experience at the Wind River Nursery. *Western Journal of Applied Forestry* 14:61–64.

Howat, D. 2000. *Acceptable Salinity, Sodicity and pH Values for Boreal Forest Reclamation*. Environmental Sciences Division, Alberta Environment, Edmonton, Alberta, Canada.

Indian and Northern Affairs Canada (INAC). 2002. Mine site reclamation policy for the Northwest Territories. Minister of Indian Affairs and Northern Development, Ottawa, Ontario, Canada.

Indian and Northern Affairs Canada (INAC) and Government of the Northwest Territories (GNT). 2010. Giant Mine remediation project developer's assessment report. Northwest Territories. Ea0809-001, 764pp. http://www.reviewboard.ca/upload/project_document/EA0809-001_Giant_DAR_1288220431.pdf (accessed April 15, 2016).

Jurgensen, M. F. 1979. Microorganisms and the reclamation of mine wastes. In *Forest Soils and Land Use*, ed. C. T. Youngberg, pp. 251–286. Oregon State University Press, Corvallis, OR.

Kimberley Process Rough Diamond Statistics. 2016. Public statistics area. https://kimberleyprocessstatistics.org/public_statistics (accessed April 1, 2016).

Koch, I., L. Wang, C. A. Ollson, W. R. Cullen, and K. J. Reimer. 2000. The predominance of inorganic arsenic species in plants from Yellowknife, Northwest Territories, Canada. *Environmental Science and Technology* 34:22–26.

Leskiw, L. A. 1998. Land capability classification for forest ecosystems in the oil sands region: Working manual. Alberta Environmental Protection, Environmental Regulatory Service, Land Reclamation Division, Edmonton, Alberta, Canada. http://www.environment.gov.ab.ca/info/library/6858.pdf (accessed July 19, 2016).

Mackenzie, D. D. and M. A. Naeth. 2010. The role of the forest soil propagule bank in assisted natural recovery after oil sands mining. *Restoration Ecology* 18:418–427.

Mertz, E. 2016. Alberta NDP's plan to phase out coal could triple power bills: Coal Association. http://globalnews.ca/news/2610760/alberta-ndps-plan-to-phase-out-coal-could-triple-power-bills-coal-association/ (accessed May 20, 2016).

Mining Association of Canada. 2015. Facts and figures of the Canadian mining industry 2015. The Mining Association of Canada, Ottawa, Ontario, Canada.

Mining Fact. 2016. What are abandoned mines? http://www.miningfacts.org/Environment/What-are-abandoned-mines/ (accessed May 3, 2016).

Mossop, G. D., P. D. Flach, S. G. Pemberton, and J. C. Hopkins. 1982. Field excursion 22: Athabasca oil sands, sedimentology and development technology. *11th International Congress on Sedimentology*, McMaster University, Hamilton, Ontario, Canada, August 22–27, 1982.

Naeth, M. A. and S. R. Wilkinson. 2004. Revegetation of disturbed sites at Diavik Diamond Mine, N.W.T. Annual Report to Diavik Diamond Mines, Inc., Edmonton, Alberta, Canada, 12pp.

Naeth, M. A. and S. R. Wilkinson. 2010. Diamond mine reclamation in the Northwest Territories: Substrates, soil amendments and native plant community development phase I final report. Final report to Diavik Diamond Mines, Inc., Edmonton, Alberta, Canada, 39pp.

Naeth, M. A. and S. R. Wilkinson. 2011. Reclamation at the Diavik Diamond Mine in the Northwest Territories substrates, soil amendments and native plant community development phase II final report. Final report to Diavik Diamond Mines, Inc., Edmonton, Alberta, Canada, 31pp.

Naeth, M. A. and S. R. Wilkinson. 2014. Establishment of restoration trajectories for upland tundra communities on diamond mine wastes in the Canadian arctic. *Restoration Ecology* 22:534–543.

Naeth, M. A., S. R. Wilkinson, D. D. Mackenzie, H. A. Archibald, and C. B. Powter. 2013. Potential of LFH mineral soil mixes for reclamation of forested lands in Alberta. Oil Sands Research and Information Network, University of Alberta, School of Energy and the Environment, Edmonton, Alberta, Canada. OSRIN Report No. TR-35, 64pp.

Natural Resources Canada. 2016. Canadian reserves of selected major metals—Re-opened mines spur increase in reserves in 2011. http://www.nrcan.gc.ca/mining-materials/publications/15745 (accessed on April 1, 2016).

Oil Sands Information Portal. 2015. Age of permanent reclamation in the mineable oil sands region: 2014. http://osip.alberta.ca/library/Dataset/Details/28 (accessed on June 1, 2016).

Oil Sands Magazine. 2016. Water usage in the oil sands. http://www.oilsandsmagazine.com/oil-sands-water-usage/ (accessed on May 1, 2016).

Plouffe, A., P. E. Rasmussen, G. E. M. Hall, and P. Pelchat. 2004. Mercury and antimony in soils and non-vascular plants near two past-producing mercury mines, British Columbia, Canada. *Geochemistry—Exploration, Environment, Analysis* 4:353–364.

Powter, C. B. 2002. *Glossary of Reclamation and Remediation Terms Used in Alberta*, 7th ed. Alberta Environment, Edmonton, Alberta, Canada.

Powter, C. B., N. R. Chymko, G. Dinwoodie, and R. Dyer. 2012. Regulatory history of Alberta's industrial land conservation and reclamation program. *Canadian Journal of Soil Science* 92:39–51.

Province of Alberta. 2014. *Environmental Protection and Enhancement Act*. Alberta Queen's Printer, Edmonton, Alberta, Canada, 164pp.

Pugh, R. E., D. G. Dick, and A. L. Fredeen. 2002. Heavy6 metal (Pb, Zn, Cd, Fe, and Cu) contents of plant foliage near the Anvil Range lead/zinc mine, Faro, Yukon Territory. *Ecotoxicology and Environmental Safety* 52:273–279.

Regional Aquatics Monitoring Program (RAMP). 2016. Hydrology of the Athabasca oil sands region. http://www.ramp-alberta.org/river/hydrology/river+hydrology.aspx (accessed June 10, 2016).

Reid, N. B. and M. A. Naeth. 2005a. Establishment of a vegetation cover on tundra kimberlite mine tailings: 1. A greenhouse study. *Restoration Ecology* 13:594–601.

Reid, N. B. and M. A. Naeth. 2005b. Establishment of a vegetation cover on tundra kimberlite mine tailings: 2. A field study. *Restoration Ecology* 13:602–608.

Rio Tinto. 2016. About Diavik-geography. http://www.riotinto.com/canada/diavik/geography-11460.aspx (accessed May 10, 2016).

Rooney, R. C., E. S. Bayley, and D. W. Schindler. 2012. Oil sands mining and reclamation cause massive loss of peatland and stored carbon. *Proceedings of the National Academy of Sciences (PNAS)* 109(13):4933–4937.

Soil Classification Working Group. 1998. *The Canadian System of Soil Classification*, 3rd ed. Agriculture and Agri-Food Canada Publication 1646, NRC Research Press, National Research Council of Canada, Ottawa, Ontario, Canada, 187pp.

Soil Quality Working Group (SQWG). 1987. *Soil Quality Criteria Relative to Disturbance and Reclamation (Revised)*. Alberta Agriculture, Food and Rural Development, Edmonton, Alberta, Canada, 56pp.

Straker, J., R. Berdusco, C. Jones. R. Jones, and S. Harrison. 2005. Elk Valley coal waste as a growth medium: Results of soil and vegetation analysis from Elk Valley Coal's Fording River, Greenhills and Line Creek mines. *British Columbia Mine Reclamation Symposium*, pp. 1–6, The University of British Columbia, Vancouver, British Columbia, Canada.

Straker, J., R. Jones, R. Berdusco. B. O'Brien, S. Woelk, and C. Jones. 2004. Reclamation of high-elevation wildlife habitat at Elk Valley Coal's Fording River and Greenhills Operations. https://open.library.ubc.ca/collections/50878/items/1.0042458 (accessed July 5, 2016).

Suncor Energy. 2016. Wapisiw lookout reclamation. http://www.suncor.com/sustainability/environment/land/wapisiw-lookout-reclamation (accessed June 1, 2016).

Syncrudc Canada Ltd. 2016a. Case study, rehabilitated sites. Syncrude Canada Ltd. http://www.businessbiodiversity.ca/documents/16-Syncrude.pdf (accessed May 1, 2016).

Syncrude Canada Ltd. 2016b. Land reclamation-our progress. http://www.syncrude.ca/environment/land-reclamation/our-progress/ (accessed May 1, 2016).

Syncrude Canada Ltd. 2016c. Database of reclamation pictures. https://www.flickr.com/photos/syncrudecanada (accessed on November 24, 2016).

Testa, B. M. 2010. Reclaiming Alberta's oil sands mines. *Earth Magazine*. American Geoscience Institute. http://www.earthmagazine.org/tags/restoration (accessed June 11, 2016).

TransAlta Corporation. 2016. Reclaiming mine land: From coal to crops and conservation areas. http://www.transalta.com/newsroom/feature-articles/2015-07-24/reclaiming-mine-land-coal-crops-and-conservation-areas (accessed July 1, 2016).

Visser, S. 1984. Management of microbial processes in surface mined land in western Canada. In *Soil Reclamation Processes: Microbiological Analyses and Applications*, eds. R. L. Tate and D. A. Klein, pp. 203–236. Marcel-Dekker, New York.

Wikipedia. 2013. Victor Diamond Mine. https://en.wikipedia.org/wiki/Victor_Diamond_Mine#cite_note-14 (last modified May 10, 2016).

Zheng, J., H. Hintelmann, B. Dimock, M. S. Dzurko. 2003. Speciation of arsenic in water, sediment, and plants of the Moira watershed, Canada, using HPLC coupled to high resolution ICP-MS. *Analytical and Bioanalytical Chemistry* 377:14–24.

17 Case Studies of Successful Mine Site Rehabilitation
Malaysia

Soon Kong Yong and Suhaimi Abdul-Talib

CONTENTS

17.1 Introduction 309
    17.1.1 Tin Minerals and Application of Tin 309
    17.1.2 Tin Mining in Malaysia 310
    17.1.3 Alluvial Mining Boom in Malaysia (1958–1987) 311
    17.1.4 Aftermath of Tin Mining Boom in Malaysia 312
17.2 Environmental Impact of Tin Mining 312
17.3 Rehabilitation of Ex-Tin Mines 318
17.4 Food Safety 321
17.5 Successfully Rehabilitated Ex-Tin Mines in Malaysia 322
    17.5.1 Sunway Resort City, Selangor 322
    17.5.2 Mines Wellness City at Serdang, Selangor 322
    17.5.3 Tin Tailings Afforestation Centre 322
    17.5.4 Perak Herbal Garden, Kampung Kepayang, Perak 325
    17.5.5 Paya Indah Wetlands, Dengkil, Selangor 325
17.6 Issues and Recent Development in Mining Regulations 328
    17.6.1 Loopholes in the Early Mining Regulations 328
    17.6.2 National Mineral Policy 328
    17.6.3 Recent Guidelines on Redevelopment of Ex-Tin Mines 328
17.7 Conclusion 329
References 329

17.1 INTRODUCTION

17.1.1 TIN MINERALS AND APPLICATION OF TIN

Tin (Sn) is a silver metal (atomic number 50) that occurs in the earth's crust, mainly as cassiterite or tinstone (SnO_2), and is the main source of commercial tin production. Cassiterite is heavy, metallic, hard, and often brown or black due to presence of iron impurities. Tin is relatively stable against corrosion as compared to iron and copper. Before the invention of stainless steel, tin was widely used for coating cans of processed foods, bakeware, biscuit containers, and culinary foil. Dense tin is also used as weight for fishing and tire balancing. Metal alloys produced from tin have been used for manufacturing coins, lead-free solder for plumbing or electronics, dental amalgam, and handicrafts from pewter and bronze. Tin metal is relatively nontoxic, but its organic derivatives (i.e., tributyltin oxide) are very toxic and have been used as antifouling agents for marine vessels.

17.1.2 TIN MINING IN MALAYSIA

Tin deposits in Malaysia were mainly discovered in alluvial mine pits, and these are primarily located on the western side of the main mountain range in Peninsular Malaysia. The so-called tin belt spans from north of the Malay Peninsula, stretching southward through the Kinta Valley in Perak to Kuala Lumpur and further down to Malacca (Figure 17.1). Alluvium is typically made up of a variety of highly weathered materials, including fine particles of silt and clay and larger particles of sand and gravel, and often contains a good deal of organic matter. Cassiterite particles can easily be liberated by hydraulic methods (i.e., gravel pumping and dredging) (Figure 17.2). Large volumes of water are used for processing alluvial deposits in both methods. In the gravel pumping

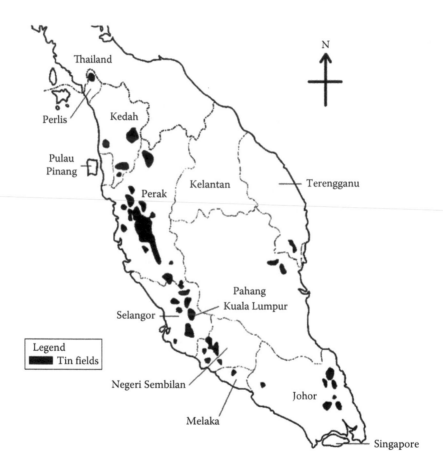

FIGURE 17.1   Tin fields in the Malay Peninsula.

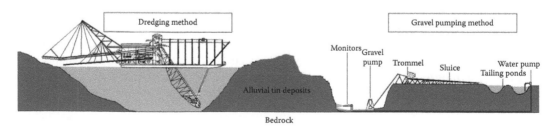

FIGURE 17.2   Dredging and gravel pumping methods for mining tin in the Malay Peninsula.

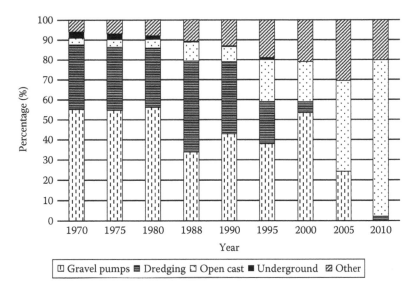

FIGURE 17.3   Tin ore mining methods in Peninsular Malaysia from 1970 to 2010.

method, alluvial deposits are mobilized by pressurized water discharged from monitors, and the slurry is pumped upward into a wooden sluice (e.g., palong) to recover cassiterite. In the dredging method, submerged alluvial is collected mechanically before employing separation processes similar to those of the gravel pumping method. During the processing of alluvial deposits, large volumes of water are used for suspending fine fractions of the soil, and the heavier cassiterite ore is separated and recovered using a wooden sluice or jig. Waste slurry containing tin tailings is dumped directly into empty pits for sedimentation, whereby pond water can be reused for processing alluvial deposits after adequate sedimentation. Opencast and underground methods are also employed on the hard rock tin deposits at Sungai Lembing, Pahang, and Klian Intan, Perak (Cope 2000, Yeap 2000). Other mining methods include the labor-intensive panning process, as well as reprocessing of tin by-products (i.e., amang). Figure 17.3 shows the methods of tin mining in Malaysia between 1970 and 2010. The gravel pump method was simple and was preferred by the Chinese-owned tin mines. At the height of the Malaysian tin mining boom, the gravel pump method was slowly replaced by the more efficient dredging method. In 2010, small-scale mining activities were conducted mainly by opencast and other methods.

17.1.3   ALLUVIAL MINING BOOM IN MALAYSIA (1958–1987)

By 1883, British Malaya had become the largest tin producer in the world, and by the end of the nineteenth century, British Malaya was supplying about 55% of the world's tin requirement. Even though the production decreased toward the mid-1980s (Figure 17.4), Malaysia still led the world in tin export. However, tin mining industry in Malaysia has endured stiff competition from the new and cheap tin mines in Brazil, Indonesia, and China. In October 1985, the price of tin dropped from RM29 to around RM15 per kg, effectively ending profitability for the tin miners. The tin mines were gradually closed, causing Malaysian tin exports to decrease to about 30% in 1992. In 1994, Malaysia became a net importer of tin. Today, the Malaysian tin industry is still active, but has not regained its past glory, since the collapse of its tin market. Major mining activities are still carried out at the Rahman Hydraulic Tin Private Limited's opencast tin mine at Klian Intan, Perak. Despite the significant but brief rise of global tin price between 2001 and 2011, tin mining activity in Malaysia remains unchanged due to low profitability. Alluvial tin mining is costly due to heavy

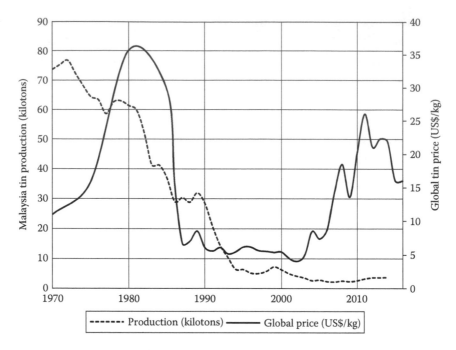

FIGURE 17.4 Malaysian tin ore production (kilotons) and global tin price (US$/kg) between 1970 and 2016.

dependence on fuel consumption (Sykes and Kettle 2013). High oil prices during that period also may have discouraged miners from returning to tin fields.

17.1.4 Aftermath of Tin Mining Boom in Malaysia

The tin mining industry has contributed enormously to the development of Malaysia as a nation. In fact, tin mining has also attracted Chinese migrant workers to Malaya during the British colonial rule. The economies in the state of Perak and Selangor (including the Federal Territory of Kuala Lumpur) were mostly built because of the lucrative tin mining industry. Vital infrastructures such as railways, power stations, and telecommunications were built for supporting the mining industry. Despite the largely successful outcome, tin mining has also left Malaysia with serious environmental problems. Immediately after the collapse of the tin mining industry, large areas of wasteland were unintentionally created and were left idle with little or no economic productivity.

17.2 ENVIRONMENTAL IMPACT OF TIN MINING

In the early stages of the mining process, exploration of vast areas of virgin forests has impacted wildlife due to loss of habitat. Removal of vegetation exposed the top soil to erosion, causing landslides, loss of nutrients, and siltation of rivers. Migration of waste slurry to nearby lakes or rivers occurred after flooding and negatively impacted the quality of water resources. Even after the collapse of the tin mining industry in October 1985 (Yap 2007), the undesirable environmental remnants from tin mining still persist today. Tin tailings, a waste from the tin mining industry, dominate the uneven and often barren landscape. Since 1990, approximately 113,700 ha of a tin tailings site was created in the entire Peninsular Malaysia, and 16,440 ha of this area was submerged under water (Chan 1990).

Tin tailings consist of mixtures of sand and slime (i.e., clay and silts) (Palaniappan 1974). The amount of sand and slime in the ex-tin mines is estimated at around 10%–20% and 90%–80%,

TABLE 17.1
Classification of Tin Tailings for Construction

Type	Composition	Drainage	Bearing Capacity	Revegetation Potential	Suitable Land Use
A	Sand	High	High	Low	Construction
B	Clay, covered with sand	Moderate	High	High	Revegetation/construction
C	Sand, covered with clay	Low	Moderate	Moderate	Recreation/revegetation
D	Clay	Low	Low	Moderate	Recreation/revegetation
E	Clay/sand, covered with water	Low	Low	Low	Recreation

Source: Japan International Cooperation Agency (JICA), The reclamation project of ex-mining land for housing development and other purpose, Feasibility Study Report, I & II, Malaysian Government, 1981.

respectively (Lim et al. 1981). Sand and slime are different in physicochemical characteristics and need to be treated before any development can take place. Tin tailings have been classified into five types (Table 17.1), based on the composition of loose sand and soft clay (e.g., slime), the ability of the tailings to support the load of buildings, and their revegetation potential. Type A tailings require less ground stabilization, and they are the most suitable for construction purposes. Other tailing types may be used for construction after ample stabilization of the tailings has taken place. However, Type A tailings are not suitable for agriculture due to their low retention of moisture and nutrients (Table 17.2). Moreover, prior usage of heavy machinery on the ex-tin mines may have compacted the sand tailings, and the compaction impedes penetration by plants roots. Hence, the Type A tailings are usually barren. The natural succession of plants is limited only to grass species, and it occurs at a significantly slower pace than other types of tailings (Palaniappan 1974). By contrast, slime is more suitable for revegetation due to its higher retention of nutrients and moisture (Ang and Ho 2002). Even so, low drainage may cause water logging and root rot in plants. Regardless of the intended new use, treatment of some sort is required to address the weakness of sandy and slimy tin tailings (Tan and Komoo 1990). The rehabilitation program for ex-tin mines depends on the intended land use as well as the physicochemical characteristics of the tin tailings. Although most of the rehabilitation practices for ex-tin mines in Malaysia have focused on agricultural uses, there is an increasing trend in redeveloping ex-tin mines for commercial and housing purposes.

Tin tailings may contain high concentrations of heavy metals and radionuclides due to the presence of arsenopyrite (FeAsS) (Teh 1981) and rooseveltite (BiAsO$_4$) (Teh and Cheng 2002). Arsenic (As), copper (Cu), zinc (Zn), lead (Pb), and cadmium (Cd) were found in tin tailings, and these could be released into the aquatic environment by dissolution and erosion (Romero et al. 2014). Excessive levels of As in soils have been reported at Kampung Kepayang, Dengkil, and Bestari Jaya (Table 17.3). Pollutants from tin tailings may migrate deep into the aquifer and contaminate groundwater. In fact, well water at an ex-tin mine in Sentul was contaminated with As, which was associated with three cases of skin cancer (Jaafar et al. 1993). Soil samples from an abandoned underground tin mine at Sungai Lembing, Pahang, contained high concentrations of heavy metals, and Cu and Zn were easily leached into water resources (Ahmad and Sarah 2014).

Tin tailings may contain traces of radioactive by-product minerals called amang (i.e., monazite, zircon, xenotime, struverite, ilmenite, and rutile) (Table 17.4). Amang was considered a waste and was disposed of in empty pits at the site of the tin mine (Wu 1999). Amang contains precious and often radioactive elements, such as tantalum (Ta), thorium (Th), niobium (Nb), uranium (U), and the rare earth elements (REE) (i.e., lanthanum [La], yttrium [Yt], cerium [Ce]). In the 1970s, amang was commercially reprocessed at Puchong, Dengkil, Kampar, and Lahat. There have been concerns

TABLE 17.2

pH, OM, CEC, and Nutrient Contents (N, Ca, K, Mg, and P) of Tin Tailings at Dengkil, Bestari Jaya, Bidor, Kampung Kepayang, Kundang, Universiti Putra Malaysia, and Kampung Pasir Semenyih

Location	Soil Texture	pH (H₂O)	OM (%)	CEC (cmol/kg)	N (%)	Ca (cmol/kg)	K (cmol/kg)	Mg (cmol/kg)	P (cmol/kg)	References
Dengkil, Selangor	Unspecified	2.8–3.0								Maesschalk and Lim (1978)
Dengkil, Selangor	Sand					0	0.47	0.10	0	Tompkins (2003)
	Sandy slime					0.03	1.04	0.44	0	
	Slime					0	3.64	1.55	0.26	
Bestari Jaya, Selangor	Sand					0.06–0.12	0.95–1.29	0.20–0.48	0	Tompkins (2003)
	Slime					0.77	6.74	5.06	0.46	
Bestari Jaya, Selangor	Unspecified	4.8–7.2	4.78–9.91	13.76–26.98						Ashraf et al. (2010)
Kampung Coldstream Bidor, Perak	Sand, sandy clay loam Sandy loam	6.08	0.66	0.77						Mender (2004)
Bidor, Perak	Sand	4.2–5.1	0.042–0.062	1.0–1.95	0.012	0.06	0.012–0.03	0.007–0.059	0	Ang and Ho (2002)
	Slime	3.2–6.5	1.4	25.27	0.20	0.17	0.19	0.30	0.048	
Bidor, Perak	Sand	6.5	0.14	3.50						Wan Asma et al. (2011)

(Continued)

TABLE 17.2 (Continued)

pH, OM, CEC, and Nutrient Contents (N, Ca, K, Mg, and P) of Tin Tailings at Dengkil, Bestari Jaya, Bidor, Kampung Kepayang, Kundang, Universiti Putra Malaysia, and Kampung Pasir Semenyih

Location	Soil Texture	pH (H₂O)	OM (%)	CEC (cmol/kg)	N (%)	Ca (cmol/kg)	K (cmol/kg)	Mg (cmol/kg)	P (cmol/kg)	References
Kampung Kepayang	Sand	6.53–8.08	0.03–0.37	1.00–1.30	0.01–0.03	0.45–0.88	0.02–0.03	0.27–0.41	0.004–0.01	Jusop et al. (1986)
Perak	Sand and slime mixture	6.40–7.30	0.15–1.86	1.20–6.90	0.01–0.20	0.71–6.09	0.03–0.05	0.19–1.34	0.006–0.04	
	Slime	5.70–8.22	0.09–0.52	1.00–7.80	0.01–5.06	0.28–5.63	0.03–0.11	0.38–1.78	0.006–0.05	
Kundang, Selangor	Sand	4.6	0.14	0.39	0.01		0.06	0.01	0.01	Wan Abdullah et al. (1992)
Universiti Putra	Sand	5.7	0.2		0.02	0.31	0.01	0.02	0.02	Radziah and Shamsuddin (1990)
Malaysia, Selangor	Sandy loam	4.8	0.36		0.02	0.17	0.01	0.02	0.01	
	Slime	4.7	2.36		0.06	12.65	0.05	0.17	0.03	
Kampung Pasir Semenyih, Selangor	Sand	4.47		1.54	0.02	1.35	0.09	0.50	0.02	Majid et al. (1998)

TABLE 17.3

Heavy Metals and Radionuclide Content in Tin Tailings at Kampung Kepayang, Dengkil, Bestari Jaya, Bidor, Puchong, and Kampung Gajah

Location	Tailing Type	Concentration (mg/kg)									Radioactivity (Bq/kg)					References
		Pb	Cu	Zn	As	Cr	Cd	Hg	Ni	Sn	U-238	Th-232	Ra-226	Ra-228	K-40	
Kampung Kepayang, Perak	Sand and slime mixture	NA	9.53–24.12	30.6–75.9	**67.2–298.3**	39.10	NA	NA	NA	NA	5.57–17.40[a]	NA	NA	NA	NA	Jusop et al. (1986)
	Sand	NA	9.21–10.51	40.3–42.8	**139.8–256.9**	9.45–10.36	NA	NA	NA	NA	5.37–8.41[a]	NA	NA	NA	NA	
	Slime	NA	34.82–47.79	115–178	**337.2–389.8**	**134.7–158.5**	NA	NA	NA	NA	19.40–20.10[a]	NA	NA	NA	NA	
Dengkil, Selangor	Sand	6–17	0.53–12.15	10–95	<0.01–53.4	NA	NA	NA	0.93–13.66	NA	1.05–3.63[a]	NA	NA	NA	NA	Tompkins (2003)
	Sandy slime	8.12	3.28	10.1	19.0				15.3	2.91	1.81[a]					
	Slime	36.3	8.63	26.5	0.65				10.5	7.64	4.50[a]					
Dengkil, Selangor	Unspecified	**5416.56**	**158.9**	6.23	NA	NA	NA	NA	NA	NA	3.37–7.88	**10.5–45.2**	NA	NA	NA	Yasir et al. (2001)
Dengkil, Selangor	Unspecified	NA	NA	NA	NA	NA	NA	NA	NA	NA	**449.2**	**37.5**	**105.8**	NA	247.3	Yasir et al. (2005)
Bestari Jaya, Selangor	Unspecified	85–115	**100–137**	100–132	**62–91**	NA	NA	NA	NA	325–498	NA	NA	NA	NA	NA	Ashraf et al. (2010)
Bestari Jaya, Selangor	Sand	4.16–5.15	4.52–5.20	3.95–4.64	**21.8–39.1**	NA	NA	NA	38.2–45.6	1.74–1.78	1.90–1.91[a]	NA	NA	NA	NA	Tompkins (2003)
	Slime	38.4	11.6	34.5	6.31	NA	NA	NA	11.5	8.35	10.2[a]	NA	NA	NA	NA	
Kampung Coldstream, Bidor, Perak	Sand, sandy clay loam, sandy loam	10.4	4.20	15.8	NA	NA	0.84	NA	3.07	NA	NA	NA	NA	NA	NA	Mender (2004)

(Continued)

TABLE 17.3 (Continued)

Heavy Metals and Radionuclide Content in Tin Tailings at Kampung Kepayang, Dengkil, Bestari Jaya, Bidor, Puchong, and Kampung Gajah

Location	Tailing Type	Concentration (mg/kg)									Radioactivity (Bq/kg)					References
		Pb	Cu	Zn	As	Cr	Cd	Hg	Ni	Sn	U-238	Th-232	Ra-226	Ra-228	K-40	
Bidor, Perak	Sand	NA	3.36–9.47	2.85–55.0	0.02–4.48	0.09–4.66	0.02–0.36	0.03–0.07	0.29–8.78	NA	NA	NA	NA	NA	NA	Ang and Ang (2000)
Bidor, Perak	Slime	NA	3.96–17.01	0.1–30.4	0.03–3.18	0.30–17.01	0.03–0.58	0.08–0.64	3.64–29.3	NA	NA	NA	NA	NA	NA	Ang and Ang (2000)
Bidor, Perak	Slime	9–11	NA	NA	NA	NA	1.5	NA	NA	NA	NA	NA	NA	NA	NA	Ang et al. (2010)
Puchong, Selangor	Lake sediment	3.57–13.55	1.0–2.47	0.38–0.51	0.32–0.35	NA	0.00	0.20–0.60	NA	0.00	**9.37–26.32**	**30.76–35.34**	NA	NA	Neg	Yasir et al. (2008)
Kampung Gajah, Perak	Lake sediment	NA	NA	NA	NA	NA	NA	NA	NA		NA	NA	**152.6–289.8**	**344.2–434.7**	**1533–2095**	Saat et al. (2014)
National background											3.75[b]	10.51[b]	64[c]	84[c]		Yasir et al. (2005)
Global average		0.1–200	1–80	3–300	0.1–20	2–100	0.01–3	0.01–1	2–50		0.01–1.0[a]					Markert (1996)

Note: The values in bold are those higher that the national background/global average values.

[a] Concentration in soil (mg/kg).

[b] Yasir et al. (2005).

[c] Yasir et al. (2008).

TABLE 17.4
Chemical Formulae for Amang
Minerals in Malaysia

Name	Chemical Formula
Ilmenite	$FeO \cdot TiO_2$
Zircon	$ZrSiO_4$
Monazite	$(Ce, La, Y, Th)PO_4$
Xenotime	YPO_4
Columbite	$(Fe, Mn)(Nb, Ta)_2O_2$
Struverite	$(Ti, Ta, Fe^{3+})_3O_6$

about high radioactivity levels at processing plants and storage facilities (Roberts 1995). Traces of amang may spread away from the source through natural weathering processes, possibly raising radioactivity to a hazardous level. Excessive exposure to ionizing radiation may adversely affect human health. Surveys have found that radioactivity levels at Dengkil (Yasir et al. 2005), Kampung Gajah (Saat et al. 2014), and Puchong (Yasir et al. 2008) were higher than at the national background levels (Table 17.3).

17.3 REHABILITATION OF EX-TIN MINES

Immediately after the collapse of the tin industry in Malaysia, many of the unproductive ex-tin mines were returned to the state government. Redevelopment or rehabilitation of the ex-tin mines is the responsibility of both the state and federal governments with active participation from private entities and nongovernmental organizations. A report in 1990 indicated that a small portion (i.e., 9.6%) of the ex-tin mines have been reclaimed and redeveloped for both agricultural and nonagricultural purposes, which almost exclusively occurred in Kuala Lumpur, Selangor, and Perak (Table 17.5). The trend is most likely influenced by several factors: (1) availability of existing infrastructures, (2) high demand for housing in densely populated and urbanized areas, and (3) increasing demand for food due to a growing population. Most of the developed ex-tin mines are already well equipped with infrastructures. Ex-tin mines near Kuala Lumpur were quickly redeveloped as residential areas, business centers, and townships. Table 17.6 shows selected ex-tin mines in Malaysia that have been successfully redeveloped.

TABLE 17.5
Area of Ex-Tin Mines (ha) and Redeveloped Area (ha) for Agricultural
and Nonagricultural Purposes in Peninsular Malaysia in 1990

State	Agricultural (ha)	%	Nonagricultural (ha)	%	Area of Tailings (ha)
Perak	3890	3.421	2300	2.023	71,850
Selangor and Kuala Lumpur	830	0.730	3920	3.448	28,250
Kedah	10	0.009	20	0.018	2,510
Others[a]	0	0	0	0	11,090
Total	4730	4.160	6240	5.488	113,700

Source: Chan, Y.K., Tin mining land—An overview of the current situation in Peninsular Malaysia, in: Paper presented at the *National Seminar on Ex-Mining Land and BRIS Soil: Prospects and Profit*, Serdang, Malaysia, 1990.

[a] Johor (5660 ha), Pahang (2910 ha), Negeri Sembilan (2090 ha), Melaka (380 ha), Terengganu (40 ha), Pulau Pinang (10 ha).

TABLE 17.6

Locations and Land Use of the Redeveloped Ex-Tin Mines in Malaysia

Location	New Land Use	References
Serendah Golf Course, Selangor	Recreation	Abdul Rasid (2008)
Clearwater Sanctuary Golf Resort, Perak	Recreation	Ibrahim (2010)
Lake Garden, Kuala Lumpur	Recreation	Ibrahim (2010)
Taiping Lake Gardens, Perak	Recreation	Ibrahim (2010)
Lake Fields, Sg Besi, Kuala Lumpur	Housing	Ibrahim (2010)
Taman Kinrara 1–5, Puchong, Selangor	Housing	Oh (2015)
Kinta Nature Park, Batu Gajah, Perak	Nature conservation	Wu (2001)
Paya Indah Wetlands, Dengkil, Selangor	Nature conservation	Ho (1998)
Tin Tailing Research Station, Kundang, Selangor	Agriculture/Research	Haron et al. (2015)
Tin Tailing Afforestation Centre, Bidor, Perak	Agroforestry/Research	Ang and Ang (2000)
Tambun, Perak	Agriculture	Ibrahim (2010)
Bidor, Perak	Agriculture	Mender (2004)
Slim River, Perak	Aquaculture	Agromatic Corporation Pty Ltd (2015)
Bestari Jaya, Selangor	Agriculture/Aquaculture/ Education	Jabatan Perancangan Bandar dan Desa Selangor (2015)
Kampar, Perak	Housing/Education Agriculture/Aquaculture	Ibrahim (2010), Ng (2011), and Leong et al. (2012)

Malaysia has been facing a shortage of land for housing purposes, due to an urban migration problem that began in the early 1970s. The rise of illegal squatters in urban areas may cause social and health problems. The Malaysian government has tried, but was unsuccessful in increasing low-cost housing during the second, third, and fourth Malaysia Plan (1970–1985) (Said 1985). Scarcity of land in the urbanized Kuala Lumpur area has pushed property prices higher, often beyond the affordability of the urban poor. In 1986, the Malaysian federal government introduced the Industrial Master Plan to diversify the Malaysian economy in order to reduce overreliance on the tin mining sector and to intensify manufacturing to boost national exports. This policy may have boosted migration and pushed demand for housing in cities. The Ministry of Housing and Local Development and state governments have been looking at redeveloping ex-tin mines, including building affordable housing for the urban poor. In the Kuala Lumpur suburbs, low-cost houses were built on the reclaimed tin mines. Since the 1990s, rising living standards in the Klang Valley has attracted investment in satellite cities around the Kuala Lumpur city center. Reclaimed mines at Serdang, Petaling Jaya, Subang Jaya, and Puchong have been redeveloped, ranging from business centers, shopping malls, recreational parks, and learning centers. In 1985, there were about 7487 ha of undeveloped tin-mined lands surrounding the city of Ipoh. About 12,141 ha of ex-tin mines has been reclaimed for residential projects, such as the Lahat Mines at Batu Gajah and Bandar Baru Kampar (Perak State Parliment 2004).

Most nonagricultural development of the ex-tin mines has been carried out with little or no treatment. For example, in Selangor, some ex-tin mine water ponds were drained and converted into open landfills, where the reclaimed land will be further developed after the closure of the landfill (Pariatamby 2014). The ex-tin mine water pond at Taman Desa, Kuala Lumpur, has been used as a storm water retention reservoir in the Storm Water Management and Road Project (SMART) in Kuala Lumpur (Abdullah 2004). Recently, an ex-tin mine water pond has been proposed for rehabilitation for storm water management (Takaijudin et al. 2012). But north of the Klang Valley lies a sizable area of idle ex-tin mine water ponds in rural Bestari Jaya. Similarly, in Perak, many ex-mine water ponds also have been left undeveloped in Kampar, Mambang di Awan, and Gopeng.

TABLE 17.7

Strategies for Conditioning Sand and Slime Tin Tailings at Tin Tailing Research Station (TTRS), Bidor, Perak, and TTRS, Kundang, Selangor

Problem of Tin Tailings	Conditioning Strategies	
	Sand	Slime
Water retention	Lowering plant position (45 cm) below soil surface	Raising planting bed
Compaction	Ploughing, and conditioning with organic matter	Ploughing and rotovation (2 rounds)
Low nutrient	Fertilizing with organic fertilizer (Oil palm waste)	—
High acidity	—	Liming with limestone

Sources: Ang, L.H. and Ho, W.M. Afforestation of tin tailings in Malaysia, in: Paper presented at the *12th International Soil Conservation Organization Conference*, Beijing, China, 2002; Haron, S. et al., Returning agricultural productivity to former tin mining land in Peninsular Malaysia, in: J. Griffiths (Ed.), *Living Land*, United Nations Convention to Combat Desertification, Leicester, U.K., 2015, pp. 166–170.

Inadequate private investment is one of the causes for slow development of ex-tin mines located in rural regions. The federal and state governments have intervened by building key infrastructures. One example is the establishment of higher education institutions, namely Universiti Tunku Abdul Rahman at Kampar, Perak, and Universiti Industri Selangor in Bestari Jaya, Selangor. Besides that, the state governments in Selangor and Perak, with the assistance of agencies from the federal government, have actively promoted agriculture and aquaculture on the ex-tin mines.

Currently, two research stations have been set up by the Malaysian Agricultural Research and Development Institute (MARDI) and the Forest Research Institute Malaysia (FRIM) at Kundang (i.e., Tin Tailing Research Station) (Haron et al. 2015) and Bidor (i.e., Tin Tailing Afforestation Centre), respectively. Research has been carried out at both research stations, and they aim to create suitable environments for raising livestock and growing plants and to improve agricultural productivity on the tin tailings. Despite difficulties of farming on the tin tailings, researchers at MARDI and FRIM have discovered strategies for treating the nutrient-devoid tin tailings (Table 17.7), and they have successfully produced at least 25% of Malaysia's vegetables and fruits (Said 1985). Successful cultivation of taro (Tong 2016), cassava (Tan and Chan 1995, Tan 2000), groundnuts (Termidi et al. 2009), pomelo, winter melon (Mansor et al. 2010), mango, guava, and papaya (Ang and Ng 2000) have been reported at orchards located in the southern part of Perak. The acidity of slime and nutrient-deficient sand tailings from ex-tin mines can be improved by liming and soil conditioning, respectively. Cheap and abundant organic materials (i.e., manures, peat, oil palm biomass, and sewage sludge) have been used for conditioning sand tailings. Due to the porous nature of sandy tin tailings, soil conditioning will only be successful if organic material is thoroughly and homogenously blended with the sandy tin tailings (Wan Asma et al. 2011). Overall, the cost for developing tin tailings for agriculture may be between RM3500 and RM5000 per ha, which covers the irrigation system, roads, land preparation, and initial operational costs (Mohd Salleh and Sabtu 2005).

Most of the ex-tin mines have been used for aquaculture, where the water ponds are being utilized for farming freshwater fish (e.g., Chinese carps) since the 1920s. After the tin market collapse, more ex-tin mine water ponds were converted to fish farms by the miners to supplement their income. Intensive fish farming in ex-tin mine water ponds also addresses the increasing demand for freshwater fish due to the growing population. Popular fish species, such as tilapia (*Oreochromis niloticus*), river catfish (*Clarias* sp.), and barramundi (*Lates calcarifer*), and high-value fish, such as jade perch (*Scortum barcoo*), marble goby (*Oxyeleotris marmorata*), and giant freshwater prawn (*Macrobrachium rosenbergii*), have been cultured in ex-tin mine water ponds. The area of the ex-tin mine aquaculture system in Malaysia has increased by 256%, from 641 ha in 1990 to 1642 ha

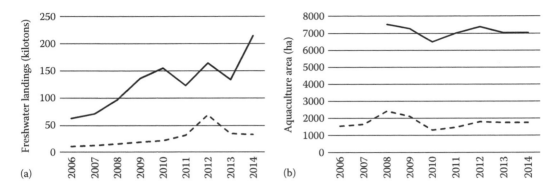

FIGURE 17.5   (a) Freshwater landings (kilotons) and (b) area of aquaculture systems (ha) between 2006 and 2014 for the entire aquaculture systems (solid) and ex-tin mine water ponds (dash).

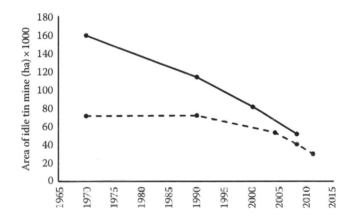

FIGURE 17.6   Area of idle tin mines (ha) in Peninsular Malaysia (solid) and Perak (dash) between 1970 and 2011.

in 2007. During the same period, freshwater landings also increased by 13.3 times (Department of Fisheries of Malaysia 1990). Between 2006 and 2014, no significant changes were recorded on freshwater landings and area coming from the ex-tin mine aquaculture system (Figure 17.5). Moreover, in the state of Selangor, the percentage of freshwater fish landings from ex-tin mine water ponds has decreased from 5.6% in 2000 to 1.0% in 2010 (Jabatan Perancangan Bandar dan Desa Selangor 2015), possibly due to the shift of interest from aquaculture to the construction sector.

According to the recent survey by the Department of Mines and Geoscience Malaysia, the area of idle ex-tin mines in general has been decreasing rapidly in both Perak and Peninsular Malaysia (Figure 17.6). It is projected that all idle ex-tin mines in Malaysia will be fully developed in about 10 years' time (Osman 2013).

17.4   FOOD SAFETY

Tin tailings are known to contain hazardous elements. It is possible that under acidic and oxic conditions, the contaminants may migrate from tin tailings and accumulate in living organisms. For example, ex-tin mine water ponds at Tronoh, Perak, have elevated Pb levels at both active and abandoned tin mines, and Pb levels in winter melon and Tilapia and *Cichla* fish samples were above the safe limit according to the Malaysian Food Acts 1983 and Regulations 1985 (Mansor et al. 2010). Native fish species, collected from an ex-tin mine at Sungai Lembing, Pahang, also were contaminated with heavy metals, where the concentrations of As and Ni exceed the permissible limits in food set by the Malaysia Food Act (MFA) 1983 and the World Health Organization (WHO) (Ahmad and Sarah 2015). At Bestari Jaya, Selangor, the mean concentrations (dry weight basis) for As, Cu, Zn, and Sn for 15 fish species collected from ex-tin mine water ponds have exceeded the permissible limits in food set by the Malaysia Food Act (MFA) 1983 (Ashraf et al. 2011).

The effects of tin mining may include widespread pollution of heavy metals in soil and water resources. Although raised levels of heavy metals and radionuclides have been detected in the soils and lake sediments of the ex-tin mines (Table 17.2), the risk to the public is largely not a concern. For example, heavy metal contents in the tin tailings may be immobile and may not be contaminating the environment (Ashraf et al. 2010, Roberts 1995). In fact, there are few reported cases of adverse health effects due to contaminants from ex-tin mines in Malaysia. Moreover, food crops have been produced from the ex-tin mines, and the levels of heavy metals and radionuclides for most of the crops (Table 17.8) and fish (Table 17.9) still adhere to the strict standards of the Malaysian Food Act and WHO. However, most pollutants exposed during mining activities still exist and may not diminish over time. Constant monitoring of the food crop is essential for safeguarding the health of consumers.

17.5   SUCCESSFULLY REHABILITATED EX-TIN MINES IN MALAYSIA

17.5.1   SUNWAY RESORT CITY, SELANGOR

Sunway Resort City or Bandar Sunway is an integrated mix of developments located in the Kuala Lumpur suburbs of Petaling Jaya and is Malaysia's first successful ex-tin mine rehabilitation project. Since early 1986, 350 ha of ex-tin mine land was transformed into a township that consists of a theme park, a shopping mall, two universities, three hotels, and several premium residential areas. The developer (i.e., Sunway Group) is trying to emulate their success by developing its second township on another ex-tin mine at Sunway City Ipoh (Ang et al. 2014b).

17.5.2   MINES WELLNESS CITY AT SERDANG, SELANGOR

The Mines Wellness City is located south of Kuala Lumpur and is also an integrated, mixed development built on the derelict ex-tin mine of the Hong Fatt Mines (Howard 1937). Development began in 1988 with the conversion of the mine pit to a water theme park. Since then, more development has taken place, including a hotel, a golf course, a business park, a man-made beach resort, and residential areas.

17.5.3   TIN TAILINGS AFFORESTATION CENTRE

The Tin Tailings Afforestation Centre (TTAC) is Malaysia's pioneer reforestation project on a 121.4 ha ex-tin mine at Bidor, Perak. The project began under a lease by the Perak State Government in 1996. It was opened in 2001 and has been managed by the Forest Research Institute of Malaysia (FRIM). The aim of TTAC is to rehabilitate the 85 ha derelict ex-tin mine into a commercial tropical forest plantation, which will supply pulp and hardwood for various applications. Under this pilot program, intensive studies have been conducted on various aspects for the rehabilitation work, including improving the structure and nutrients of tin tailings and developing tree planting techniques on tin tailings. Tin tailings at TTAC are sandy, slime, or a mixture of both. The site was initially colonized by weeds (i.e., *Imperata cylindrica* and *Melastoma malabathricum*, *Muntingia calabura*, *Vitex pubescens*, and *Mallotus* sp.), which have been removed before the tree planting process. Sandy slime tailings are suitable for planting without much treatment. Slime tailings require proper drainage and liming. Sand tailings require more treatment for addressing low contents of water and nutrients (see Section 17.3). Two important strategies have been employed for increasing

TABLE 17.8

Concentration of Heavy Metals in Vegetables and Fruits Produced from Ex-Tin Mines in Malaysia

Crop	Location of Source	Metal (Concentration/mg/kg Dry Weight Basis)									Reference
		Cd	Hg	Pb	As	Zn	Cu	Cr	Ni	Sn	
Chinese radish						15.33					Mender (2004)
Tapioca				0.85							
Sweet potato		0.12									
Yam bean		0.07				2.41	0.63			0.56	Mender (2004)
Lotus seed	Kampar			163.2[a]				358.3[a]			Leong et al. (2012)
Lotus root	Perak			4.2[a]				12.3[a]			
Guava	Bidor, Perak	0.12	0.02	8.71	0	8.74	3.74	0.39	0.55		Ang and Ng (2000)
Mango		0.06	0.26	0.63	0	7.63	3.90	0.52	1.07		
Papaya		0.34	0.08	1.89	0	5.20	2.01	0	0		
Pineapple		0	0.003	0	0	0	0	0	0		Ang et al. (2011)
Permissible limit		0.05–0.1	0.5	0.1–0.3	1.4	60	40	0.05		230	FAO/WHO Codex Alimentarius (2001)
		1	0.05	2	1	40	30			40	Malaysian Food Act (1983)

Note: The values in bold are those higher than the levels permitted in the Malaysian Food Act 1983/FAOWHO.

[a] denotes that the metal contents were calculated on wet sample weight basis.

TABLE 17.9

Concentration of Heavy Metals in Fishes Farmed from the Ex-Tin Mine Water Ponds in Malaysia

Fish Species	Location of Source	Metal Concentration (mg/kg Dry Weight Basis)												Reference
		Cd	Hg	Pb	As	Zn	Cu	Cr	Ni	Sn	Ra-226	Ra-228	K-40	
Assortment of fishes[a]	Puchong, Selangor	0.00	**0.11–0.53**	0.00	0.06–0.11	0.5–3.81	0.13–2.27	NA	NA	0.00				Yasir et al. (2008)[b]
Assortment of fishes[c]	Bestari Jaya, Selangor	NA	NA	0.07–1.78	0.00–0.83	**16.25–104.23**	1.51–36.31	NA	NA	**56.34–153.45**				Ashraf et al. (2011)
Assortment of fishes[d]											3.24–5.67	1.42–4.48	161.9–239.7	Saat et al. (2014)
Oreochromis mossambicus	Glami Lemi,	0.22	NA	0.0048	**4.7**	17.5	1.2	NA	NA	NA				Low et al. (2011)
	Negeri Sembilan	0.05	0.0081	0.011	**1.07**	4.0	0.27							Low et al. (2015)[b]
Permissible		0.05–0.1	0.5	0.1–0.3	1.4	60	40	0.05	NA	230				FAO/WHO Codex Alimentarius (2001)
Limit		1	0.05	2	1.0	100	30	NA	NA	40				Malaysian Food Act (1983)

Note: The values in bold are those higher that the national background/global average values.

a   *Oreochromis mossambicus, Clarias batrachus, Pangasius sutchi, Ophiocephalus micropeltes, Ophiocephalus striatus, Notopterus notopterus.*

b   Wet weight basis.

c   *Hampala macrolepidota, Rasbora sumatrana, Rasbora elegans, Cyclocheilichthys apogon, Osteochilus vittatus, Trichogaster trichopterus, Pristolepis fasciatus, Acantopsis choirorhynchos, Oxyeleotris marmorata, Labiobarbus cuvieri, Chela anommalura, Osteochilus hasseltii, Mastacembelus armatus, Macrobrachium rosenbergii, Channa striatus.*

d   *Leptobarbus hoevenii, Puntius schwanenfeldii, Pristolepis fasciatus, Cichla monoculus, Osteochilus hasseltii, Oreochromis mossambicus, Channa micropeltes, Aristichthys nobilis.*

the availability of water for plants: (1) artificially introducing organic matter for increasing nutrient content and water holding capacity of sand tailings and (2) planting tree seedlings 45 cm below the surface of sand tailings to reduce evapotranspiration. First, the compacted sand tailings are ploughed to allow penetration and healthy growth of the tree roots (Tompkins 2003). Agricultural wastes from the palm oil industry (i.e., empty fruit bunch) and peat are homogenously mixed or alternately layered with sand tailings to increase organic matter. Planting leguminous crops (i.e., *Calopogonium mucunoides*, *Centrosema pubescens*, *Indigofera tinctoria*, *Mucuna cochinchinensis*, *Vigna radiata*, *Glycine max*, *Arachis pintoi*, and *Pueraria javanica*) helps to increase biomass and soil nitrogen (Haron et al. 2015, Majid et al. 1994). The initial stage of reforestation has been conducted using fast-growing local and foreign tree species (Table 17.10). Leguminous acacia species are most suitable due to their high survival and growth rate on sand tailings (Ang et al. 2014a). These fast-growing species are enriching the soil nitrogen and providing adequate shade from the heat for the enrichment of delicate tropical rainforest species. Trees are more likely to survive and regenerate under the canopies of acacia trees. By 2014, there were 60 forest tree species in TTAC and more than 70 wildlife species that had been observed in the area. The presence of wild animals and birds has contributed to natural regeneration because they scatter seeds through their digestive tracts. Tree species not dispersed by birds and bats have been introduced through enrichment planting (Ang n.d.).

17.5.4 Perak Herbal Garden, Kampung Kepayang, Perak

The 22 ha ex-tin mine was set aside by the Perak Department of Agriculture for agricultural research by the Soil Science Department, Universiti Putra Malaysia (Jusop et al. 1986). Since 1987, the ex-tin mine has been redeveloped as a center for vegetable crops, and, in 2009, Perak Herbal Garden was opened and inaugurated by the late Sultan Azlan Shah. Various herb species of local and international origins have been planted. The collection of herbs in Perak Herbal Garden has reached almost 500 species. The famous local herb species such as the Misai Kucing (*Orthosiphon stamineus*), Tongkat Ali (*Eurycoma longifolia*), Hempedu Bumi (*Andrographis paniculata*), Kaduk (*Piper sarmentosum*), Kacip Fatimah (*Labisia pumila*), and dan Mas Cotek (*Ficus deltoidea*) have been planted at the Perak Herbal Garden. In addition, Perak Herbal Garden is also known to the public for a variety of recreational activities.

17.5.5 Paya Indah Wetlands, Dengkil, Selangor

Paya Indah Wetlands (PIW) is a 3100 ha rehabilitated ex-tin mine located near the Kuala Lumpur International Airport and Putrajaya. It comprises 14 ex-tin mine water ponds (1050 ha), a disturbed peat swamp forest (825 ha), and an undisturbed peat swamp forest (2800 ha). Tin fields at Dengkil and PIW were actively mined since the 1950s. In 1997, the "Sri Banting" dredging ceased operation, marking the end of tin mining activities in Dengkil. The Selangor State Government named Paya Indah as a wetland reserve in February 1998. The initial rehabilitation work was carried out by the Malaysian Wetland Foundation in December 1997, and then it was taken over by the Malaysian Department of Wildlife and National Parks in August 2005. So far, the Malaysian federal government has spent a total of RM85 million for rehabilitation of PIW (New Straits Times 2010, The Sunday Daily 2005).

Since the inception of PIW in 1999, a rapid rehabilitation process has resulted in about a 30% increase in wildlife population, with 130 species of birds, 40 species of fish, 25 species of mammals and reptiles, and 220 species of aquatic and terrestrial plants and rare herbs (Ho 1998). Tailings at PIW are sandy or a mixture of sand and peat. Presence of peat at PIW contributes to a high organic matter content and a low pH value in the pond water. Floating aquatic weeds (i.e., *Salvinia molesta*, *Nelumbo nucifera*) dominate the surface of the ex-tin mine water ponds, and it has been populated with a variety of fish species such as haruan (*Channa striata*), selat (*Notopterus notopterus*), sepat

TABLE 17.10

List of Tree Species Planted in the Tin Tailings Afforestation Centre [TTAC], Bidor, Perak

Cultivated Plant Species	Local Name	Area (ha)	References
Dalbergia pinnata			Ang and Ho (2004)
Khaya ivorensis	Khaya	15	
Swietenia macrophylla	Mahogany	1	
Peronema canescens	Sungkai	1	
Fagrae crenulata	Malabera	4	
Dyera costulata	Jelutong	6	
Intsia palembanica	Merbau		
Tectona grandis	Teak		
Maesopsis eminii			
Dryobalanops oblongifolia	Keladan		
Hopea odorata	Merawan siput jantan	13	
Acacia crassicarpa	Akasia	5	
Acacia mangium	Akasia		
Acacia hybrid[a]	Akasia	5	
Acacia auriculiformis	Akasia		
Acacia aulacocarpa	Akasia		
Morinda spp.	Mengkudu		
Anisoptera sp.	Mersawa		
Melia excels	Setang		
Palaquium spp.	Nyatoh		
Shorea platyclados	Meranti bukit		
Shorea roxburghii	Meranti temak nipis		
Neobalanocarpus heimii	Chengal		
Dryobalanops aromatica	Kapur		Forest Research Institute
Shorea parvifolia	Meranti sarang punai		Malaysia (2014) and AEON
Shorea leprosula	Meranti Tembaga		Cheers Club (2014)
Shorea curtisii	Seraya		
Shorea ovalis	Meranti Kepong		
Shorea macroptera	Melantai		
Shorea assamica	Meranti pipit		
Hopea pubescens	Merawan bunga		
Garcinia hombroniana	Beruas		
Melaleuca leucadendron	Gelam		
Sindora coriacea	Sepetir licin		
Careya arborea	Putat kedang		
Cananga odorata	Kenanga		
Palaquium gutta	Teban merah		
Aquilaria malaccensis	Karas		
Pentaspadon motleyi	Pelong		
Ardisia elliptica	Mata pelanduk		
Syzygium cumini	Salam		
Sterculia parvifolia	Kelumpang burung		
Sandoricum koetjape	Santol		
Ficus hispida	Hairy fig		
Adenanthera pavonina	Saga		
Bridelia tomentosa	Bredelia		
Bouea macrophylla	Kundang		
Callerya atropurpurea	Tulang diang		

[a] denotes hybrid of *Acacia mangium* and *Acacia auriculifomris*.

Kedah (*Trichogaster tricopterus*), sepat Siam (*Trichogaster pectoralis*), river catfish (*Clarias batrachus, Clarias teijsmanni, Mystus nemurus*), fighting fish (*Beta pugnax*), tapah (*Wallago leeri*), and puyu (*Anabas testudineus*). Overall, the ex-tin mine water pond has had no trouble in sustaining life, but it has low plant species diversity.

Compared to other ex-tin mines in Malaysia, natural regeneration of sandy tailings at PIW may have occurred rapidly due to: (1) a possible higher content of peat, and (2) close proximity to an undisturbed peat swamp forest. To the best of the authors' knowledge, a diversity study has been conducted only on the bird species, and no report on plant species diversity has ever been published. However, many trees have been planted on selected plots at PIW. Private corporations and nongovernmental organizations have also contributed to the rehabilitation effort by planting about 2400 tree seedlings from 20 species at the Malaysia–Japan Friendship Forest AEON Woodland at PIW.

There are five types of habitat in PIW: (1) dryland; (2) shrub patches; (3) marsh swamp, a treeless shallow water body; (4) lotus swamp, shallow water body; and (5) open water body, deep large lake. A total of 67 tree species were recorded along the lake edges in the marsh swamp area, 17 in the lotus swamp, and 6 in the open water body. The dominant plant species in these five habitats is shown in Table 17.11. Heterogeneous vegetation at PIW provides diverse food resources, suitable perching locations, safe foraging, and breeding sites for a variety of bird species (Rajpar and Zakaria 2013). Moreover, the lotus swamp is a habitat for aquatic invertebrates

TABLE 17.11

Dominant Plant Species at Dryland, Shrub Patches, Marsh Swamp, Lotus Swamp, and Open Water Body at Paya Indah Wetlands [PIW], Dengkil, Selangor

Dryland	Shrub Patches	Marsh Swamp	Lotus Swamp	Open Water Body
Acacia auriculiformis	*Melastoma malabathricum*	*Eleocharis dulcis*	*Nelumbo nucifera*	*Potamogeton* spp.
Acacia mangium	*Dillenia suffruticosa*	*Stenochlaena palustris*	*Nymphaea pubescens*	*Utricularia* spp.
Ficus rubiginosa	*Acacia auriculiformis*	*Philydrum lanuginosum*	*Nymphaea lotus*	*Salvinia molesta*
Ficus benjamina	*Acacia mangium*	*Scleria purpurascens*	*Eleocharis dulcis*	*Scirpus olneyi*
Syzygium grande	*Imperata cylindrica*	*Lepironia articulata*	*Nymphaea nouchali*	*Eleocharis dulcis*
Syzygium polyanthum	*Stenochlaena palustris*	*Acacia auriculiformis*	*Elodea* sp.	*Somatina purpurascens*
Caryota mitis	*Gleichenia linearis*	*Acacia mangium*	*Phragmites karka* reeds	*Phragmites australis*
Delonix regia	*Salvinia molesta*	*Macaranga tanarius*	*Typha angustifolia*	*Nymphaea odorata*
Fragraea fragrans		*Peltophorum pterocarpum*		*Myriophyllum spicatum*
Imperata cylindrical		*Cinnamomum iners*		*Scirpus sylvaticus*
Cynodon dactylon		*Melicope glabra*		*Schoenoplectus californicus*
Distichlis spicata		*Melastoma malabathricum*		*Schoenoplectus mucronatus*
Cinnamonum iners				*Scirpus maritimus*
Melicope glabra				*Sagittaria latifolia*
Melastoma malabathricum				*Hydrilla* sp.

Sources: Salari, A. et al., *Wetlands* 34(3), 565, 2014; Rajpar, M.N. and Zakaria, M., *Int. J. Zool.*, 2011, 1.

(i.e., worms, crustaceans, and molluscs), which serve as food resources for water birds, fish, amphibians, and reptiles (Rajpar and Zakaria 2011). A total of 100 bird species were recorded in PIW; 25 are water bird species, and 75 are terrestrial bird species. The total bird density at PIW is 83.92 ± 4.53 birds/ha. The terrestrial birds have the highest species diversity (i.e., Shannon's index, $N_1 = 20.83$), species richness (i.e., Margalef's index, $R_1 = 7.97$), and species evenness (i.e., McIntosh's index, $E = 0.73$). This indicates that the rehabilitated PIW is a bird habitat for both water and terrestrial birds (Rajpar and Zakaria 2010).

17.6   ISSUES AND RECENT DEVELOPMENT IN MINING REGULATIONS

17.6.1   LOOPHOLES IN THE EARLY MINING REGULATIONS

Though mining activities were subjected to numerous laws and regulations, loopholes existed that did not mandate that miners fulfil a land restoration obligation after mining activities ceased. As of 1985, the State Mining Enactment stated that (1) all mined areas should be filled and levelled to a maximum of five percent slope and (2) disturbed lands or slopes should be stabilized through revegetation and construction of "silt traps." A bond is payable by the miner as a guarantee of filling and levelling pits in mined sites. Because the cost for land filling and levelling was greater than their bond, miners could choose not to fulfil their obligation (Said 1985). Moreover, there was no provision for the restoration of topsoil to the depth that originally existed or for a future land use plan (Ang 1994) or for rehabilitation of mined lands after mining operations ended. All of these weaknesses are the result of the high cost of rehabilitating mined lands.

17.6.2   NATIONAL MINERAL POLICY

Mining rehabilitation is part of the nine thrusts of Malaysia's National Mineral Policy, which recognizes the need for research and development and encourages environmental stewardship. The environmental aspects of mine development are regulated by the Environmental Quality (Prescribed Activities) (Environmental Impact Assessment) Order 1987, which was an amendment to the Environmental Quality Act of 1974. The State Mineral Enactments require the following items before any large-scale mining project is approved:

1. Environmental Impact Assessment (EIA) report
2. Mine Rehabilitation Plan (MRP)
3. Annual contribution for the Mine Rehabilitation Fund/Common Rehabilitation Fund

For a large-scale mine, the miner is responsible for conducting rehabilitation work as the mining operations progress, not after closure.

The Minister in charge of mining is empowered by the Mineral Development Act (1994) to make regulations concerning environmental protection and safety. New rules and regulations have since been enacted, such as the Mineral Development (Operational Mining Scheme, Plans and Record Books) Regulations 2007 that deal with information needed in the proposed mining scheme. Regulation 3(i)(q) requires environmental protection measures including pollution control, monitoring, and contingency plans; Regulation 3(i)(s) states that progressive rehabilitation and post mine closure plans need to be addressed (Termidi et al. 2009).

17.6.3   RECENT GUIDELINES ON REDEVELOPMENT OF EX-TIN MINES

In 2009, the Department of Environment Malaysia has developed and published three series of contaminated land management and control guidelines that cover the following: (1) site screening levels (SSL) for contaminated land, (2) assessing and reporting contaminated sites, and (3) remediation on contaminated sites. The SSL was adopted from the United States Environmental Protection Agency (USEPA), and, at the time when this article was written, the Malaysian government was still working on its first standards for contaminated land (Department of Environment 2009a,b,c). In 2012, the Ministry of Housing and Local Government of Malaysia has published a planning guide on the redevelopment of brownfields in Malaysia (Federal Department of Town and Country Planning Peninsular Malaysia 2012). Ex-tin mines were classified in the A category brownfield that requires proper risk assessment and remediation measures. Two Malaysian governmental agencies (i.e., Solid Waste Management Companies of Malaysia and Department of Environment of Malaysia) will monitor and issue clearance for further development.

17.7 CONCLUSION

Since the collapse of the global tin market, tin mining has left unproductive and sometimes hazardous ex-tin mines. Natural attenuation has largely reduced the health risk associated with potential pollutants from tin tailings. Human intervention can accelerate this process for specific land use, as illustrated by the case studies reported in this chapter. The federal and state governments have conducted rehabilitation projects with crucial investment from private entities, and they have created successful models for future reference. More can be done, especially in making information about ex-tin mines available to the public. To make this happen, a database of the ex-tin mines must be compiled and constantly updated. This database is currently being developed by a federal government agency.

Despite the many changes in laws and regulations related to mining, Malaysian law still holds the land titleholders responsible for cleaning up land contamination. Information of prior land use is usually not disclosed to potential homebuyers by land developers. As a result, homebuyers may not be able to bear this responsibility if this rule is enforced. Malaysia has seen another boom in housing, and a large portion of the housing project is indeed built on ex-tin mines. It is time that the federal and state governments emend existing regulations and pass the remediation cost of the contaminated ex-tin mines on to the real polluters.

In sum, Malaysia is on a course of "solving" the derelict ex-tin mine problems by 2020. The problems may have been prevented if ample laws and enforcement were in place for ensuring proper closure of the mining projects. Malaysia still has plenty of exploitable mineral deposits, one of which is gold, with the associated proposed gold mining project at Bau, Sarawak. This latest mining project will test the effectiveness of Malaysian mining laws in safeguarding the wellbeing of the environment and public health.

REFERENCES

Abdul Rasid, F. 2008. Brownfields (ex-mining land) in Malaysia. National Institute of Valuation, Valuation and Property Services Department, Ministry of Finance, Kajang, Selangor, Malaysia.

Abdullah, K. 2004. Stormwater Management and Road Tunnel (SMART) a lateral approach to flood mitigation works. In Paper presented at the *International Conference on Bridge Engineering and Hydraulic Structures*, Selangor, Malaysia.

AEON Cheers Club. 2014. Bidor Forest Tree Diversity Planting Campaign. Last Modified September 14, 2014. Accessed June 24, 2016. http://aeoncheersclub.com/event-bidor-forest-tree.php.

Agromatic Corporation Pty Ltd. 2015. AutoPot™ Project—Slim River. Accessed June 20, 2016. http://autopot.com.au/slimriver.

Ahmad, A., and A. Sarah. 2014. Assessment of abandoned mine impacts on concentrations and distribution of heavy metals in surface sediments of catchments around Sungai Lembing Abandoned Tin Mine. *Iranica Journal of Energy & Environment* 5(4):453–460.

Ahmad, A., and A. Sarah. 2015. Human health risk assessment of heavy metals in fish species collected from catchments of former tin mining. *International Journal of Research Studies in Science, Engineering and Technology* 2(4):9–21.

Ang, L. H. 1994. Problems and prospects of afforestation on sandy tin tailings in Peninsular Malaysia. *Journal of Tropical Forest Science* 7(1):87–105.

Ang, L. H. n.d. Environmental and ecological benefits of growing acacias on problematic soils. Tin Tailings Afforestation Centre, FRIM, Bidor, Perak, Malaysia.

Ang, L. H., and T. B. Ang. 2000. Greening the tin tailing areas in Malaysia. In *Proceedings of the International Conference on Forestry and Forest Products Research*, S. Appanah, M. Y. Yusmah, W. J. Astinah, and K. C. Khoo (Eds.), pp. 195–205. Kepong, Malaysia: Forest Research Institute Malaysia.

Ang, L. H., and W. M. Ho. 2002. Afforestation of tin tailings in Malaysia. In Paper presented at the *12th International Soil Conservation Organization Conference*, Beijing, China.

Ang, L., and W. Ho. 2004. A demonstration project for afforestation of denuded tin tailings in Peninsular Malaysia. *Cuadernos de la Sociedad Española de Ciencias Forestales* 17:113–118.

Ang, L. H., W. M. Ho, and L. K. Tang. 2014a. A model of greened ex-tin mine a lowland biodiversity depository in Malaysia. *Journal of Wildlife and Parks* 29:61–67.

Ang, L. H., and L. T. Ng. 2000. Trace element concentration in mango (*Mangifera indica* L.), seedless guava (*Psidium guajava* L.) and papaya (*Carica papaya* L.) grown on agricultural and ex-mining lands of Bidor, Perak. *Pertanika Journal of Tropical Agricultural Science* 23(1):15–22.

Ang, S. L., D. Subramaniam, J. M. Lim, Y. K. M. Ng, and S. H. Yap. 2014b. Growing together with our communities. *Berita Sunway*, July–September 2014: 20–21.

Ang, L. H., L. K. Tang, W. M. Ho, H. T. Fui, and M. Ramli. 2011. Bio-accumulation of mercury, lead, arsenic and cadmium by pineapple grown as an agroforestry crop for ex-tin mines in peninsular Malaysia. *Acta Horticulturae* 902:313–318.

Ang, L. H., L. K. Tang, W. M. Ho, T. F. Hui, and G. W. Theseira. 2010. Phytoremediation of Cd and Pb by four tropical timber species grown on an ex-tin mine in peninsular Malaysia. *International Journal of Environmental, Chemical, Ecological, Geological and Geophysical Engineering* 4(2):70–74.

Ashraf, M. A., M. Maah, and I. Yusoff. 2011. Bioaccumulation of heavy metals in fish species collected from former tin mining catchment. *International Journal of Environmental Research* 6(1):209–218.

Ashraf, M. A., M. J. Maah, I. Yusoff, and M. M. Gharibreza. 2010. Heavy metals accumulation and tolerance in plants growing on ex-mining area, Bestari Jaya, Kuala Selangor, Peninsular Malaysia. In Paper presented at the *International Conference on Environmental Engineering and Applications*, Singapore.

Chan, Y. K. 1990. Tin mining land—An overview of the current situation in Peninsular Malaysia. In Paper presented at the *National Seminar on Ex-Mining Land and BRIS Soil: Prospects and Profit*, Serdang, Malaysia.

Cope, L. W. 2000. Malaysian tin. *Engineering and Mining Journal* 201(12):32.

Department of Environment. 2009a. Contaminated land management and control guidelines no. 1: Malaysian recommended site screening levels for contaminated land. Ministry of Natural Resource and Environment, Putrajaya, Malaysia. http://www.doe.gov.my/portalv1/wp-content/uploads/Contaminated-Land-Management-and-Control-Guidelines-No-1_Malaysian-Recommended-Site-Screening-Levels-for-Contaminated-Land.pdf. Accessed 26 May, 2016.

Department of Environment. 2009b. Contaminated land management and control guidelines no. 2: Assessing and reporting contaminated sites. Ministry of Natural Resource and Environment, Putrajaya, Malaysia. http://www.doe.gov.my/portalv1/wp-content/uploads/Contaminated-Land-Management-and-Control-Guidelines-No-2_Assessing-and-Reporting-Contaminated-Sites.pdf. Accessed 7 June, 2016.

Department of Environment. 2009c. Contaminated land management and control guidelines no. 3: Remediation of contaminated sites. Ministry of Natural Resource and Environment, Putrajaya, Malaysia. http://www.doe.gov.my/portalv1/wp-content/uploads/Contaminated-Land-Management-and-Control-Guidelines-No-3_Remediation-of-Contaminated-Sites.pdf. Accessed 7 June, 2016.

Department of Fisheries of Malaysia. 1990. Annual fisheries statistics. http://www.dof.gov.my/dof2/resources/user_1/UploadFile/Usahawan%20Perikanan/Sumber/1990/Foreword.pdf. Accessed 13 July, 2016.

FAO/WHO Codex Alimentarius. 2001. Maximum levels for cadmium in cereals, pulses and legumes. Joint FAO/WHO Standards. FAO/WHO, Rome, Italy.

Federal Department of Town and Country Planning Peninsular Malaysia. 2012. Garis Panduan Perancangan—Pengenalpastian Bagi Pembangunan Semula Kawasan Brownfield. Ministry of Urban Wellbeing, Housing and Local Government. http://www.mpn.gov.my/sites/default/files/gp023_-_garis_panduan_perancangan_pengenalpastian_bagi_pembangunan_semula_kawasan_brownfield.pdf. Accessed 20 March, 2016.

Forest Research Institute Malaysia. 2014. Volunteers plant 8,000 trees at SPF Bidor. Last Modified September 18, 2014. Accessed June 24, 2016. http://www.frim.gov.my/volunteers-plant-8000-trees-at-spf-bidor/?wppa-album=113&wppa-photo=430&wppa-cover=0&wppa-occur=1.

Haron, S., M. R. Mohd Noor, W. A. Wan Yusoff, and R. Md. Yon. 2015. Returning agricultural productivity to former tin mining land in Peninsular Malaysia. In *Living Land*, J. Griffiths (Ed.), pp. 166–170. Leicester, U.K.: United Nations Convention to Combat Desertification.

Ho, S. C. 1998. The role of Paya Indah: The Malaysian wetland sanctuary in raising public awareness and understanding of wetlands and their conservation through its WATER programme. In Paper presented at the *International Workshop on Wetland Conservation and Management: The Role of Research and Education in Enhancing Public Awareness*, Penang, Malaysia.

Howard, S. June 1937. Hong fatt mine. *Wonders of World Engineering* 29:535–541.

Ibrahim, Y. 2010. Sustainable reuse of ex-mining land. *Malaysian Townplan Journal* 7(1):13–21.

Jaafar, R., I. Omar, A. Jidon, B. Wan-Khamizar, B. Siti-Aishah, and S. Sharifah-Noor-Akmal. 1993. Skin cancer caused by chronic arsenical poisoning—A report of three cases. *The Medical Journal of Malaysia* 48(1):86–92.

Jabatan Perancangan Bandar dan Desa Selangor. 2015. Laporan Tinjauan—Kajian Rancangan Struktur Negeri Selangor 2035. Selangor State Government B3: Pertanian, perhutanan & perlombongan. http://www.jpbdselangor.gov.my/Laporan/RSN_Selangor/laporan-tinjauan/B3.0-Pertanian-Perhutanan-Perlombongan.pdf. Accessed 20 March, 2016.

Japan International Cooperation Agency (JICA). 1981. The reclamation project of ex-mining land for housing development and other purpose, Feasibility Study Report, I & II, Malaysian Government.

Jusop, S., N. Wan, N. Mokhtar, and S. Paramananthan. 1986. Morphology, mineralogy and chemistry of an ex-mining land in Ipoh, Perak. *Pertanika* 9(1):89–97.

Leong, E. S., S. Tan, and Y. P. Chang. 2012. Antioxidant properties and heavy metal content of lotus plant (*Nelumbo nucifera Gaertn*) grown in ex-tin mining pond near Kampar, Malaysia. *Food Science and Technology Research* 18(3):461–465.

Lim, K. H., L. Maene, G. G. Maesschalck, and W. H. W. Sulaiman. 1981. Reclamation of tin tailings for agriculture in Malaysia. Technical Bulletin Faculty of Agriculture. UPM, Serdang, Malaysia, p. 61.

Low, K. H., S. M. Zain, and M. R. Abas. 2011. Evaluation of metal concentrations in red tilapia (*Oreochromis* spp) from three sampling sites in Jelebu, Malaysia using principal component analysis. *Food Analytical Methods* 4(3):276–285.

Low, K. H., S. M. Zain, M. R. Abas, K. Md. Salleh, and Y. Y. Teo. 2015. Distribution and health risk assessment of trace metals in freshwater tilapia from three different aquaculture sites in Jelebu Region (Malaysia). *Food Chemistry* 177:390–396.

Maesschalk, G. G., and K. H. Lim. 1978. The effect of organic waste materials, fertilization and mulching on the physical properties and yield of mungbean on tin-tailings. In Paper presented at the *Workshop on Research and Activities of Soil Science Department*, Universiti Putra Malaysia, Serdang, Malaysia.

Majid, N. M., A. Hashim, and I. Abdol. 1994. Rehabilitation of ex-tin mining land by agroforestry practice. *Journal of Tropical Forest Science* 7(1):113–127.

Majid, N. A., N. Muhamad, B. K. Paudyal, and Z. Shebli. 1998. Survival and early growth of *Acacia mangium*, *Ceiba pentandra* and *Casuarina equisetifolia* on sandy tin tailings. *Pertanika Journal of Tropical Agricultural Science* 21(1):59–65.

Malaysian Food Act. 1983. Ministry of Health Malaysia, Regulation 38, Fourteenth Schedule.

Mansor, N., M. A. Z. M. Shukry, and H. Afif. 2010. Investigation of Pb dispersal and accumulation around untreated former tin mines in Perak, Malaysia. In Paper presented at the *Second International Conference on Chemical, Biological and Environmental Engineering*, Cairo, Egypt.

Markert, B. 1996. *Instrumental Element and Multi-Element Analysis of Plant Samples: Methods and Applications*. Chichester, U.K.: John Wiley & Sons.

Mender, K. 2004. Assessment of heavy metals in soils and tuber crops on ex-mining land of Southern Perak, Malaysia. MSc thesis (FP 2004 26), Faculty of Agriculture, Universiti Putra Malaysia, Selangor, Malaysia.

Mohd Salleh, S., and M. Sabtu. 2005. Asian experience in land reclamation and rural development with special reference to Malaysia. In Paper presented at the *International Workshop on Land Reclamation and Rural Development: Policies, Strategies and Practices*, Cairo, Egypt.

New Straits Times. 2010. Malaysian wetlands: Resurrection of Paya Indah. *New Straits Times*, July 16, 2010. http://www.nst.com.my/nst/articles/13mino/Article#ixzz0tq1FLD60. Accessed June 16, 2016.

Ng, W. L. 2011. Diversity studies of fish and shrimp species in disused tin-mining ponds of Kampar, Perak. Thesis, UTAR, Perak, Malaysia.

Oh, I. Y. 2015. Puchong: Former mining town's growing pains. *The Star Online*, Metro Focus, December 8, 2015, Accessed June 25, 2016. http://www.thestar.com.my/metro/focus/2015/12/08/former-mining-towns-growing-pains-from-floods-to-congestion-progress-in-puchong-has-not-come-without/.

Osman, R. M. 2013. Rate of development of ex-mining land in Peninsular Malaysia. In Paper presented at the *National Geoscience Conference 2013, Kinta Riverfront Hotel & Suites*, Ipoh, Malaysia.

Palaniappan, V. 1974. Ecology of tin tailings areas: Plant communities and their succession. *Journal of Applied Ecology* 11: 133–150.

Pariatamby, A. 2014. MSW management in Malaysia—Changes for sustainability. In *Municipal Solid Waste Management in Asia and the Pacific Islands: Challenges and Strategic Solutions*, A. Pariatamby and M. Tanaka (Eds.), pp. 195–232. Singapore, Singapore: Springer.

Perak State Parliment. 2004. Jawapan-jawapan kepada soalan-soalan bertulis mesyuarat kedua, penggal pertama, Dewan Negeri Yang Ke-Sebelas, Perak Darul Ridzuan. Perak State Parliment. http://epla.perak.gov.my/dewan/other/2004/KERTAS_MESYUARAT_BERTULIS_(JUN%202004).pdf. Accessed 23 February, 2016.

Radziah, O., and Z. H. Shamsuddin. 1990. Growth of *Sesbania rostrata* on different components of tin tailings. *Pertanika* 13(1):9–15.

Rajpar, M. N., and M. Zakaria. 2010. Density and diversity of water birds and terrestrial birds at Paya Indah Wetland Reserve, Selangor Peninsular Malaysia. *Journal of Biological Sciences* 10(7):658–666.

Rajpar, M. N., and M. Zakaria. 2011. Bird species abundance and their correlationship with microclimate and habitat variables at Natural Wetland Reserve, Peninsular Malaysia. *International Journal of Zoology* 2011:1–17.

Rajpar, M. N., and M. Zakaria. 2013. Avian density in different habitat types at Paya Indah Natural Wetland Reserve, Peninsular Malaysia. *Journal of Animal and Plant Sciences* 23(4):1019–1033.

Roberts, P. D. 1995. Radiometric measurements, soil and water sampling in tin mining areas of Malaysia. BGS technical report. British Geological Survey, Nottingham, U.K.

Romero, F. M., C. Canet, P. Alfonso, R. N. Zambrana, and N. Soto. 2014. The role of cassiterite controlling arsenic mobility in an abandoned stanniferous tailings impoundment at Llallagua, Bolivia. *Science of the Total Environment* 481:100–107.

Saat, A., N. M. Isak, Z. Hamzah, and A. K. Wood. 2014. Study of radionuclides linkages between fish, water and sediment in former tin mining lake in Kampung Gajah, Perak, Malaysia. *Malaysian Journal of Analytical Sciences* 18(1):170–177.

Said, I. 1985. Development of rehabilitation techniques to reclaim tin-mined lands for low-cost housing in Malaysia. Masters thesis, Department of Landscape Architecture, Kansas State University, Manhattan, KS.

Salari, A., M. Zakaria, C. C. Nielsen, and M. S. Boyce. 2014. Quantifying tropical wetlands using field surveys, spatial statistics and remote sensing. *Wetlands* 34(3):565–574.

Sykes, J. P., and P. Kettle. 2013. Tin Mine Cost, Greenfield Research Ltd, Curtin University & ITRI Ltd. Accessed June 16, 2016. http://www.slideshare.net/JohnSykes/tin-mine-costs-june-2013. Accessed 16 June, 2016.

Takaijudin, H., A. Shahir, and A. Hashim. 2012. The utilization of an abandoned mining pond as a retention pond in dealing with stormwater runoff. *Flood Recovery, Innovation and Response III* 159:27–37.

Tan, B. K., and I. Komoo. 1990. Urban geology: Case study of Kuala Lumpur, Malaysia. *Engineering Geology* 28(1):71–94.

Tan, S. L. 2000. Cassava breeding and agronomy research in Malaysia during the past 15 years. In Paper presented at the *Sixth Regional Workshop Cassava's Potential in Asia in the 21st Century: Present Situation and Future Research and Development Needs*, Ho Chi Minh City, Vietnam.

Tan, S. L., and S. K. Chan. 1995. Recent progress in cassava varietal improvement and agronomy research in Malaysia. In Paper presented at the *Fourth Regional Workshop on Cassava Breeding, Agronomy Research and Technology Transfer in Asia*, Trivandrum, Kerala, India.

Teh, G. 1981. The Tekka tin deposit, Perak, Peninsular Malaysia. *Geological Society of Malaysia, Bulletin* 14:101–118.

Teh, G. H., and K. K. Cheng. 2002. EPMA characterisation and geochemistry of cassiterites from the Kuala Lumpur area. *Bulletin Geological Society of Malaysia* 45:363–368.

Termidi, H., S. Aliman, and M. Mohd Lip. 2009. Sustainable development initiatives in the Malaysian Mineral Industry. In Paper presented at the *APEC Conference on Sustainable Development of Mining Sector in APEC*, Singapore.

The Sunday Daily. 2005. Paya Indah wetlands to reopen 2006. *The Sunday Daily*, March 28, 2005. Accessed June 16, 2016. http://www.thesundaily.my/node/178976.

Tompkins, D. S. 2003. Sandy tin tailings in Malaysia: Characterization and rehabilitation. PhD thesis, University of Plymouth, Plymouth, England.

Tong, P. S. 2016. *Colocasia esculenta* (taro, yam, keladi) as a small farm crop in the Kinta Valley of Malaysia. *UTAR Agriculture Science Journal* 2(1):49–56.

Wan Abdullah, W. Y., M. H. Ghulam, and A. B. Othman. 1992. Physical and chemical properties of tin tailings in MARDI Station, Kundang, Selangor. MARDI, Kuala Lumpur, Malaysia.

Wan Asma, I., K. Wan Rasidah, A. Rosenani, H. Aminuddin, and A. Rozita. 2011. Effects of mulching and fertiliser on nutrient dynamics of sand tailings grown with *Acacia* hybrid seedlings. *Journal of Tropical Forest Science* 23: 440–452.

Wu, J. C. 1999. The Mineral Industry of Malaysia. United States Geological Survey. http://minerals.usgs.gov/minerals/pubs/country/1999/9318099.pdf. Accessed 19 February, 2016.

Wu, J. C. 2001. The mineral industry of Malaysia. In *2001—USGS Mineral Resources Program*, USGS (Ed.). Reston, VA: USGS.

Yap, K. M. 2007. Tin Mining in Malaysia. *Jurutera* 12:12–18.

Yasir, M. S., A. Ab Majid, and R. Yahaya. 2005. Heavy metals migration beneath unlined waste disposal site at Dengkil, Selangor, Malaysia. In Paper presented at the *Sixth ITB-UKM Joint Seminar on Chemistry*, Denpasar, Indonesia.

Yasir, M. S., N. Ahmad Kabir, R. Yahaya, and A. A. Majid. 2008. Kandungan logam berat dan radionuclide tabii di dalam ikan, air, tumbuhan dan sedimen di bekas lombong. *The Malaysia Journal Analytical Science* 12(1):172–178.

Yasir, M. S., I. Bahari, A. Sahibin, D. S. A. Rahim, and H. A. Rahman. 2001. The concentration of natural radionuclides and heavy metals in soils from tin mining and its surrounding area. *Jurnal Sains Nuklear Malaysia* 19:50–56.

Yeap, E. B. 2000. The prospects for hardrock gold and tin deposits in Malaysia. In Paper presented at the *Annual Geological Conference 2000*, Shangri-La Hotel, Penang, Malaysia.

18 Mine Rehabilitation in New Zealand
Overview and Case Studies

Robyn C. Simcock and Craig W. Ross

CONTENTS

18.1 History of NZ Mine Rehabilitation, Legislation, and Overview.................................... 336
    18.1.1 Mining before 1971.. 337
    18.1.2 Mining since 1971.. 338
    18.1.3 Mining Public Conservation Lands... 338
18.2 Mine Rehabilitation to Pastoral Land Uses.. 340
    18.2.1 Establishing a Suitable Topography.. 340
    18.2.2 Selecting Surrogate Soils 340
    18.2.3 Establishing a Dense Vegetative Cover 342
    18.2.4 Maintaining Rehabilitated Pasture 343
18.3 Mine Rehabilitation to Native Ecosystems 344
    18.3.1 Selecting Root Zones and Plant Species 344
    18.3.2 Seeding Techniques... 348
    18.3.3 Natural Regeneration 349
    18.3.4 Direct Transfer of Wildings and Salvaged Sods... 350
    18.3.5 Maintenance 354
18.4 Conclusion.. 354
References... 357

New Zealand has relatively abundant minerals and energy deposits such as coal, natural gas, and petroleum. Mining currently focuses mainly on alluvial, epithermal, and orogenic gold deposits, lignite, subbituminous, and bituminous coal deposits, and Fe-rich mineral sand along with aggregates for construction (Pope and Craw 2015). Rehabilitation of mined land is required by national statutes, because as NZ's Parliamentary Commissioner for the Environment (2010) noted, environmental offsets, or enhancing the value of one part of Public Conservation Lands to compensate for damage to another, requires great care. Effective rehabilitation is fundamental to reducing the extent, duration, and degree of such damage. It is also fundamental to mining's "social license to operate" in New Zealand (Pope and Craw 2015), as in other countries (Tibbett 2015).

New Zealand trades heavily on "clean, green," "100% pure," and "100% natural" brands to support its land-based export industries and tourism (Frohlick and Johnston 2011). Tourism activities and mining both use New Zealand's public conservation lands, which are primarily managed to protect biodiversity and cover nearly one third of the country, some 8.5 million ha in 2012 (Parkes et al. 2017). In the year ending March 2015, mining contributed $NZ 3.6 billion (1.8% of GDP), while tourism directly contributed 4.9% to GDP. Total tourism expenditure was $NZ 29.8 billion, of which about one third was from international tourists (www.stats.govt.nz). A University of Victoria economic analysis of minerals exploration highlighted the need for New Zealand mining to avoid creating negative economic spillover effects to this dominant tourism industry (Bertram 2011).

Unlike Australia, "remote" New Zealand mines are still only a few hours drive from major cities, and some of the largest operations are literally on the back doorsteps of rural towns (e.g., Waihi and Huntly).

This chapter provides an overview of mining history and legislation as it relates to rehabilitation activities, divided into pre-1971, 1971–1991, and post-1991 periods. However, the focus of this chapter is New Zealand rehabilitation practice as it has been informed by case studies involving published research. Findings and practices for the two most common rehabilitated land uses, these being pastoral land and native ecosystems, are discussed separately. Nearly half of New Zealand's land mass is still dominated by native ecosystems, with the majority of "nonnative" ecosystems being grazed pastures based on European species. Rehabilitation practices for such farmland are largely consistent with those in international literature for areas with relatively low moisture stress. Effective practices were developed in the 1970s and 1980s as part of permitting large mines and have changed little since. However, mine rehabilitation practices and research on New Zealand native ecosystems only started in the 1990s, and are still relatively undeveloped, particularly when compared with the extensive literature and experience in Australia (e.g., Koch 2015; Spain et al. 2015), Europe, and North America.

It is difficult to translate specific research from Australia (or Europe or North America) to New Zealand as tens of millions of years of isolation and the absence of mammals created a diversity of ancient and unique plants, fungi and animals, and ecosystems. New Zealand's 839 islands typically have a moist, temperate, maritime climate that supported evergreen, long-lived forests with little resilience to fire. Shortly after Polynesian settlement in the thirteenth century, repeated burning reduced the original closed forest cover by nearly 40% in one of the most rapid and complete landscape transformations anywhere in the world (McWethy et al. 2010). Forest soils are typically shallow and acidic, with low available phosphorus and moderate to high organic matter. Forests tend to regenerate from long-lived seedlings rather than seed banks. Forest structures typically resemble rainforest, with epiphytes and multiple layers, in contrast to their northern hemisphere latitudinal equivalents. A relatively high proportion of New Zealand's native bird and insect species are ground dwelling, flightless, unusually large, slow moving, and long lived. Many birds are "utterly vulnerable" to introduced mammals (Young 2004), as are many large lizards, insects, and plants. Combined with typically slow breeding rates, this means many species are highly vulnerable to predation-induced population collapse. Dominant plants, from trees to tussocks, may also propagate irregularly, in "mast" seasons.

Rats arrived in New Zealand with Polynesian settlers. Europeans introduced two more rat species, mustelids, and Australian possums; a total of 29 warm-blooded species were introduced over a period of about 150 years ending with chamois in 1907 (Parkes et al. 2017). Rehabilitation of native New Zealand ecosystems therefore necessarily requires killing to reduce predation and browse (Brown et al. 2015; Parliamentary Commissioner for the Environment 2011); rehabilitation also requires identification and control of a naturalized weed flora that exceeds the number of native species. Consequently, New Zealand conducts large-scale, multiple-pest mammal control (Brown et al. 2015) and biocontrol of weeds programs. New Zealand citizens, philanthropists, NGOs, and corporate businesses are increasingly enthusiastic to be involved with and fund pest management, with the central government adopting a "Predator Free vision" in 2016 (Parkes et al. 2017).

18.1 HISTORY OF NZ MINE REHABILITATION, LEGISLATION, AND OVERVIEW

Mining in New Zealand began with extraction of manganese from Waiheke Island in 1841 (Isdale 1981). In 1861, a 50-year gold rush era was an important catalyst in the exploration and settlement of Coromandel and more remote areas of New Zealand, such as Otago and Westland (Bagley 1980; May 1962; Weston 1991). Revenues from gold were a major source of income for New Zealand during this era, providing funds for development of early agriculture and industry

(Taylor and Walker 1987). By 1870, all of New Zealand's main gold and coal fields had been discovered (Nathan 2006). Large reserves of black iron sands were also identified, but were not profitable to smelt to produce iron and steel until technological developments in the 1960s.

18.1.1 MINING BEFORE 1971

Little rehabilitation occurred before 1971 as the Mining Act 1926 allowed payment of a levy to the government in lieu of rehabilitation; many mining operations left unsightly, unproductive land. Most such "legacy" areas were small; the largest of these were "herringbone" piles of boulder dredge tailings and water-sluiced river terraces left by alluvial gold miners in Westland and Otago (Simcock et al. 2004). The remnants of some of the water races that delivered water to hydraulic hoses and incised channels are now tourist attractions in these regions; both the races and access roads are of particular value as well-graded cycling and walking tracks. Some historic placer deposits in central Otago have naturally developed distinctive, relatively rare ecosystems and are now managed as reserves to protect the unusual flora and their saline substrates (Druzbicka et al. 2015). In Westland, larger areas of herringbone tailings were recontoured, fertilized, and underplanted with introduced legumes in the 1970s to regain productivity by growing *Pinus radiata* plantations (Fitzgerald 1987).

The most serious and long-term environmental impact of early mining in New Zealand is streams contaminated by water issuing from adits, tailings, or waste rock stacks with pyritic and arseno-pyritic rocks. In the South Island, such sites are concentrated in the Reefton and north Otago gold fields (Craw et al. 2000; Haffert 2009; Malloch et al. 2015) or coal mines associated with Brunner coal measures in the Buller Coal Field and near the Paparoa National Park (Pope et al. 2010; Trumm 2007). Many of these sites are now covered with dense forest and largely hidden from public view, although some streams remain severely impacted (e.g., Cannel Creek) (Trumm and Cavanagh 2006). In the North Island, legacy sites include historic gold mines on the Coromandel such as the Maratoto and Tui Mines. The latter, on the flanks of Mount Te Aroha, is a base metal underground mine abandoned in 1973. A $NZ 21.7M remediation program was largely completed in 2013 (Figure 18.1). It focused on chemical and physical stabilization of about 4 ha and 90,000 m³ of tailings and waste rock together with treatment of adit discharges to reduce loads of metal contaminants entering the Tui and Tunakahoia Streams (Environment Waikato, n.d.).

FIGURE 18.1 Legacy Tui Mine tailings dam and processing plant foundations after stage 1 rehabilitation surrounded by native forest, Te Aroha maunga (mountain).

18.1.2 MINING SINCE 1971

In 1971, two key acts provided the first legislative powers allowing conditions to be attached to a mining license, including restoration or rehabilitation. Under the Mining Act 1971 and Coal Mines Act 1979, the primary responsibility for setting conditions lay with the former Ministry of Energy (now part of the Ministry of Business, Innovation and Employment), operating through the Mines/Coal Mines Inspectorate. However, additional conditions were imposed by the appropriate land-administrating department, for example, the NZ Forest Service or Department of Lands and Survey before 1987 and the Department of Conservation after 1987. Catchment Boards, until their envelopment into Regional Councils, were also involved in imposing license conditions.

Licenses had a general set of imposed conditions. These conditions and associated bonds drove the process of land rehabilitation. A common condition was to maintain land use capability after mining. This requirement triggered rehabilitation research to develop guidance for rehabilitation, particularly for large mines permitted during the 1980s within (valuable) grazed pastures. Research usually focused on reclaimed soil profiles, particularly whether and how much topsoil was necessary to maintain land use capability and the potential for mine overburden and/or tailings to act as surrogate soils or subsoils under specific fertilizer regimes.

When environmental legislation changed again in 1991, mining operations permitted before 1991 maintained conditions set under the old legislation. Hence all, or parts, of the largest opencast mines currently operating in New Zealand operate under pre-1991 conditions, including the largest coal mine (Stockton, started opencast in 1944); Huntly coal fields (including Rotowaro, started in 1978); and the two most productive gold mines, Martha Mine (Waikato, started in 1988) and Macraes Flat (Otago, started in 1990). A 2009 report on Stockton Mine explains the complexity of the regulatory approach and enforcement for such mines (Parliamentary Commissioner for the Environment 2009); about 1400 ha of land at Stockton was covered by at least 55 old mining licenses.

Newer mines operate under conditions set under the Resource Management Act 1991 (RMA 1991) and Crown Minerals Act 1991. The RMA 1991 was designed to provide for sustainable management of natural and physical resources, stating all persons have a duty to "avoid, remedy, or mitigate" adverse effects on the environment. This underpins current requirements to rehabilitate mined land (Gregg et al. 1998) as, although mineral extraction is excluded from the sustainability provision, all other parts of the Act apply (Nathan 2006). Discharge and land use consents granted under the RMA 1991 are typically region and/or site specific as they are imposed by district councils (responsible for noise, light, and land use) and regional councils (responsible for discharges to air and water, water takes, and erosion control). There are no national standards for land rehabilitation. Within a region, conditions placed on smaller mines may be relatively generic, depending on the intended use of rehabilitated land, but those for larger mines are typically highly complex and site specific.

18.1.3 MINING PUBLIC CONSERVATION LANDS

New Zealand's 8.5 million ha of conservation lands are a major crown asset on many levels. In May 2010, tens of thousands of New Zealanders protested government plans to boost the economy by opening an additional c. 7000 ha of the 3.4 million ha of conservation land closed to mine exploration (referred to as "Schedule 4" lands under the Crown Minerals Act 1991). The areas included parts of Paparoa National Park, Great Barrier Island, and the Coromandel Peninsula. More than 37,000 written submissions were received by the Ministry of Economic Development on the proposal. This was an overwhelming display of public sentiment for a country of just 4.5 million people (Nippert 2011; Rudzitis and Bird 2011). The Parliamentary Commissioner for the Environment, an independent role, also criticized the proposal (Parliamentary Commissioner for the Environment 2010). The government backed down. However, mining is permitted on 5.1 million ha (not in Schedule 4) administered by the Department of Conservation under the Conservation Act 1987 if an Access

Arrangement is agreed. The Parliamentary Commissioner for the Environment considers mining in these areas has a privileged permitting position compared with other concessions (Parliamentary Commissioner for the Environment 2010) as the removal of ecosystems over large areas, other than for mining, would generally be considered contrary to the Conservation Act. Other commercial activities require a concession issued under the Conservation Act that can only be granted if the activity is not contrary to the provisions of the Act or the purposes for which the land concerned is held. The purposes for which most conservation land is held include protecting natural and cultural heritage, retaining areas of wilderness, and enabling recreational opportunities (Parliamentary Commissioner for the Environment 2010).

Any mines or prospecting activities that may affect most indigenous birds, all indigenous mammals (being two bat species and marine animals), reptiles, amphibians, and specific invertebrates also require Wildlife Permits issued under Wildlife Act 1953. Permits may also specify mitigation and management actions. There is no Act in New Zealand equivalent to the "Endangered Species Act" in the United States. Mining has been permitted in areas with acutely threatened plant and invertebrate species, perhaps most notably the endemic, carnivorous *Powelliphanta* land snails (Warne and Morris 2012). A feature of the *Powelliphanta* genus is the large number of species and subspecies in relatively small geographic areas, based largely on shell morphology (Buckley et al. 2014).

The majority of the approximately 15 opencast coal and 70 alluvial gold mines with active Access Arrangements on Public Conservation Lands in 2016 are in Westland, South Island. These include operations in "ecological areas" created by (former) NZ Forest Service to be specifically set aside for their ecological values as representative examples of ecosystems (Parliamentary Commissioner for the Environment 1993). Most of these operations are relatively small, 5–10 ha, and the extraction short-lived, compared with the 100–1000 ha operations of the aforementioned coal and hard rock gold mines in other areas of New Zealand. The Department of Conservation sets land access, compensation, rehabilitation, and bond requirements. The latter are forfeited if rehabilitation requirements are not met. Typical conditions require compensation for residual damage and rehabilitation and rehabilitated land "to be stable and free from erosion." This allows for natural regeneration, often through an intermediate period in woody weeds for smaller sites. Compensation may include monetary compensation based on the area and value of disturbed vegetation (and may include a biodiversity offset approach) and for "intrusion of an industrial operation." Compensation may also include specific biodiversity and/or historic mine site enhancements on conservation land.

These revenues from mining in conservation lands are mainly used to help protect New Zealand's unique biodiversity against the preeminent threats of pest plants and animals. Smaller mines contribute to a mining compensation fund used for work that improves biodiversity conservation outcomes, particularly outcomes that can achieve multiyear benefits without an ongoing funding commitment. From 2005/6 to 2015/16, the mining compensation fund supported 48 projects totaling $NZ 2.5M, of which the majority targeted plant pests, for example, killing willow trees invading Lake Brunner wetlands. The Mining Compensation Fund has also had a role in assisting with sudden needs, such as in 2014/15 when "mast" (abundant) seeding of beech trees triggered an increased requirement for pest mammal control as part of the "Battle for our Birds" program (Department of Conservation 2016b). This was an extensive program, much broader and larger than the environmental programs within individual mining permit regions that were recipients of past funds.

The largest mining compensation package to date in New Zealand was the May 2013 deed of agreement for access to conservation land allowing Buller Coal Limited access to an estimated 3 million tonnes of coking coal on the Denniston Plateau (Anon 2013). This compensation was specifically targeted to achieve biodiversity benefits through pest animal and weed control in the Heaphy and Denniston areas to the value of $NZ 21M. The Denniston program was to last 50 years and covers a minimum 500 ha, of which at least 200 ha was within the range of the endemic land snail *Powelliphanta patrikensis*. A further $589,000 was tagged for projects enhancing recreational and tourism values of nearby built mine heritage.

18.2   MINE REHABILITATION TO PASTORAL LAND USES

Many of New Zealand's largest mines are in agricultural areas created by felling native forests during the nineteenth and early twentieth centuries. These mines include titano-magnitite extraction from sand dunes with weakly developed soils and opencast coal mining in the older, ash-mantled Waikato, North Island, and loess- and alluvium-covered areas within Otago and Southland, near the bottom of the South Island. A government proposal to develop large lignite deposits for a liquid fuels plant in Southland in the 1980s initiated the most detailed New Zealand research into pasture rehabilitation (Ross and Widdowson 1985, 1987). This was followed by research on pasture after alluvial mining in Nelson (McQueen 1983), Westland (Parker 1991; Ross 1988), and Manawatu (Simcock 1993) regions. Rehabilitation trials were also associated with consenting of large epithermal gold mines (Martha and Golden Cross Mines) at the base of the Coromandel Peninsula, North Island (Gregg et al. 1990, 2003; Gregg and Stewart 1991), and mesothermal gold Macraes Mine in north Otago (Cossens and Keating 1990).

Despite New Zealand generally having much thinner topsoils (typically just 10–20 cm) than Europe or North America, research outcomes are consistent with international literature, particularly with areas having relatively low seasonal moisture stress and temperate climates. New Zealand research identifies four key influences on success of mine rehabilitation to pasture for grazing: creating a suitable topography; creating a chemically and physically favorable surrogate soil profile; rapidly establishing a dense vegetative cover to protect surfaces from erosion; and maintenance of vegetation density to rebuild organic matter and minimize weed competition. The research that underpins each of these findings is discussed below.

18.2.1   Establishing a Suitable Topography

Establishing slopes gentle enough to resist water erosion and retain adequate moisture yet steep enough to provide for adequate drainage is site and overburden specific. For example, Connolly et al. (1981) reported amelioration of excessive drainage on parts of a dredged site with 900 mm rainfall by reducing slope angles. Gregg et al. (2003) identified slope as the driver for an average 20% lower pasture production (measured over 11 years) between rehabilitated embankments (16°–18°) and the unmined control area (at about 5°). However, in most years the reduction was less than predicted using New Zealand models of hill country pasture productivity of a 340 kg DM/ha reduction with every degree of slope increase over 10° (Lambert et al. 1983). Conversely, inadequate slope combined with shallow subsoils and relatively low-permeability, clay-rich overburden substantially increased tree mortality at a Waikato mine.

18.2.2   Selecting Surrogate Soils

In nearly all cases in New Zealand, reconstructed soil profiles that use salvaged topsoil have the highest productivity in the short and medium term, the greatest resilience to climate variation and maintenance, and require the least ongoing inputs. Surrogate subsoils selected from overburden or tailings are widely used to create the lower parts of reconstructed "soil" profiles, particularly in wetter climates and where natural subsoils are poorly drained. The risk of adverse outcomes is inevitably increased where topsoils and subsoils are mixed together, or topsoils are mixed with overburden, as the dilution of organic matter damages the physical structure of the topsoil and increases nitrogen requirements (Mew and Ross 1991; Ross and Widdowson 1985). Surrogate subsoils are used at many mines, rather than reusing original subsoil layers. At larger mines, surrogate soils are based on site-specific glasshouse pot trials using waste rock overburdens and tailings to assess chemical, fertilizer, and lime requirements. Subsequent field trials are critical to inform physical limitations of slope, hydrology, compaction, erosion, crusting, and layering. Adequate drainage to achieve root

FIGURE 18.2 Rehabilitated pasture on the lower levels of the tailings storage waste rock embankment of the Martha Mine, Waihi.

oxygen supply is balanced against water supply and root zone depth to sustain plant growth through seasonal droughts.

At Martha Mine, Waihi, the original volcanic ash subsoil was retained under the tailings dam to assist attenuation of any leachate (Gregg et al. 1995, 2003). Extensive trials identified the alkaline tailings had few limitations to plant growth and supported high rates of pasture growth (e.g., 11–14 T dry matter/ha in the first year), with or without amendment of phosphate(P)-supplying compost (Mason 1996). In contrast, lime applications were needed to raise the pH of nonacid-forming overburden above 5.5 (Gregg et al. 1995). Under the average 2300 mm rainfall, pasture rehabilitation on the 16°–18° slopes of the tailings embankments was successful using a minimum 100 mm of stockpiled topsoils placed on about 500 mm of nonacid-forming overburden (surrogate subsoils) amended with lime and 1400 kg/ha potassic superphosphate (Figure 18.2). If the surrogate subsoil was not limed, pasture yields were initially suppressed but recovered over several years. This recovery was attributed to initial phytotoxic effects of soluble aluminum being mitigated over time as organic leachates from topsoil complexed aluminum (Gurung et al. 1996).

Not enough topsoil was salvaged at Rotowaro coal mine to provide a complete cover over rehabilitated areas. This led to a "lack of nutrients and organic matter in the (rehabilitated) soils and poor physical properties" (Solid Energy 2010). Pastures in areas rehabilitated with subsoils and basal inorganic fertilizers were also vulnerable to invasion by the windblown weed pampas (*Cortaderia selloana* and *C. jubata*) when grasses exhausted the limited nitrogen (N) and available P; the natural volcanic ash subsoils are high in aluminum and iron oxides which sorb phosphate. Additions of municipal biosolids markedly improved pasture production, whether incorporated into ash to about 300 mm depth or spread over the surface. Incorporating 10-year-old biosolids at 125 dry T/ha provided about 930 kg N/ha and 625 kg P/ha, with beneficial effects on pasture biomass lasting at least 8 years. Significant improvement in soil physical properties was only related to the highest biosolids application rate of 400 dry T/ha.

Both topsoils and subsoils were largely absent at the abandoned Tui Mine. Tailings with high lead, cadmium, mercury, and sulfide minerals that continually oxidized created low pH conditions and mobilization of these heavy metals into two streams. After 30 years, no vegetation had established on the tailings. Pot and field trials identified high rates of lime and/or sewage sludge amendments overcome a surface pH of 2.8–3.9, allowing revegetation. Metal-tolerant varieties of fescue (*Festuca rubra*) and browntop (*Agrostis capillaris*) outperformed their nonmetal-tolerant counterparts by as much as 4 and 10 times, respectively (Morrell 1997). Professor RR Brooks pioneered such phytoremediation research in New Zealand (Robinson and Anderson 2007). Pot trials were followed by a successful field trial using four native species, which showed low levels of bioaccumulation into leaves. Fieldwork identified the importance of avoiding exposure of shallow, largely unweathered tailings present below 200 mm (Morrell 1997).

King (2008) conducted similar trials on tailings from Macraes hard rock gold mine, where a considerably drier, colder environment with short growing season and desiccating winds provides additional challenges. Although rye corn could be established directly into fertilizer-amended tailings, plant uptake of arsenic (As) exceeding 50 mg/kg meant topsoil capping was needed to reduce As uptake. Mosses, grasses, and native vegetation have naturally established on historic processed wastes of the Reefton gold fields with very high As (Malloch et al. 2015); the potential impact on native invertebrates and animals in these ecosystems is beginning to be quantified. In contrast, some rehabilitated pastures have benefited from micronutrient additions to overcome localized deficiencies in animal health. For example, Connolly et al. (1981) report copper and molybdenum are required amendments for areas mined for iron sands in the North Island.

In line with international experience, soil structure typically deteriorates under stripping and storage, with fine-textured soils being particularly vulnerable, hence minimizing or ameliorating compaction is often a critical feature of rehabilitation. Managing the transition between layers of overburden, subsoil, and topsoil is also important to ensure adequate rooting depth and water supply, or adequate drainage. Sharp textural interfaces between layers are often detrimental. Keating (1988) reported sharp textural boundaries between fine soil and underlying coarse tailings exacerbated inadequate drainage in gold dredged areas in Westland. Ripping or subsoiling can be effective at breaking up such layers, alleviating structural degradation, and improving drainage. Ripping applied to rehabilitated loess soils in Southland was effective at increasing hydraulic conductivity (a measure of drainage). The effect, which lasted for 5 years, improved soil physical conditions and increased pasture yields (Ross and Orbell 1986). Horne et al. (1990) reported bearing strength and resistance to pugging were important considerations for root zone selection at Martha Mine, which was to be grazed by cattle. Their experience was supported by Morton and Harrison (1990) for alluvial dredge materials in Westland and by Longhurst and O'Conner (1990) in the Waikato coal fields.

Topsoil is widely recognized as the most valuable surface material to facilitate plant growth through its beneficial drainage, water and nutrient supply (and storage), and ability to resist compaction. However, in the late 1970s and the early 1980s, an unusual approach was developed by operators of large alluvial gold dredges in Westland frustrated with the cost of managing the aggressive weeds gorse and broom that sprouted from a long-lived seed bank in the replaced topsoil. The young silty topsoils with low organic matter also tended to form surface seals or crusts that increased vulnerability to rill and sheet erosion (Keating 1990). Topsoils were buried. Evenly spread rainfall and low seasonal moisture deficits allowed use of dredge fines, being coarse sands and fine gravels (less than about 13 mm diameter) to create a root zone over coarser dredged gravels and boulders. The dredge fines were compacted by heavy rollers to improve water-holding capacity. High rates of basal superphosphate fertilizers were incorporated to support legumes. These were followed by frequent applications of inorganic N fertilizers (typically urea) to support high-fertility grazed ryegrass (*Lolium perenne*) pastures. The high bearing strength of the dredge fines allowed larger, heavier cattle to be grazed with little damage (Morton and Harrison 1990).

18.2.3 ESTABLISHING A DENSE VEGETATIVE COVER

On most mine sites that rehabilitate to pasture, rapid development of a dense sward is desirable to suppress erosion. Mulches, tackifiers (sticking agents), and hydroseeding are not generally used in New Zealand, as a productive pasture can be quickly established given adequate basal rates of fertilizers and nonacid-forming overburden. Exceptions are erosive environments combined with erosion-prone materials low in organic matter, such as coastal iron sands and north Otago tailings. At such sites, barley, oats, and triticale (*Hordeum vulgare*, *Avena sativa*, *Triticosecale* hybrid) have been used to rapidly stabilize sands against wind erosion. Connolly et al. (1981) report such a crop being direct-drilled with pasture when the grain crop was 150–200 mm tall. Munro (1980) reports this technique generally returned land to agricultural use less than 6 months after completion of mining. King (2008) reported using oxidized schist to anchor seeds of rye corn sown directly into fertilizer-amended tailings at Macraes Mine. At Stockton Mine, newly rehabilitated slopes are vulnerable to erosion from the extreme rainfall, which is both high (over 5000 mm/year) and intense (>200 mm/day about once a month). Hydroseeding, straw mulching, rock mulching, and soil–plant sod transfer techniques have been developed at the site to minimize soil losses (Kingsbury and Cunningham 2007; Rodgers et al. 2011).

18.2.4 Maintaining Rehabilitated Pasture

The importance of post-establishment management to minimize physical damage to rehabilitated soil structure by maintaining post-grazing plant biomass and avoiding stock pugging damage was highlighted by Ross and Mew (1990). Ross and Widdowson (1985) showed ongoing (maintenance) applications of nitrogen fertilizers were important to maintain pasture production if topsoil was either absent or diluted with overburden. Such fertilizers are typically required until adequate organic matter levels are developed. New Zealand's climate is well known to be favorable for pasture growth. Gregg et al. (1995) reported rapid soil development at Martha Mine, Waihi. The total carbon content of mine tailings to 7.5 cm depth rose from 0.3% to 2.0% over 6 years under a maintenance regime using 60 kg inorganic N/ha annually. Trials indicated application of organic sources of nitrogen such as biosolids can greatly shorten both the carbon recovery time and avoid the need for regular nitrogen applications in the medium term. In contrast, 200 kg N/ha/year as inorganic N was applied on loess soils mixed with subsoil or overburden to raise productivity to undisturbed levels (Ross and Orbell 1986). However, the high inorganic N applications reduced clover abundance and earthworm numbers (Widdowson and McQueen 1990). These authors suggested that earthworm introduction would benefit restored soils if less than 2 worms/m² survived soil movement. Nonnative earthworms have been seeded to benefit soil recovery at Martha Mine. Gregg et al. (2003) reported earthworm densities on 5–11-year-old rehabilitated embankments were higher than in the control, but controls had greater worm densities than 3-year-old rehabilitation.

Gold and iron sand dredges have produced relatively large areas of rehabilitated land with higher pasture productivity and easier contour than the original farmland (Simcock and Ross 2014; Taylor and Walker 1987) (Figure 18.3). This allowed higher stock carrying capacity and more intensive

FIGURE 18.3 Rehabilitated pasture with a patch of remnant native forest on Westland river valley after gold dredging.

farm management. Such gains are most likely where pre-mining production is limited by poor drainage, low-fertility pasture species, or regular flooding. However, a wide variation in the quality of reclaimed pastoral land is reported for smaller operators, particularly where gorse reversion, inadequate drainage, or inadequate ongoing fertilization occurs (Mew and Ross 1991; Simcock and Ross 2014). This considerable success of pasture (and exotic forest) establishment after alluvial mining in Westland has been contrasted with the outcomes of indigenous rehabilitation (Mew et al. 1997).

18.3 MINE REHABILITATION TO NATIVE ECOSYSTEMS

Native forest rehabilitation before 1990 for both gold and coal mining was "virtually nonexistent" (Gregg et al. 1998). Only the largest mines had adequate baseline vegetation assessment or clear specification of vegetation outcomes expected after rehabilitation (Mew et al. 1997). Most sites were "essentially left to nature." As a consequence, a cover of weeds was common: gorse (*Ulex europeus*), broom (*Cytisus scoparius*), blackberry (*Rubus fruticosus*), and/or Himalayan honeysuckle (*Leycesteria formosa*) (Mew and Ross 1989). Little specific information was available to guide miners. Miners routinely discarded (buried) topsoils, along with all stumps and vegetation, following practices developed for establishing *Pinus radiata* plantations on old dredge tailings with gorse and broom seed banks. Burying topsoils reduced competition from these woody weeds.

A study of natural regeneration on old gold workings in Westland and Nelson concluded forest composition could regenerate to that approaching natural forest "within a hundred years or so" and it would take several hundred years for full species complement and natural structure of the forest to be achieved (Fitzgerald 1987). Fitzgerald considered the main barriers to satisfactory regeneration were fire, grazing, and absence of forest duff, topsoil, and fines. Changes in post-mining drainage conditions created by coarse, bouldery tailings at the surface, steep slopes, and/or lack of topsoil were also identified as key impediments by Mew et al. (1997).

In 1990, a research program was started to provide guidance for miners rehabilitating native forest. The research was tasked with identifying the most effective means of ensuring a return to tall native forest on Westland mines in as short a time as possible. The first research trials were established in 1990 at Giles Creek opencast coal mine, a site at which mining license conditions required restoration to indigenous rainforest (Gregg et al. 1998). These Giles Creek rehabilitation trials established performance of native forest seedlings on a variety of surrogate soil profiles, including raw overburdens.

Planting of native nursery-grown and wilding native seedlings is considered the lowest-risk approach to achieve predictable, short-term outcomes. However, there are two important caveats. First, the root zone must be favorable. Second, the effects of pest animal browsers, fire, and pest plants must be mitigated. Results from the Giles Creek trials highlighted the importance of suitable drainage and fertility; however, the site has no acidic overburdens. The importance of mitigating acid-generating materials in root zones was quantified over 10 years at Wangaloa coal mine, Otago, South Island. Wangaloa, together with the Golden Cross Mine in the North Island, were New Zealand's earliest "closed" mines with completed native rehabilitation projects. In this section, rehabilitation learnings at these mines are presented alongside those of mines such as Stockton that have many hundreds of ha still to revegetate. Revegetation methods explored included regeneration of seeds and propagules from mixed, salvaged topsoils that are not stockpiled, direct transfer of intact plant–soil sods, various seeding techniques, and natural regeneration from adjacent or remnant areas.

18.3.1 SELECTING ROOT ZONES AND PLANT SPECIES

The Giles Creek mine operates in lowland, evergreen beech forest on alluvial terraces that form a wildlife corridor. The temperate climate features mild, short moisture deficits under an annual rainfall of about 2900 mm. Dense silver and red beech forests (*Lophozonia menziesii* and *Fuscospora fusca*) at the site were cutover in the early 1980s for timber, but a forest structure remained with a shrub understorey (<5 m height) that contained seedlings of all canopy species and a diverse ground cover with an abundance of ferns. In general, soils are naturally free draining. An average 30 cm of organic and topsoil layers overlie alluvial gravels and coal seams associated with mudstones and sandstones. The bird-dispersed weed, Himalayan honeysuckle, formed a dominant cover up to 3 m high in disturbed areas, but gorse, broom, and blackberry were generally absent.

The first Giles Creek trial in 1991 confirmed the inadequacy of the 1980s "soilless" technique developed for dredge tailings. Nursery-grown seedlings were planted into fine, weathered, uncultivated, granitic gravels overburden. Despite herbicide sprays to control competing pasture and rush species, over the following 3 years, 7 of the 10 species had more than 50% mortality and slow growth (≤50 mm/year up to April 1994). This was attributed to poor drainage and low chemical fertility, particularly low nitrogen. Broadleaf (*Griselinea littoralis*) had growth rates exceeding 100 mm/year but was highly palatable; plants unprotected by fencing were severely browsed by hares and deer, as was karamu (*Coprosma robusta*). Transplants of "wilding" silver beech seedlings (up to 1 m tall) taken from adjacent forest fared almost as well as nursery stock (Langer et al. 1999). Kahikatea (*Dacrycarpus dacrydiodes*), totara (*Podocarpus totara*), koromiko (*Hebe salicifolia* syn. *Veronica salicifolia*), and karamu performed equally well (or poorly), whether established as nursery-raised or bare-rooted seedlings (Langer et al. 1999). Given the advantages of bare-root transplants—lower costs of production, transportation, and transplanting—it is perhaps surprising very few nurseries grow bare-rooted native seedlings.

This first trial led to investigations of the response of beech and broadleaf seedlings to fertilizer in glasshouse (pot) and fenced field trials (Langer et al. 1999). As is now common practice, fertilizers were placed in a spade slot at about 5–10 cm depth and 10–20 cm from the stem base to minimize stimulating growth of competing pasture weeds. After 18 months, karamu and kahikatea growth was about 25% greater for treatments containing nitrogen but not for treatments with only phosphorus. Red beech and marble leaf (*Carpodetus serratus*) responded more strongly to phosphorus. The pot trial showed the beneficial buffering effect provided by topsoil compared with gravel or siltstone overburdens. This buffering effect was greatest in the absence of nitrogen fertilizers. As shown internationally, organic soil horizons are important providers of nitrogen for plant growth. The pot trial also identified one overburden that supplied adequate P, highlighting the value of such screening tests for identification of surrogate subsoils, as they identify forms of plant-available P, such as calcium-bound P, not detected by standard agronomic tests. Foliar analyses in a linked field trial showed nitrogen (alone) was greater in plants grown in layered topsoil over subsoil compared with mixed soil horizons and more than double that of plants on bare overburden.

These Giles Creek results have been confirmed at other mine sites where nitrogen levels are very low, either due to absence of topsoil or to dilution of topsoil by mixing with overburden. For example, a series of unpublished trials at Stockton Mine from 1994 clarified N limitations influenced height and width of red tussock (*Chionochloa rubra*), koromiko, and rata (*Metrosideros umbellata*) grown in overburdens. However, at this highly exposed, mid-altitude site, the extent of aboveground shelter was also critical; 20%–40% positive growth responses to fertilizer were in the most sheltered areas. At Stockton, mitigating exposure by deliberately creating a rough "hummocky" surface had a greater impact on growth than applying fertilizer for koromiko and mountain beech (*Fuscospora cliffortioides*), but not for the more wind-tolerant monocotyledons flax (*Phormium cookianum*) and toetoe (*Austroderia toetoe*) (Theinhardt 2003). The Stockton trials also showed the danger of broadcast fertilization, as this technique caused the European weed *Juncus squarrosus* to grow vigorously, smothering smaller statured, slower growing native plants.

An early Giles Creek soil-layering trial quantified the value of separately stripping and replacing topsoil and subsoil horizons, compared with mixing these horizons (Figure 18.4). The former treatment aimed to restore soils as closely as possible to original conditions, the latter to produce a surrogate soil profile that was simpler to create (Mew et al. 1997). The trial was planted

FIGURE 18.4   Giles Creek Mine Rehabilitation Trial at 10 years old. Native seedlings were planted into replaced topsoil over subsoil (foreground) and into overburden gravels (background). (Photo taken in 2002.)

in September 1992. After 4.5 years, 6 of the 11 species had greater than 90% survival. However, inadequate drainage at this high rainfall, relatively flat site, meant that mixing topsoil and subsoil approximately halved growth rates of planted native seedlings (mean height growth 49 mm/year in mixed soil horizons treatment and 96 mm/year in the layered treatment). Tensiometer and oxygen diffusion measurements showed that the mixed soil treatment remained saturated and deficient in oxygen for long periods after rainfall (Jackson 1994). In contrast, the same depth of free-draining topsoil over subsoil to a total depth of 0.6 m, combined with cutoff drains, created adequate drainage for the native species. In other trials, ripping and surface contouring were used to enhance root zone drainage. In practice, the expense of subsoil replacement means topsoils are generally placed directly on adequately drained overburden.

The impact of acid-generating rocks in plant root zones on native plant establishment and regeneration has been researched at Wangaloa coal mine in southeast Otago, led by Professor Dave Craw and Dr Cathy Rufaut, Department of Geology at University of Otago. The Wangaloa Mine operated between the 1940s and 1989. Mining left a pit lake and 75 ha within a 252 ha site affected by sulfur- and boron-containing coal and near-coal overburdens including pyrites (FeS_2) (Begbie et al. 2007). From 2000 to 2010 the site was rehabilitated to a recreational area, mainly by planting native, nursery-grown seedlings. First, overburden and areas with naturally regenerated vegetation (mainly gorse) were bulldozed into more stable slopes. This created a highly variable rooting environment into which 50,000 native seedlings were planted in 2003 and 2004. A further 50,000 seedlings were planted over the following 7 years, many as replacement for dead plants (Rufaut et al. 2015). Natural revegetation and planted native seedlings growth was dense on Quaternary soils developed in 5 m of loess (Figure 18.5), with plantings assessed as self-sufficient and self-propagating after 7 years. However, areas of acidic (coal-rich) overburden were hostile to seedling establishment (Todd et al. 2009). Most hydroseeded grasses and planted seedlings in these areas died, and surviving plants grew slowly. By 2013, canopy cover of planted areas ranged between 50% and 100% on plots in loess and an average of 20% cover on waste rock where plants were largely established from seed (Rufaut et al. 2015). At both Wangaloa and Stockton, native nursery-grown plants successfully establish in substrates down to about pH 4. They are generally highly tolerant of soluble Al, particularly in the presence of organic matter, unlike many nonnative pastures (grasses and legumes).

FIGURE 18.5 Wangaloa Mine rehabilitation in areas with Quaternary soil substrate where native shrubs were planted into windrows of gorse, at about 2 years after planting. (Photo taken in 2006.)

Both Wangaloa and the more benign Giles Creek Mines also show the value and importance of measuring revegetation over at least 5 years, and preferably 10 years. At Wangaloa, plant species that had the highest initial survival rates were koromiko and the large grass, toetoe. However, growth rates of koromiko were about 100 mm/year (2003–2006), reaching about 450 mm after 3 years—a height at which plants are still vulnerable to many browsers (Todd et al. 2009). In contrast, survival of planted mānuka (*Leptospermum scoparium*) seedlings was low, but surviving plants reached over 1 m height after 3 years. The long-term Wangaloa research has also shown plants can have an important role in mitigating the effects of acid drainage. The pH of the mine pit lake, which receives most surface and groundwater runoff and has a water residence time of 1–2 years, increased from a consistent 4.6–4.8 before rehabilitation (1998–2002) to near pH 6 during rehabilitation. Begbie et al. (2007) considered these changes in lake water composition from year to year might be a result of increased input of rainwater that has had less interaction with acid substrate because of increasing vegetation cover. Neither finding was obvious for up to 10 years.

Favorable outcomes at Wangaloa sites with benign root zones mirror those of the more modern Golden Cross Mine where a similar number of nursery-raised seedlings were planted into replaced ash soils between 1998 and 2001, at a similar density (about 10,000 plants/ha). The 130 ha Golden Cross site is at the base of the Coromandel Peninsula. It has a mild climate with a long-term average 2900 m rainfall, more than double that of Wangaloa, although both sites have similar low levels of drought stress, as evapotranspiration is substantially lower at the cooler, more southern Wangaloa. Over 100,000 native shrubs and trees of over 60 species were planted on and around the Golden Cross site. Ross and Crequer (2006) reported most areas reached canopy closure after 6–8 years. The authors, an independent peer reviewer, and an Environmental Waikato compliance officer, considered in 2006 that the aim of leaving a self-sustaining environment consistent with the adjacent Forest Park had been largely achieved. Success has required ongoing pest control, exclusion of grazing animals, and removal of wilding radiata pine seedlings (Ross and Crequer 2006).

The Giles Creek soil-layering trial delivered an unexpected result. After 4.5 years, the two beech species were almost eliminated from both soil treatments, but more than 60% survived in the overburden gravel plots (Davis et al. 1997), and more than 80% survived in nearby mudstone overburden after 3 years (Langer et al. 1999). Davis et al. (1997) and Johnston et al. (2003) suggested mortality was due to the transfer of *Phytophthora cinnamomi* in topsoils from the surrounding forest. From 2002 to 2004, Dr Peter Johnston and Dr Ross Beever compared seedling disease development in soils with *P. cinnamomi* into which beech seedlings with/without mycorrhizas were placed, with a further treatment of a phosphorous acid-based, oomycete-specific fungicide. The fungicide has no impact on mycorrhizal fungi but kills *Phytophthora*.

All New Zealand beech species are obligately ectomycorrhizal. *P. cinnamomi* was almost certainly introduced to New Zealand by humans, possibly by early Polynesian settlers (Johnston et al. 2003), and now has a widespread, patchy distribution throughout New Zealand that includes habitats modified by people as well as remote, apparently undisturbed forest (Podger and Newhook 1971). Johnston and Beever attempted to manipulate mycorrhizas of beech seedlings in the nursery to increase seedling survival in restoration sites where *P. cinnamomi* is present. Although nursery trials were not successful, a field trial at Giles Creek showed the fungicide provided effective protection for the seedlings (Johnston et al. 2003). An initial fungicide application at planting might assist field establishment of nursery-raised beech seedlings by allowing time for seedlings to become colonized by protective mycorrhizal species from adjacent mature forest and/or leaf litter applied to the planting site. The trial showed ectomycorrhizal status (high or low) of beech seedlings had no effect; however, Johnston and Beever considered this might be due to changes in the structure of the ectomycorrhizal community caused by growing the seedlings (taken from natural forest) in shade houses and low organic material for 2–3 years. In practice, highly mycorrhizal wildings used in mine rehabilitation at the site exhibit low mortality, and the transplanting of seedlings, up to 3–4 m tall, has been successful over decades at Giles Creek.

18.3.2 SEEDING TECHNIQUES

Few New Zealand mines use "traditional" seeding treatments such as hydroseeding, broadcast seeding, brush layering (fascining), or mulch spreading. Direct seeding is underutilized in NZ mine sites mostly due to lack of available seed, because the native seed industry is very small and key species may only seed heavily in "mast" years. Native plants established from seed are also relatively slow to provide a dense, erosion-resistant cover compared with pasture and cereal species (Simcock and Ross 1997), and native seedling germination is reduced under compacted conditions (Bassett et al. 2005). However, native seeding provides two specific advantages compared with planting nursery-raised seedlings. First, seedlings develop root systems that are less prone to toppling and with an unhindered taproot (Douglas et al. 2007). Second, the seedlings can tolerate more hostile conditions, as reported at Wangaloa, where mānuka and kānuka (*Kunzea ericoides*) naturally established (from seeds blown from adjacent remnant vegetation) onto coal-rich and coal-bearing substrates too hostile for nursery-grown plants to survive (pH < 4 and/or high boron). The two species established from seed on areas with pH between 2.5 and 5 (Rufaut et al. 2015; Todd et al. 2009).

Hand-broadcasting of native seeds onto overburden gravels with some soils and scattered branches was trialed at Giles Creek in the spring of 1991. Karamu, koromiko, and mānuka were successful using seeding rates of 0.1 g seed/m^2 for karamu and 0.01 g/m^2 for the finer mānuka. Although mortality over the subsequent 2 years was generally high, after 5 years, about 15 seedlings/m^2 of karamu and mānuka survived. Seedlings that survived the first winter responded strongly to fertilizer (nitrogen and phosphorus), but fertilizer also stimulated the growth of competitive herbaceous species such as Yorkshire fog (*Holcus lantanus*), lotus, and Himalayan honeysuckle. Langer et al. (1999) concluded that broadcasting of seed had potential as a revegetation technique for sites with low weed competition. In mine sites within undisturbed native forests competition from groundcovers can be very low. However, nonnative grasses are sometimes seeded to control erosion. In such cases, success of native seeding is likely to be limited by intensive competition, as reported in trials seeding native species into farmland (Douglas et al. 2007).

In the mid-2000s, Solid Energy invested in large-scale collection of native seeds from areas of wild, rehabilitated, and purpose-planted seed and "stool bed" nurseries at Stockton Mine. Collection was driven by the need for millions of locally sourced (eco-sourced) native plants adapted to the low fertility, highly exposed site conditions to rehabilitate over 1000 ha. The most common and most successful species were mānuka, flax, koromiko, and olearia. Aerial seeding by helicopter was used for several years to enhance planted areas, but lack of early results and probable low seed viability contributed to the technique being halted (Kingsbury 2008).

Small trials at three Stockton sites in 1998 demonstrated the efficacy of fascining using branches of mānuka. This technique was more efficient at this wet, relatively cold site at 600–1000 m ASL, as mānuka capsules could take months to open and release seed, allowing storage of material. However, on exposed sites the branches needed to be held down with rocks, wire pegs, or netting, which increased the cost of the technique. Growth rates on most substrates were relatively slow, for example, mānuka reached 50–150 mm height after 3 years on mixed soil substrates. Such growth rates meant seedlings were vulnerable to smothering from *Juncus squarrosus* and grasses where fertility levels (N and P) were raised.

Extensive trials from 2001 to 2003 developed techniques for establishing mosses, lichens, and herbs on a variety of rock overburdens using hydro-mulching (Ross et al. 2003). This technique targeted sites too inhospitable or steep for planting seedlings, such as high walls. It was most successful for bare rocky surfaces with a high proportion of protected, stable microsites that faced south (i.e., lowest drought stress). Techniques were also developed for nursery propagation of the most successful mosses used in the mixtures, *Campylopus clavatus* and *Racomitrium pruinosum* (Buxton et al. 2005; Ross et al. 2003; Stanley et al. 2003). Moss cover was highest using higher application rates of moss (2 L/m²), adding fertilizers and soils, and covering mosses over the summer months. Propagation within confined areas helped keep source material free of weed grass and rush species.

18.3.3 Natural Regeneration

There is ongoing debate in New Zealand about the efficacy and benefits of natural regeneration of native forest through exotic species. Fitzgerald (1987) reported successful revegetation of beech forest after nineteenth-century alluvial gold mining. However, this occurred in conditions with much lower weed propagule pressure and pest animal pressure (deer and goats) than are currently present and in relatively small sites within relatively larger areas of more intact forest. Mew and Ross (1992) noted that open bare areas on mine sites present particularly difficult microclimatic conditions for the establishment of some species of native seedlings planted in a similar manner to exotic trees. Natural regeneration is hindered where a dense sward of exotic grasses is present as these prevent regeneration, at least in the medium term. Effects may be exacerbated in the presence of deer or hare browse (as noted at Giles Creek).

The effect of natural regeneration through pasture and gorse was investigated in early trials (Langer et al. 1999). An area of reshaped gravel-tailings-mixed topsoil associated with an alluvial gold mine was either unplanted or established in pasture for 18 months (with a large basal N and P fertilizer application) before being sprayed with herbicide and planted with native shrubs in 1996. Herbicide was used to control gorse in half of each plot from the time native shrubs were planted and the area was fenced. Initial survival and growth of native species established through grass was lower, possibly due to moisture competition from the grass or due to damage by grass grubs that built up in the pasture. Gorse invasion was reported as limited in December 1997 when monitoring was stopped; observations about 5 years later indicated that gorse dominated all treatments and had smothered native seedlings.

The soil-layering trial at Giles Creek also monitored self-establishing (adventive) plants. Unplanted groundcover reached 40% in both soil treatments after 4.5 years. The mixed-soil plots were dominated by tall (>50 cm) native rushes, whereas the layered plots were dominated by small (<5 cm) herbs, the native *Nertera depressa*, and the exotic catsear (*Hypochoeris radicata*). Spontaneous koromiko and mānuka seedlings were present by the end of the third year, but only in the layered soil treatment.

FIGURE 18.6   Strongman Mine rehabilitation where native seedlings were planted into salvaged respread soil and coarse wood. (Photo taken in 2012.)

High risks of erosion and weed competition that are typical at New Zealand mines mean native dominance at canopy closure is a key measure of success, as it is associated with low risk of erosion and low ongoing weed control costs. Once rehabilitated root zones have been identified, plant density trials are useful to establish trade-offs between planting density, maintenance inputs, and time to canopy closure. For example, a trial at Stockton Mine, established in 2005, compared the effects of initial planting density (3, 5, 7, 9, and 11,000 plants/ha) with and without an annual fertilizer application. An unexpected result was the value of precocious nursery-plant species which self-propagate from seed and wind-broken stem fragments within 1–3 years. Such plants boost the density of native seedlings under conditions of low weed pressure, reducing the time to canopy closure. They include *Olearia arborescens, Chionochloa conspicua,* and *Hebe moorea.*

At other areas of Stockton and at Strongman Mine, plant performance was also quantified in permanent belt transects 50 m long and 2 m wide (100 m^2). Transects were initially used to audit density and mortality of planted areas, but have proven invaluable to track natural regeneration over 5–10 years. At Strongman Mine, 17 permanently marked transects were established in planted areas between April 2005 and 2009 and monitored until 2012 or 2015. Areas were planted at a mean density of 4700/ha nursery-raised seedlings (Figure 18.6). By 2011 the density of native seedlings greater than 100 mm height had increased to 11,400/ha on average. In 2015, the density in three remaining transects had reached over 50,000/ha (Table 18.1). The dramatic increase was due both to precocious species (koromiko, toetoe, and mānuka) seeding and colonization of the site by wind-blown ferns and kamahi (*Weinmannia racemosa*). Establishment rates increased as the variety of sheltered microhabitats increased. The average increase in density has been around 4500 stems per hectare per year since 2005 (despite the initial oversowing of browntop grass at about 3 kg/ha), but surged between 2011 and 2015 to about 9900 stems per hectare per year. Plant density would be expected to be reduced when canopy closure is reached as high vegetation cover will suppress establishment of shade-intolerant species. The effect of the increase in adventive seedlings is a decrease in relative frequency of plant species that have not propagated, that is, karamu and flax (Figure 18.7).

18.3.4   DIRECT TRANSFER OF WILDINGS AND SALVAGED SODS

Native "wilding" seedlings are usually sourced from areas about to be stripped for mining. This requires rehabilitation to be explicitly built into mine planning to deliver favorable mine schedules and costs. The cost of hand salvaging seedlings (or trees) is specific to species, size, and site. At the relatively small Giles Creek site, mine staff salvaged beech seedlings at "quiet" times for minimal cost. At the much larger Stockton site, wilding salvage was a specific contract to an external operator.

TABLE 18.1

**Number of Plants Recorded in the Three Oldest 100 m²
Belt Transects at Strongman Mine**

Species	Site 3 2005	Site 3 2015	Site 11 2008	Site 11 2015	Site 15 2008	Site 15 2015
Hebe salicifolia	43	103	2	2	27	131
Leptospermum scoparium	33	461	15	232	2	56
Coriaria arborea	0	0	0	0	2	0
Coprosma robusta	29	53	10	4	13	21
Austroderia richardii	2	12	20	22	3	225
Phormium species	19	34	7	7	12	5
Weinmannia racemosa	4	71	3	2	1	6
Olearia species	0	12	0	2	0	4
All fern species	15	36	11	0	47	46

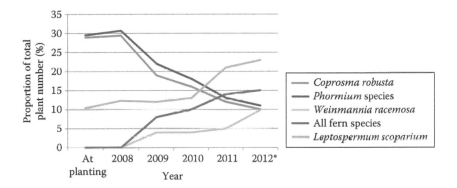

FIGURE 18.7 Relative change in abundance of native species at Strongman Mine over 6 years. *Note*: * indicates transect number reduced.

In 1992, such sourcing and establishing wildings were estimated to be at least three times the cost of equivalent nursery-grown root trainers, as the much larger wilding plants were slower to plant. Despite this, the first large areas of rehabilitation at Stockton during the mid-2000s used about 2000 wildings and 5000 nursery-grown root trainers per hectare. The wildings, which were mainly flax, tussocks, and small trees, were considered essential to delivering an immediate positive visual impact (Kingsbury 2008). Transplants of individual wildings were also used at Stockton to create intensively managed areas for seed and cutting production where plants were fertilized to boost seed quality. Transplants were also used to conserve genetic material from a wide range of subalpine herbs, shrubs, and grasses from the Mt Augustus ridgeline.

   Direct transfer is a specific form of wilding salvage in which intact mats or sods of vegetation and attached soil are removed and relocated in a way that preserves the potential for natural regeneration. The technique was pioneered in the early 1990s at Giles Creek using stumps and root plates of beech trees with attached seedlings and underlying soils to rehabilitate lowland beech/podocarp forest (Ross et al. 2000; Simcock and Ross 1995). In 1993, direct transfer was evaluated for rehabilitating pakihi wetland, dominated by tangle fern, wire rush (*Empodisma minus*), and sedges, at Giles Creek. After 4 years, 90% vegetative cover was achieved. The proportion of adventive species had increased from about 3% to between 30% and 40%, depending on whether initial rehabilitation had achieved a complete or partial vegetation cover (Ross et al. 1998, 2000). A key factor in the success

FIGURE 18.8 Alluvial gold mine site in lowland beech forest rehabilitated using direct transfer of spaced beech tree saplings and replaced soils, about 5 years after initial treatment.

was retention of the limited soil permeability under which the plants of pakihi ecosystems are competitive. Direct transfer has since been used successfully in other Westland alluvial gold mines (Figure 18.8) and has become the preferred method for rehabilitating native forest after mining in New Zealand. It has particular value in advancing rehabilitation outcomes by decades to hundreds of years and is highlighted in rehabilitation guidance (Simcock and Ross 2014) and under consent conditions of Westland mines consented from about 2000.

Direct transfer techniques have been tailored for a variety of ecosystems at Stockton Mine, from tussock wetland to alpine herb field and beech forest. The Stockton mining operation is a mosaic of multiple pits and overburden landforms over 2500 ha. The shallow soils with restricted, dense root mats within the surface organic-enriched horizons (typically 100–400 mm) combined with low stature (<3 m) of most plants, allow highly proficient removal of intact soil/plant sods. The soils are difficult to handle conventionally, as they are often saturated. Stockton trials started in the mid-1990s using techniques adapted from Giles Creek and unspecialized machinery. Face shovels, excavators, and articulated dump trucks were used, and sods unloaded using a "running spread" technique that caused substantial damage and significant gaps between sods. Despite this, plant and invertebrate diversity was far greater than in conventionally rehabilitated areas and more similar to undisturbed areas; trees up to 4 m high and 70 years old were successfully relocated (Simcock et al. 1999).

The Stockton experience of applying direct transfer over tens of hectares showed ecosystems differ in their suitability for direct transfer (DT). The best DT outcome (highest vegetation cover and least bare ground) is associated with source vegetation <1.5 m in height with dense root mats and low rock or boulder content. DT is the only technique used to introduce plant species not propagated in nurseries and is particularly successful at establishing plants that are poor at dispersing, such as beeches, pines (*Halocarpus* and *Lepidothamnus* sp.), and some herbs. The transfer of near-intact soils helps maintain both the chemical and physical conditions in which the native plants are competitive.

In the mid to late 2000s, Solid Energy refined direct transfer techniques using a range of specialized digger buckets (Figure 18.9), transportation trays, and auto-unloading transporters. Strips of direct transfer were used for erosion control on steep slopes adjacent to water courses and the tops of backfill slopes. The instant, dense ground cover and dense root mat reduces total suspended solids in runoff to levels comparable with hydroseeding and straw mulching (Rodgers et al. 2011). Direct transfer preserves the organic layers, while the living vegetation binds and physically protects soils against erosive rains that typically exceed 200 mm/day, at least monthly.

(a)   (b)

FIGURE 18.9   A method of direct transfer for shrubland at Stockton Mine (a) uses specialized buckets (b).

FIGURE 18.10   A *Powelliphanta* snail release site on Stockton ridgeline about a year after rehabilitation using the direct transfer method has achieved a high level of ecosystem intactness.

Progress was driven by large-scale use at Cypress Mine (12 ha of red tussock wetland) and Mt Augustus where *Powelliphanta* snails were present (Rodgers et al. 2011). Direct transfer has been successful at relocating the habitat of the *Powelliphanta* snails (Figure 18.10). Some snails inadvertently "piggybacked" on sods; others have been released into areas rehabilitated using direct transfer once it had been assessed as providing favorable habitat for the snails. Monitoring confirmed the presence of snails (Hamilton 2015). Large flightless invertebrates such as native crickets (weta), beetles (Simcock et al. 1999), and, most important, earthworms (Boyer et al. 2011), the main food of *Powelliphanta*, are also conserved in DT. Boyer et al. (2011) report direct transfer sods contained an earthworm biomass and diversity slightly lower and more variable than in undisturbed ecosystems. In comparison, conventionally stripped and spread soils had very low earthworm biomass and diversity.

Powelliphanta augusta was discovered in 2004 within about 6 ha of its total remaining habitat at Stockton Mine (Walker et al. 2008). Under conditions of a Wildlife Permit issued in early 2006, over 6100 of these snails and 1100 eggs were brought into captivity in cool-stores before the stripping of about 5 ha of their c. 6 ha total habitat. The permit required the snails and eggs to be translocated and a captive breeding colony maintained for 10 years as a safeguard against failure of the translocated populations (James et al. 2012). By late 2007, about 4000 snails had been translocated to other sites, including rehabilitated habitat about 800 m north of the original sites. A successful captive rearing and management program was developed for these long-lived, relatively slow-maturing and relatively large snails (Department of Conservation 2016a), 2400 of which were housed individually in moss-lined containers and fed with nonnative earthworms (Allan 2010). A major component of the project was the use of high-quality, innovative rehabilitation methods to create suitable habitat for the release of snails hatched in captivity. Hamilton (2015) developed a "mark and recapture" monitoring method, recording annual survival exceeding 80% at most locations, sexual maturity at about age 8 years, and average shell growth rate of 2.6 mm/year. Five years of this monitoring indicates most of the *P. augusta* populations appear to be stable and many appear to be increasing, although it will take decades to confirm success of translocations (Hamilton 2015). Given infrastructure construction, urbanization, and quarrying have been identified as a risk for land snails in Europe (Cuttelod et al. 2011) continuation of monitoring of translocated *P. augusta* over the next 20 years could be of international value.

18.3.5 MAINTENANCE

Indigenous New Zealand ecosystems are typically vulnerable to pest animals as well as pest plants. Isolated areas of Stockton, Golden Cross, and Strongman Mines all showed high mortality of planted native seedlings and retardation of native establishment from seed where dense lotus and/ or pasture grasses established. At Stockton, areas of dense pasture and lotus growth were variously associated with trial importation of topsoil, use of biosolids seeded with pasture grasses (which was effective at suppressing erosion), and with early rehabilitation trials that established lotus using fertilizer and lime amendments. Survival and growth of palatable native species has also been reduced at all three mine sites, mirroring results of the early fenced and unfenced treatments at Giles Creek. At Strongman Mine, selective browsing is considered the main factor behind elimination of the nitrogen-fixing plant tutu (*Coriaria arborea*) from most monitored transects sites over 6 years (unpublished data). At Tui Mine, karamu, wineberry (*Aristotelia serrata*), and koromiko (*Hebe stricta* syn *Veronica stricta*) were removed from most of the site by goats; at the Stockton Mine, growth of *Coprosma propinqua* and broad-leafed tussock *Chionochloa conspicua* was severely suppressed by hares in plots that were outside the fenced, earlier mentioned density trial. Mew et al. (1997) also report planted nursery-grown native seedlings on Westland sites being impacted by stock where fences were ineffective.

18.4 CONCLUSION

New Zealand requirements for successful rehabilitation of mined land to pastoral uses are well established. The technical capacity to deliver favorable outcomes has been demonstrated since the mid- to late 1980s across a range of mine sizes and mine types. Pasture production similar or greater than pre-mining has been achieved after dredging of alluvial and coastal deposits for gold or iron sands and opencast coal and gold mining from the loess-covered landscapes of Southland and Otago to the ash-mantled Waikato. Rehabilitation outcomes for pasture are straightforward to quantify and, with suitable identification of surrogate subsoils, usually straightforward to achieve over relatively short time frames of less than 5 years. The plant species commonly used are readily available and adapted to rapid growth rates in open, exposed environments. Well-researched foliar and soil fertility requirements can be optimized using conventional fertilizers. New Zealand guidelines are available and international guidelines have relatively high relevance.

The main determinants of successful pasture rehabilitation in New Zealand, being suitable slopes and root zones, have been consistently highlighted over 30 years of research and guidance. The selection of suitable slopes must strike a balance between gradients steep enough to ensure adequate drainage yet gentle (and short) enough to resist erosion from rainfalls that generally exceed 1000 mm per annum and may reach 6000 mm per annum. Most New Zealand open-pit mines create conservative bench and batter backfill forms. These enable runoff to be run along benches to ponds that provide water quality management and peak flow management. More natural landforms are uncommon. Suitable root zones for pastures are easily specified in terms of minimum rooting depth, drainage, water supply, and fertility. New Zealand mines create these root zones, even in the absence of topsoil, by selecting overburden textures, depths, and slopes that provide a suitable physical environment. Inorganic fertilizers (and lime, where acidity is too low) and, more recently, municipal biosolids are used to achieve suitable pH and chemical fertility, which are both much higher than those needed to establish native species. Pastures and plantation forests generally use fast-growing, nonnative species bred to respond strongly to fertile substrates; metal-tolerant or hyperaccumulator species are not used. Each large, hard-rock mine has developed rehabilitation techniques specific to its particular waste rock and tailings, based on pot and field trials.

Uniformity of plant growth within single paddocks or timber plantations is highly desirable and enhanced by uniformity of slope and root zone. Uniformity and fertility requirements of these traditional productive systems contrast with native New Zealand ecosystems. Where native New Zealand ecosystems are rehabilitated, uniformity and fast growth are not necessarily desirable—highly fertile conditions usually increase competition with nonnative plant species and increase the palatability of native species to introduced herbivores. Specific rehabilitation methods for native New Zealand ecosystems were developed following the poor outcomes of techniques used in plantation forestry and began relatively recently in the early 1990s. This research focused on the properties of surrogate and replaced soil profiles needed to achieve acceptable survival and growth rates of nursery-grown or wild-salvaged native shrub and tree species, and such trials are repeated with new mines (e.g., Norton et al. 2013). Research shows that while providing suitable soils underpins native plant survival and growth, adequate shelter provided by rough microtopography (aided by return of coarse wood, boulders) and control of pest plants and pest animals are also vital to achieve sustained growth.

Although planting native seedlings is a common approach, a range of establishment options are available if native vegetation is present before mining. These include regeneration of seeds and propagules from mixed, salvaged topsoils that are not stockpiled, direct transfer of intact plant–soil sods, various seeding techniques, and natural regeneration from adjacent or remnant areas. Very little native seed is commercially produced in New Zealand, so sites need to be self-sufficient. Further, although the seral species used across New Zealand for forest rehabilitation are similar, recent genetic work indicates local genotypes are present (e.g., kowhai, mānuka) so the use of local, "eco-sourced" seed sources is strongly encouraged. Natural regeneration to supplement, or replace, planting has been highly successful at sites with low weed competition, as shown at Wangaloa, Strongman, and Stockton Mines.

The greatest variety of outcomes, and the greatest uncertainty and controversy, are associated with mines in areas with native ecosystems. This reflects the location of many of these mines on public land administered for conservation and/or mines affecting rare or threatened native animals, plants, and/or ecosystems. From about 2000, New Zealand research has increasingly focused on the effectiveness of different rehabilitation methods to deliver self-sustaining native ecosystems. In the past, self-sustaining rehabilitation was considered to be a native plant cover that was dense and tall enough to suppress common weeds and resist pest animal browse. The expectation was that natural regeneration and colonization of plants and animals from nearby remnant native ecosystems would deliver a natural outcome over the long-term with minimal input. However, more recently there has been an increased focus on delivering a greater initial diversity of plant species and much faster ecosystem recovery. This change has probably been driven by the need to reduce uncertainty for rehabilitation of habitat for slow-moving invertebrates of high conservation value, for example, *Powelliphanta* species within the Stockton and Denniston coal mines. There is a paucity of published data in this area. Notable exceptions are research at Wangaloa comparing invertebrates in areas either naturally seeded or planted with nursery seedlings (Rufaut et al. 2006, 2010), invertebrates in very young direct transfer at Stockton Mine (Simcock et al. 1999), and research on native earthworms on mines within Westland coal measures (Boyer and Wratten 2010). Research indicates a critical component for fast rehabilitation of native ecology is the salvage and immediate reuse of living profiles with intact plant/soil sods, as this can conserve plants not cultivated in nurseries, poorly dispersing invertebrates, and complex mycorrhizal interactions. Here, international guidance is of limited relevance due to the extreme endemism of flora, fauna, and ecosystems and the unique pressures from pest plants and animals.

The underlying importance of habitat heterogeneity to support biodiversity heterogeneity is emphasized internationally but is only beginning to be implemented in New Zealand. International guidance on effective methods to create and underpin ecosystem diversity is useful, particularly examples of dendritic and S-shaped landforms under climates with similar erosion potentials. Methods to create variation in rooting environments by manipulating root zones and surface mulches are also useful. One method that has generally been highly successful at conserving and enhancing heterogeneity has been direct transfer or sod translocation. Results appear to have been much more successful in New Zealand than overseas, although published results are sparse and generally short term (less than 5 years). Success of the technique in New Zealand is attributed to features that limit plant stress, that is, shallow root systems and relatively high rainfall with low drought stress deficits, and development of machinery and methods that reduce root disturbance. Areas of direct transferred wetland, shrubland, and subalpine herb fields between 10 and 20 years old are available to assess medium-term outcomes, but the multidisciplinary research required is expensive.

The area available for direct transfer is determined by mine scheduling, that is, areas of final landform must be created before direct transfer can occur, unless space for temporary storage is possible. Where direct transfer is not possible, two rehabilitation issues are commonly debated. First, there is a tension between planting fast-growing seedlings to achieve certainty of cover and slower or more variable techniques relying on seeding and natural regeneration. The second key rehabilitation issue is to what extent and over what time frame rehabilitation techniques can be used to replace native ecosystems that are removed by mining.

With respect to the tension between planting and natural regeneration or seeding, planting is often preferred by miners and regulators and may be the only option where competition from non-native species and/or erosion risk is intense. The latter, seed-based techniques are more risky, slower and more variable, whether using deliberate seeding or natural seeding near patches of planted, direct transfer or nearby remnant areas. Research at Wangaloa coal mine (Rufaut and Craw 2010; Rufaut et al. 2015) and Central Otago gold mines (Craw et al. 2006) indicates seeding can allow native plants to establish onto mine overburdens that are hostile to both nursery-raised native plants and nonnative species such as pasture grasses and legumes. Data from Strongman and Stockton Mines show the density and diversity of ferns and shrubs that can self-establish under favorable conditions. In practice, both methods may be suited for mine sites that operate for more than 5–10 years. Nursery-raised plants are suited to areas with weedy topsoils, or where erosion risk is extremely high, as they can be planted through a surface armored with mulch (as observed at Wangaloa, where pebble mulch protected underlying finer sediment). Many mine sites have at least some areas where there is a deficit of topsoil. The placement of topsoil and favorable different overburdens should be able to be manipulated to achieve areas favoring seeding and natural regeneration techniques, particularly if such areas include islands of direct transfer, pest plant competition is absent or can be controlled, microtopography creates stable, sheltered sites, and suitable native seed is available. Direct transfer and seeding techniques may also be preferred where biosecurity risks are high, where nursery-grown plants are not wanted, and/or where longer-term maintenance to remove competing weeds is feasible. The latter approach is being used to revegetate small sites within relatively intact forest at Tui mine. Further research is needed to reduce uncertainty by documenting outcomes.

The second key rehabilitation issue is to what extent and over what time frame rehabilitation techniques can be used to replace native ecosystems that are removed by mining. The result impacts the quantum of off-site mitigation or "environmental compensation." There are few New Zealand studies that offer information beyond initial plant growth rates and mortality to canopy closure. This is not surprising, given rehabilitation using native plants only started in the early 1990s, and large areas of native rehabilitation were not achieved until the 2000s. Further, until about 2000, closure criteria and consequent release of mining bonds were tied to native canopy cover, not fauna or specific ecosystems. More long-term, multidisciplinary studies over at least 10 years are needed, such as those at Wangaloa by the University of Otago. Such studies could be based on specific rehabilitation data required to be collected and publically reported under resource consent and wildlife permit conditions for mines permitted since 2000.

REFERENCES

Allan, T.E. 2010. Husbandry of the carnivorous land snail *Powelliphanta augusta*. MSc thesis, Victoria University of Wellington, Wellington, New Zealand.

Anon. 2013. Deed of agreement between the minister for conservation and Buller coal limited. Compensation for access agreement under S61 of the Crown Minerals Act. http://www.doc.govt.nz/Documents/about-doc/news/issues/compensation-deed-for-denniston-coal-mine.pdf. Accessed November 2016.

Bagley, S. 1980. Regional report. Nelson Marlborough. *Goldfields Seminar, Cromwell 1980*. KL Jones, New Zealand Historic Places Trust, Publication no. 14, Wellington, New Zealand.

Bassett, I.E., Simcock, R.C., Mitchell, N.D. 2005. Consequences of soil compaction for seedling establishment: Implications for natural regeneration and restoration. *Austral Ecology* 30(8):827–833.

Begbie, M., Craw, D., Rufaut, C., Martin, C.E. 2007. Temporal and spatial variability of acid rock drainage in a rehabilitated coal mine, Wangaloa, South Otago, New Zealand. *New Zealand Journal of Geology and Geophysics* 50:227–238.

Bertram, G. 2011. Mining in the New Zealand economy. *Policy Quarterly* 7(1):13–19.

Boyer, S., Wratten, S. 2010. The potential of earthworms to restore ecosystem services after opencast mining—A review. *Basic and Applied Ecology* 11:196–203.

Boyer, S., Wratten, S., Pizey, M., Weber, P. 2011. Impact of soil stockpiling and mining rehabilitation on earthworm communities. *Pedobiologia* 54(Suppl. 29):S99–S102.

Brown, K., Elliott, G., Innes, J., Kemp, J. 2015. Ship rat, stoat and possum control on mainland New Zealand. An overview of techniques, successes and challenges. Department of Conservation, Wellington, New Zealand, September 2015. https://www.trc.govt.nz/assets/Documents/Environment/biodiversity/ship-rat-stoat-possum-control.pdf. Accessed November 2016.

Buckley, T.R., Whaite, D.J., Howitt, R., Winstanley, T., Ramon-Laca, A., Gleeson, D. 2014. Nuclear and mitochondrial DNA variation within threatened species and subspecies of the giant New Zealand land-snail genus *Powelliphanta*: Implications for classification and conservation. *Journal of Molluscan Studies* 80(3):291–302.

Buxton, R.P., Stanley, J., Alspach, P., Morgan, C., Ross, C. 2005. Mosses and other early colonizers: Pioneers for revegetating mine waste rock dumps and rock walls. In Paper presented at the 2005 *National Meeting of the American Society of Mining and Reclamation*, Breckenridge, CO, June 19–23, 2005. American Society of Mining and Reclamation, Lexington, KY.

Connolly, L.P., Wein, V.S., Denize, P.R. 1981. Land rehabilitation practice at Waipipi Ironsands Ltd. In Paper presented at the *Australian Institute of Mining and Metallurgy New Zealand Branch Annual Conference*, Thames, U.K., August 19–September 21, 1981.

Cossens, G.G., Keating, R.D. 1990. Mine waste reclamation studies at Macraes Flat, Otago. 1990. In *Proceedings of the Workshop Issues in the Restoration of Disturbed Land*, Occasional report no. 4, pp. 120–131. Fertilizer and Lime Research Centre, Massey University, Palmerston North, New Zealand.

Craw, D., Chappell, D., Reay, A., Walls, D. 2000. Mobilisation and attenuation of arsenic around gold mines, east Otago, New Zealand. *New Zealand Journal of Geology and Geophysics* 43(3):373–383.

Craw, D., Rufaut, C.G., Hammitt, S., Clearwater, S., Smith, C.M. 2006. Geological controls on natural ecosystem recovery on mine waste in southern New Zealand. *Environmental Geology* 8:1389–1400.

Cuttelod, A., Seddon, M., Neubert, E. 2011. European red list of non-marine mollsucs. IUCN and the Natural History of Bern, Switzerland, Luxembourg Publications Office of the European Union, Luxembourg.

Davis, M.R., Langer, E.R., Ross, C.W. 1997. Rehabilitation of native forest species after mining in Westland. *New Zealand Journal of Forestry Science* 27:51–68.

Department of Conservation. 2016a. *Powelliphanta augusta* recovery strategy 2015–2020. *P. augusta* Recovery Working Group, Department of Conservation. Wellington, New Zealand.

Department of Conservation. 2016b. The science behind the Department of Conservation's predator control response. Battle for our Birds. http://www.doc.govt.nz/Documents/conservation/threats-and-impacts/battle-for-our-birds-2016/battle-for-our-birds-science-2016.pdf. Accessed November 2016.

Douglas, G.B., Dodd, M.B., Power, I.L. 2007. Potential of direct seeding for establishing native plants in to pastoral land in New Zealand. *New Zealand Journal of Ecology* 31(2):143–153.

Druzbicka, J., Rufaut, C., Craw, D. 2015. Evaporative Mine water controls on natural revegetation of placer gold mines, Southern New Zealand. *Mine Water and the Environment* 34(4):375–387.

Environment Waikato. n.d. Tui Mine remediation project. https://www.waikatoregion.govt.nz/services/regional-services/waste-hazardous-substances-and-contaminated-sites/tui-mine. Accessed November 2016.

Fitzgerald, R.E. 1987. Natural regeneration after early gold mining, Westland and North-west Nelson. A report for the Ministry of Energy, Wellington, New Zealand, 52pp.

Frohlick, S., Johnston, L. 2011. Naturalising bodies and places. *Annals of Tourism Research* 38:1090–1109.

Gregg, P.E.H., Stewart, R.B. 1991. Land reclamation at Waihi: Principles. In *Proceedings of the Australian Institute of Mining and Metallurgy, New Zealand Branch Annual Conference*, Wellington, New Zealand, pp. 78–84.

Gregg, P.E.H., Stewart, R.B., Brodie, K.M., Stewart, F. 2003. From field trials to reality: Rehabilitation of the tailings storage embankment at the Waihi Gold Mine. In *Proceedings of the Workshop Environmental Management Using Soil-Plant Systems*, Occasional report no. 16, pp. 141–148. Fertiliser and Lime Research Centre, Massey University, Palmerston North, New Zealand.

Gregg, P.E.H., Stewart, R.B., Mason K. 1995. Pastoral reclamation of tailings and overburden material from a New Zealand opencast gold mine. In *Proceedings Sudbury 95, Conference on Mining and the Environment*, pp. 1249–1258. Sudbury, Ontario, Canada, May 28–June 1, 1995.

Gregg, P.E.H., Stewart, R.B., Ross, C.W. 1998. Land reclamation practices and research in New Zealand. In *Land Reclamation: Achieving Sustainable Benefits*, H.R. Fox, H.M. Moore, and A.D. McIntosh (Eds.), pp. 365–372. Blakerna, Rotterdam, the Netherlands.

Gregg, P.E.H., Stewart, R.B., Stewart, G. 1990. Determination of a suitable plant growth medium on gold and silver mining waste, Martha Hill, Coromandel Peninsula. I. Agronomic and soil aspects. In *Proceedings of the Workshop Issues in the Restoration of Disturbed Land*, Occasional report no. 4, pp. 101–108. Fertilizer and Lime Research Centre, Massey University, Palmerston North, New Zealand.

Gurung, S.R., Stewart, R.B., Loganathan, P., Gregg, P.E.H. 1996. Aluminium organic matter fluoride interactions during development on oxidised mine waste. *Soil Technology* 9:273–279.

Haffert, L.T. 2009. Metalloid mobility at historic mine and industrial processing sites in the South Island of New Zealand. PhD thesis, University of Otago, Dunedin, New Zealand.

Hamilton, M. 2015. Monitoring *Powelliphanta* land snails: An assessment of the current technique and the development of a new mark-recapture technique. MSc thesis, Lincoln University, Lincoln, New Zealand.

Horne D.J., Stewart, R.B., Gregg, P.E.H. 1990. Determination of a suitable plant growth medium on gold and silver mining waste, Martha Hill, Coromandel Peninsula II Physical properties. In *Proceedings of the Workshop Issues in the Restoration of Disturbed Land*, Occasional report no. 4, pp. 109–113. Fertiliser and Lime Research Centre, Massey University, Palmerston North, New Zealand.

Isdale, A.M. 1981. New Zealand historic places regional report. Thames Coromandel, *Goldfields Seminar, Cromwell 1980*. KL Jones, New Zealand Historic Places Trust, Publication no. 14, Wellington, New Zealand.

Jackson, R.L. 1994. Land rehabilitation after mining: Soil drainage studies. Landcare Research contract report LC94/9530, Lincoln, New Zealand.

James, A.F., Brown, R., Weston, K.A., Walker, K. 2012. Modelling the growth and population dynamics of the exiled Stockton coal plateau land snail, *Powelliphanta augusta*. *New Zealand Journal of Zoology* 40(3):175–185.

Johnston, P.R., Horner, I.J., Beever, R. 2003. *Phytophthora cinnamomi* in New Zealand's indigenous forests. In *Phytophthora in Forests and Natural Ecosystems. Second International IUFRO Working Party 7.02.09 Meeting*, Albany, Western Australia, Australia, September 30–October 5, 2001, J.A. McComb, G.E.St.J. Hardy, and I.C. Tommerup (Eds.), pp. 41–48. Murdoch University Print, Perth, Western Australia, Australia.

Keating, R.D. 1988. Mining and land reclamation in New Zealand. In *Proceedings of the 13th Canadian Land Reclamation Association Convention*, Ottawa, Ontario, Canada, August 7–10, pp. 9–17.

Keating, R.D. 1990. Trends in land rehabilitation practices 1976–1990. In *Proceedings of the Workshop Environmental Management Using Soil-Plant Systems*, Occasional report no. 16, pp. 42–48. Fertiliser and Lime Research Centre, Massey University, Palmerston North, New Zealand.

King, A.R. 2008. Revegetation of mine tailings impoundments and associated environmental issues. MSc thesis, University of Otago, Dunedin, New Zealand.

Kingsbury, M. 2008. Revegetation of the Stockton Coal Mine, Buller. *International Plant Propagators Society Combined Proceedings* 2007(57):114–119.

Kingsbury, M., Cunningham, H. 2007. Establishment of temporary exotic grass covers on a highly disturbed sub alpine mine site. In Paper presented at the *International Erosion Control Conference*, New Plymouth, New Zealand, November 21, 2007.

Koch, J.M. 2015. Mining and ecological restoration in the Jarrah forest of Western Australia. In *Mining in Ecologically Sensitive Landscapes*, M. Tibbett (Ed.), pp. 111–140. CSIRO Publishing, Clayton South, Australia. CRC Press/Bakema.

Lambert, M.G., Clark, D.A., Costell, D.A., Fletcher, R.H. 1983. Influence of fertiliser and grazing management on North Island moist hill country 1. Herbage accumulation. *New Zealand Journal of Agricultural Research* 26:95–108.

Langer, L., David, M.R., Ross, C. 1999. Rehabilitation of lowland indigenous forest after mining in Westland. Science for Conservation 117. Department of Conservation, Wellington, New Zealand.

Longhurst, R.D., O'Conner, M.B. 1990. Pasture establishment to rehabilitate land after open-cast mining at Rotowaro. In *Proceedings of the Workshop Issues in the Restoration of Disturbed Land*, Occasional report no. 4, pp. 159–162. Fertiliser and Lime Research Centre, Massey University, Palmerston North, New Zealand.

Malloch, K., Craw, D., Trumm, D. 2015. Arsenic forms and distribution at the historic Alexander gold processing site, West Coast. In *Proceedings of the Australian Institute of Mining and Metallurgy New Zealand Branch 2015 Annual Conference*, Dunedin, New Zealand, pp. 241–250. http://www.crl.co.nz/downloads/geology/2015/Malloch_et_al_2015_AusIMM_Arsenic.pdf.

Mason, K.A. 1996. Rehabilitation of unoxidised pyritic waste rock and tailings at Martha Hill Gold Mine New Zealand. MSc Hort. thesis, Massey University, Palmerston North, New Zealand.

May, P.R. 1962. *The West Coast Gold Rushes*. Pegasus Press, Christchurch, New Zealand.

McQueen, D.J. 1983. Land reclamation after gravel extraction on Ranzau soils, Nelson, New Zealand. New Zealand Soil Bureau Scientific Report 58. Department of Scientific and Industrial Research, Upper Hutt, New Zealand.

McWethy, D.B., Whitlock, C., Wilmshurst, J.M., McGlone, M.S., Fromont, M., Li X., Dieffenbacher-Krall, A., Hobbs, W., Fritz, S.C., Cook, E. 2010. Rapid landscape transformation in South Island New Zealand following initial Polynesian settlement. *Proceedings of the National Academy of Sciences of the United States of America* 107(50):21343–21348.

Mew, G., Ross, C.W. 1989. Assessment of land rehabilitation after mining within the Department of Conservation Estate, Westland. DSIR division of land and soil sciences contract report 89/12, DSIR, Wellington, New Zealand.

Mew, G., Ross, C.W. 1991. Beech forests after mining. *Terra Nova* 4:52–53.

Mew, G., Ross, C.W. 1992. From beech to coal, and back. *New Zealand Forestry* 37(2):18–20.

Mew, G., Ross, C., Davis, M.R., Langer, L. 1997. Rehabilitation of indigenous forest after mining, West Coast. Science for Conservation 54. Department of Conservation, Wellington, New Zealand. http://www.doc.govt.nz/Documents/science-and-technical/sfc054.pdf. Accessed November 2016.

Morrell, W.J. 1997. An assessment of the revegetation potential of base-metal tailings from the Tui Mine, Te Aroha, New Zealand. PhD thesis, Massey University, Palmerston North, New Zealand.

Morton, J.D., Harrison, R.W. 1990. Restoration of alluvial soils for pasture production on the West Coast, South Island. In *Proceedings of the Workshop Issues in the Restoration of Disturbed Land*, Occasional report no. 4, pp. 63–74. Fertiliser and Lime Research Centre, Massey University, Palmerston North, New Zealand.

Munro, A.S. 1980. Some aspects of Waipipi Ironsands. Post Graduate Diploma of Science thesis, University of Otago, Dunedin, New Zealand.

Nathan, S. 2006. Mining and underground resources—History of mining. Te Ara—the Encyclopedia of New Zealand. http://www.TeAra.govt.nz/en/mining-and-underground-resources/page-2 (accessed October 20, 2016).

Nippert, M. 2011. Biggest protest in a generation. *New Zealand Herald*, May 2.

Norton, D.A., Creedy, S., Keir, D. 2013. Substrate modification for enhanced native forest restoration, Reefton. *Ecological Management and Restoration* 14(2):147–150.

Parker, R.W. 1991. Rehabilitation guidelines for land disturbance by alluvial gold mining in Nelson and Westland. Resource allocation report 2. Energy and Resources Division, Ministry of Commerce, New Zealand, 60pp.

Parkes, J.P., Nugent, G., Forsyth, D.M., Byrom, A.E., Pech, R.P., Warburton, B., Choquenot, D. 2017. Past, present and two potential futures for managing New Zealand's mammalian pests. *New Zealand Journal of Ecology* 41(1):151–161.

Parliamentary Commissioner for the Environment. November 1993. Environmental management of coal mining at Fletcher Creek, Reefton: An enquiry. Parliamentary Commissioner for the Environment, Wellington, New Zealand.

Parliamentary Commissioner for the Environment. October 2009. Stockton revisited: The mine and the regulatory minefield, Parliamentary Commissioner for the Environment, Wellington, New Zealand. http://www.pce.parliament.nz/media/1336/stockton_mine.pdf.

Parliamentary Commissioner for the Environment. September 2010. Making difficult decisions: Mining the conservation estate, Parliamentary Commissioner for the Environment, Wellington, New Zealand. http://www.pce.parliament.nz/media/1301/making-difficult-decisions.pdf.

Parliamentary Commissioner for the Environment. June 2011. Evaluating the use of 1080: Predators, poisons and silent forests, Parliamentary Commissioner for the Environment, Wellington, New Zealand. http://www.pce.parliament.nz/media/1294/evaluating-the-use-of-1080.pdf.

Podger, F.D., Newhook, F.J. 1971. *Phytophthora cinnamomi* in indigenous plant communities in New Zealand. *New Zealand Journal of Botany* 9(4):625–638.

Pope, J., Craw, D. 2015. Current research on mine water and the environment in New Zealand. *Mine Water and the Environment* 34(4):363.

Pope, J., Newman, N., Craw, D., Trumm, D., Rait, R. 2010. Factors that influence coal mine drainage chemistry West Coast, South Island, New Zealand. *New Zealand Journal of Geology and Geophysics* 53(2):115–128.

Robinson, B., Anderson, C. 2007. Phytoremediation in New Zealand and Australia. In *Methods in Biotechnology (23): Phytoremediation: Methods and Reviews*, N. Willey (Ed.), pp. 455–468. Humana Press Inc., Totowa, NJ.

Rodgers, D., Simcock, R., Bartlett, R., Wratten, S., Boyer, S. 2011. Benefits of vegetation direct transfer as an innovative mine rehabilitation tool. In Paper presented at *2011 Workshop on Australian Mine Rehabilitation*, Adelaide, South Australia, Australia, August 17–19, 2011.

Ross, C.W. 1988. West Coast alluvial gold mining rehabilitation study. *Guidelines for Land Rehabilitation*. Department of Scientific and Industrial Research, Division of Land and Soil Sciences contract report 88/24. Wellington, New Zealand.

Ross, C., Buxton, R., Stanley, J., Alspach, P., Morgan, C. 2003. Revegetation of steep batters of mine rock dumps using pioneer plants. In *Proceedings of the Workshop Environmental Management Using Soil–Plant Systems*, Occasional report no. 16, L.D. Currie, R.B. Stewart, and C.W.N. Anderson (Eds.), pp. 160–167. Fertilizer and Lime Research Centre, Massey University, Palmerston North, New Zealand.

Ross, C.W., Crequer, D. 2006. Land rehabilitation at Golden Cross—8 years after mine closure. In Paper presented at the *Australian Institute of Mining and Metallurgy New Zealand Branch Conference*. Waihi, New Zealand.

Ross, C., Mew, G. 1990. Land restoration after mining on the West Coast, South Island. In *Proceedings of the Workshop Issues in the Restoration of Disturbed Land*, Occasional report no. 4, pp. 51–62. Fertiliser and Lime Research Centre, Massey University, Palmerston North, New Zealand.

Ross C.W., Orbell, G.E. 1986. Land restoration after mining. In *Proceedings of the Third National Drainage Seminar*, Hamilton, Ontario, Canada, August 18–21, K.W. McAuliffe and J.M.C. Boag (Eds.), pp. 184–209. Massey University Occasional report no. 7.

Ross, C.W., Widdowson, J.P. 1985. Restoration research after opencast mining in Southland. *New Zealand Soil News* 33(5):163–169.

Ross, C.W., Widdowson, J.P. 1987. Ripping it out, then putting it back. Open cast coal mining and land rehabilitation. *Soil and Water* (Autumn):8–12.

Ross, C., Williams, P., Simcock, R. 1998. Rehabilitation of pakihi wetland after mining in North Westland. In Paper presented at *Restoring the Health & Wealth of Ecosystems*, Landcare Research, Christchurch, New Zealand, September 28–30.

Ross, C., Williams, P., Simcock, R. 2000. Can pakihi fernland be restored after mining? In Poster paper *Society for Ecological Restoration International Conference*, Liverpool, U.K., September 2000.

Rudzitis, G., Bird, K. 2011. The myth and reality of sustainable New Zealand: Mining in a pristine land. *Environment Magazine*, November–December 2011. http://www.environmentmagazine.org/Archives/Back%20Issues/2011/November-December%202011/Myths-full.html. Accessed November 2016.

Rufaut, C.G., Clearwater, S., Craw, D. 2010. Recolonisation and recovery of soil invertebrate assemblages at an inactive coal mine in southern New Zealand. *New Zealand Natural Sciences* 35:17–30.

Rufaut, C.G., Craw, D. 2010. Geoecology of ecosystem recovery at an inactive coal mine site, New Zealand. *Environmental Earth Sciences* 60:1425–1437.

Rufaut, C.G., Craw, D., Foley, A. 2015. Mitigation of acid mine drainage via a revegetation programme in a closed coal mine in southern New Zealand. *Mine Water Environment* 34:464–477.

Rufaut, C.G., Hammitt, S., Craw, D., Clearwater, S.G. 2006. Plant and invertebrate assemblages on waste rock at Wangaloa coal mine, Otago, New Zealand. *New Zealand Journal of Ecology* 30:311–319.

Simcock, R.C. 1993. Reclamation of aggregate mines in the Manawatu, Rangitikei and Horowhenua Districts, New Zealand. PhD thesis, Massey University, Palmerston North, New Zealand.

Simcock, R., Ross, C. 1995. Methods of restoring New Zealand native forests after mining. In *Proceedings of the 1995 PACRIM Congress*, Auckland, New Zealand. Australasian Institute of Mining and Metallurgy Publication Series No. 9/95, pp. 533–537.

Simcock, R., Ross, C. 1997. Hydro seeding with New Zealand native species. In *Proceedings of a Workshop on Scientific Issues in Ecological Restoration 1995*, Christchurch, New Zealand, M.C. Smale and C.D. Meurk (Eds.), pp. 42–48. Landcare Research Science Series 14.

Simcock, R., Ross, C. 2014. *Guidelines for Mine Rehabilitation in Westland*. Envirolink Advice grant:937-WCRC83. Landcare Research, West Coast Regional Council, Lincoln, New Zealand. http://envirolink. govt.nz/assets/Envirolink-reports/937-WCRC83-Guidelines-for-mine-rehabilitation-in-Westland.pdf. Accessed November 2016.

Simcock, R., Ross, C., Pizey, M. 2004. Rehabilitation of alluvial gold and opencast coal mines from 1904 to 2004. In *Proceedings of the 37th Annual Conference of the Australasian Institute of Mining and Metallurgy New Zealand Branch*, Nelson, New Zealand, August 29–September 1, pp. 77–82.

Simcock, R., Toft, R., Ross, C., Flynn, S. 1999. A case study of the cost and effectiveness of a new technology for accelerating rehabilitation of native ecosystems. In *Minerals Council of Australia 24th Annual Environmental Workshop*, Townsville, Queensland, Australia, pp. 234–251.

Solid Energy. November 2010. Solid energy New Zealand Ltd application for resource consent: Discharge to land. Rotowaro Mine, Biosolids Operational Field trial 2010/2011.

Spain, A.V., Tibbett, M., Hinz, D.A., Ludwig, J.A., Tongway, D.J. 2015. The mining-restoration system an ecosystem development following bauxite mining in a biodiverse environment of the seasonally-dry tropics of Australia. In *Mining in Ecologically Sensitive Landscapes*, M. Tibbett (Ed.), pp. 159–228. CSIRO Publishing, CRC Press/Bakema.

Stanley, J., Buxton, R., Alspach, P., Morgan, C., Ross, C. 2003. Mine site revegetation using indigenous mosses: Progress over the past five years and future opportunities. In *Proceedings of the International Meeting on the Preservation and Ecological Restoration in Tropical Mining Environments*, Noumea, New Caledonia, July 15–20, 2003.

Taylor, P.C., Walker, B.V. 1987. The development of mining in New Zealand. In *Proceedings of the Pacific Rim Congress 1987*, pp. 679–687. The Australasian Institute of Mining and Metallurgy, Parkville, Victoria, Australia.

Theinhardt, N.I. 2003. Plant restoration at an open cast coal mine, West Coast, New Zealand. MSc Forestry thesis, University of Canterbury, Christchurch, New Zealand.

Tibbett, M. 2015. Mining in ecologically sensitive landscapes: Concepts and challenges. In *Mining in Ecologically Sensitive Landscapes*, M. Tibbett (Ed.), pp. 1–6. CSIRO Publishing, Clayton South, Australia. CRC Press/Bakema.

Todd, A.J., Rufaut, C.G., Craw, D., Begbie, M.A. 2009. Indigenous plant species establishment during rehabilitation of an opencast coal mine, South-East Otago, New Zealand. *New Zealand Journal of Forest Science* 39:81–98.

Trumm, D. 2007. Acid mine drainage in New Zealand. *Reclamation Matters* 1:23–28.

Trumm, D., Cavanagh, J. 2006. Investigation of acid-Mine-Impacted waters at Cannel Creek. Landcare Research contract report LC0506/169, CRL report 06-41101, Lincoln, New Zealand. http://www.wcrc.govt.nz/ Documents/Environmental%20Management/BellvueMineRemediation.pdf. Accessed November 2016.

Walker, K.J., Trewick, S.A., Barker, G. 2008. *Powelliphanta augusta*, a new species of land snail, with a description of its former habitat, Stockton coal plateau, New Zealand. *Journal of the Royal Society of New Zealand* 38(3):163–186.

Warne, K., Morris, R. 2012. The black and the green. *New Zealand Geographic*, July–August, 116. https:// www.nzgeo.com/stories/the-black-and-the-green/. Accessed November 2016.

Weston, R. 1991. The Macraes gold mine. Development of a major resource. In *Proceedings of the 25th Annual Conference of Australasian Institute of Mining and Metallurgy New Zealand Branch*, Waihi, New Zealand, pp. 32–53.

Widdowson, J.P., McQueen, D. 1990. Rehabilitation after open cast mining in Southland. In *Proceedings of the Workshop Issues in the Restoration of Disturbed Land*, Occasional report no. 4, pp. 51–62. Fertiliser and Lime Research Centre, Massey University, Palmerston North, New Zealand.

Young, D. 2004. *Our Islands Ourselves: A History of Conservation in New Zealand*. University of Otago Press, Dunedin, New Zealand.

Index

A

Abandoned mine sites, biosolids
  commercial mining, 241
  flint dust, 242
  materials and methods
    fixed split-plot treatment factor, 248
    greenhouse benches, 247
    ICP-AES, 246
    mine waste materials, 243, 246
    nitric-perchloric acid digest, 246
    soil moisture meters, 245
    statistical analyses, 248
    sudex, 244
    suspended solids, 247
    tap water, 246
    treatments, 243
  microbial activity, 242
  mine waste materials, 243–255
  Pb and Zn contamination, 241
  plant growth, 242
  results
    concentrations, heavy metals, 251, 253
    diammonium phosphate fertilizer, 255
    digested biosolids, 253
    electrical conductivities, waste materials, 248, 252
    heavy metals, 249
    metal availability, contaminated soils, 253
    salinity effects, 248
    soil organic matter, 252
    sudex growth, with and without biosolids, 250
    total and extractable concentrations, 254
Abiotic methylation, 209, 272
Acid-base accounting (ABA) procedure, 40–41, 43–44, 51
Acid mine drainage (AMD), 96, 217, 232, 260, 266, 279
  abandoned open pit coal mine, 33–34
  accretion and migration, 37–38
  by-products
    acidity, 39
    aluminum, 40
    metal hydroxides, 38
    sulfate salts, 38–39
  environmental impact, 19–20
  heavy metal(loid)s, (im)mobilization of, 272–274
  mitigating acid and, 181–182
  poor water quality, 175
  predictive techniques
    acid-producing potential, 40
    evaluation, 42–44
    kinetic tests, 41–42
    mathematical and geochemical modeling, 40
    mineralogical and petrologic studies, 40
    remote sensing, 40
    static test, 40–41
  prevention and control
    AMD generation, 44
    bactericides treatment, 45

in situ mobilization and subsequent
      migration, 44
    oxidant infiltration barriers, 46–47
    preventive coatings, 45
    selective handling, 45
  pyrite oxidation
    abiotic and biotic processes, 35
    acidophilic bacteria, 35
    alkaline material, 37
    biochemistry, 36
    ferric sulfate hydrolysis, 36
    protons, 36
    reaction kinetics, 35
    sulfide minerals, 35, 37
  remediation techniques
    classification, 21–23
    net acidity, 22
  sources, 34–35
  treatment
    active treatment, 49–50
    chemical treatment, 48
    in situ neutralization, 48–49
    passive treatment, 50
Acid-neutralizing capacity (ANC), 40–41
Acid-producing potential (APP), 40–41, 43–44
Adsorption process, 104, 272
Aerial seeding, 349
Alberta (Canada), mine site reclamation,
    see Canada
ALDs, *see* Anoxic limestone drains
Algal biochar, 24
Alkaline amendments, 231
Alluvial gold mine site, 352
Alluvial mine pits, 310
Alternative covers assessment project (ACAP), 226
Amang (tin tailings), 313, 318
AMD, *see* Acid mine drainage
ANC, *see* Acid-neutralizing capacity
Animal remediation, 99
Anoxic limestone drains (ALDs), 50, 158
Anthropogenic process, 265
AOSR, *see* Athabasca oil sands region
APP, *see* Acid-producing potential
Appalachian Regional Reforestation Initiative (ARRI)
  Core Team, 138
  GFW, 140–141
  goals, 138
  mechanism, 138
  reforestation guidelines, 140
  Science Team, 139
Approximate original contour (AOC), 127, 129
Area mining method, 124, 126
ARRI, *see* Appalachian Regional
    Reforestation Initiative
Arsenic trioxide, 225, 304
Assimilatory redox reaction, 268
Athabasca oil sands region (AOSR), 296–297

B

Bactericides, 44–45, 51
Beef cattle production
    pasture and grazing research, Queensland,
        Australia, 115
    pasture and grazing trials, Hunter Valley, Australia
        carcass comparison, 120–121
        cattle weight, 119
        pasture establishment, 116–118
        pasture quality, 119–120
        stock type, pasture, and soils, 119
        unmined grazing land, 118
Best management practices (BMP), 305
Bioaccumulation, 102, 342
Biochar, 9, 231
    characteristics, 23
    engineered biochar, 27
    in mine site remediation
        algal biochar, 24
        amendments, 26–27
        applications, 24–26
        bioavailability and leachability, 26
        metal immobilization, 24
        microbial community composition, 24, 26
        sludge-derived biochar, 24
        softwood-and hardwood-derived
            biochars, 24
        surface complexation/coprecipitation, 24
BIOLOG® system, 84
Biomagnification, 102
Biomethylation process, 209, 272
Biomining technology, 21
BioNeutro-GEM (BNG), 277–278
Bioreactor-wetland treatment system, 158–159
Biosolids
    abandoned mine sites
        commercial mining, 241
        concentrations, heavy metals, 251, 253
        diammonium phosphate fertilizer, 255
        digested biosolids, 253
        electrical conductivities, waste
            materials, 248, 252
        fixed split-plot treatment factor, 248
        flint dust, 242
        greenhouse benches, 247
        heavy metals, 249
        ICP-AES, 246
        metal availability, contaminated soils, 253
        microbial activity, 242
        mine waste materials, 243–255
        nitric-perchloric acid digest, 246
        Pb and Zn contamination, 241
        plant growth, 242
        salinity effects, 248
        soil moisture meters, 245
        soil organic matter, 252
        statistical analyses, 248
        sudex growth, with and without
            biosolids, 250
        suspended solids, 247
        tap water, 246
        total and extractable concentrations, 254
        treatments, 243

    soil carbon sequestration, 61, 68
    soil organic carbon, 61
        data sources, 63, 65
        dynamics, 66–68
        meta-analysis, 65–66, 68
        organic matter, source of, 62
        types/fraction, 62–63
    sustainable strategy, 61
    treatments, 116–118
Biowastes
    biological characteristics, 10
    biosolids
        soil carbon sequestration, 61, 68
        soil organic carbon, 61–63, 65–68
        sustainable strategy, 61
    chemical characteristics, 9–10
    mine waste rehabilitation and potential benefits, 7–8
    municipal solid waste (MSW), 61
    organic matter, 62–64
    physical characteristics, 8
    types, 7
Blastofiltration, 206
Border area (Bd), 274
Brunisolic soil, 297

C

Canada
    coal, *see* Coal mine reclamation
    distribution, 292
    environmental impacts, 294–295
    metal and diamond, *see* Metal and diamond mine
        reclamation
    oil sands reclamation
        Gateway Hill, 297–299
        surface mining, 296
        Wapisiw tailings reclamation, 299–300
    regulations, 293–294
    sustainable resource development, 295–296
Canadian Environmental Protection Act, 291
Capping remediation methods, 20
Carbon nanotubes (CNTs)
    inorganic contaminants, 105
    organic contaminants, 102
Carbon sequestration
    amendment factors, 196
    carbon credits, 195
    land management decisions, 195
    mined, disturbed soils, 190
    quantity of C, 194
    rates of
        biosolids application rate, 194
        efficacy of amendments, 191, 194
        organic residuals *vs.* soils reclaimed, 194
        SOC accumulation, 191–193
    restoration methods, 190
    short-term carbon cycle, 189
    SOC storage, 195
Cassiterite, 309–311
Caulofiltration, 206
Chemical characterization, mine sites
    environmental impact
        AMD, 19–20
        mine tailings, 17–19

geochemical dynamics, 17
hydrogeochemistry, 17
remediation techniques
AMD, 21–23
biochar applications, 23–27
mine tailings, 20–21
Chemical fixation, 97–98
Chemical leaching, 97
Cluff Lake uranium mine, 303
CNTs, *see* Carbon nanotubes
Coal mine reclamation
drainage systems, 300
Fording River mine, 301–302
underground and surface mining, 300
Whitewood mine, 301
COCs, *see* Contaminants of concern
Colorimetric procedure, 246
Compensation, 339
Composite cover system, 47
Composted biosolids, 242
Compost treatments, 116–118
Conservation Act, 338–339
Contaminants of concern (COCs), 180–181
Contaminated site management policy, 294
Contour mining method, 126
Coupled nutrient cycling processes, 79
Covering, *see* Capping remediation methods
Covers with capillary barrier effects
(CCBE), 232
Cowarra mine, 217
Crown Minerals Act, 338
Cypress mine, 353

D

Dabaoshan mine, 217
Degraded mine sites, environmental
impacts, 59–61
Dekalb-Marowbone-Lantham complex, 153
Denniston coal mine, 355
Dense tin, 309
Derelict metalliferous/pyritic mine sites
abandoned metalliferous mines, 216
capping approaches
contaminated wastes, 222
Mole River arsenic mine, 225–226
Webbs Consols mine, 222–224
issues, 216–218
landfill covers, 216
metal toxicants, 215
remediation and management options
contaminated sites, risk assessment of, 218
cost-effectiveness, 219
electrolysis-based cleanup methods, 221
electroremediation technology, 220
ex situ techniques, 219
HHRA, 219
high risk sites and areas, 219
in situ techniques, 219
long-term contaminated soils, 221
organic contamination, 219
phytoremediation technology, 220–221
RBLM, 217, 219
soil washing process, 221

technology
cap-waste interactions, 228–229
conventional MSW sites, 226
irrigation, 227–228
landfill capping systems, 226
plant species, 229–230
soil amendments, 230–232
Derelict Mines Program (DMP), 225
Diavik Diamond mine, 302–303
Diethylenetriaminepentaacetic acid (DTPA), 98, 246
Dimethyl selenide (DMSe), 272
Dissimilatory redox reaction, 268
Dredging method, 310–311

E

Eastern Coal Fields (ECF), 148
Ecological risk assessments (ERA), 218
Efficient nutrient cycling, 191
Electrical conductivity (EC), 297, 299
Electrokinetic remediation, 98
Electroremediation technology, 220
Endangered Species Act, 339
Engineered tailing-soil, 86–87
Environmental Impact Assessment (EIA), 176–177, 328
Environmental Protection and Enhancement Act (EPEA), 294
Environmental Quality Act, 328
Ephemeral channels, 157
Ephemeroptera, Plecoptera, and Trichoptera
(EPT taxa), 148, 168, 170
Extensive trials, 341
Ex-tin mines
aquaculture, 320
area of, 318
freshwater landings, 321
inadequate private investment, 320
infrastructures, 318
reclaimed tin mines, 319
redeveloped locations and land use of, 319
rehabilitation
Mines Wellness City, 322
Perak Herbal Garden, 325
PIW, 325, 327–328
Sunway Resort City/Bandar Sunway, 322
TTAC, 322, 325–326
sand and slime tin tailings, 320
storm water management, 319
Extracellular polymeric substances (EPS), 210

F

Falling Rock (FR), 160
Field emission scanning electron microscopy/
energy dispersive x-ray spectroscopy
(FESEM/EDS), 273
Fording River mine, 301–302
Forest Research Institute Malaysia (FRIM), 320, 322
Forestry Reclamation Approach (FRA), 149
Bent Mountain project
hydrology and water chemistry, 133–134
soil, 132
spoil types, 131
tree survival and growth, 132
understory recruitment, 133

forest vegetation, 129
global application, 142
ground cover vegetation, 130–131
impacts, 141–142
loose soil, noncompacted soil growth medium, 130
proper tree planting techniques, 131
Starfire project
    bare-root tree species, 135
    growth, 136–140
    mountaintop-removal operation, 135
    subsurface compaction treatments, 135
    survival, 136
successional species, crop trees, 131
suitable rooting medium creation,
    tree growth, 129–130
Forest vegetation, 129; *see also* Forestry
    Reclamation Approach
FRIM, *see* Forest Research Institute Malaysia
Frozen block method, 304

G

Gahcho Kué Diamond mine, 180–181
Gaseous wastes, 96
Geochemical kinetic tests, 41–42
Geochemical static tests, 40–41
Geogenic process, 265
Giant Gold mine, 304
Giles Creek rehabilitation trials, 344, 346
Golden Cross mine, 344, 347, 354
Gravel pumping method, 310–311
Green Forests Work (GFW), 140–141
Ground freezing technique, 304
Guy Cove valley fill, 149–150

H

Halophytic plants, 206
Hardpan, 47
Hardwood-derived biochar, 24
HDPE, *see* High density polyethylene
Headwater stream system, Appalachia
    amphibians and fish, 148
    bioreactor-wetland treatment
        system, 164–165, 168
    colonization of macroinvertebrates, 170
    design components
        geomorphology, 153–155
        hydrologic modifications, 155–156
        stream dimension, pattern, and
            profile, 156–159
        valley reconfiguration, 155–156
    EPT taxa, 148, 168
    FRA, 149
    groundwater connection, 169
    herbivory, 170
    invasive species, 170
    large populations of macroinvertebrates, 148
    monitoring
        geomorphology, 159–160
        habitat, 161–162
        hydrology, 160
        vegetation, 161
        water quality, 160–161

native broomsedge (*Andropogon virginicus*), 168
post-construction monitoring period are, 164
pre-restoration conditions
    habitat, 153
    soils, 153
    vegetation, 153
    water quality, 150–153
pre-restoration sampling period, 164
problematic spoils, isolation of, 169
project site
    Guy Cove valley, 149–150
    presence of ponded water, 149, 151
    sediment-laden pond, 149, 151
    spring-fed channel, 150, 152
RBP values, 168
riffle bankfull cross section dimensions, 162
riparian forest census, 166–167
storm discharge volumes, 163
upland forest census, 166–168
wetland and bioreactor treatment
        system, 164, 166
Heavy metal(loid)s, mine soils
    AMD and (im)mobilization, 272–274
    Brazil, 274–276
    essential and nonessential elements, 259
    Lead and Zn mines, 259
    microbial/biotransformation process, 260
    mining-impacted soils, 260
        sink of, 266
        source of, 265–266
        transformation of, 267–272
    potential metal(loid) mobility, 264
    recalcitrant fraction, 264
    redox-sensitive elements, 260
    Republic of Korea, 276–278
    sulfide oxidation/neutralization process, 260
    tailing deposits, 259
    total metal(loid) concentrations and
        minerals, 261–264
Henrietta Creek, 301
Herringbone tailings, 337
High density polyethylene (HDPE), 46, 226
HoldAll Moisture Meter, 245
Horne Cu smelter, 265
Human health risk assessments (HHRA), 219
Hydrogeochemistry, 17
Hydro-mulching, 349
Hydroseeding, 342, 352

I

Ignition method, 246
Ilkwang mine, 276–277
Imgi mine, 276–277
Inductively coupled plasma-atomic emission
        spectroscopy (ICP-AES), 246
Inductively coupled plasma-optical emission
        spectrometry (ICP-OES), 246
Intergovernmental Panel on Climate Change
        (IPCC), 190

K

Korea Mine Reclamation Corporation, 260

L

Labor-intensive panning process, 311
Land degradation, 3, 75
Landfill covers, 216, 226, 231–232
Landfill leachate, 227–228
Land reclamation
    biowastes, 68
    mitigation activities, 293
    municipal and industrial organic
            by-products, 231
    plant species selection, 296
LEACHM, 228
LECO TruSpec CN Carbon/Nitrogen combustion
            analyzer, 246
LFH, *see* Litter-fibric-humic
Lime-stabilized biosolids, 242
Litter-fibric-humic (LFH), 297, 299
Little Millseat (LMS) watershed, 154
Long-term vertical mixing regimes, 179–180
Long-term water quality, 180–181
Loss of weight method, 246
Luvisolic soil, 297

M

Macraes mine, 340, 343
Malaysia Food Act (MFA), 322
Malaysian Agricultural Research and Development
            Institute (MARDI), 320
Malaysian rehabilitation practice
    alluvial mining boom, 311–312
    environmental impact
        ex-tin mines, 312–313
        radioactive by-product minerals, 313
        tin tailings, 312–313
    ex-tin mines, *see* Ex-tin mines
    food safety
        farmed fish, heavy metals in, 324
        pollutants, 322
        tin tailings, 321
        vegetables and fruits, heavy metals in, 323
    lucrative tin mining industry, 312
    regulations
        loopholes, 328
        National Mineral Policy, 328
        redevelopment, ex-tin mines,
                328–329
    tin minerals and application, 309
    tin mining in, 310–311
    wasteland, 312
Mambira formation, 274
Martha mine, 338, 341–343
Mehlich 3 test, 246
Metal and diamond mine reclamation
    Cluff Lake uranium mine, 303
    Diavik Diamond mine, 302–303
    Giant Gold mine, 304
Metal hyperaccumulation, 208
Metal oxide particles, 101–102
Methylation mechanism, 209, 272
Microbial consortia, 21
Microbial leaching, 99, 273
Microbial remediation, 99

Mine land rehabilitation, beef cattle production
    classification, 112
        growing media, 112–113
        infrastructure, 112
        landform establishment, 112
        maintenance revegetation, 113–114
        revegetation, 113
    goal, 111
    pasture and grazing research, Queensland, Australia, 115
    pasture and grazing trials, Hunter Valley, Australia
        carcass comparison, 120–121
        cattle weight, 119
        pasture establishment, 116–118
        pasture quality, 119–120
        stock type, pasture, and soils, 119
        unmined grazing land, 118
    stocking rate, 114–115
Mineral Development Act, 328
Mine Rehabilitation Plan (MRP), 328
Mine site reclamation, Canada, *see* Canada
Mine site remediation
    biological remediation, 98–99
    chemical remediation, 97–98
    nanoscale materials
        advantages and disadvantages, 105–106
        and applications, 101–105
        and processing, 99–101
    physical remediation, 97
Mine soils, 260
    microorganisms, 269–271
    sink of, 266
    source of, 265–266
    transformation of
        biological processes, 268–272
        physicochemical processes, 267–268
Mine tailings
    biological characteristics
        biogeochemical changes, OA, 84–86
        microbial-mediated bioweathering, 77
        native microbial inoculum, 86–87
        organic matter, 77–80
        soil microbial communities, 77, 80–84
    definition, 75–76
    environmental impact
        geochemical reactions, 17–19
        hydrological changes, 19
        sulfide minerals, 17
    fresh tailings, 76
    mineralogical profile, 76
    organic matter
        coupled nutrient cycling processes, 79
        organic carbon, 77–78
        organo-mineral interactions and aggregation, 80
        SOM, 77, 79–80
        TOC, 77
    phytostabilization, 76
    remediation techniques, 20–21
    sulfide-rich tailings, 76
    toxic constituents, 76
Mine wasteland space, 96
Mine wastes
    characteristics
        biological properties, 6–7
        biowastes, 7–10

chemical properties, 5–6
organic/inorganic amendments, 7
physical properties, 4–5
rehabilitation strategy, 7–8
revegetation strategy, 7
materials, 96
metal contamination, 294
mine spoil, 4
mine tailings, 4
overburden soil, 4
Mini Moisture Tester, 245
Mole River arsenic mine, 225–226
Mountaintop surface mining, 124–126
Mulches, 342
Municipal solid waste (MSW), 226–229, 231–232
Mycorrhizal fungi, 209

N

Nanocomposite materials, 99
Nanofibers, 99
Nano iron oxide particles, 104
Nano iron sulfide particles, 104–105
Nanometal particles, 101–102
Nanoparticles (NP), 99–101
Nano-remediation
    inorganic contaminants
        bioaccumulation, 102
        biomagnification, 102
        CNTs, 105
        heavy metals contamination, 102
        nano iron oxide particles, 104
        nano iron sulfide particles, 104–105
        nano zero-valent iron particles, 103–104
        phosphate-based nanoparticles, 104
        zeolites, 104
    organic contaminants
        CNTs, 102
        disadvantages, 101
        metal oxide particles, 101–102
        nanometal particles, 101–102
        nitroaromatic compounds, 101
        PAHs, 101
Nanoscale materials
    advantages and disadvantages, 105–106
    nanocomposite materials, 99
    nanofibers and nanotubes, 99
    nanoparticles, 99–101
    nano-remediation
        inorganic contaminants, 102–105
        organic contaminants, 101–102
    nanostructured surfaces, 99–100
    organic, 101
    properties, 100
    synthesis methods, 100–101
Nanostructured surfaces, 99–100
Nanotubes, 99
Nano zero-valent iron (nZVI) particles, 103–104
National Institute of Standards and Technology (NIST), 246
Native ecosystems
    maintenance, 354
    native nursery-grown and wilding native
        seedlings, 344
    natural regeneration, 349–351

post-mining drainage, 344
root zones and plant species selection
    acid-generating rocks, 346
    fungicide, 348
    Giles Creek mine, 344–345
    nursery-raised beech seedlings, 348
    pot trial, 345
    soilless technique, 345
    Stockton trials, 345
    tensiometer and oxygen diffusion
        measurements, 346
    vegetation cover, 347
    Wangaloa research, 347
seeding techniques, 348–349
wildings and salvaged sods
    adventive species, 351
    direct transfer techniques, 352
    earthworm biomass and diversity, 353
    living vegetation, 352
    mark and recapture monitoring method, 354
    mine schedules and costs, 350
    shrubland, 353
    transplants, 351
Natural channel design (NCD)
    techniques, 148–149, 168
New Zealand rehabilitation practice
    effective rehabilitation, 335
    history
        mining before 1971, 337
        mining public conservation lands, 338–339
        mining since 1971, 338
    native ecosystems
        maintenance, 354
        native nursery-grown and wilding native
            seedlings, 344
        natural regeneration, 349–350
        post-mining drainage, 344
        root zones and plant species selection, 344–348
        seeding techniques, 348–349
        wildings and salvaged sods, 350–354
    pastoral land and native ecosystems, 336
    pastoral land uses
        dense vegetative cover, 342–343
        maintaining rehabilitated pasture, 343–344
        surrogate subsoils, 340–342
        topography, 340
    Polynesian settlement, 336
    tourism activities and mining, 335
Nitroaromatic compounds, 101, 204
Nonnative earthworms, 343, 354
Nonspecific adsorption, 267

O

Oil sands reclamation
    Gateway Hill, 297–299
    surface mining, 296
    Wapisiw tailings reclamation, 299–300
Old tailings dam, 223
Open-pit uranium mining, 302
Ore processing area (Pr), 274
Organic amendments (OAs)
    advantages, 84
    Cu mobilization, 276

engineered pedogenesis, 85
hydro-geochemical stabilization, 85–86
impacts of, 84–85
physicochemical properties, 84
selection, 85–86
Organic carbon (OC), 77–78
Organic covers, 46
Organic matter, mine tailings
coupled nutrient cycling processes, 79
OC, 77–78
organo-mineral interactions and
aggregation, 80
SOM, 77, 79–80
TOC, 77
Organic nanoscale materials, 101

P

PAHs, *see* Polycyclic aromatic hydrocarbons
Pakihi ecosystems, 352
Paya Indah Wetlands (PIW), 325, 327–328
Peat mineral soil, 297
Phosphate-based nanoparticles, 104
Phytocapping, derelict mine sites
abandoned metalliferous mines, 216
capping approaches, 222–226
issues, 216–218
landfill covers, 216
metal toxicants, 215
remediation and management options
contaminated sites, risk
assessment of, 218
cost-effectiveness, 219
electrolysis-based cleanup methods, 221
electroremediation technology, 220
ex situ techniques, 219
HHRA, 219
high risk sites and areas, 219
in situ techniques, 219
long-term contaminated soils, 221
organic contamination, 219
phytoremediation technology, 220–221
RBLM, 217, 219
soil washing process, 221
technology
cap-waste interactions, 228–229
conventional MSW sites, 226
irrigation, 227–228
landfill capping systems, 226
plant species, 229–230
soil amendments, 230–232
Phytodegradation, 204
Phytodesalination, 206
Phytoextraction, 99, 206
Phytofiltration, 206
Phytoremediation, 98–99, 205
phytodegradation, 204
phytodesalination, 206
phytoextraction, 206
phytofiltration, 206
phytostabilization, 204
phytovolatilization, 204, 206
rhizodegradation, 206
technology, 220

Phytostabilization, 20, 98, 204
Phytotechnology
phytoremediation, 205
phytodegradation, 204
phytodesalination, 206
phytoextraction, 206
phytofiltration, 206
phytostabilization, 204
phytovolatilization, 204, 206
rhizodegradation, 206
process
contaminant removal, 209
hydraulic control, 208–209
rhizosphere modification, 209–210
soil cover, 207–208
Phytovolatilization, 98–99, 204, 206
Pit lakes mine closure issues
AMD, 175
anticipating and meeting stakeholder
expectations, 177–178
closure objectives and criteria, 175–177
health and safety issues, 183–184
identifying COCs, 180–181
liquid and solid mine waste, disposal of, 182–183
long-term vertical mixing regimes, 179–180
mine water predictions reliability and accuracy
of, 184–185
mitigating AMD, 181–182
predicting and managing water balances, 177–179
PIW, *see* Paya Indah Wetlands
Plant density trials, 350
Polycyclic aromatic hydrocarbons (PAHs), 6, 101–102
Polyvinyl chloride plastics (PVC), 46, 226
Poor soil characteristics, 59
Poor water quality prediction, 184–185
Potentially acid forming (PAF), 182
Potentially acid generating (PAG) materials, 181–182
Pot trials, 340–342, 345
Powelliphanta snail, 353
Precipitation process, 63, 97, 267, 272
Pre-Surface Mining Control and Reclamation
Act (SMCRA), 195
Processed feedstock, 23
PROC GLIMMIX, 248
Progressive reclamation, 295
Pump-and-treat method, 221
PVC, *see* Polyvinyl chloride plastics
Pyrite oxidation, AMD
abiotic and biotic processes, 35
acidophilic bacteria, 35
alkaline material, 37
biochemistry, 36
ferric sulfate hydrolysis, 36
oxygen and microorganisms, 20
protons, 36
reaction kinetics, 35
sulfide minerals, 35, 37

R

Rapid bioassessment protocols (RBP), 153, 161, 168
Rapid Flow Analyzer (RFA), 246
Reclamation certification process, 296, 299
Reclamation permit, 296

Rehabilitated pasture, 119, 121, 341–344
Remediation and management options
    contaminated sites, risk assessment of, 218
    cost-effectiveness, 219
    electrolysis-based cleanup methods, 221
    electroremediation technology, 220
    ex situ techniques, 219
    HHRA, 219
    high risk sites and areas, 219
    in situ techniques, 219
    long-term contaminated soils, 221
    organic contamination, 219
    phytoremediation technology, 220–221
    RBLM, 217, 219
    soil washing process, 221
Resource Management Act (RMA), 338
Rhizodegradation (phytostimulation), 206
Rhizofiltration, 206
Richmond Mountain mine, 217
Ripping, 342
Risk-based land management (RBLM), 217, 219
Robinson Forest, 149
Rock mulching, 343
Rotowaro coal mine, 341

S

Salicylic-sulfuric acid digestion, 246
Sam's Creek, 225
Sandy slime tailings, 322, 325
Sao Joaquim formation, 274
Selenium biomethylation mechanism, 272
Self-sustaining native ecosystems, 355
Shelocta-Gilpin-Hazleton complex, 153
Silt traps, 328
Silty clay loam soils, 66–67
Site hydrology, 297
Site screening levels (SSL), 328–329
Slags, 66, 96, 101
Sludge-derived biochar, 24
Smelting process, 265
SOC, *see* Soil organic carbon
Softwood-derived biochar, 24
Soil depth, 63, 67, 208, 229, 296
Soil-layering trial, 349
Soil microbial communities, mine tailings, 77
    composition and activities
        BIOLOG approach, 84
        measurement methods, 82–83
        quantification methods, 82–83
        respiration, 84
    environmental factors and optimum conditions, 80
    extremophiles-dominant microbial communities, 81–82
    phytostabilization, 81
    SOM dynamics, 80
Soil organic carbon (SOC), 61, 191
    data sources, 63, 65
    meta-analysis, 65–66, 68
    organic matter, source of, 62
    soil carbon dynamics
        vs. biosolids land application, 66–68
        silty clay loam soils, 66–67

    soil depth, 67
    soil texture, 66
    types/fraction, 62–63
Soil organic matter (SOM)
    biological activities, 77
    decomposition, 79
    dynamics, microbial community, 80
    stabilization, 79–80
Soil replacement methods, 97
Soil washing process, 221
Specific adsorption, 267
SSL, *see* Site screening levels
Statistical Analysis System (SAS), 248
Stockton mine, 338, 343, 350, 352, 354–355
Storm Water Management and Road Project (SMART), 319
Straw mulching, 343, 352
Strongman mine rehabilitation, 350, 354
Subaqueous disposal controls, 47
Subsoiling, 342
Successive alkalinity producing systems (SAPS), 158
Sulfide oxidation process, 273
Surface coal mining and reclamation, Appalachia
    Appalachian coalfield, 124–125
    area mining method, 124, 126
    ARRI
        Core Team, 138
        GFW, 140–141
        goals, 138
        mechanism, 138
        reforestation guidelines, 140
        Science Team, 139
    contour mining method, 126
    FRA
        Bent Mountain project, 131–134
        forest vegetation, 129
        global application, 142
        ground cover vegetation, 130–131
        impacts, 141–142
        loose soil, noncompacted soil growth medium, 130
        proper tree planting techniques, 131
        Starfire project, 135–140
        successional species, crop trees, 131
        suitable rooting medium creation, tree growth, 129–130
    mountaintop surface mining, 124–126
    post-SMCRA reforestation, 127–129
    pre-SMCRA reforestation, 126–127
Surface Mining Control and Reclamation Act (SMCRA), 124
Syncrude Canada Ltd., 297–298
Synthetic geofabrics, 46

T

Tackifiers (sticking agents), 342
Tailings
    biological characteristics
        biogeochemical changes, OA, 84–86
        microbial-mediated bioweathering, 77
        native microbial inoculum, 86–87
        organic matter, 77–80
        soil microbial communities, 77, 80–84

definition, 75–76
dumps, 277–278
environmental impact
    geochemical reactions, 17–19
    hydrological changes, 19
    sulfide minerals, 17
fresh tailings, 76
mineralogical profile, 76
organic matter
    coupled nutrient cycling processes, 79
    OC, 77–78
    organo-mineral interactions and aggregation, 80
    SOM, 77, 79–80
    TOC, 77
phytostabilization, 76
remediation techniques, 20–21
sulfide-rich tailings, 76
toxic constituents, 76
Thermal desorption, 97
Throckmorton Greenhouse Temperature Tracking
        System, 245
Time-of-flight secondary ion mass spectrometry
        (TOF-SIMS), 273
Tin Tailing Research Station (TTRS), 320
Tin tailings
    amang, 313, 318
    classification of, 313
    ex-tin mines, 313
    heavy metals and radionuclide content in, 316–317
    pH, OM, CEC, and nutrient contents, 314–315
Tin Tailings Afforestation Centre (TTAC), 322, 325
Total organic carbon (TOC), 77, 274
TransAlta coal mine, 197–198
Trimethylarsine oxide (TMAO), 272
Tui mine tailings dam, 337, 341, 354

U

Ubari formation, 274
United States Environmental Protection Agency
        (USEPA), 242, 329
Unprocessed feedstock, 23

V

Vitrification technique, 98

W

Wangaloa mine, 344, 346–347
Waste rock (Wr), 96, 274
Water balances, 177–178
Webbs Consols mine, 222–224
Weeping tiles, 300
Whitewood mine, 301
Wildlife Act, 339
Wildlife Permit, 353
World Health Organization (WHO), 322

X

X-ray diffraction (XRD), 273
X-ray fluorescence (XRF), 222, 274
X-ray near edge structure (XANES) spectra, 276

Z

Zeolites, 104